ANIMAL BIOCHEMISTRY

ANIMAL BIOCHEMISTRY

By

Dr. Lata Bhattacharya

Professor

School of Studies in Zoology & Biotechnology

Vikram University

Ujjain (M.P.)

(India)

DISCOVERY PUBLISHING HOUSE PVT. LTD.

NEW DELHI-110 002

Published by:
Tilak Wasan
DISCOVERY PUBLISHING HOUSE PVT. LTD.
4383/4B, Ansari Road, Darya Ganj
New Delhi-110002 (India)
Phone : +91-11-23279245; 23253475; 43596065
E-mail: discoverybooksindia@gmail.com
discoverypublishinghouse@gmail.com
namitwasan9@gmail.com
web : www.discoverypublishinggroup.com

***Reprinted:* 2019**

***First Edition:* 2010**

ISBN: 978-81-8356-641-4

Animal Biochemistry

Printed at:
Infinity Imaging Systems
Delhi

Preface

Biochemistry today has made spectacular progress in unraveling the mysteries of animate nature. This progress has allowed us to gain deeper insight into the principles of vital activity and has to a very significant extent stimulated the development of applied disciplines, especially medicine. The present title *"Animal Biochemistry"* is intended for those who wish to understand living organisms, especially man. Biochemistry is essential for this purpose, but it would be almost impossible for a student to survey on his own the massive body of existing knowledge, constantly augmented by a remarkable torrent of brilliat discoveries. The purpose of the book, then is to organize our knowledge into something that can be comprehended in a relatively short time and still convey a reasonable complete picture of the chemical structure and function of man. Readability without sacrifice of coverage has been a prime goal. An important device in gaining that goal is to keep attention constantly focused on function, with repeated use of rationalization to show that the chemical facts are not isolated, but part of a whole.

Biochemistry has two major goals as a fundamental science, it treats the vital functions from the standpoint of physical chemistry, and as an applied discipline, it points out practical applications for the wealth of scientific knowledge it has acquired. The dual purpose has been, as far as possible, take into account in the writing of this book. We have also tried to summarize our pedagogical experience in teaching biochemistry to students specializing in medicine and pharmacy. The material of this book has been organized according to the principle of functionally in order to trace the close relationship between the functions of the living organism and the structures of its constituents molecules as well as the chemical and physico-chemical process in which they are involved.

The aim of this book is to present a core of biochemical knowledge that is desirable for undergraduate and postgraduate students and also those involved in the field of medical, microbiology, biotechnology and pharmaceutical. Every attempt has been made to keep abrest of the advances in the subject and at the same time to include the fundamentals.

To make the work more comprehensive and informative, the author has consulled many authoritative books, research journals, abstracts, monographs etc. He is grateful to all those great scholars whose work are cited or substantially reproduced.

There can be no claim to originality except in the manner of treatment and much of the information has been obtained from the books and scientific journals available in the different libraries.

The author expresses his thanks to his friends and colleagues whose continue inspirations have initiated him to bring out this book.

The author expresses his gratitude to Mr. Wasan and Staff of M/s Discovery Publishing House Pvt. Ltd., for their whole hearted co-operation in the publication of this book.

In the mean time, the author will remain sincerely responsible for any shortcomings of the book and be grateful to the readers for their suggestions and constructive criticism for the continuous betterment of the book. He takes this opportunity to appeal to the readers to send their suggestions straightway to his publisher.

Author

CONTENTS

CHAPTER

1 Introduction

Proteins, nucleic acids, and carbohydrates all contain acidic or basic functional groups. The strengths of the acidic groups vary over a broad range from that of the strongly acidic phosphate and sulfate esters to that of the very weakly acidic alcoholic - OH group. The net electrical charge, as well as the spatial distribution of the charged groups, affects the properties of a macromolecule greatly. Therefore, it will be worthwhile for us to consider some aspects of acid-base chemistry before discussing other topics.

Strengths of Acids and Bases: the pK_a's

The strength of an acid is usually described by the acid dissociation constant K_a

$$HA = A^- + H^+ \quad ...(1.1)$$

$$Ka = [A^-][H^+]/[HA] \quad ...(1.2)$$

For strong acids K_a is high and for weak acids it is low. Since the values of K_a vary by many orders of magnitude it is customary to use as a measure of the acid strength pK_a. This is the negative logarithm of K_a (p$K_a = -\log K_a$).

For very strong acids pK_a is less than zero, while very weak acids have pK_a values as high as 15 or more. In the biochemical literature *the strength of a base is nearly always given by the* pK_a *of the conjugate acid.* Thus, A^- in Eq. 1.1 base and HA its conjugate acid. The base could equally well be uncharged A and its conjugate acid HA^+. In both cases Eq. 1.2 would hold. This defines the strength of both the acid HA and the base A^-. It follows that strong bases have weak conjugate acids with high pK_a values, while weak bases have strong conjugate acids with low pK_a values.

For a compound containing several acidic groups we define a series of consecutive dissociation constants K_{1a}, K_{2a}, K_{3a}, etc. For the sake of simplicity we will omit the a's and call these K_1, K_2, K_3....

$$H_3A \xrightarrow{K_1} H_2A \xrightarrow{K_2} HA \xrightarrow{K_3} A \quad ...(1.3)$$

If there are n consecutive dissociation constants there will be $n + 1$ ionic species H_3A, H_2A, etc. Notice that H_3A could be a neutral molecule with H_2A, HA, and A bearing changes of -1, -2, and -3, respectively. Alternatively, H_3A might carry 1, 2, or 3 positive changes. In every case the mathematical expressions will be the same. For this reason the charges have been deliberately omitted from Eq. 1.3 and others that follow. Each constant in Eq. 1.3 defined as in Eq. 1.2 *i.e.,*

K_2 = [HA][H^+]/[H_2A], etc. Keep in mind that there are significant differences between the *apparent equilibrium constants* (concentration equilibrium constants) that we ordinarily use and thermodynamic equilibrium constants that are obtained by extrapolation to zero ionic strength. A related complication is the uncertainity associated with the measurement of pH. This is often considered a measurement of hydrogen ion activity but this is not a correct statement. However, for all practical purposes, in the range of about pH 4-10 the pH can be equated with -log [H^+]. Often only one of the ionic forms of Eq. 1.3 will be important in a biochemical reaction.

A particular ionic species may be the substrate for an enzyme. Likewise, an enzyme-substrate complex in only a certain state of protonation may react to given products. In these cases, and whenever pH affects an equilibrium, it is useful to relate the concentration [A_1] of a given ionic form of a compound to the total of all ionic forms $[A]_t$ using Eqs. 1.4 and 1.5

$$[A_i] = [A]_t/F_i \qquad ...(1.4)$$

$$[A]_t = [A] + [HA] + [H_2A] + \ldots [H_nA] \qquad ...(1.5)$$

For Eq. 1.4 $A_1 = H_nA$, $A_2 = H_{n-1}A$, etc. and F_1, F_2, etc. are the *Michaelis pH functions, which* were proposed by L. Michaelis in 1914. For the case represented by Eq. 1.3 there are four ionic species and therefore four Michaelis pH functions which have the following form Eq. 1.6. Here, K_1, K_2, etc. are the usual consecutive acid dissociation constants.

$$F_1 = 1 + K_1/[H^+] + K_1K_2/[H^+]^2 + K_1K_2K_3/[H^+]^3$$

$$F_2 = [H^+]/K_1 + 1 + K_2/[H^+] + K^2K_3/[H^+]^2$$

$$F_3 = [H^+]^2/K_1K_2 + [H^+]/K_2 + 1 + K_3/[H^+]$$

$$F_4 = [H^+]^3/K_1K_2K_3 + [H^+]^2/K_2K_3 + [H^+]/K_3 + 1 \qquad ...(1.6)$$

If there are only three ionic forms the first three of these equations will apply if the final term is dropped from each. The student should be able to verify these equations and to write the appropriate pH functions for other cases. Since these relationships are met so often in biochemistry it is worthwhile to program a computer to evaluate the Michaelis pH functions and to apply them as needed. From Eq. 1.4 can be seen that *the reciprocal of the Michaelis pH function for a given ionic form represents the fraction of the total compound in that form* and that the sum of these reciprocals for all the ionic forms is equal to one.

Notice that in Eq. 1.3 single arrows have been used rather than the pairs ($\rightleftarrows$) that are often employed to indicate reversible equilibria. This is done so that *the direction of the arrow indicates whether we are using a dissociation constant or an association constant.* The use of single arrows in this manner to indicate how the equilibrium constants are to be written is a good practice when dealing with complex equilibria.

Titration Curves

When a neutral amino acid is titrated with acid, the carboxylate groups become protonated and acid is taken up. Likewise titration with base removes protons from the protonated amino groups and base is taken up. If we plot the number of equivalents of acid or base that have reacted with the neutral amino acid versus pH, a titration curve such as that generated. The curve for histidine contains three steps; the first corresponds to the titration of the carboxylate group with acid, the second to the titration of the protonated imidazole of the side chain, and the third to the titration of the protonated amino group with base. Each step is characterized by a midpoint that is equal to the pK_a value for the group being titrated.

The ends of the curve at low and high pH, which are drawn with a dashed line, are obtained only after corrections have been applied to the data. If only the equivalents of acid or base *added* rather than the number *reacted* are plotted, we obtain the solid line shown in Fig. 1.1. We see that at the low pH end there is no distinct end point. As we add more acid to try to complete the titration, the correction that must be applied to the data becomes increasingly greater. The difference between the dashed and solid lines reflects the fact that at the low pH end much of the acid added is used to simply lower the pH. Therefore, we have a large free [H^+]. Similarly, at the high pH end we have a high free [OH^-]. The exact shapes of the ends of the titration curve depend heavily on the total concentration.

Likewise, the magnitude of the correction required to obtain a plot of equivalents of acid or base reacted varies with the concentration and is smaller the higher the concentration of the substance being titrated. An important rule in doing acid-base titrations, especially when very small samples are available, is to use the highest possible concentration of sample in the smallest possible volume and to titrate with relatively concentrated acid or base. Because of the difficulty of adequately correcting titration curves at low pH it is hard to estimate the pK_a values of the carboxyl groups of amino acids accurately from titration.

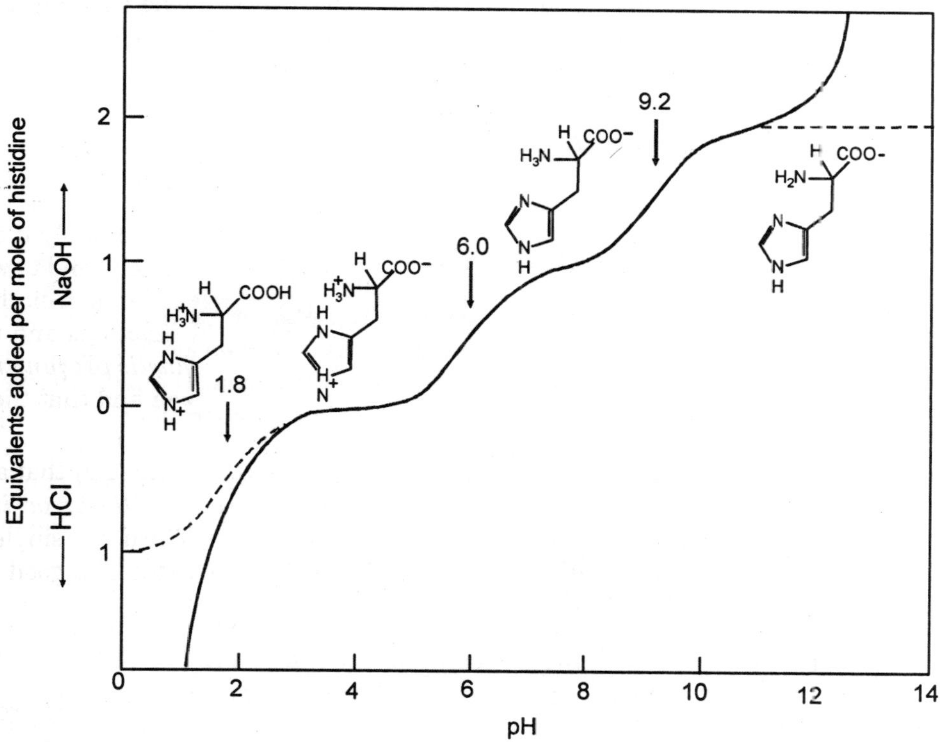

Fig. 1.1. Titration curve for histidine. The solid line represents the uncorrected titration curve for 3 mM histidine monohydrochloride titrated with 0.2 M HCL to lower pH or with 0.2 M NaOH to higher pH assuming pK_a values of 1.82, 6.00, and 9.17. The dashed line represents the corrected curve showing the number of protons bound or lost per mole of histidine monohydrochloride.

An additional experimental difficulty exists at the high pH end because of the tendency for basic solutions to absorb CO_2 from the air. Sometimes formaldehyde is added to shift the apparent pK_a of the amino groups to lower values and make the titration more satisfactory. Despite the difficulties, real proteins can be titrated successfully to estimate the numbers of carboxyl, histidine, tyrosione, and other groups. An example is shown in Fig. 1.2.

The experimental data have been fitted with a theoretical curve based on the pK_a's of the carboxyl, histidine, tyrosine, and amino groups as determined by NMR measurements. Account has been taken of the effect of the electrical field created by the many charged groups in distorting the curve from that obtained by summing the theoretical titration curves of the component groups. However, when there are multiple acid-base groups that are close together in a protein, a more complex situation involving tatutomerism arises. Titration curves based on plots of light absorption versus pH or of NMR chemical shifts versus pH often useful. They have the important advantage that no special correction for free acid or base is needed at low or high pH.

Buffers

A mixture of a weak acid HA and of its *conjugate base* A constitutes a buffer which resists changes in pH. This can be seen most readily by taking logarithms of both sides if Eq. 1.2. By replacing log K with $-pK_a$ and log $[H^+]$ with -pH and rearranging we obtain Eq. 1.7. Henderson-Hasselbalch equation). It is sometimes useful to rewrite this as Eq. 1.8 where α is the fraction of the acid that is dissociated at a given pH.

$$\text{PH} - pK_a = \log([\text{A}]/[\text{HA}]) \qquad \text{...(1.7)}$$

$$\text{pH} - pK_a = \log[\alpha/1\text{-}\alpha];$$

$$\alpha = [\text{A}]/[\text{HA}]+[\text{A}] = \frac{10^{(\text{pH}-\text{pK}_a)}}{1+10^{(\text{pH}-\text{pK}_a)}} \qquad \text{...(1.8)}$$

Logarithms to the base 10 are used in both equations. These equations are useful in preparing buffers and in thinking about what fraction of a substance exists in given ionic form at a particular value of pH. From Eq. 1.7 it is easy to show that when the pH is near the pK_a relatively large amounts of acid or base must be added to change the pH if the concentrations of the buffer pair A and HA are high. Buffers are often added to maintain a constant pH in biochemical research[6] and naturally occurring buffer systems within body fluids and cells are very important.

Among the most important natural buffers are the proteins themselves, with the imidazole groups of histidine side chains providing much of the buffering capacity of cells around pH 7 (Eq. 1.1 and 1.2). Table 1.1 lists some useful bio-chemical buffers and their pK_a values. Here are a few practical hints about buffer preparation. Buffers containing monovalent ions tend to change pH with dilution less than do those with multivalent ions such as HPO_4^{2-} or $HP_2O_7^{3-}$. The pK_a values of carboxylic acids and of phosphoric acid, or of its organic derivatives, change very little with change in temperature. The pH of a buffer prepared with such components is nearly independent of temperature (Table 1.1).

However, the pK_a of the $-NH_3^+$ group changes greatly with temperature. Buffer composition can be calculated readily from Eq. 1.8 and pK_a values from Table 1.1. It is convenient to keep in the taboratry standardized (to ~1% error) 1 M HCl and 1M NaOH for use in buffer preparation. Compositions calculated from Eq. 1.8 will usually yield buffers of pH glucose to those expected.

Final adjustment value may be needed if the pH is more than one use may from a pK_a value. When two buffering materes are present, the composition should be calculated independently for each. The measurement of pH should always be done with great care because it is easy to make errors. Everything depends upon the reliability of the standard buffers used to calibrate the pH meter.

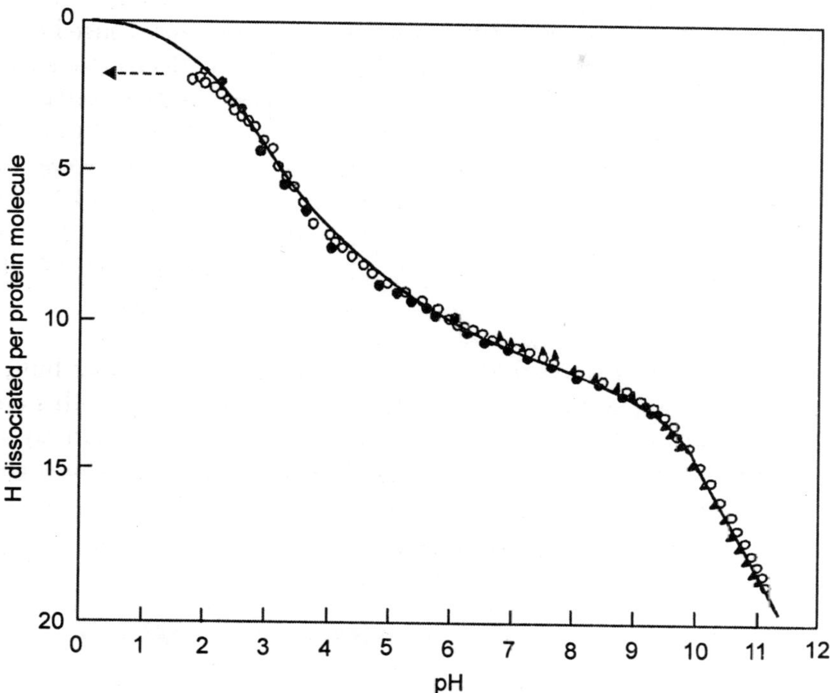

Fig. 1.2. Acid-base titration curve for hen lysozyme at 0.1 ionic strength and 25°C. O, initial titration from the pH attained after dialysis; •, back titration after exposure to pH 1.8; ▲, back titration after exposure to pH 11.1. The solid curve was constructed on the basis of "intrinsic" pK_a values based on NMR data.

Often, especially during isolation of small compounds, it is desirable to work in the neutral pH region with volatile buffers, *e.g.*, trimethylamine and CO_2 or ammonium bicarbonate, which can be removed by vacuum evaporation or lyophilization.

ISOLATING COMPOUNDS

Before structural work can begin, pure substances must be separated from the complex mixtures in which they occur in cells and tissues. Often, a substance must be isolated from a tissue in which it is present in a very low concentration. After it is isolated in pure form, if it is a large molecule, it must often be cut up into smaller pieces which are separated; purified, and identified. Accurate quantitative analysis is required to determine the ratios of these fragments. Considerable ingenuity may then have to be exercised in putting the pieces of the jigsaw puzzle back together to determine the structure of the "native" molecule. Many books, a few of which are cited here, provide instructions. There are also journals and other periodicals dedicated to biochemical methods.

Fractionation of Cells and Tissues

A fresh issue or a paste of pack cells of a microorganism, usually collected by centerlization, may be the starting material. Tissue the ground in a kitchen-type blender or, for genration environment in a special *hemogenizer.* The purpuse televehijem homogenizer is a small appartus may glass or plastic pestel rotates inside a tite-first stablertar tube (see standal laboratory equpment catalisis for pictures). Microbial cells (*sonecation*) or inficial pressure cells. It is important to pay attention to the pH, buffer contribution, and, if subcellular organelles are tube secrded, the osmotic pressure. To preserve the organelles, 0.25 M sucrose if frequently frii suspending medium, and $MgCl_2$ as well as complexing agent such as ethylendediamine state (EDTA) may be added Solution enzymes are often extracted without addition of secrose, but reducing compounds such as glutathione, mercaptoethanol, or dithiothreitol Eq. 1.23 may be added.

The crude *homogenate* may be strained and is usualy centrifuged briefly to remove cell fregments and other "debris". Large-scale purification of proteins is often initiated with such a crude homogenate. Cell organelles are also often separated by centrifugation. In one procedure a homogenate in 0.25 M sucrose (*isotonic* with most cells) is centrifuged for 10 min at a field of 600-1000 times the force of gravity (600-1000 *g*) to sediment nuclei and whole cells. The supernatant fluid is then centrifuged another 10 min at - 10,000 *g* to sediment mitochondria and lysosomes. Finally, centrifugation at ~100,000 *g* for about an hour yields a pellet of microsomes which contains both membrane fragments and ribosomes. Each of the separated components can be resuspended and recentrifuged to obtain cleaner preparations of the organelles.

Table 1.1. Practical pK_a Values for Some Useful Buffer Compounds at 25°C and Ionic Strength 0.1[a]

Compound	pK_a	*Grams per mole*	$d(pK_a)/dT$	*Charge on conjugate base*
Citric acid (pK_1)		192		−1
Formic acid	3.7	0	−1	
Citric acid (pK_2)	4.45	192	−0.0016	−2
Acetic acid	4.64	60	0.0002	−1
Succinic acid (pK_2)	5.28	118	0	−2
Citric acid (pK_3)	5.80	192	0	−3
3,3-Dimethylglutaric acid	5.98	160	0.006	−2
Piperazine (pK_1)	(5.68) 6.02	86		0
Cacodylic acid (dimethylarsinic acid)	6.1	138		−1
MES[b] 6.1	195	−0.011		
BIS-TRIS[b]	6.41	209	−.017	
Carbonic acid (pK_1)	6.4c			
Pyrophosphoric acid (pK_3)	6.76	178		−2
Phosphoric acid (pK_2)	6.84	98	−.0028	−2
PIPES[b]	6.90	353	−.0085	−2
Imidazole	7.07	68	−.020	0

Compound	*pK_a*	*Grams per mole*	*d(pK_a)/dT*	*Charge on conjugate base*
BES[b]	7.06	213	–.016	–1
Diethylamalonic acid	7.2	136		
TES[b]	7.37	279	–.020	–1
HEPES[b]	7.46	238	–.014	–1
N-Ethylmorpholine	7.79	115	–.022	0
Triethanolamine	7.88	149	–.020	0
TRICINE[b]	8.02	178	–.021	–1
TRIS[b] 8.16	121	–.031	0	
Glycylglycine	8.23	132	–.028	–1
BICINE[b]	8.26	163	–.028	–1
4-Phenolsulfonic acid	8.70	174	–.013	–2
Dethanolamine	9.00	105	–.024	0
Ammonia	9.2		–0.031	0
Boric acid	9.2		–0.008	mixed
Pyrophosphoric acid (pK_4)	9.41	178		–3
Ethanolamine	9.62	61	–.029	0
Glycine (pK_2)	9.8	75	–0.025	–1
Piperazine (pK_2)	9.82	86		0
Carbonic acid (pK_2)	10.0		–0.009	–2
Piperidine	11.1	85		0

The Good buffers have dipolar ionic constituents. Since no form is without electrically charged groups they are unlikely to enter and disrupt cells.

[b] Abbreviations used:

BES	*N,N*-Bis(2-hydroxyethyl)-2-aminoethanesulfonic acid
BICINE	*N,N*-Bis(2-hydroxyethyl)glycine
BIS-TRIS	Bis(2-hydroxyethyl)iminotris(hydroxymethyl)methane
HEPES	*N*-2-Hydroxyethylpiperazine-N'2-ethanesulfonic acid
MES	2-(*N*-Morpholino)ethanesulfonic aicd
PIPES	Piperazine-*N,N'*-bis(2-ethanesulfonic acid)
TES	*N*-Tris(hydroxymethyl)methyl-2-aminoethanesulfonic acid
TRICINE	*N*-Tris(hydroxymethyl)methylglycine
TRIS	Tris (hydroxymethyl) aminomethane

[c] For CO_2 (solid) + $H_2O \rightarrow H^+$ HCO_3^-, apparent pK_a.

The sedimented particles can often be solubilized by chemical treatment, for example, by the addition of either ionic or non-ionic detergents. Membrane proteins can be isolated following solubilization in this way. The soluble supernatant fluid remaining after the highest speed centrifugation provides the starting material for isolation of soluble enzymes and many small molecules.

Separations Based on Molecular Size, Shape, and Density

The simplest way to separate very large dissolved molecules is to let the small ones pass through a suitable sieve which may be a membrane with holes or a bed of gel particles. If the size of the particles approaches that of the holes in the sieve, the rate of passage will depend upon shape as well as size.

Dialysis, Ultrafiltration, and Perfusion Chromatography

In dialysis and ultrafiltration, a thin membrane, *e.g.*, made of cellulose acetate (cellophane) and containing holes 1-10 nm in diameter (typically 5 nm), is used as a semipermeable barrier. Small molecules pass through but large ones are retained. Dialysis depends upon diffusion and can be hastened by adequate stirring. Ultrafiltration requires a pressure difference across the membrane. The more sophisticated procedures of gel filtration and perfusion chromatography were introduced in 1959. A column is packed with material such as the crosslinked dextran Sephadex, polyacrylamide gels (such as the Bio-Gel P Series), or agarose gels (*e.g.*, Bio-Gels A and Sepharoses). These come in the form of soft beads, the interior network of which is a three-dimensional network of polymer strand.

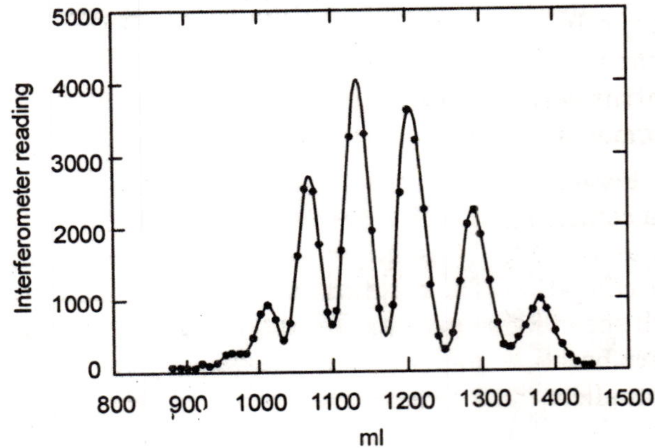

Fig. 1.3. Separation of oligosaccharides by gel filtration. The sugars dissolved in distilled water were passed through a column of Sephadex G-25. The peaks contain (right to left) glucose, cellobiose, cellotriose, etc.

Recently porous beads of hard crosslinked polystyrene, glass, or various other silicate materials have been employed. The interstices between strands, whose size depends upon the degree of crosslinking introduced chemically into the gel, are small enough to exclude large molecules but to admit smaller ones. If a mixture of materials of different molecular size is passed through such a column the smaller molecules are retarded because of diffusion into the gel, while the larger molecules pass through unretarded (Fig. 1.3). Sephadex G-25 excludes all but salts and compounds no larger than a simple sugar ring. Sephadex G-200, which is much less crosslinked, permits separation of macromolecules in the range of 5-200 kDa. As explained already get filtration also provides an important way of estimating M_r for proteins and other macromolecules.

Centrifugation

Centrifuges of many sizes and speeds are used in the laboratory to remove debris as well as to collect precipitated proteins and other materials at various steps in a purification scheme. The most remarkable are the *ultracentrifuges* which produce forces greater than 4×10^5 times that of gravity. They can be used both for separation of molecules and for determination of M_r. When macromolecules in a solution are subjected to an ultracentrifugal field they are accelerated rapidly to a constant velocity of sedimentation. This is expressed as a *sedimentation constant* s, which is the rate (cm/s) per unit of centrifugal force.

The unit of s is the second but it is customary to give it in Svedberg units, S (1S = 10^{-13}s). Sizes of particles are often cited by their S values. The sedimentation constant is affected by the sizes, shapes, and densities of the particles. If carried out at constant velocity an equilibrium will eventually be attained in which sedimentation is just balanced by diffusion and a smooth concentration gradient forms from the top to the bottom of the centrifuge cell or tube. Concentration gradients can also be formed by centrifuging a concentrated solution of small molecules. Such a concentration gradient is also a *density gradient* which can be made very steep. For example, a gradient with over a 10% increase in density from top to bottom can be created using 6 M *cesium chloride* (CsCl) and is widely used in DNA separations. If DNA is added prior to centrifugation it will come to rest in a narrow band or bands determined by the *buoyant densities* of the species of DNA present. After centrifugation, which is usually done in a plastic tube, a hypodermic needle is inserted through the bottom of the tube and the contents are pumped or allowed to flow by gravity into a fraction collector. Another type or gradient centrifugation *(zone centrifugation)* utilizes a preformed gradient to stabilize bands of cell fragments, organelles, or macromolecules as they sediment.

For example, RNA can be separated into several fractions of differing sedimentation constants in a centrifuge tube that contains *sucrose* ranging in concentration from 25% at the bottom to 5% at the top. This is prepared by a special mixing device or "gradient maker" prior to centrifugation. The solution of RNA is carefully layered on the top, the tube is centrifuged at a high speed for several hours, and the different RNA fractions separate into slowly sedimenting sharp bands. A 20-60% gradient of sucrose or glycerol may be used in a similar way to separate organelles.

Separations Based on Solubility

Some fibrous proteins are almost insoluble in water and everything else can be dissolved away. More often soluble proteins are precipitated from aqueous solutions by adjustment of the pH or by addition of large amounts of salts or of organic solvents. The solubility of any molecule is determined both by the forces that hold the molecules together in the solid state and by interactions with solvent molecules and with salts or other solutes that may be present. Proteins usually have many positively and negatively charged groups on their surfaces. If either a positive or a negative charge predominates at a given pH the protein particles will tend to repell each other and to remain in solution.

However, near the isoelectric point (pI), the pH at which the net charge is zero, the solubility will usually be at a minimum. The pH of a tissue extract may be adjusted carefully to the PI of a desired protein. Any protein that precipitates can be collected by centrifugation and redissolved to give a solution enriched in the protein sought. Some proteins, such as those classified (by an old system) as *globulins,* are insoluble in water but are readily *"salted in"* by addition of low concentrations (*e.g.,* up to 0.1 M) of salts.

Low concentrations of salts increase the solubility of salts increase the solubility of most proteins because the salt ions interact with the charged groups on the protein surfaces and interfere with strong electrostatic forces that are often involved in binding protein molecules together in the solid state. Some salts, including $CaCl_2$ and NaSCN, which bind to proteins, are especially effective in salting in. Addition of *high* concentrations of salt causes precipitation of most proteins from aqueous solutions. The most effective and most widely used materials for this *"salting out"* of proteins are $(NH_4)_2SO_4$ and Na_2SO_4. Because the salt ions interact so strongly with water, the protein molecules interact less with water and more with each other.

A similar intramolecular effect may cause the stabilization of proteins of halophilic bacteria by 1-4 M KCl. Different proteins precipitate at different concentrations of an added salt. Hence, a fraction of proteins precipitating between two different concentrations of salt can be selected for further purification Fig. 1.4.

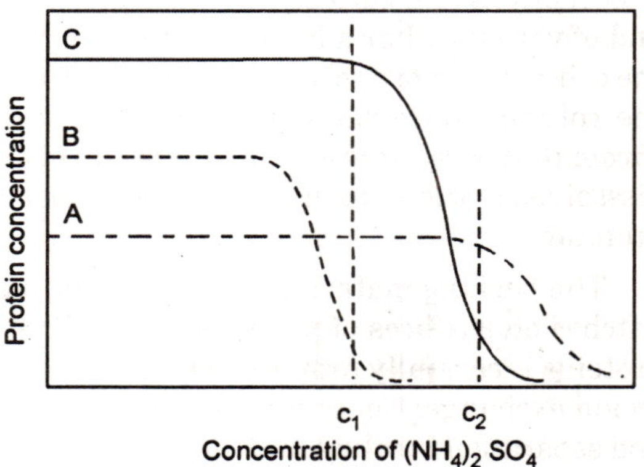

Fig. 1.4. Hypothetical behaviour of a solution containing three proteins, A, B, and C, upon ammonium sulfate fractionation. The concentration of protein remaining in the solution is plotted against ammonium sulfate concentration (usually expressed as % saturation). Addition of ammonium sulfate to concentration C_1 will precipitate largely protein B, which can be removed by centrifugation. Addition of additional salt to C_2 will precipitate largely protein C, while A remains in solution.

Precipitation methods are popular first steps in purification of proteins because they can be carried out on a large "batch" scale. The amounts of ammonium sulfate used are often expressed as percentage saturation, *i.e.,* as a percentage of the amount required-to saturate the solution (4.1 M at 25°). Convenient tables are available that permit one to weigh out the correct amount of solid ammonium sulfate to give a desired percentage saturation or to go from one percentage saturation to a higher one. Proteins are otten stabilized by low concentrations of simple alcohols or ketones and by higher concentrations of polyhydroxy alcohols, such as glycerol and sucrose, and also by certain inert, synthetic polymers such as *polyethyleneglycol* (PEG). The latter is a widely used precipitant.

The polyhydroxy-alcohols and PHG are all hydrated but tend not to interact strongly with the protein molecules. On the other hand, simple alcohols may denature proteins by their interaction with nonpolar regions.

Separation by Partition

Many of the most important separation methods are based on repeated equilibration of a material between two separate phases, at least one of which is usually liquid. Small molecules may be separated by *countercurrent distribution* in which a material is repeatedly partitioned between two immiscible liquid phases, one more polar than the other. New portions of both liquids are moved by a machine in a "counter-current manner" between the equilibration steps or are moved continuously through coiled tubes. A similar result is accomplished by using as one phase a solid powder or fine "beads" packed in a vertical column or spread in a thin layer on a plate of glass.

The methods are usually referred to as *chromatography,* a term proposed by Tswett to describe separation of materials by colour. In 1903 Tswett passed solutions of plant leaf pigments (chlorophylls and carotenes) in nonpolar solvents such as hexane through columns of alumina and of various other adsorbents and observed separation of coloured bands which moved down the column as more solvent was passed through. Individual pure pigments could be *eluted* from the column by continued passage of solvent. This important method is called *adsorption chromatography.* It is assumed that the pigments are absorbed on the surface rather than being disssolved in the solid material. A related method is *hydrophobic interaction chromatography* of proteins.

The packing material bears long-chain alkyl groups that can interact with the hydrophobic patches on surfaces of proteins. A very different adsorbent that is very useful in separation of proteins is carefully prepared microcrystalline *hydroxylapatite.* It presumably functions in part by ion exchange. Column packing materials such as silica gel contain a large amount of water, and separation involves partition between an immobilized aqueous phase in the gel and a mobile, often organic, solvent flowing through the column.

Usually materials elute sooner when they are more soluble in the mobile phase than in the aqueous phase. These methods are closely related to perfusion chromatography. Aromatic amino acids, lipids, and many other materials can be separated on *reversed-phase* columns in which nonpolar groups, usually long-chain alkyl groups, are covalently attached to silica gel, alumina, or other inert materials.

The mobile phase is a more polar solvent, often aqueous, and gradually made less polar by addition of an organic solvent. In reversed-phase chromatography more polar compounds migrate faster through the system than do nonpolar materials, which experience hydrophobic interaction with the solid matrix. Many traditional chromatographic methods including reversed-phase chromatography have been adapted for use in automatic systems which employ columns of very finely divided solid materials such as silica, alumina, or ion exchange materials coated onto fine glass beads. These *high-performance liquid chromatographic (HPLC)* systems often utilize pressures as high as 300 atmospheres. Separations are often sharper and faster than with other chromatographic methods.

Reversed-phase columns in which the solid matrix may carry long (*e.g.* C_{18}) hydrocarbon chains have been especially popular for separation both of peptides and of proteins. Proteins may also be separated by gel filtration, ion exchange, or other procedures with HPLC equipment.

A sheet of high-quality filter paper containing adsorbed water serves as the stationary phase in *paper chromatography.* However, *thin-layer chromatography,* which employs a layer of silica gel or other material spread on a glass or plastic plate, has often supplanted paper chromatography because of its rapidity and sharp separations (Fig. 1.5).

An approach that requires no stationary phase at all is *field flow fractionation.* Here a suitable external field (*e.g.,* electrical, magnetic, and centrifugal) or a thermal gradient is imposed on the particles flowing through a narrow channel. For volatile materials *vapor phase chromatography* (gas chromatography) permits equilibration between the gas phase and immobilized liquids at relatively high temperatures. The formation of volatile derivatives, *e.g.,* methyl esters or trimethylsilyl derivatives of sugars, extends the usefulness of the method. A method which makes use of neither a gas nor a liquid as the mobilephase is *supercritical fluid chromatography.* A gas above but close to its critical pressure and temperature serves as the solvent.

The technique has advantages of high resolution, low temperatures, and ease of recovery of products. Carbon dioxide, N_2O, and xenon are suitable solvents.

Ion Exchange Chromatography

Separation of molecules that contain electrically charged groups is often accomplished best by ion exchange chromatography. The technique depends upon interactions between the charged groups of the molecules being separated and fixed ionic groups on an immobile matrix. Separation depends upon small differences in pK_a values and net charges and upon varying interactions of nonpolar parts of the molecules being separated with the matrix. Since changes in pH can affect both the charges on the molecules being separated and those of the ion exchange material, the affinities of the molecules being separated are strongly dependent on PH.

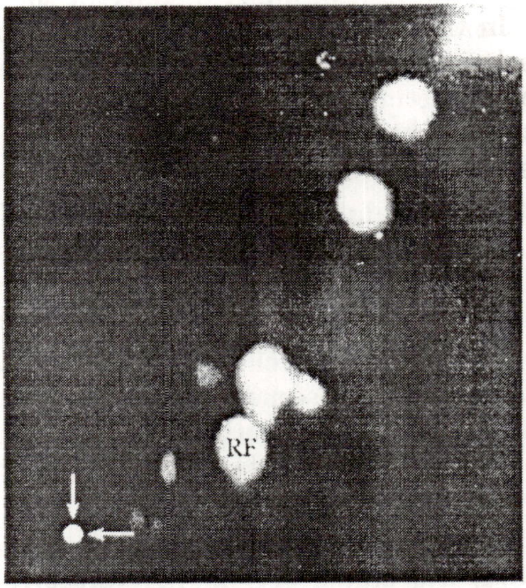

Fig. 1.5. Photograph of a two-dimensional thin layer (silica gel) chromatogram of a mixture of flavins formed by irradiation of ~10 mg of the vitamin riboflavin. The photograph was made by the fluorescence of the compounds under ultraviolet light. Some riboflavin (RF) remains. The arrows indicate the location of the sample spot before chromatography. Chromatography solvents: a mixture of acetic acid, 2-butanone, methanol, and benzene in one direction and *n*-butanol, acetic acid, and water in the other.

For example, proteins and most amino acids are held tightly by cation exchangers at low pH but not at all at high pH. Aqueous solutions are usually employed and the columns are packed with beads of *ion exchange resins,* porous materials containing bound ionic groups such as $-SO_3^-$, $-COO^-$, $-NH_3^+$, or quaternary nitrogen atoms. Synthetic resins based on a cross-linked polystyrene are usually employed for separation of small molecules. For larger molecules chemical derivatives of cellulose or of crosslinked dextrans (Sephadex), agarose, or polyacrylamide are more appropriate. Positively charged ions, such as amino acids in a low pH solution, are placed on a cation exchange resin such as Dowex 50, which contains dissociated sulfonic acid groups as well as counter ions such as Na^+, K^+, or H^+.

The adsorbed amino acids are usually eluted with buffers of increasing pH containing sodium or lithium ions. The procedure, which was developed by Moore and Stein, is widely used for automatic quantitative analysis of amino acid mixtures obtained by hydrolysis of a protein or peptide (Fig. 1 6). Ion exchange chromatography of proteins and peptides is often done with such ion exchange materials as carboxymethyl-Sephadex and phosphocellulose, which carry negatively charged side chains or diethylaminoethylcellulose (DEAE-cellulose), which carries positively charged amino groups. These materials do not denature proteins or entrap them and have a large enough surface area to provide a reasonable absorptive capacity.

The mobile phase is usually buffered. For anion exchangers the pH should be above ~4.4 to keep most carboxylate side chains on the proteins ionized. The pH may be increased to pH 7-10, where most histidine imidazolium ions have dissociated, increasing the mobility of many proteins. For cation exchange the pH is usually buffered below pH 6 or 7 (Fig. 1.6).

Affinity Chromatography

In this technique the chromatographic absorbent is designed to make use of specific biochemical interactions to "hook" selectively a particular macromolecule or group of macromolecules. Affinity chromatography is used in many ways, including the purification of enzymes, antibodies, and other proteins that bind tightly to specific small molecules. Because of their open gel structure agarose derivatives in bead form provide a good solid support matrix.

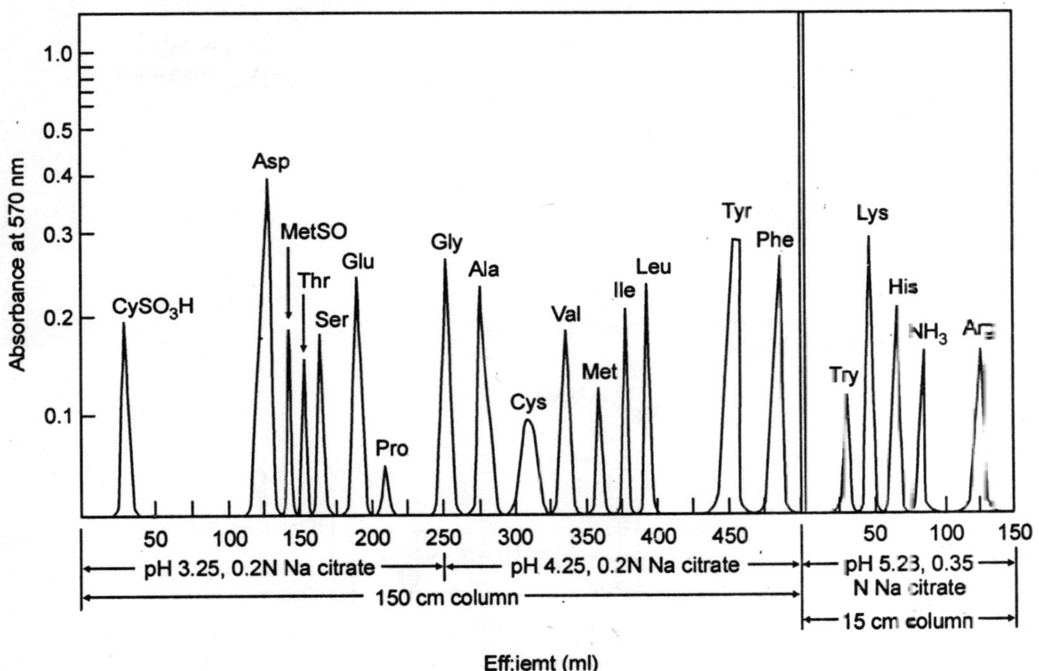

Fig. 1.6. Separation of amino acids by cation-exchange chromatography on a sulfonated polystyrene resin in the Na^+ form by the method of Moore and Stein. The amino acids were detected by reaction with ninhydrin, areas under the peaks are proportional to the amounts. Two buffers of successively higher pH are used to elute the amino acids from one column, while a still higher pH buffer is used to separate basic amino acids on a shorter column.

The hydroxyl groups of the agarose are often linked to amino compounds. In one widely used procedure the agarose is treated with cyanogen bromide ($Br - C \equiv N$) in base to "activate" the carbohydrate (Eq. 1.9 setp a). Then the amino compound is added (step-*b*). The isourea product shown is the major one with agarose gels but dextran-based matrices tend to react by steps *c-e*. Absorbents containing a large variety of R groups of shapes specifically designed to bind to the desired proteins can be made in this way. If the coupling is done with a diamine [$R = (CH_2)_H—NH_2$], the resulting ω-aminoalkyl agarose can be coupled with other compounds by reaction with *carbodiimide* (Eq. 1.10).

For reaction in nonaqueous medium dicyclohexylcarbodiimide (R" = cyclohexyl) is often used, but for linking groups to agarose a water-soluble reagent such as l-ethyl-3-(3-dimethylaminopropyl) carbodiimide is recommended. Carbodiimides are widely used for forming amide or phosphodiester linkages in the laboratory. The formation of an amide with a side chain of ω-aminoalkyl agarose can be pictured as in Eq. 1.10.

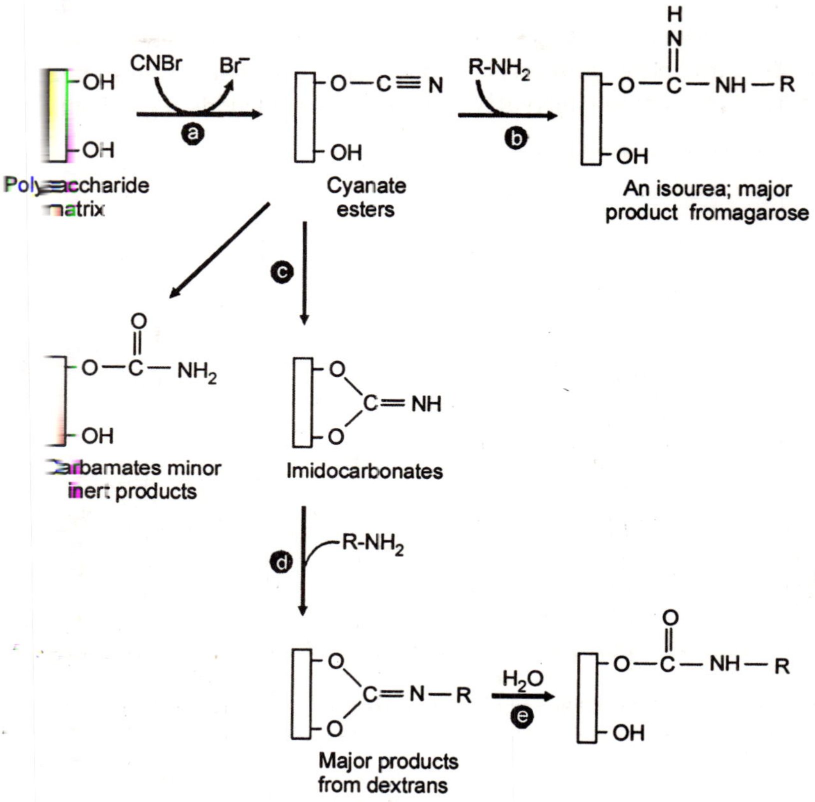

CNBr, Br⁻ (a); R-NH₂ (b); (c); R-NH₂ (d); H₂O (e)

Polysaccharide matrix

Cyanate esters

An isourea; major product fromagarose

Carbamates minor inert products

Imidocarbonates

Major products from dextrans

...(1.9)

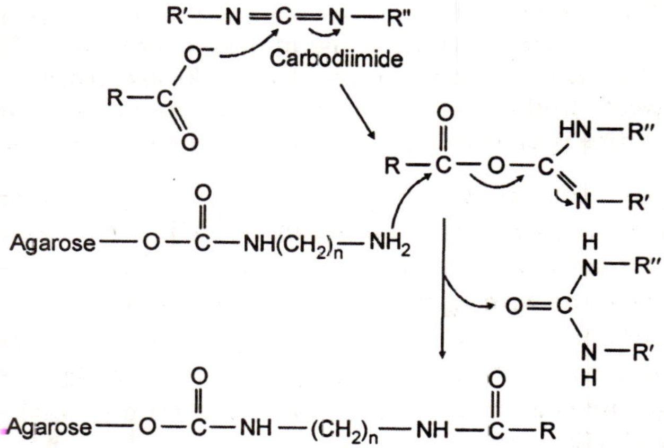

$R'-N=C=N-R''$ Carbodiimide

Agarose—O—C(=O)—NH(CH₂)ₙ—NH₂

Agarose—O—C(=O)—NH—(CH₂)ₙ—NH—C(=O)—R ...(1.10)

Many other means of preparation of adsorbents for affinity chromatography are also available. For example, 1,1'-carbonyl-diimidazole can be used to couple a diamine to the matrix Fig. 1.11. This reagent has the advantage that it does not depend upon the relatively unstable isourea linkages formed by Eq. 1.9 to hold the specific affinity ligands.

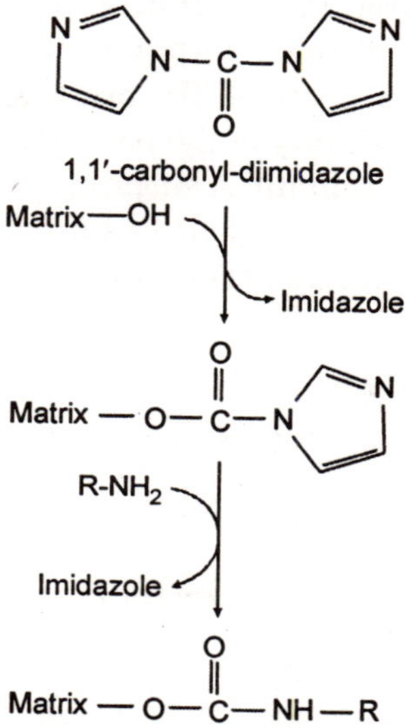

1,1'-carbonyl-diimidazole

Matrix—OH → (− Imidazole) → Matrix—O—C(=O)—N(imidazole)

R-NH₂ → (− Imidazole) → Matrix—O—C(=O)—NH—R ...(1.11)

An example of a successful application of affinity chromatography is the isolation of the enzyme cytidine deaminase from cells of *E. coli*. Cytidine was linked covalently via long *spacer* arms to the agarose beads as in the following diagram:

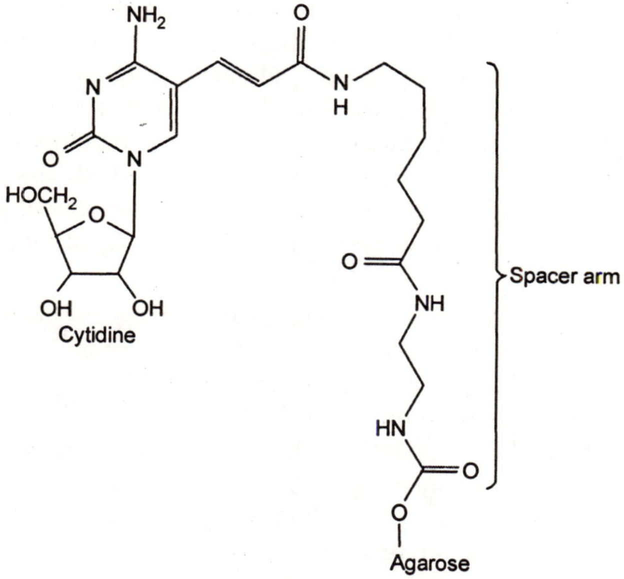

Cytidine

Spacer arm

Agarose

A cell extract was subjected to ammonium sulfate fractionation and the dialyzed protein was then poured through the affinity column which held the cytidine deaminase molecules because of their affinity for the cytidine structures that were bound to the agarose. The protein was eluted with a borate buffer; the borate formed complexes with the adjacent hydroxyl groups of the cytidine and thereby released the protein?

After passage of the protein through an additional column of DEAE-Sephadex the deaminase had been purified 1700-fold compared to the crude extract. Another technique is to engineer genes to place a polyhistidine "tag" at the C terminus of a protein chain. Commercial cloning vehicles and kits are available for this purpose. The protein produced when the engineered gene is expressed can be captured by the affinity of the polyhistidine tag for Cu^{2+}, Ni^{2+}, Co^{2+}, or Zn^{2+} held in chelated form on an affinity column. A surprising discovery was that certain dyes, for example Cibachron blue, when covalently coupled to a suitable matrix, often bind quite specifically to proteins that have a nonpolar binding pocket near a positive charge. This includes many enzymes that act on nucleotides.

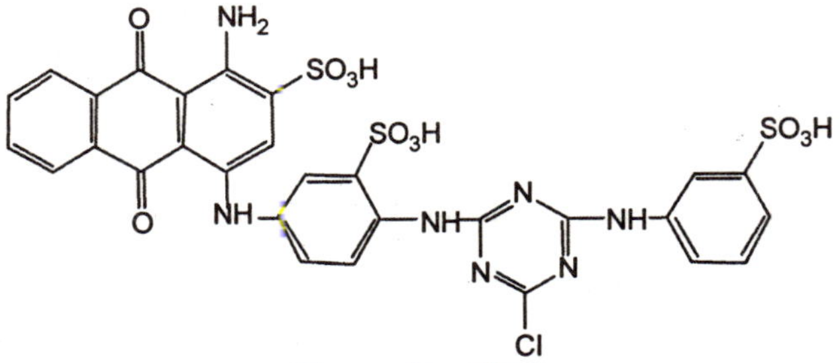

Cibacron blue F3GA

Electrophoresis and Isoelectric Focusing

The methods considered in this section make use of movement of molecules in an electrical field. Separation depends directly upon differences in the net charge carried by molecules at a fixed pH. The net charge for compounds containing various combinations of acidic and basic groups can be estimated by considering the pK_a of each group and the extent to which that group is dissociated at the selected pH using Eqs. 1.3 to 1.5. At some pH, the *isoelectric point* (pI), a molecule will carry no net charge and will be immobile in an electric field. At any other pH it will move toward the anode (+) or cathode (–). The pH at which the protein carries no net charge in the complete absence of added electrolytes is called the *isoionic point. Electrophoresis,* the process of separating molecules, and even intact cells by migration in an electrical field, is conducted in many ways.

In *zone electrophoresis,* a tiny sample of protein solution, *e.g.,* of blood serum, is placed in a thin line on a piece of paper or cellulose acetate. The sheet is moistened with a buffer and electrical current is passed through it. An applied voltage of a few hundred volts across a 20-cm strip suffices to separate serum proteins in an hour. To hasten the process and to prevent diffusion of low-molecular-weight materials, a higher voltage may be used. Two to three thousand volts may be applied to a sample cooled by water-chilled plates. Large-scale electrophoretic separations may be conducted in beds of starch or of other gels. One of the most popular and sensitive methods for separation of proteins is electrophoresis in a column filled with *polyacrylamide* or *agarose gel* or on a thin layer of gel on a plate.

The method depends upon both electrical charge and molecular size and has been referred to as *electrophoretic molecular sieving.* Polyacrylamide gel electrophoresis is often carried out in the presence of ~1% of the denaturing detergent sodium docecyl sulfate which coats the polypeptide chain rather evenly. This method, which is often referred to as *SDS-PAGE,* has the advantage of breaking up complex proteins composed of more than one subunit and sorting the resultant monomeric polypeptide chains according to molecular mass.

A disulfide-reducing reagent (Eq. 1.23) such as ~1% 2-mercaptoethanol is usually present but may be omitted to permit detection of crosslinked peptides. *Capillary electrophoresis* is increasingly popular and can be used to separate attomole amounts. It can be used not only for separation of proteins but also for rapid estimation of the net charge on a protein. The separation is conducted in tubes with internal diameters as small as 10-15 μm and as short as 1 cm. Multiple channels cut into a glass chip can be used. Whereas in conventional zone electrophoresis most of the electrical current is carried by the buffer, in *isotachophoresis* the ions being separated carry most of the current. In *isoelectric focusing,* a pH gradient is developed electrochemically in a vertical column or on a thin horizontal plate between an anode and a cathode. The pH gradient in a column is stabilized by the presence of a density gradient, often formed with sucrose, and the apparatus is maintained at a very constant temperature.

Proteins within the column migrate in one direction or the other until they reach the pH of the isoelectric point where they carry no net charge and are "focused" into a narrow band. As little as 0.01 pH unit may separate two adjacent protein bands which are located at positions corresponding to their isoelectric points. A newer development is the use of very narrow pH gradients that are immobilized on a polyacrylamide matrix. With this technique some hemoglobin mutants differing only in substitution of one neutral amino acid for another have been separated. Special techniques are needed for highly basic proteins. A two-dimensional method in which proteins are separated by isoelectric focusing (preferably with an immobilized pH gradient) in the first dimension and by SDS-gel electrophoresis in the second has become a popular and spectacularly successful method for studying complex mixtures of proteins.

Over 2000 proteins can be separated on a single plate. A similar procedure but without SDS can be used to examine undenatured proteins. Computer-assisted methods are being developed to catalog the thousands of proteins being identified in this way and also to allow rapid identification of spots by mass spectrometry. The technique can be applied to intact proteins in subpicomole quantities, even in whole cell lysates, or an enzyme such as trypsin can be used to cut the proteins into pieces on the gel plate and the mixtures of peptides can be analyzed by mass spectrometry. Capillary electrophoresis or capillary isoelectric focusing can be applied before samples are sent to the mass spectrometer.

DETERMINING THE RELATIVE MOLECULAR MASS, M_R

The evaluation of M_r is often of critical importance. Minimum values of M_r can often be computed from the content of a minor constituent, *e.g.,* the tryptophan of a protein or the iron of hemoglobin. However, physicochemical techniques provide the basis for most measurements. Observations of osmotic pressure or light scattering can also be used and provide determinations of M_r that are simple in principle, but which have pitfalls.

Ultracentrifugation

Some of the most reliable methods for determining M_r depend upon *analytical ultracentrifuges.* These instruments, capable of generating a centrifugal field as much as 4×10^5

times that of gravity, were developed in the 1920s and 1930s by T. Svedburg and associates in Uppsala, Sweden. Driven by oil turbines, the instruments were expensive and difficult to use, but by 1948 a reliable electrically driven machine, the Beckman Model E ultracentrifuge, came into widespread use. It has had a major impact on our understanding of proteins, on methods of purification of proteins, and on our understanding of interactions of protein molecules with each other and with small molecules. Nevertheless, it was not until 1990 that a truly "user-friendly" analytical untracentrifuge became available.

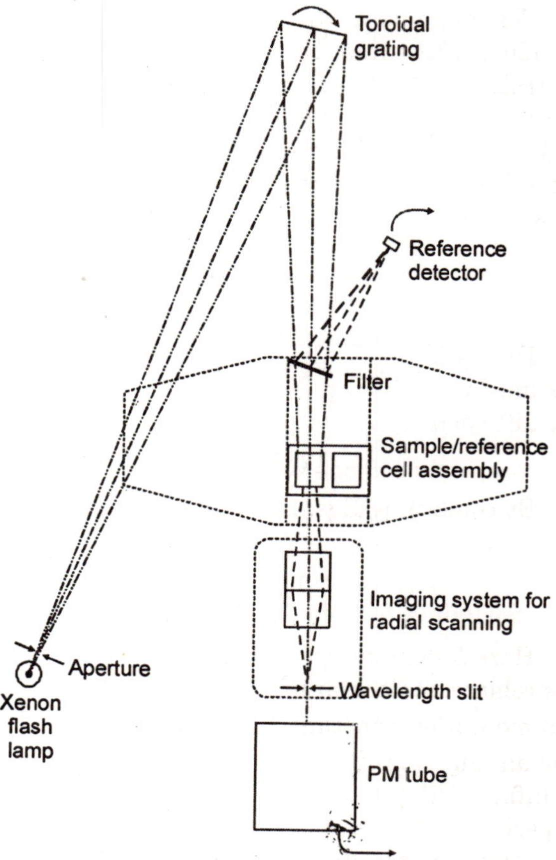

Fig. 1.7. The scanning absorption optical system of the Beckman Optima XL-A ultracentrifuge.

The Beckman Model XL-A centrifuge has a very small rotor driven by an air-cooled induction motor and is computer controlled. Data are recorded automatically in digital form and computer programs are available to carry out the necessary computations. The instrument can record ultraviolet-visible spectra at multiple radial positions in the sample cell. A straightforward determination of M_r is obtained by centrifuging until an equilibrium distribution of the molecules of a protein or other macromolecular material is obtained and by recording the variation in concentration from the center to the periphery of the centrifuge cell. Using short cells, this *sedimentation equilibrium* can be attained in 1-5 hours instead of the 1-2 days needed with older instruments. For a single component system the concentration distribution at equilibrium is given by Eq. 1.12

$$c(r) = c(a) \exp [M (1-\bar{v}\ p)\ \omega^2 (r^2-a^2) / 2RT] \qquad ...(1.12)$$

Here $c(r)$ is the concentration c at the radial position r (measured from the centrifuge axis), a is the radial distance of the meniscus, M is the molecular mass in daltons, and $\bar{v}$ is the partial specific volume in ml /gram. For most proteins $\bar{v}$ varies from 0.69-0.75. It is the reciprocal of the density of the particle. Rho (ρ) is the density (g/ml) of the solvent. A plot of log $c(r)$ against r^2 is a straight line of slope M $(1-\bar{v}\rho)$ / 2RT. The computer can also accommodate mixtures of proteins of differing molecular masses, interacting mixtures, etc.

Sedimentation Velocity

The relative molecular mass M_r can also be measured from observation of the velocity of movement of the boundary (or boundaries for multicomponent systems) between solution and solvent from which the macromolecules have sedimented.

This boundary, which can be visualized by optical methods, is quite sharp initially, but it broadens with time, because of diffusion, as the macromolecules sediment. A molecule in a centrifuge is acted upon not only by the applied centrifugal force but also by an opposing *buoyant force* that depends upon the difference in density of the sedimenting particles and the solvent and by a frictional drag, which is proportional to a *frictional coefficient f*. Setting the sum of these forces to zero for the hydrodynamic steady state yields 3-13, which defines the *sedimentation constant s*.

$$s = \frac{v}{\omega^2 r} = \frac{M(1-\bar{v}\rho)}{Nf} \qquad ...(1.13)$$

Here f is the frictional coefficient which is difficult to predict or to measure but is often assumed to be the same as the frictional coefficient that affects *diffusion*. It can be obtained from the diffusion coefficient D using Eq. 1.14.

$$D(\text{cm}^2\ \text{s}^{-1}) = k_B T / f = RT / Nf \qquad ...(1.14)$$

By combining Eqs. 1.13 and 1.14 we obtain the Svedberg equation:

$$M = \frac{RTs}{D(1-\bar{v}\rho)} \qquad ...(1.15)$$

Here R is in the cgs (cm-gram-second) unit of 8.31×10^7 erg mol^{-1} deg^{-1}. Using this equation the relative molecular mass M_r, which is numerically the same as M, can be evaluated from the sedimentation constant s. Since s, D, and $\bar{v}$ must all be measured with care, the method is demanding. It is often necessary to measure s and D at several concentrations and to extrapolate to infinite dilution. It is also customary to correct the data to give the values $s°_{20,w}$ and $D°_{20,w}$ expected at 20°C in pure water at infinite dilution. To the extent that we can regard protein molecules as spherical we can substitute for f frictional coefficient of a sphere:

$$f_{sphere} = 6\pi\eta r_h \qquad ...(1.16)$$

Here r_h is the hydrated radius or *Stokes radius* of the protein. On this assumption s will be expected to increase with the relative molecular mass approximately as $M_r^{2/3}$. A plot of log s against log M_r should be a straight line. Figure 1.8 shows Such a plot for a number of proteins.

The plots for nucleic acids, which can often be approximated as rods rather than spheres fall on a different line from those of proteins. Furthermore, the sedimentation constant falls off more rapidly with increasing molecular mass than it should for spheres. From analysis of a variety of well-characterized proteins, Squire and Himmel observed that if proteins are assumed to contain 0.53 g H_2O per gram of protein and to have a mean value for $\bar{v}$ of 0.730 g/cm³ the value of M_r can be predicted by Eq. 1.17 with the standard deviation indicated. Here, S is the sedimentation constant in Svedberg units (10^{-13}s).

$$M_r = 6850\ S^{3/2})\ 0.090\ M_r \qquad ...(1.17)$$

For proteins with various values of $\bar{v}$ Eq. 1.18, applies.

$$M_r = 922\ [S/(1-\bar{v}\ \rho)]^{3/2}) \pm 0.066\ M_r \qquad ...(1.18)$$

Gel Filtration and Gel Electrophoresis

Several newer methods of molecular mass determination were developed in the 1960s-1980s. One is gel filtration. A column of gel beads such as Sephadex is prepared carefully and is calibrated by passing a series of protein solutions through it. The volume V_e at which a protein peak emerges from the column can be expressed as the sum of two terms Eq. 1.19 in

$$V_e = V_o + \sigma V_i \qquad ...(1.19)$$

Which V_o is the *void volume, i.e.,* the elution volume that is observed for very large particles that are completely excluded from the gel, and V_i is the internal volume within the beads of gel. The value of σ is inversely related to the diffusion constant D, which for a spherical particle D is related by Eq. 1.20 to the Stokes radius r_h. This equation comes directly from Eqs. 1.16 and 1.14.

$$r_h = \frac{k_B T}{6\pi\eta D} \qquad ...(1.20)$$

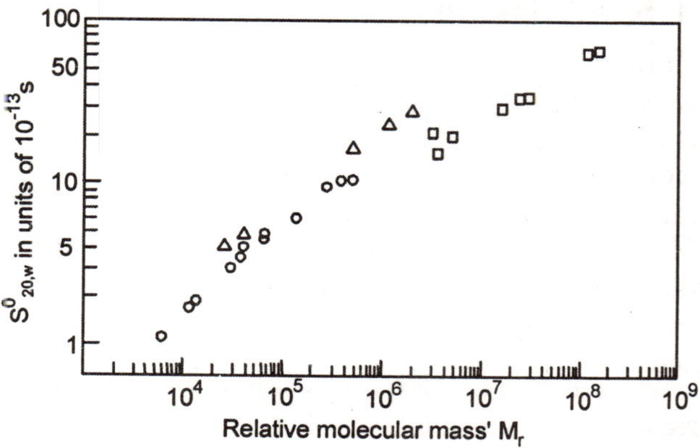

Fig. 1.8. Plots of the logarithm of the sedimentation constant *s* against the logarithm of the molecular weight for a series of proteins and nucleic acid: (O) globular proteins, (□) RNA, and (□) DNA. Proteins include lipase (milk), cytochrome *c*, ribonuclease (pancreatic), lysozyme (egg white), follicle-stimulating hormone, bacterial proteases, human hemoglobin, prothrombin (bovine), malate dehydrogenase, g-globulin (horse), tryptophanase (*E. coli*), glutamate dehydrogenase (chicken), and cytochrome *a*. Double-stranded DNA molecules are those of bacteriophage Φ × 174 (replicative form), T_7, λ_{b2}, T_2, and T_4, and that of a papilloma virus. The RNA molecules are tRNA, rRNA, and mRNA of *E. coli*, and that of turnip yellow mosaic virus.

This suggests a proportionality between σ and the molecular radius. In fact, which a and b are constants provides a fairly good approximation for σ and V_e is correlated approximately with log M_r as shown in Fig. 1.9.

$$\sigma = a \log r_h + b \qquad ...(1.21)$$

A series of reference proteins of known molecular masses are used to calibrate the column and M_r for an unknown protein is estimated from its position on the graph. Another modification of the method depends upon chromatography in a high concentration of the denaturing salt guanidinium chloride. The assumption is made that proteins are denatured into random coil conformation in this solvent. Probably the most widely used method for determining the molecular mass of protein subunits is gel electrophoresis in the presence of the denaturing detergent sodium dodecyl sulfate (SDS).

The protein molecules are not only denatured but also all appear to become more or less evenly coated with detergent. The resulting rodlike molecules usually show a uniform dependence of electrophoretic mobility on molecular mass (plotted as log M_r). Again, the molecular mass of the protein under investigation is estimated by comparison of its rate of migration with that of a series of marker proteins.

Mass Spectrometry

Mass spectrometry has played a role in biochemistry since the early 1940s when it was introduced for use in following isotopic labels during metabolism. However, it was not until the 1990s that suitable commercial instruments were developed to permit mass spectrometry using two new methods of ionization. The techniques are called *matrix-assisted laser desorption/ ionization time-of-flight (MALDI-TOF)* and *electrospary ionization (ESI)* mass spectrometry. In the MALDI technique a pulsed laser beam strikes a solid sample and heats, vaporizes, and ionizes compounds with little decomposition. Proteins or other biopolymers are mixed with a "matrix" that absorbs the heat of the laser beam. The protein sample together with the matrix is dried.

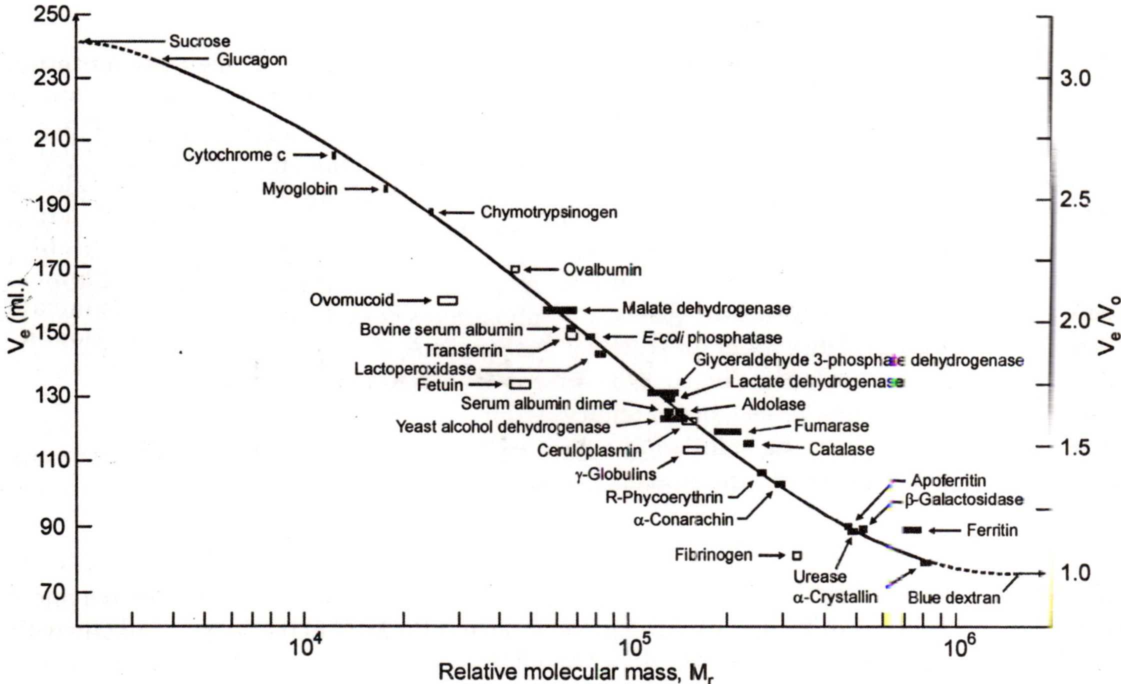

Fig. 1.9. Elution volume of various proteins on a column of Sephadex G-200 as a function of molecular mass. The right-hand vertical axis shows the ratio of the elution volumes to that of blue dextran, a high-molecular-mass polysaccharide that is excluded from the internal volume.

Most proteins form crystals and the laser beam is directed toward individual protein crystals or aggregates. Various materials are used for the matrix. Compounds as simple as glycerol, succinic acid, or urea can be used with an infrared laser. For proteins an ultraviolet nitrogen laser tuned to 337 nm is usually employed with an ultraviolet light-absorbing matrix such as hydroxybenzoic acid, 2,5-dihydroxybenzoic acid, α-hydroxy cinnamic acid, or sinapinic acid. The matrix ionizes, desorbs from the surface, and transfers energy to the crystalline protein, causing it to ionize and desorb from the surface. Oligosaccharides and oligonucleotides can be ionized in a similar way. MALDI spectra are relatively simple, often containing a single major peak corresponding to the singly charged molecular ion $[M + H]^+$ of mass $m + I$ and perhaps a doubly charged molecular ion $[M + 2H]^{2+}$ of mass $(m + 2)/2$. For oligomeric proteins the major peak is often that of the monomer with weaker peaks for oligomers.

The instrument can also be adjusted to generate negative ions whose detection is useful for study of phosphorylated peptides, many oligosaccharides and oligonucleotides. With a TOP spectrometer there is no upper limit to the mass range and masses of over 100 kDa can be measured to about ±0.1%. Femtomole quantities can be detected. The MALDI method is especially useful for complex mixtures of peptides and can be utilized in peptide sequencing. The technique is also appropriate for studying mixtures of glycoproteins. Negative-ion MALDI can be applied to oligonucleotide mixtures. Further improvements in resolution in both MALDI and ESI methods are anticipated as a result of development of Fourier transform mass spectrometers. In ESI dissolved in an appropriate solvent (usually a 50:50 mixture of methanol and water for proteins), is infused directly into the ionization chamber of the spectrometer through a fused silica capillary. At the end of the capillary the solution is subjected to electrical stress created by a voltage difference of about 5 kV between the electrospray neerale and the sampling orifice (the counter-electrode).

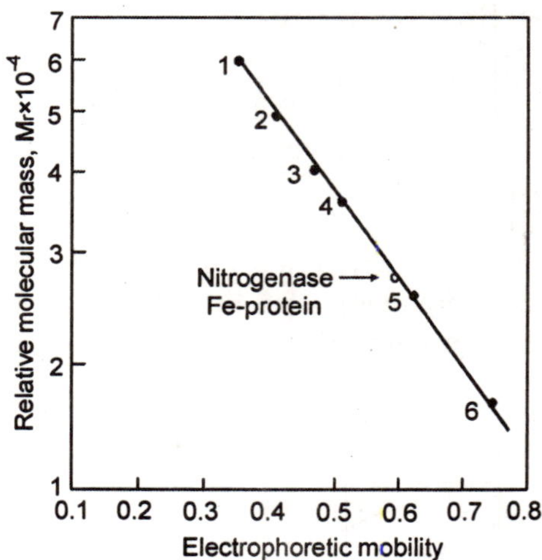

Fig. 1.10. Estimation of the molecular mass of the polypeptide chain of the nitrogenase Fe-protein using SDS-poly-acrylamide electrophoresis; from a set of four standard curves. The marker proteins are (1) catalase, (2) fumarase, (3) aldolase, (4) glyceraldehyde-phosphate dehydrogenase, (5) cc-chymotry-psinogen A, and (6) myoglobin. (o) indicates position of azoferredoxin.

The process results in the formation of singly and/or multiply charged molecular ions which are guided into the analyzer for mass analysis. For proteins every arginine, lysine, and histidine may bind a hydrogen ion to form a variety of positive ions. A 100-kDa protein may easily bind 100 protons bringing the m/z ratio for the fully charged protein to 1000, well below the maximum m/z ratio of ~/2400 for a typical quadrupole mass spectrometer.

Since not all basic groups are protonated the spectra consist of families of peaks of differing m/z (Fig. 1.12). For a single pure protein the molecular mass can be calculated from the ratio of m/z values from any pair of adjacent peaks. It is better, especially if there is a mixture of proteins, to use a computer to extract M_r. The accuracy can be quite high, typically ± 0.01%: one mass unit in 10 000. Some complexity arises from the fact that each carbon atom in a protein contains about 1% of ^{13}C. This means that for a protein of mass > 10 kDa there will be a confusing array of peaks in the mass spectrum and it may be difficult to pick out the relatively minor "monoisotopic" peak that arises from molecules containing only ^{12}C, $^{1}H_{L}$, ^{14}N, ^{16}O, and ^{32}S.

In fact, the peak representing the most abundant mass will be a few mass units higher than the monoisotopic peak. New computer programs have been devised to assist in the analysis. Use of ^{13}C and ^{15}N-depleted nutrients also extends the applicability of mass spectrometry.

Electrospray mass spectrometry utilizes a soft ionization technique at nearly atmospheric pressure. As a result, intact molecular ions are formed in high yield. The instrument can be interfaced readily to HPLC or capillary electrophoresis columns and subfemtomole amounts of proteins can be detected. A disadvantage is that salt concentrations must be kept low (< mM) and that the protein tends to bind Na^+, K^+, and anions that may confuse interpretation of spectra.

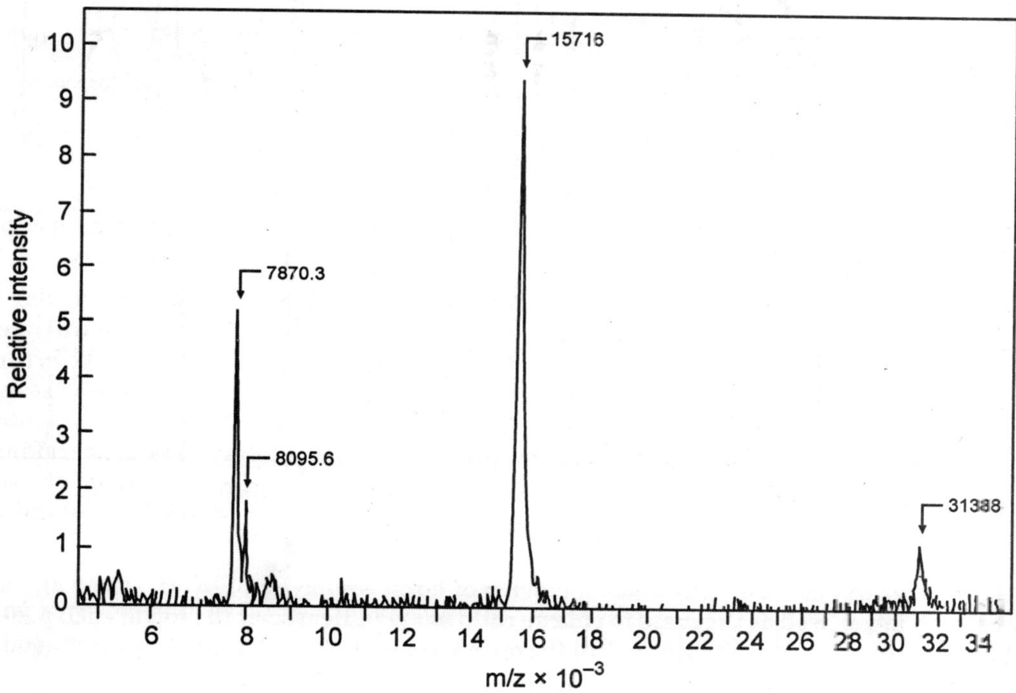

Fig. 1.11. Matrix-assisted laser desorption/ionization time-of-flight (MALDI-TOF) mass spectrum of bovine erythrocyte Cu-Zn superoxide dismutase averaged over ten shots with background smoothing. One-half μl of solution containing 10 pmol of the enzyme in 5 mM ammonium bicarbonate was mixed with 0.5 μl of 50 mM α-cyanohydroxycinnamic acid dissolved in 30% (v/v) of acetonitrile-0.1% (v/v) of trifluoroacetic acid. The mixture was dried at 37° C before analysis. The spectrum shows a dimer of molecular mass of 31,388 Da, singly charged and doubly charged molecular ions at 15,716, and 7870 Da, respectively. The unidentified ion at mass 8095.6 may represent an adduct of the matrix with the doubly charged molecular ion.

DETERMINING AMINO ACID COMPOSITION AND SEQUENCE

When they are first isolated, proteins are usually characterized by M_r, isoelectric point, and other easily measured properties. Among these is the amino acid composition which can be determined by completely hydrolyzing the protein to the free amino acids. Later, it is important to establish the primary structure or amino acid sequence. This has been accomplished traditionally by cutting the peptide chain into smaller pieces that can be characterized easily. However, most protein sequences are now deduced initially from the corresponding DNA sequences, but further chemical characterization is often needed.

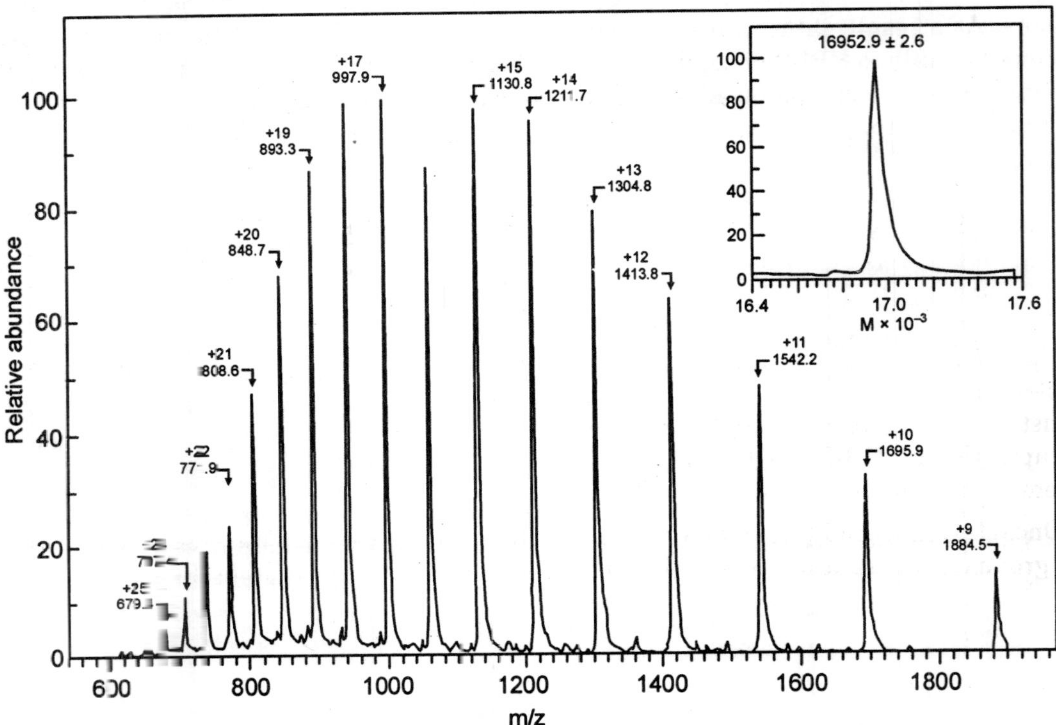

Fig. 1.12. Positive ion electrospray mass spectrum of horse apomyoglobin (M_r 16,950.4). The net charge on each ion as well as the mass to charge ratio *m/z* is indicated at the top of each peak. The inset shows a computer "deconvolution" of the spectrum with the calculated value of molecular mass.

Cleavage of Disulfide Bridges

Before a polypeptide chain can be degraded it is usually necessary to break any disulfide bridges. For some proteins such as the keratins of hair, these linkages must be broken even to get the protein into solution. Oxidation with performic acid (Eq. 1.22) has been used on ribonuclease but is not often employed because the performic acid also oxidizes tryptophan residues.

$$P_1-CH_2-S-S-CH_2-P_2 \xrightarrow{H-C(=O)-O-OH} P_1-CH_2-SO_3^- + P_2-CH_2-SO_3^- \qquad ...(1.22)$$

Reduction by dithiothreitol or dithioerythritol usually successful.

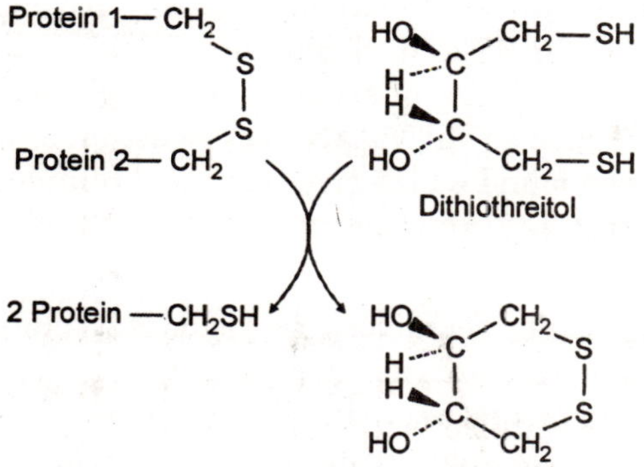

Protein 1—CH_2–S–S–CH_2—Protein 2 + Dithiothreitol (HO–CH(CH$_2$–SH)–CH(OH)–CH$_2$–SH) → 2 Protein —CH_2SH + cyclic disulfide

...(1.23)

Upon oxidation these dithiols cyclize to form stable disulfides, driving the reaction to completion. These same compounds are also widely used to protect SH groups in enzymes against accidental oxidation by oxygen and to dissolve highly crosslinked insoluble proteins. Mercaptoethanol, HS-CH_2-CH_2-OH, may be used for the same purposes but requires higher concentrations and has a disagreeable odor.

Once cleaved, disulfide bridges may be prevented from reforming by conversion of the resulting thiol groups to stable derivatives, *e.g.*, with iodoacetate and iodacetamide (Eq. 1.24).

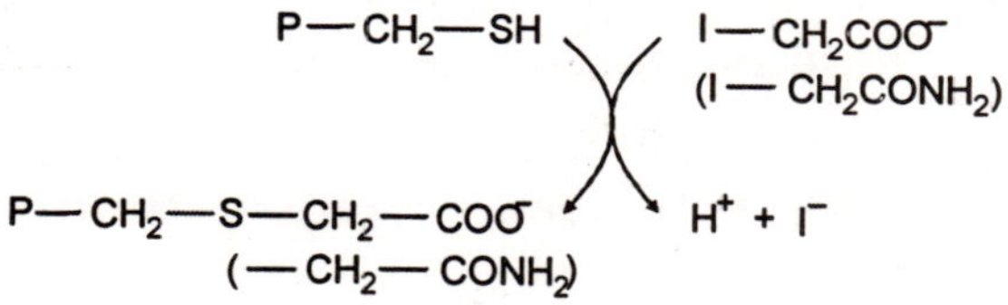

$$P—CH_2—SH + I—CH_2COO^- \ (I—CH_2CONH_2) \rightarrow P—CH_2—S—CH_2—COO^- \ (—CH_2—CONH_2) + H^+ + I^-$$

...(1.24)

A better reagent is acrylonitrile (Eq. 1.25).

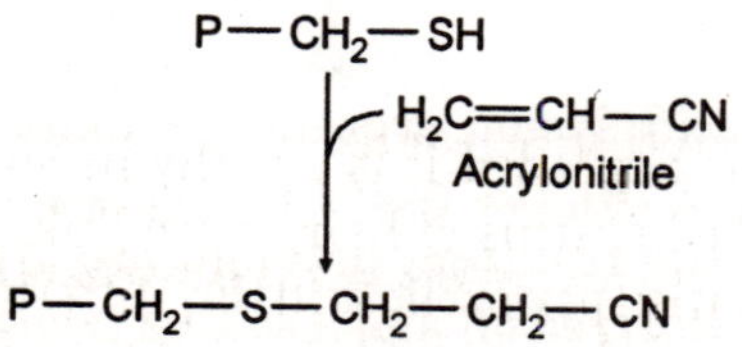

$$P—CH_2—SH \xrightarrow{H_2C{=}CH—CN \text{ (Acrylonitrile)}} P—CH_2—S—CH_2—CH_2—CN$$

...(1.25)

In modern sequencing methods vinylpyridine, which reacts in a similar way, is often used. It can be detected during amino acid analysis or sequencing after derivatization with phenylisothiocyanate (Eq. 1.30).

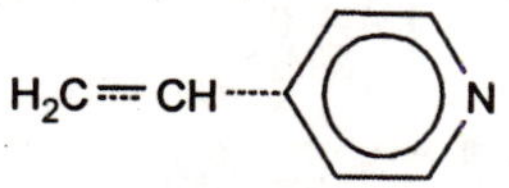

$H_2C{=}CH$–(pyridine ring, N)

Vinylpyridine

Hydrolysis and Other Chain Cleavage Reactions

Most biopolymers are inherently unstable with respect to cleavage to monomer units by reaction with water. Hydrolysis can be catalyzed by protons, by hydroxyl ions, or by the protein-hydrolyzing enzymes. Complete hydrolysis of proteins is usually accomplished by heating under nitrogen with 6 M HC1 at 150°C for 65 min. Some amino acids, especially tryptophan, are destroyed and the amides in the side chains of asparagine and glutamine are converted to the free acids. Some peptide linkages such as Val-Val are very resistant and tend to be incompletely hydrolyzed. No procedure has been found which gives the ideal complete hydrolysis. Use of 4 M methanesulfonic acid containing 3-(2-aminoethyl)indole instead of 6 M HC1 gives less decomposition of tryptophan.

Base-catalyzed hydrolysis of proteins also gives good yields of tryptophan but causes extensive racemization of amino acids. Complete enzymatic digestion of proteins can be accomplished with a mixture of enzymes including proteases produced by fungi (Pronase). However, the enzymes attack each other making quantitative analysis difficult. The problem can be circumvented by immobilizing the hydrolytic enzymes in a column of agarose gel. The protein to be hydrolyzed is passed through the gel and the constituent amino acids emerge from the bottom of the column.

Selective Enzymatic Hydrolysis

The traditional strategy in sequence determination is to cut protein chains into smaller pieces which can be separated by chroma tography or electrophoresis and sequenced individually. Enzymatic cleavage is especially useful because of its specificity. *Trypsin,* a so-called *endopeptidase*, cleaves peptide chains at a rapid rate only if the carbonyl group of the amide linkage cleaved is

$$\underset{\text{Lysine side chain}}{R-NH_3^+} + \text{(maleic anhydride)} \xrightarrow{\;-2H^+\;} R-\overset{H}{N}-\underset{\underset{O}{\|}}{C}-CH{=}CH-COO^- \qquad ...(1.26)$$

contributed by one of the basic amino acids lysine, arginine, or aminoethylcysteine. If the protein is treated with maleic anhydride (Eq. 1.26), the lysine residues are protected and trypsin will cleave only at the Arg-X positions. If the resultant peptides are separated and held at pH 3.5 overnight, the blocking groups are hydrolyzed off and a second trypsin treatment can be used to cleave at the Lys-X positions. The number of cleavage sites for trypsin can be increased by converting an -SH group to a positively charged one by aminoethylation (Eq. 1.27). The reaction can be accomplished either with ethyleneimine (caution: carcinogen) as shown in this equation or by bromoethylamine, which eliminates Br ~ to form ethyleneimine.

$$-CH_2-SH + \underbrace{CH_2-CH_2-NH_2^+}_{\text{ring}} \longrightarrow -CH_2-S-CH_2CH_2NH_3^+ \qquad ...(1.27)$$

Because of its specificity for basic residues, trypsin converts a protein into a relatively small number of tryptic peptides which may be separated and characterized. Trypsin acts primarily on denaturated proteins, and to obtain good results the disulfide bridges must be broken first.

Chymotrypsin is less specific than trypsin and pepsin is even less specific (Table 1.2). Nevertheless, they can be used to cut a peptide chain into smaller fragments whose sequences can be determined. To establish the complete amino acid sequence for a protein, "overlapping" peptide fragments must be found that contain sequences from ends of two different tryptic fragments.

Table 1.2. Specificities of Commonly Used Protein-Hydrolyzing Enzymes

Trypsin	Lys-X Arg-X	X not Pro
Chymotrypsin		
Rapidly:	Phe-X, Tyr-X, Trp-X	X not Pro
Slowly:	Y-X	X not Pro
	Y = Leu, Asn, Gln, His, Met, Ser, Thr	
Staphylococcus aureus		
protease V-8	Glu-X	X not Pro
Clostripain	Arg-X	
Pepsin		
preferentially:	X-Phe-X, X-Tyr-X, X-Leu-X	
less so:	X-Ala-X	
Thermolysin		
Rapidly:	X-Y	
	Y = Ile, Leu, Val, Ala, Phe, Met	
Slowly:	X-Y	
	Y = Tyr, Gly, Thr, Ser	

In this way the tryptic peptides can be placed in the order in which they occurred in the native protein. This tedious procedure is rarely used today. Peptide sequencing is still important but is usually coordinated with gene sequencing, X-ray structure determination, or mass spectroscopy which minimize the need for overlapping fragments. While trypsin cuts the peptide linkages Lys-X and Arg-X, a fungal protease cleaves only X-Lys.

A protease from the submaxillary glands of mice cleaves only Arg-X, one from *Staphylococcus* specifically at Glu-X, and one from kidneys at Pro-X. Several enzymes catalyze stepwise removal of amino acids from one or the other end of a peptide chain. *Carboxypeptidases* remove amino acids from the carboxyl-terminal end, while *aminopeptidases* attack the opposite end. Using chromatographic methods, the amino acids released by these enzymes may be examined at various times and some idea of the sequence of amino acids at the chain ends may be obtained.

A *dipeptidyl aminopeptidase* from bovine spleen cutsTdipeptides one at a time from the amino terminus of a chain. These can be converted to volatile trimethylsilyl derivatives and identified by mass spectrometry. If the chain is shortened by one residue using the Edman degradation the dipeptidyl aminopeptidase is again used, a different set of dipeptides that overlaps the first will be obtained and a sequence can be deduced. Carboxypeptidase Y can be used with MALDI mass spectrometry to deduce the C-terminal amino acid sequence for a peptide. However, He and Leu cannot be distinquished.

Nonenzymatic Cleavages

Of the various nonenzymatic methods that have been proposed, one has been outstandingly useful. Cyanogen bromide, N s C - Br, cleaves pep-tide chains adjacent to methionine residues. The sulfur of methionine displaces the bromide ion Eq. 1.28 and because of a favourable spatial relationship, the resulting sulfonium compound undergoes C-S bond cleavage through participation of the adjacent peptide group, (Eq. 1.28 step *b*). The C=N of the product is then hydrolyzed with cleavage of the peptide chain in step *c*. The linkage Asp-Gly can often be cleaved specifically by treatment with hydroxylamine at high pH.

Procedures for specific cleavage of tryptophanyl bonds have been devised. The Asp-Pro linkage is susceptible to cleavage by trifluoroacetic acid, which is used in the automated Edman degradation employed in peptide sequencing (Eq. 1.30). Cleavage by trifluoroacetic acid can be used to generate peptides for subsequent sequence determination.

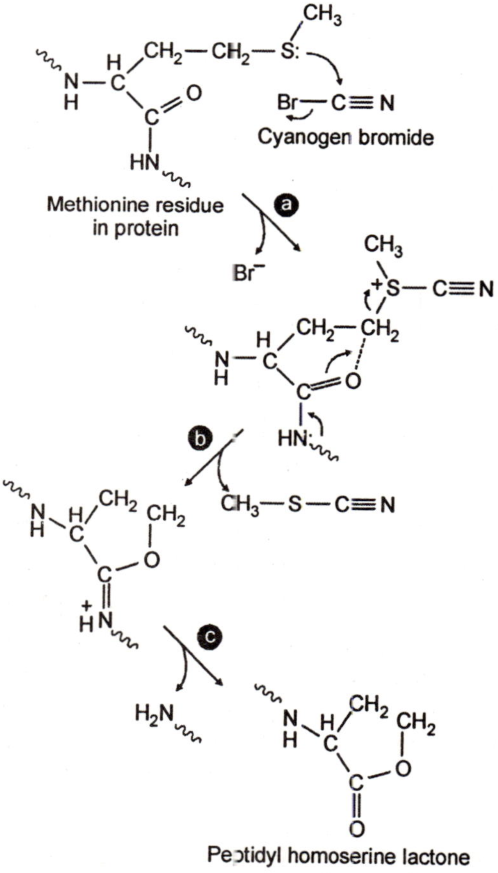

Methionine residue in protein

Cyanogen bromide

Br^-

$CH_3-S-C\equiv N$

H_2N

Peptidyl homoserine lactone

...(1.28)

Separating the Peptides

A procedure that has been very important in the development of protein chemistry is *peptide mapping* or "fingerprinting." The procedure begins with cleavage of the disulfide linkages, denaturation, and digestion with trypsin or some other protease. The sizes and amino acid compositions of the resulting series of peptides are characteristic of the protein under study.

The mixture of peptides is placed on a thin layer plate and subjected to chromatography in one direction, then to electrophoresis in the other direction, with the peptides separating into a characteristic pattern or fingerprint.

Fingerprinting has been especially useful in searching for small differences in protein structure, for example, between genetic variants of the same protein. Currently, the peptides are usually separated on ion exchange or gel filtration columns, by reversed-phase HPLC, or by capillary electrophoresis and are then often passed, in subpicomole amounts, into a mass spectrometer.

Determining Amino Acid Sequence

The covalent structure of insulin was established by Frederick Sanger in 1953 after a 10-year effort. This was the first protein sequence determination. Sanger used partial hydrolysis of peptide chains whose amino groups had been labeled by reaction with 2,4-dinitrofluorobenzene to form shorter end-labeled fragments. These were analyzed for their amino acid composition and labeled and hydrolyzed again as necessary.

Many peptides had to be analyzed to deduce the sequence of the 21rresidue and 30-residue chains that are joined by disulfide linkages in insulin. The Sanger method is mainly of historic interest, although end-labeling may still be used for various purposes. A more sensitive labeling reagent than was used by Sanger is *dansyl chloride*. It reacts to form a sulfonamide linkage that is stable to acid hydrolysis and is brilliantly fluorescent (Eq. 1.29). The related reagent dimethylaminoazobenzene-4′-sulfonyl chloride gives highly coloured derivatives easily seen on thin-layer chromatography plates.

The Edman Degradation

One of the most important reagents for sequence analysis is *phenylisothiocyanate*, whose use was developed by P. Edman. This reagent also reacts with the N-terminal amino group of peptides (Fig. 1.30 step *a*). The resulting adduct undergoes cyclization with cleavage of the peptide linkage (Eq. 1.30 step *b*) under acidic conditions. After rearrangement (step *c*) the resulting *phenylthiohydantoin* of the N-terminal amino acid can be identified. The procedure can then be repeated on the shortened peptide chain to identify the amino acid residue in the second position.

With careful work the Edman degradation can be carried down the chain for several tens of residues. Ingenious protein *sequenators* have been devised to carry out the Edman degradation automatically. Each released phenylthiohydantoin is then identified by HPLC or other techniques. Commercial sequenators have often required 5-20 nmol of peptide but new microsequenators can be used with amounts as low as 5-10 picomoles or less.

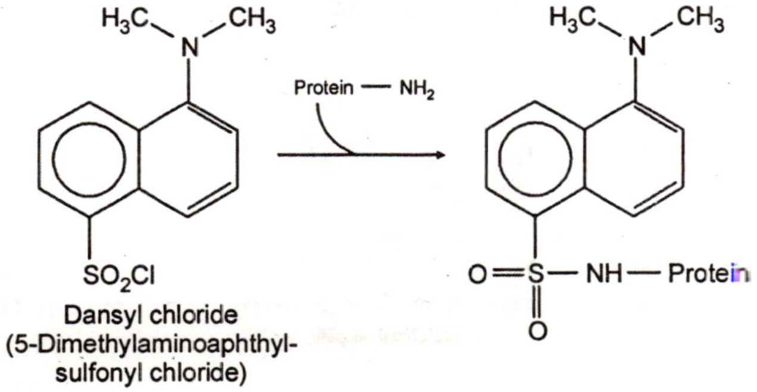

H_3C–N–CH_3 (naphthalene)–SO_2Cl $\xrightarrow{\text{Protein}-NH_2}$ H_3C–N–CH_3 (naphthalene)–$O{=}S({=}O)$–NH–Protein ...(1.29)

Dansyl chloride
(5-Dimethylaminoaphthyl-sulfonyl chloride)

Microsequencers permit sequence analysis on minute amounts of protein. Microsequencing can *be* used in conjunction with two-dimensional electrophoretic separations of proteins such as that shown in Box matter. The proteins in the polyacrylamide gel are electrophoretically transferred onto a porous sheet (membrane) of an inert material such as polyvinyl difluoride.

After staining a selected spot is cut out and placed into the sequencer. To avoid the problems associated with blocked N termini, the protein may be treated with proteases on the membrane and the resulting peptide fragments may then be separated on a narrowbore HPLC column and sequenced. Because many proteins are modified at the N terminus, blocking application of the Edman degradation, it would be useful to have a similar method for sequencing from the C terminus. It has been difficult to devise a suitable strategy, but there has been some success.

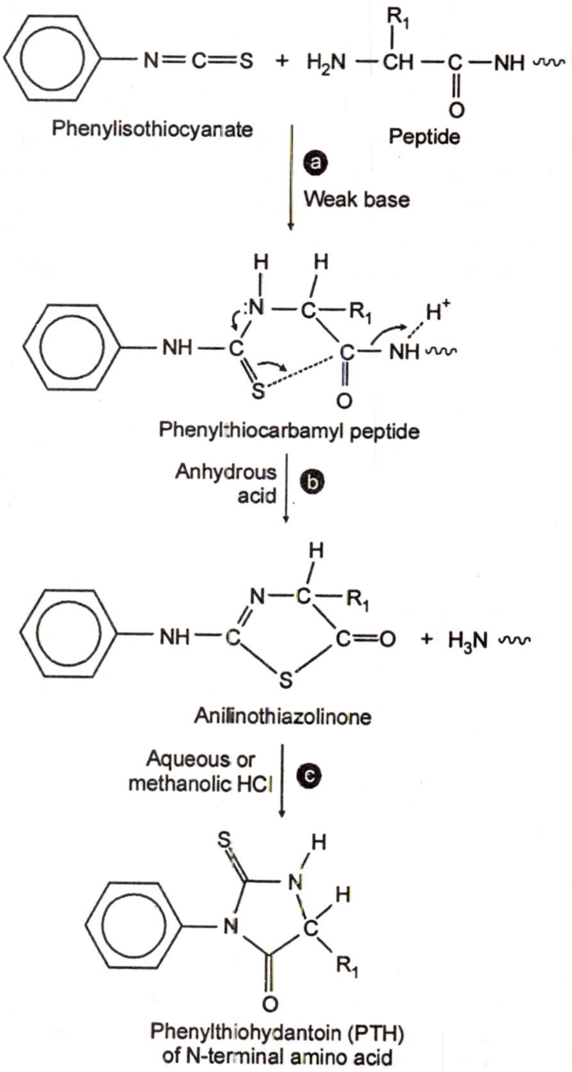

Phenylisothiocyanate + Peptide

(a) Weak base

Phenylthiocarbamyl peptide

(b) Anhydrous acid

Anilinothiazolinone + H_3N

(c) Aqueous or methanolic HCl

Phenylthiohydantoin (PTH) of N-terminal amino acid ...(1.30)

Protein Sequences from the Genes

Complete sequences of large numbers of genes have been determined and the corresponding sequences of proteins can be read directly from those of the corresponding genes. One method for sequencing a gene is to isolate a specific messenger RNA, which does not contain intervening sequences. A DNA copy (cDNA) is made from the mRNA and is used to ascertain the sequence of the encoded protein. The genomic DNA is also often sequenced.

Introns are recognized by the nucleotide sequences at their ends and the correct amino acid sequence for the encoded protein is deduced. In many instances, however, a gene can be identified only after part of the protein, often an N-terminal portion has been sequenced. This knowledge permits synthesis of an oligonucleotide probe that can be used to locate the gene.

Nucleotide sequences can be verified by comparison with sequences of tryptic or other fragments of a protein. Similarly, protein sequences are often checked by sequencing the corresponding genes as well as by study of X-ray structures. Substantial numbers of errors are made in sequencing of both DNA and protein so that checking is important.

Mass Spectrometry in Sequencing

Proteins can also be sequenced by mass spectrometry or by a combination of Edman degradation and mass spectrometry. Until recently the peptides had to be converted to volatile derivatives by extensive methylation and acetylation or by other procedures. However, newer ionization methods including MALDI (Fig. 1.11) and ESI (Fig. 1.12) have made it possible to obtain mass spectra on unmodified peptides.

In one procedure a nonspecific protease cleaves a peptide chain into a mixture of small oligopeptides which are separated by HPLC into 20-40 fractions, each of which may contain 10-15 peptides but which can be sent directly into the ionization chamber of the mass spectrometer. ^^'"Peptides can be generated from a protein using immobilized enzymes, separated on a chromatographic column, and introduced sequentially into the mass spectrometer. Examination of peptide mixtures by mass spectrometry provides a way of verifying sequences deduced from DNA sequencing. Mass spectrometry is also used widely to study covalently modified proteins. As a rule, these cannot be recognized from gene sequences.

Locating Disulfide Bridges

A final step in sequencing is often the location of S-S bridges. The reduced and alkylated protein can be cleaved enzymatically (*e.g.,* with elastase, pepsin, or thermolysin) to relatively small fragments, each of which contains no more than one modified cysteine. The same enzymatic cleavage can then be applied to the unreduced enzyme. Pairs of peptide fragments remain linked by the S-S bridges. These crosslinked pairs can be separated, the disulfide bridges cleaved, and the resulting peptides identified, each as one of the already sequenced fragments. Mass spectrometry provides a rapid method for their identification. Another elegant way of locating S-S bridges, employs *diagonal electrophoresis.*

Electrophoresis of the digest containing the crosslinked pairs is conducted in one direction on a sheet of filter paper. Then the paper is exposed to performic acid vapor to cleave the bridges according to Eq. 1.22 electrophoresis is conducted in the second direction and the paper is sprayed with ninhydrin. The spots falling off the diagonal are those that participated in S-S bridge formation. They can be associated in pairs from their positions on the paper and can be identified with peptides characterized during standard sequencing procedures. Diagonal electrophoresis and its relative diagonal chromatography are useful for other purposes as well.

After electrophoresis or chromatography is conducted in one direction, the paper or thin-layer plate may be sprayed with a reagent that will react with some comsprayed or may be irradiated with light before the separation is repeated in the second direction Fig. 1.5.

Detecting Products

Important to almost all biochemical activity is the ability to detect, and to measure quantitatively, tiny amounts of specific compounds. "Colour reagents," which develop characteristic colours with specific compounds, are especially popular. For example, *ninhydrin* can be used as a "spray reagent" to detect a small fraction of a micromole of an amino acid or peptide in a spot on a chromatogram. It can also be used for a quantitative determination, the colour being developed in a solution. More sensitive than absorption of light (colour) is fluorescence. *Fluorescamine* (Eq. 1.31) reacts with any primary amine to form a highly fluorescent product. As little as 50 pmol of amino acid can be determined quantitatively. A yet more sensitive fluorogenic reagent for detection of amino acids, peptides, and amines of all types is *o*-phthaldialdehyde.

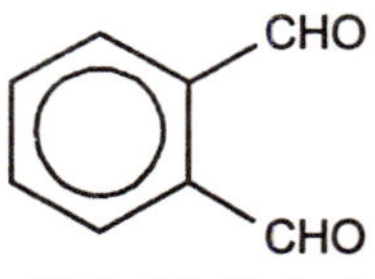

o-Phthaldialdehyde

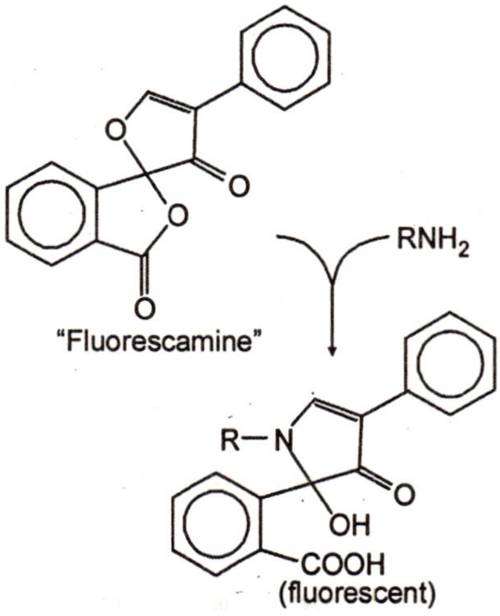

...(1.31)

Reaction with naphthalene 2,3-dicarboxaldehyde increases (Eq. 1.32) increase as the limit of detection 100-fold or more.

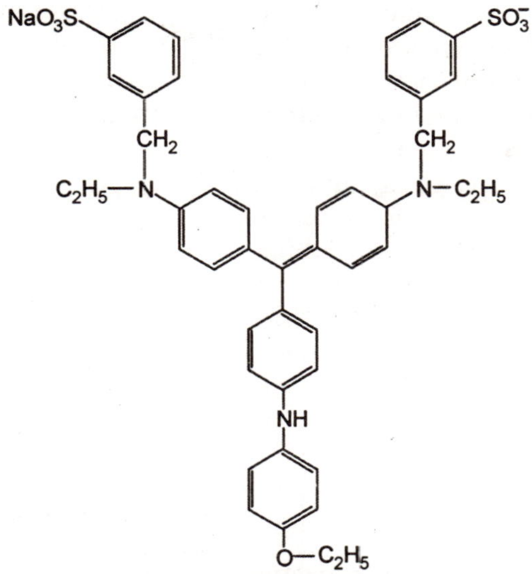

Coomassic brilliant blue ...(1.32)

Detection of proteins on thin-layer plates, gel slabs, or membranes is often accomplished by staining with a dye, the most widely used being Coomassie brilliant blue. Various silver-containing stains may also be used. After separation of a protein mixture by electrophoresis and transfer to an inert membrane, the resulting protein "blots" can be stained with specific antibodies. Flame ionization detectors can measure as little as a few picomoles of almost any column.

The importance of developing new, more sensitive analytical methods by which the quantity of material investigated can be scaled down can hardly be overemphasized. Increasingly sensitive methods of detection, including mass spectrometry, now permit measurement of fmol (10^{-15} mol) quantities in some cases. With this ability the output of neurotransmitters from a single neuron in the brain can be measured and the contents of single cells can be analyzed.

Absorption of Light

Side chains of the three aromatic amino acids phenylalanine, tyrosine, and tryptophan absorb ultraviolet light in the 240- to 300-nm region, while histidine and cystine absorb to a lesser extent. Absorption spectrum of a "reference compound" for tyrosine. There are three major absorption bands, the first one at 275 nm being a contributor to the well-known 280-nm absorption band of proteins. There is a much stronger absorption band at about 240 nm. Sensitive methods for estimating protein concentration depend upon the measurement of this absorption together with that from other side chains at around 280 or 230 nm. There is an even stronger absorption band at 192 nm.

However, at these wavelengths even air absorbs light and experimental difficulties are extreme. At 280 nm, and even more at 230 nm, it is easy to contaminate samples with traces of light-absorbing material invisible to the eye. Therefore, most estimations of protein concentration from light absorption depend upon the 280-nm band.

The spectra of N-acetyl ethyl esters of all three of the aromatic amino acids and of cystine. To a first approximation, the absorption spectra of proteins can be regarded as a summation of the spectra of the component amino acids.

However, the absorption bands of some residues, particularly of tyrosine and tryptophan, are shifted to longer wavelengths than those of the reference compounds in water. This is presumably a result of being located within nonpolar regions of the protein. Notice that the spectra for tyrosine, phenylalanine, and cystine in Fig. 1.14 have been multiplied by factors of 2 to 20. It is evident that if all of the light-absorbing side chains were present in equal numbers tryptophan would dominate the absorption band and that phenylalanine would contribute little except some small wiggles. The molar extinction coefficient e can be estimated from the numbers of residues of each type per molecule as follows:

$$\varepsilon_{280}\,(M^{-1}cm^{-1}) = 5500\,(\text{no. Trp}) + 1490\,(\text{no. Tyr}) + 125\,(\text{no. cystine}) \quad ...(1.33)$$

For proteins of unknown composition, a useful approximation is that a solution containing 1 mg/ml of protein has an absorbance at 280 nm of about 1.0.

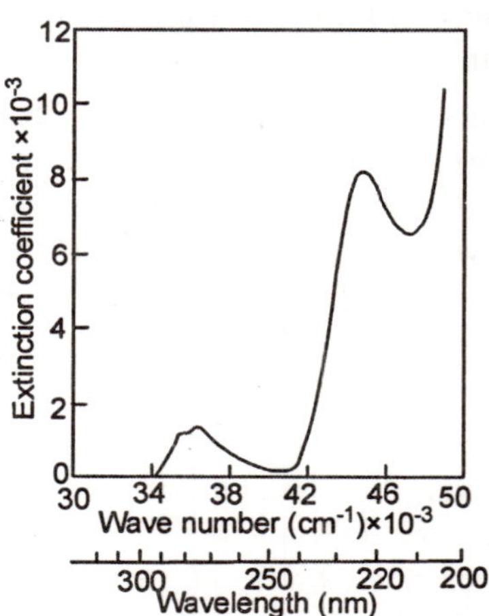

Fig. 1.13. The absorption spectrum of N-acetyltyrosine ethyl ester in an aqueous phosphate buffer of pH 6.8. Absorbance (as molar extinction coefficient, plotted against increasing energy of light quanta in units of wave number. The more commonly used wavelength scale is also given. Spectra are most often presented with the low wavelength side to the left. In the convention adopted here the energy of a quantum increases to the right. There are three π–π* electronic transitions that give rise to absorption bands of increasing intensity. The third π–π* transition of the aromatic ring is at ~ 52,000 cm^{-1} (192 nm) and reaches a molar extinction coefficient of ~ 40,000. The *n*–π* and rt–rt* transitions of the amide group in this compound also contribute to the high energy end of the spectrum.

QUANTITATIVE DETERMINATIONS AND MODIFICATION REACTIONS OF SIDE CHAIN GROUPS

The functional groups present in the side chains of proteins include $-NH_2$, –SH, S—S, –OH, $-COO^-$, the imidazole group of histidine, the guanidine group of arginine, the phenolic group of tyrosine, the indole ring of tryptophan, and the $-S-CH_3$ group of methionine. These are able to enter into a great variety of chemical reactions, most of which make use of the nucleophilic properties of these groups. The reactions are most often those of nucleophilic *addition* or nucleophilic *replacement*.

In may instances, the reactions are nonspecific; amino, thiol, and hydroxyl groups may all react with the same reagent. The usefulness of the reactions depends to a large extent on the discovery of conditions under which there is some selectivity. It is also important that the reactions be complete. Only a few reactions will be considered here; these and others have been reviewed by Glazer *et al.*

Reactions of Amino Groups

The numerous amino groups of lysine residues and of the N termini of peptide chains usually protrude into the aqueous surroundings of a protein. Chemical modification can be done in such a way as to preserve the net positive charge which amino groups carry at most pH values, to eliminate the positive charge leaving a neutral side chain, or to alter the charge to a negative value. Alterations of these charges can greatly affect interactions of the protein molecules with each other and with other substances. Amino groups react reversibly with carbonyl compounds to form Schiff bases.

Reduction of the latter by sodium borohydride or sodium cyanoborohydride causes an irreversible change (Eq. 1.34 steps *a* and *b*). Cyanoborohydride is specific for Schiff bases and does not reduce the carbonyl compound. However, side products may cause problems. Depending upon which carbonyl compound is used, the net positive charge on the amino group may be retained or it may be replaced with a different charge by this "reductive alkylation" sequence. Formaldehyde will react according to Eq. 1.34 two steps to give a dimethyl amino group with no change of net charge.

Pyridoxal phosphate converted by Eq. 1.34 into a fluorescent label. With a limited amount of pyridoxal phosphate only one or a few lysine residues may be labeled, often at active centers of enzymes. Schiff bases formed from glyceraldehyde in Eq. 1.34 can undergo the Amadori rearrangement stable products which, however, can be reconverted to the original amino groups upon acid hydrolysis. The borohydride reduction product of Eq. 1.34 with glyceraldehyde can be reconverted to the original amine by periodate oxidation.

$$R-NH_2 \underset{(a)}{\overset{R'-CHO}{\rightleftharpoons}} R-N=CH-R' \text{ (Schiff base)}$$

$$\xrightarrow[(b)]{NaBH_4 \text{ or } NaCNBH_4} R-NH-CH_2-R' \xrightarrow[(c)]{R'-CHO} R-\overset{+}{N}(=CH-R')(CH_2-R') \xrightarrow[(d)]{NaBH_4} R-\overset{+}{N}H(CH_2-R')_2$$

For formaldehyde R+H ...(1.34)

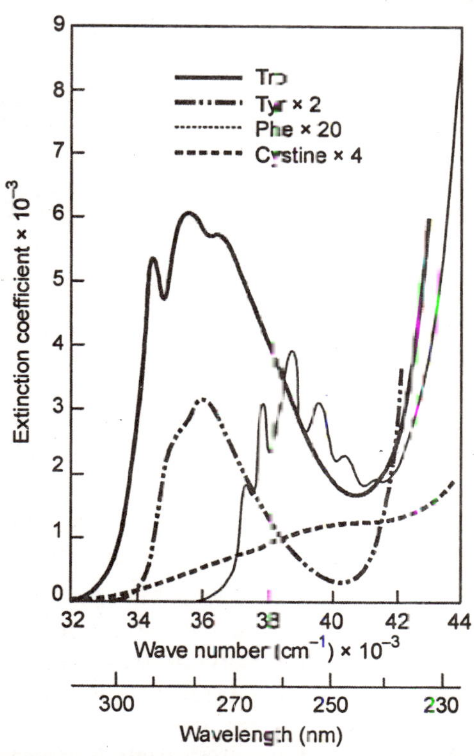

Fig. 1.14. The spectra of the first electronic transitions of the N-acetyl derivatives of the ethyl esters of phenylalanine, tyrosine, and tryptophan together with that of the dimethyl ester of cystine in methanol at 25°C. The spectra for the Tyr, Phe, and cystine derivatives have been multiplied by the factors given on the graph.

Another addition reaction of amino groups is *carbamoylation* with sodium cyanate (Eq. 1.35). A displacement reaction by an amino group on an acid anhydride such as acetic anhydride (Eq. 1.36) leads to *acylation,* a

$$R-NH_2 + \underset{\text{Cyanate}}{{}^{-}O-C\equiv N\cdots H^+} \longrightarrow R-NH-\overset{O}{\overset{\|}{C}}-NH_2$$

...(1.35)

nonspecific reaction which is also undergone by thiol, hydroxyl, and other groups. When acetic anhydride is used the net positive charge of an amino group is

$$R{-}NH_2 + \underset{\text{Acetic anhydride}}{H_3C{-}\overset{\overset{O}{\|}}{C}{-}O{-}\overset{\overset{O}{\|}}{C}{-}CH_3} \longrightarrow R{-}\overset{H}{N}{-}\overset{\overset{O}{\|}}{C}{-}CH_3 + CH_3COOH \qquad ...(1.36)$$

lost. However, the product obtained with succinic anhydride or maleic anhydride (Eq. 1.26) carries a negative charge. In the latter case, the modification can readily be reversed by altering the pH.

Both *amidination* (Eq. 1.37) and *guanidination* (Eq. 1.38) lead to retention of the positive charge.

$$R{-}NH_2 + \underset{\text{Methyl acetimidate}}{H_3C{-}\overset{\overset{NH_2^+}{\|}}{C}{-}OCH_3} \longrightarrow R{-}\overset{H}{N}{-}\boxed{\overset{\overset{NH_2^+}{\|}}{C}{-}CH_3} + CH_3OH \qquad ...(1.37)$$

$$R{-}NH_2 + \underset{\text{O-Methylisourea}}{H_2N{-}\overset{\overset{NH_2^+}{\|}}{C}{-}O{-}CH_3} \longrightarrow R{-}\overset{H}{N}{-}\boxed{\overset{\overset{NH_2^+}{\|}}{C}{-}CH_3} + CH_3OH \qquad ...(1.38)$$

Bifunctional imidoesters such as *dimethylsuberimidate* may be used to establish whether or not two different proteins or subunits are close together in a complex or in a supramolecular structure such as a membrane or ribosome.

$$H_3C{-}O{-}\underset{\underset{NH_2^+}{\|}}{C}{-}(CH_2)_6{-}\underset{\underset{NH_2^+}{\|}}{C}{-}O{-}CH_3$$

Dimethylsuberimidate

Another useful reaction of amino side chains is that with dansyl chloride (Eq. 1.29). Many lysine derivatives can be determined quantitatively by amino acid analysis.

Reactions of SH Groups

In addition to the alkylation with iodoacetate (Eq. 1.24), sulfhydryl groups can react with *N*-ethylmaleimide (Eq. 1.39). This reaction blocks the SH groups irreversibly and has often been used in attempts to establish whether or not a thiol group plays a role in the functioning of a protein. Loss of function in the presence of this "sulfhydryl reagent" may mean that an SH group has an essential role or it could be a result of the bulk of the group added. The *N*-ethylmaleimide group is large and could prevent proper contact between an enzyme and substrate or between two proteins.

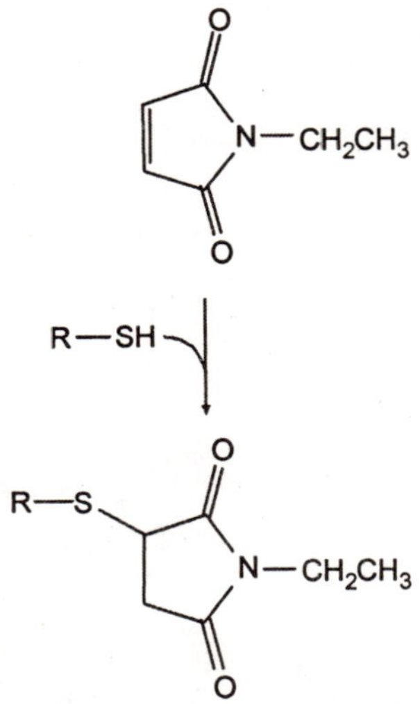

O

N—CH_2CH_3

O

R—SH

O

R—S

N—CH_2CH_3 ...(1.39)

O

To avoid the possible effect of excessive bulk, it is useful to convert the SH to the small thiocyanate group – SCN. (Eq. 1.40) Ellman's reagent 5,5′-dithiobis(2-nitrobenzoic acid; DTNB) reacts quantitatively with –SH groups to form mixed disulfides with release of a thiolate anion

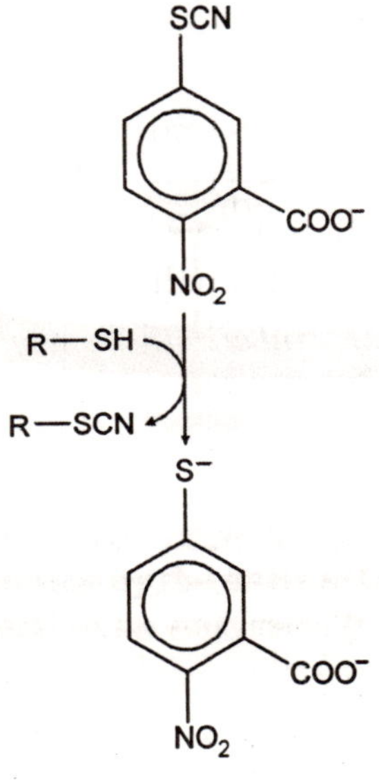

SCN

COO^-

NO_2

R—SH

R—SCN

S^- ...(1.40)

COO^-

NO_2

that (Eq. 1.41) absorbs light at 412 nm with a molar extinction coefficient of 14,150 M^{-1} cm^{-1}. While DTNB has been widely used to determine the content of -SH groups in proteins, there are some disadvantages.

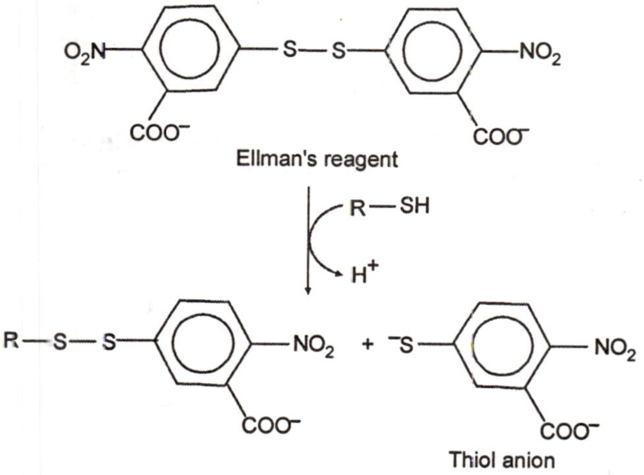

O_2N—C₆H₃(COO⁻)—S—S—C₆H₃(COO⁻)—NO_2

Ellman's reagent

↓ R—SH, H^+

R—S—S—C₆H₃(COO⁻)—NO_2 + ⁻S—C₆H₃(COO⁻)—NO_2 ...(1.41)

Thiol anion

Pyridyldisulfides such as 2-pyridyldisulfide or the isomeric 4-pyridyldisulfide react more completely and with greater selectivity.

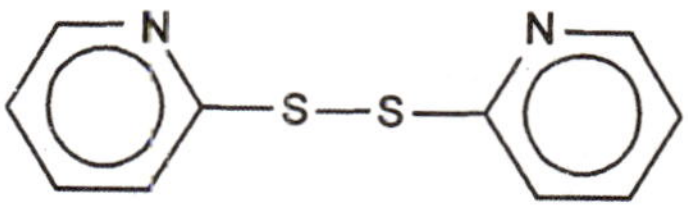

(2-pyridyl)—S—S—(2-pyridyl)

Thiol groups have a high affinity for mercury ions including organic mercury derivatives, which are widely used in the determination of protein structures by X-ray crystallography. Titration of SH groups in proteins is often accomplished with *p*-mercuribenzoate (Eq. 1.42). The reaction may be followed spectrophotometrically at 250 - 255 nm, a region in which the mercaptide product absorbs strongly.

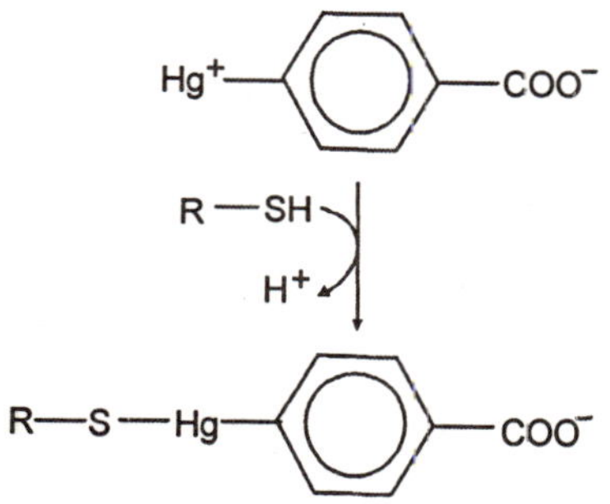

Hg^+—C₆H₄—COO^-

↓ R—SH, H^+ ...(1.42)

R—S—Hg—C₆H₄—COO^-

Reactions of Other Side Chains

There are no highly selective reactions for –OH, –COO^-, or imidazole groups. However, some hydroxyl groups in active sites of enzymes are unusually reactive in nucleophilic addition or displacement and can be modified by acylation, phosphorylation, or in other ways. Carboxyl groups, which are exceedingly numerous on protein surfaces, can be modified by treating with a water-soluble carbodimide (Eq. 1.10) the presence of a high concentration of an amine such as the ethyl ester of glycine.

The imidazole groups of residues of histidine can often be selectively destroyed by dye-sensitized photooxidation or can be acylated with ethoxyformic anhydride. Compounds with two adjacent carbonyl groups such as 1,2-cyclohexanedione (Eq. 1.34) react selectively with guanidinium groups from arginine residues in proteins. Under certain conditions the product indicated in Eq. 1.43 predominates. Related reagents are derived from camphorquinone.

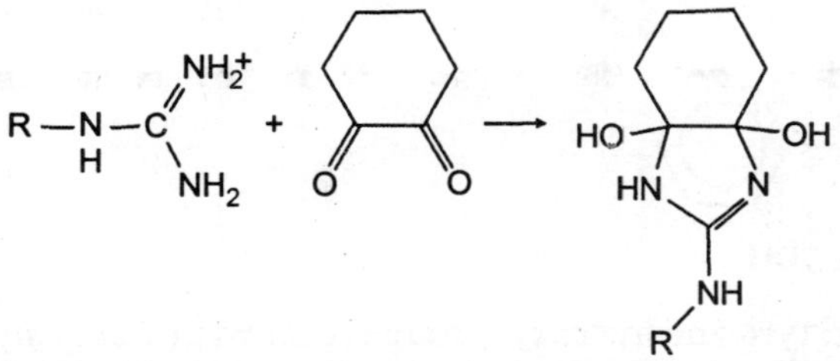

...(1.43)

The phenolic group of tyrosine undergoes iodination (Eq. 1.44), acylation, coupling with diazonium compounds, and other reactions.

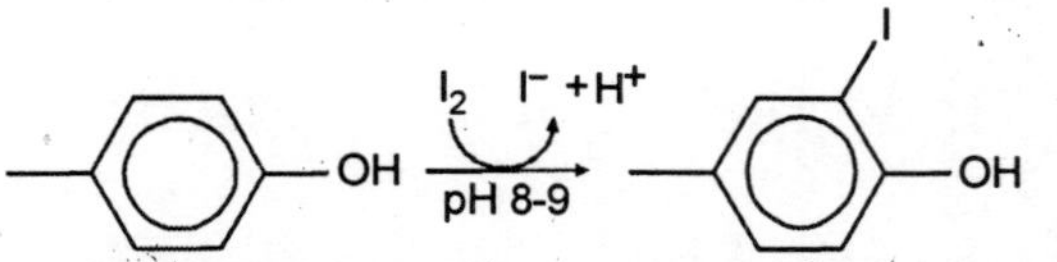

...(1.44)

The following sparingly soluble chloroamide together with I^- will also iodinate tyrosine and can be used to incorporate radiolabeled iodine into proteins.

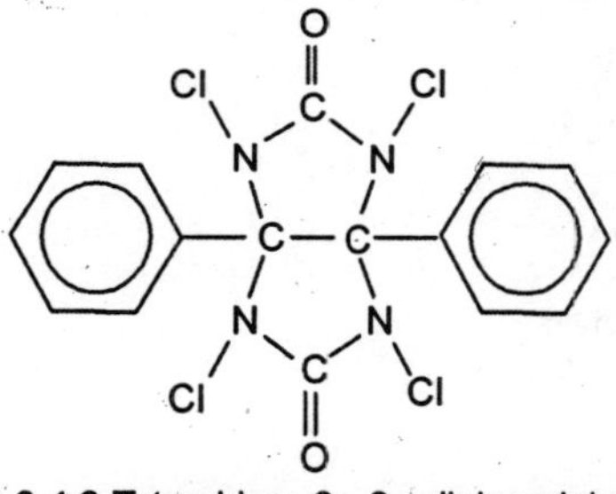

1,3,4,6-Tetrachloro-3α,6α-diphenylglycouril

Tetranitromethane reacts slowly with tyrosyl groups to form 3-nitrotyrosyl groups (Eq. 1.45). The by-product *nitroform* is intensely yellow With $\varepsilon_{350} = 14{,}400$. The reagent also oxidizes SH groups and reacts with other anionic groups.

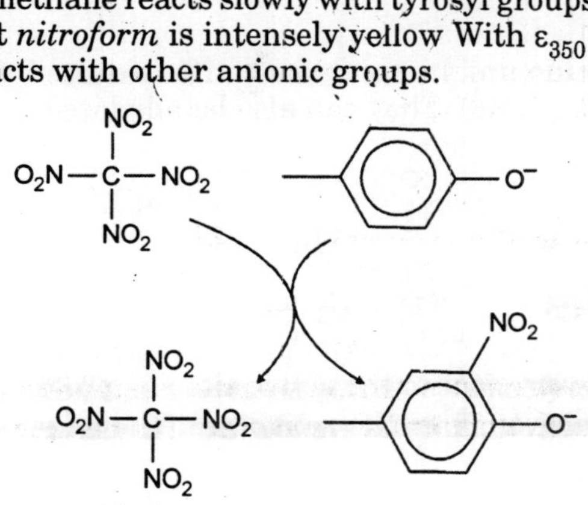

...(1.45)

Nitroform

Koshland devised the following reagent for the indole rings of tryptophan residues (Eq. 1.46)

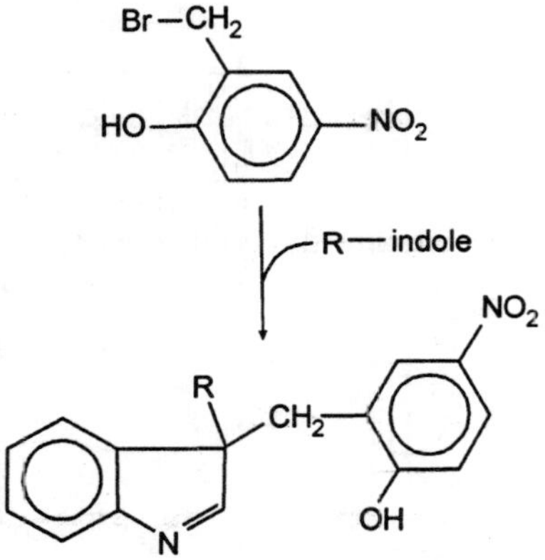

...(1.46)

Imidazole, lysine amino groups, and tyrosine hydroxyl groups react with *diethylpyrocarbonate* (Eq. 1.47) at low enough pH (below 6) that the reaction becomes quite selective for histidine. Reactivity with this reagent is often used as an indication of histidine in a protein. The reaction may be monitored by observation of NMR resonances of imidazole rings.

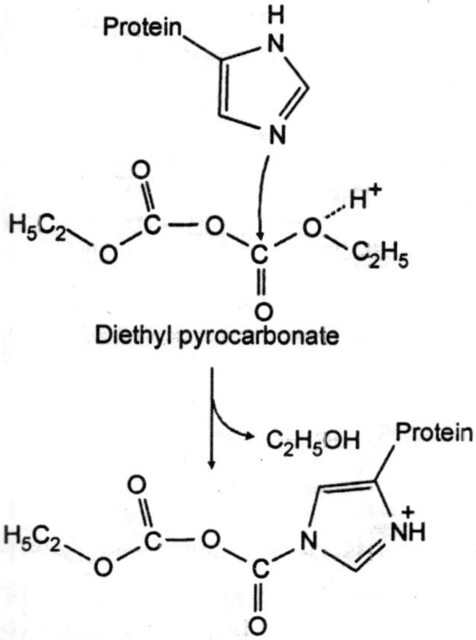

Diethyl pyrocarbonate

...(1.47)

The thioether side chains of methionine units in proteins can be oxidized with hydrogen peroxide to the corresponding sulfones (Eq. 1.48). They can also be alkylated, *e.g.*, by CH_3I to form $R—S^+(CH_3)_2$.

$$R—S—CH_3 + 2H_2O_2 \longrightarrow R—\overset{O}{\underset{O}{\overset{\|}{\underset{\|}{S}}}}—CH_3 + 2H_2O \qquad ...(1.48)$$

Affinity Labeling

To identify groups that are part of or very near to the active site of a protein, reagents can be designed that carry a reactive chemical group into the active site. The related *photoaffinity labeling* is also widely used.

SYNTHESIS OF PEPTIDES

The synthesis of peptides of known sequence in the laboratory is extremely important to biochemical research. For example, we might want to know how the effects of a peptide hormone are altered by replacement of one amino acid in a particular position by another. The synthetic methods must be precise and because there are so many steps the yield should be 98% or better for every step. Even so, it is still impractical to synthesize very large peptides. Those that have been made, such as the hormone insulin and the enzyme ribonuclease, have been obtained in low yields and have been difficult to purify. It is usually more practical to obtain large peptides from natural sources. It is often practical to clone a suitable piece of DNA in a bacterial plasmid and to set up biological production of the desired peptide.

On the other hand, for smaller peptides, laboratory synthesis is feasible. Even for large peptides it is useful because it permits incorporation of unnatural amino acids as well as isotopic labels. The general procedure for making a peptide in the laboratory is to "block" the amino group of what will become the N-terminal amino acid with a group that can be removed later. The subsequent amino acid units "activated" at their carboxyl end are then attached one by one.

The chemical activation is often accomplished by conversion of the carboxyl group of the amino acid to an anhydride. At the end of the synthesis, the blocking group must be removed from the N terminus and also from various side chain groups such as those of cysteine and lysine residues. In many respects, this procedure is analogous to the biological synthesis of proteins.

Solid-Phase Peptide Synthesis

Modern methods of peptide synthesis began with the solid-phase method introduced by Merrifield in 1962 (Fig. 1.15). To begin the synthesis a suitably protected amino acid is covalently linked to a polystyrene bead. The blocking *t*-butoxycarbonyl (Boc) group is removed as isobutene by an elimination reaction to give a bound amino acid with a free amino group. This can then be coupled to a second amino acid with a blocked amino group using dicyclohexylcarbodiimide (Eq. 1.10).

The removal of the blocking group and addition of a new amino acid residue can then be repeated as often as desired. The completed peptide is removed from the polystyrene by action of a strong acid such as HF. Advantages of the Merrifield procedure are that the peptide is held tightly and can be washed thorougly at each step. Problems arise from repeated use of trifluoroacetic acid and the need to use HF or other strong acid to cleave the peptide from the matrix and also to remove blocking benzyl groups that must be present on many side chain groups.

Newer variations of the procedure include a more labile linkage to a polyamide type of polymer and use of blocked amino acids. These "active esters" will spontaneously condense with the free amino group of the growing peptide and with suitable catalysis will eliminate pentafluorophenol. The fluorenylmethoxycarbonyl (Fmoc) blocking group is removed under mildly basic conditions. The whole procedure has been automated in commercially available equipment. Smaller peptides may be joined to form longer ones. Also useful is enzymatic synthesis. Protein-hydrolyzing enzymes under appropriate conditions will form peptide linkages, for example, joining together oligopeptides.

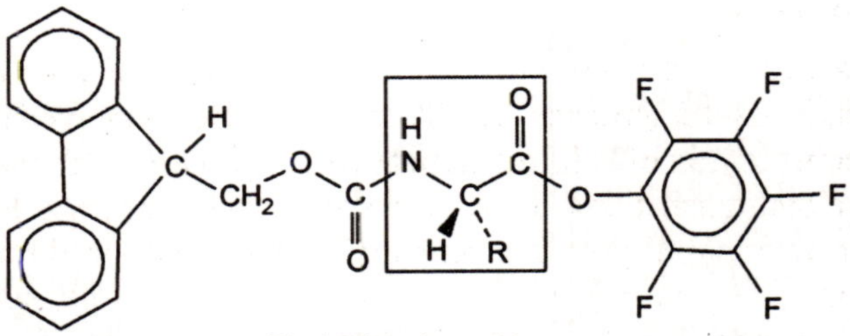

Blocked amino acid

Other new methods have been devised to join unprotected peptides. Semisynthetic approaches can also be used to place unnatural amino acids into biologically synthesized proteins through the use of suppressor transfer RNAs.

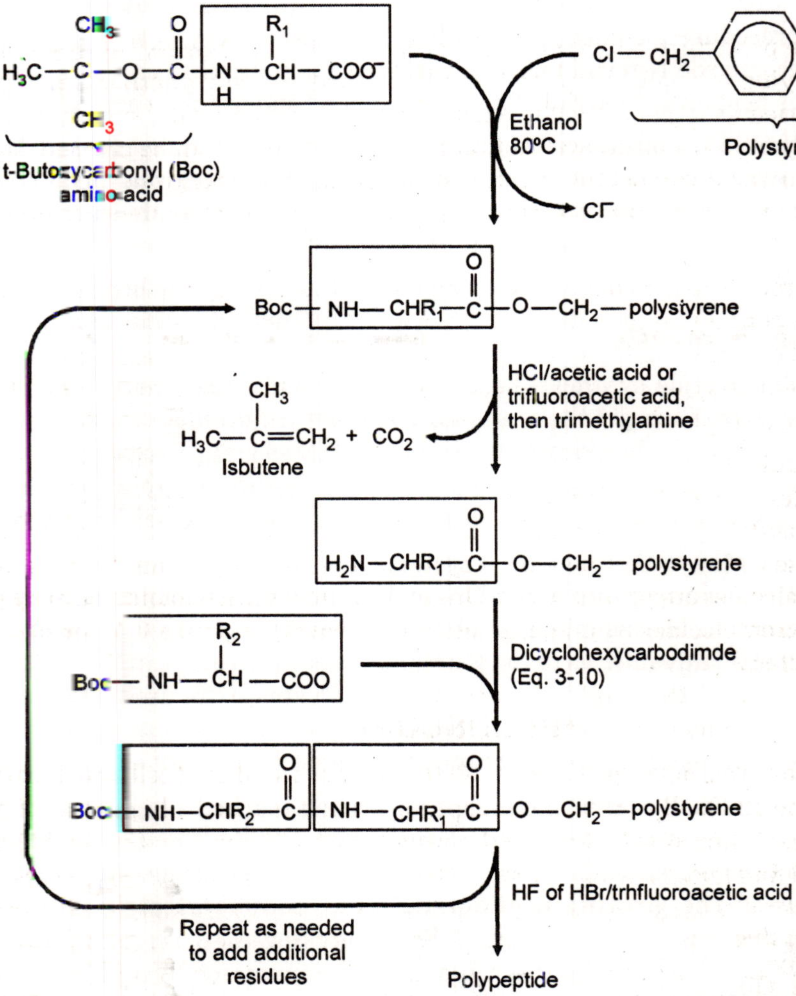

Fig. 1.15. Procedure for solid-phase peptide synthesis.

Combinatorial Libraries

Many chemists devote all of their efforts to the synthesis of new compounds, including polypeptides, that might be useful as drugs. Traditionally, this has involved the tedious preparation of a large number of compounds of related structure which can be checked individually using various biochemical or biological tests. In recent years a new approach using "combinatorial chemistry" has become very popular and is continually being adapted for new purposes. There are several approaches to creating a combinatorial library. In "split synthesis" procedures a solid-phase synthesis is conducted on beads. For example, a family of peptides, each with the same C terminus, can be started on a large number of beads. After the first amino acid residue is attached the beads are divided into up to 20 equal portions and different amino acids are added to each portion. The beads can then be mixed and again subdivided. The third residue will again contain many different amino acids attached to each of the different amino acids in the second position.

By repeating the procedure again, perhaps for many steps, a "library" of random peptides with each bead carrying a single compound will be formed. To test whether a polypeptide or other compound carried on a given bead has a derived biological activity, such as the ability to inhibit a certain enzyme, various assays that require only one bead can be devised. However, if a particular bead carries a compound of interest, how can it be identified? The bead carries only a small amount of compound but it may be possible using microsequencing procedures to identify it.

An alternative procedure is to use an encoding method to identify the beads. An alternative to the "one bead-one peptide" approach is to incorporate random sequences of a DNA segment into a gene that can be used to "display" the corresponding peptide sequence. A protein segment (which may be a random sequence) can be displayed either on the major coat proteins along the shaft or on the minor coat proteins at the end of the bacterial virus fd.

In the case of random insertions, each virus particle may display a different sequence (as many as 10^8). Peptides may be selected by binding to a desired receptor or monoclonal antibody and the DNA encapsulated in the virus particle can be used to produce more peptides for identification purposes. Many other ingenious systems for constructing and testing libraries of peptides and other molecules are being devised. One of these involves a photolithography procedure for immobilizing macromolecules in a regular addressable array, *e.g.*, in a 0.5-mm checkerboard pattern, on a flat surface.

MICROSCOPY

The light microscope was developed around 1600 but serious studies of cell structure (histology) did not begin until the 1820s. By 1890, microscope lenses had reached a high state of perfection but the attainable resolution was limited by the wavelength of light. For 450 nm blue light the limit is about 300 nm and for ultraviolet light, viewed indirectly, about 200 nm. By the 1940s the electron microscope with its far superior resolving power had overshadowed the light microscope.

For both light and electron microscopy, the preparation of thin sections of cells is a very important technique. Only with very thin sections is the image sufficiently focused. However, *confocal scanning optical microscopy,* invented in the 1950s but not used commercially until much later, provided an alternative solution to the focusing problem.

A conical beam of light focused to a point is scanned across the sample and the transmitted light (or light emitted by fluorescence) passes through a small "pinhole" aperture located in the primary image plane to a photomultiplier tube where its intensity is recorded. The illuminating beam is moved to scan the entire field sequentially. A series of pinholes in a spinning disk may accomplish the same result. The focal plane can be varied so that an image of a thick object such as a cell can be optically sectioned into layers of less than 1 μm thickness. Stereoscopic pairs can also be generated Fig. 1.17. A newer development in confocal microscopy is the use of two-photon and three-photon excitation of the fluorescent molecules that occur naturally within cells using short pulses of short-wavelength high-energy laser light.

Distribution of such compounds as NADH, DNA, and the neurotransmitter serotonin can be observed without damaging cells. Individual storage granules, each containing ~5 × 10^8 molecules of serotonin in a concentration of ~50 mM, can be seen. Another new instrument, the *near-field scanning optical microscope* (NSOM), is a lensless instrument in which the illuminating beam passes through a very small (*e.g.*, 100 nm diameter) hole in a probe that is scanned in front of the sample. It may extend the limit of optical microscopy to ~ 1/50 the wavelength of the light. Since it first became commercially available in 1939, the electron microscope has become one of the most important tools of cell biology. The practical resolution is about 0.4 nm, but recent developments in scanning electron microscopy have resulted in resolution of 0.14 nm.

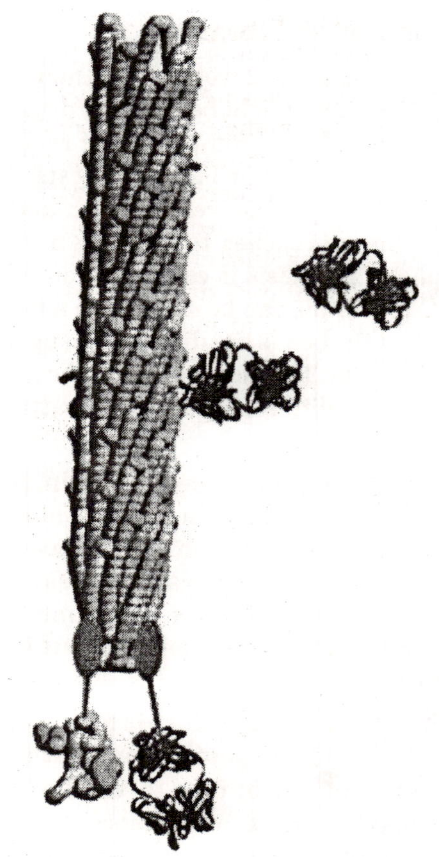

Fig. 1.16. Model of bacteriophage fd engineered to display peptides as inserts in the coat proteins of the virus. The native virus structure is shown in gray; proteins not present in the native virus are shown black or green. Inserted near the N-termini of some major coat proteins is a 6-residue peptide. To one of these peptides a specific Fab antibody fragment (green) has bound from solution, and a second Fab is shown nearby. The N-terminal region of a minor coat protein at the end of the virion has been engineered to display a (different) Fab fragment. Steric constraints are less stringent for inserts in the minor proteins, but fewer copies per virion are possible.

Of major importance was the development around 1950 of microtomes and knives capable of cutting thin (20-200 nm) sections of tissues embedded in plastic. A bacterium such as *E. coli* can be sliced into as many as 10 thin longitudinal slices a eukaryotic cell of 10 um diameter into 100 slices. Serial sections can be examined to determine three-dimensional structures. If a slice of fresh (frozen) tissue is examined directly, little is seen because most of the atoms found in cells are of low atomic mass and scatter electrons weakly and uniformly.

Therefore, thin sections must be "stained" with atoms of high atomic mass, *e.g.,* by treatment with potassium permanganate or osmium tetroxide. Tissues must also be "fixed" to prevent disruption of cell structure during the process of removal of water and embedding in plastic. Fixatives such as formaldehyde react with amino groups and other groups of proteins and nucleic acids. Some proteins are precipitated in place and digestive enzymes that otherwise would destroy much of the fine structure of the cell are inactivated. Glutaraldehyde (a five-carbon dialdehyde) is widely used to fix and crosslink protein molecules in the tissue. The methods continue to be improved. Small particles, including macromolecules, may be "shadowed." Chromium or platinum can be evaporated in a vacuum from an angle onto the surface of the specimen, Individual DNA molecules can be "seen" in this way.

In fact, only the "shadows" are seen and they are 2-3 times wider than the DNA molecules. In the *negative contrast* method a thin layer of a solution containing the molecules to be examined, together with an electron-dense material such as 1% sodium phosphotungstate, is spread on a thin carbon support film. Upon drying, a uniform electron-dense layer is formed. Where the protein molecules lie, the phosphotungstate is excluded, giving an image of the protein molecule. Surfaces of cells, slices, or intact bacteria can be coated with a deposit of platinum or carbon.

The coating, when removed, provides a "negative" *replica* which can be examined in the microscope. Alternatively, a thin plastic replica can be made and can be shadowed to reveal topography. In "freeze fracturing" and "freeze etching," fresh tissue, which may contain glycerol to prevent formation of large ice crystals, is frozen rapidly. Such frozen cells can often be revived; hence, they may be regarded as still alive until the moment that they are sliced! The frozen tissue is placed in a vacuum chamber within which it is sliced or fractured with a cold knife. If desired, the sample can be kept in the vacuum chamber at about –1.00°C for a short time, during which some water molecules evaporate from the surface. The resultant etching reveals a fine structure of cell organelles and membranes in sharp relief.

After etching, a suitable replica is made and examined. Fracturing tends to take place through lipid portions of cell membranes. Small viruses, bacterial flagella, ribosomes, and even molecules can be seen by electron microscopy. However, to obtain a clear image in three dimensions requires a computer-based technique of *image reconstruction or electron microscope tomography,* which was developed initially by Aaron Klug and associates. A sample is mounted on a goniometer, a device that allows an object to be tilted at exact angles. Electron micrographs are prepared with the sample untilted and tilted in several directions at various angles, *e.g.,* up to 90° in 10° increments. The micrographs are digitized and a computer is used to reconstruct a three-dimensional image. In *electron crystallography* micrographs of two-dimensional crystalline arrays of molecules or larger particles are prepared.

A Fourier transform of the micrograph gives a diffraction pattern which can be treated in a manner similar to that usual for X-ray diffraction to give a three-dimensional image. An important milestone in use of this technique was the determination of the structure of bacteriorhodopsin at 0.3-nm resolution. Bear in mind that X-ray crystallography can also be viewed as a form of microscopy. Invention of the *scanning tunneling microscope* (STM) by Binnig and Roher initiated a new revolution in microscopy. The STM and similar scanned probe microscopes examine surfaces by moving a fine probe mounted in a piezoelectric x,y,z-scanner across the surface to be examined. The tiny tungsten probe of the STM is so fine that its tip may consist of a single atom.

When a small voltage is applied a minuscule quantum mechanical tunnelling current flows across the small gap between the probe and the surface and a high-resolution image, sometimes at atomic resolution, is created from the recorded variation in current. The STM theoretically responds only to surfaces that conduct electrons, but nonconducting samples have been imaged at high humidity; presumably by conductance of electrons or ions through the surface water layer. The success of the STM spurred the development of many other types of scanned probe microscopes.

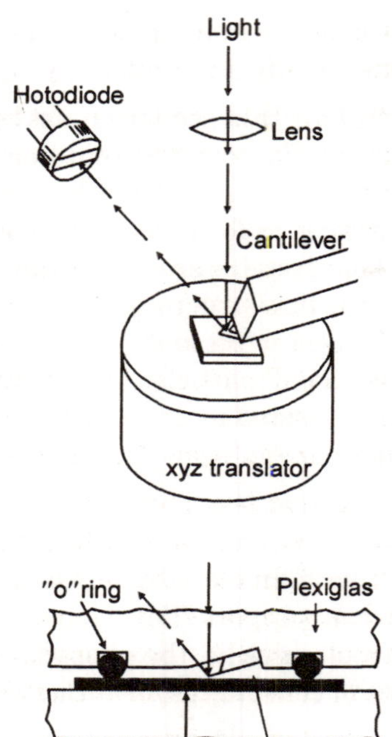

Fig. 1.17. Schematic diagram of the atomic force microscope.

Among these, the *atomic force microscope* (Fig. 1.17) has been especially useful for biological materials, including proteins and nucleic acids. The AFM moves a fine-tipped stylus directly across the sample surface or, alternatively, vibrates the probe above the surface. The small up-and-down movements of the stylus are recorded and thereby create a topographic or force-field map of the sample. AFM images contain three-dimensional information and can be used to view individual molecules (Fig. 1.18).

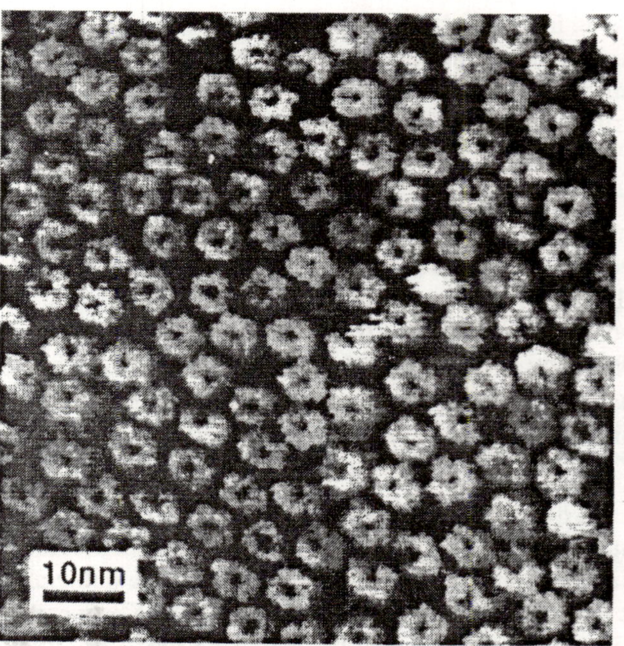

Fig. 1.18. AMF images of cholera toxin.

Chemical force microscopy is sensitive to adhesion and friction as a function of the interaction between defined chemical groups on the tip and sample. An emerging field is force spectroscopy, in which the AFM measures interaction forces between and within individual molecules. Under development are NMR microscopes. There is continual effort to see small objects more directly and more clearly!

X-RAY AND NEUTRON DIFFRACTION

One of the most important techniques by which we have learned bond lengths and angles and precise structures of small molecules is X-ray diffraction. Today, this technique, which involves measurements of the scattering of X-rays by crystalline arrays of molecules, is being used with spectacular success to shady macromolecules of biochemical and medical importance. X-rays were described by Rontgen in 1896 but there was uncertainty as to their wave nature. It was not until 1912 that the wavelengths of X-rays had been measured and it was recognized that they were appropriate for the use of X-rays in structure determination.

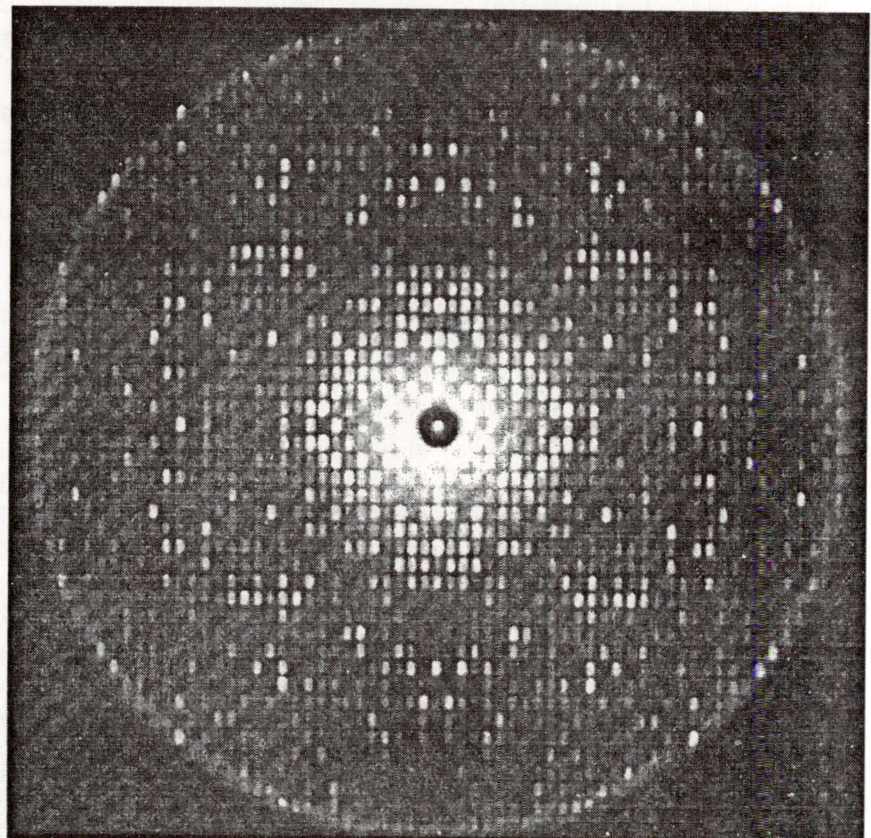

Fig. 1.19. An X-ray diffraction photograph such as was used to determine the structure of hemoglobin. This precession photograph was obtained from a crystal of human deoxyhemoglobin by rotating the crystal along two different axes in a defined manner before a narrow X-ray beam. The film also was moved synchronously. The periodicity observed in the photograph is a result of a diffraction phenomenon arising from the periodic arrangement of atoms in the crystal. The distances of the spots from the origin (center) are inversely related to the distances between planes of atoms in the crystal. In this photograph (which shows only two dimensions of the three-dimensional diffraction pattern) the spots at the periphery represent a spacing of 0.28 nm. If the intensities of the spots are measured, and if the phases of the harmonic functions required for the Fourier synthesis can be assigned correctly, the structure can be deduced to a resolution of 0.28 nm from a set of patterns of this type. In the case of human deoxyhemoglobin, a complete set of data would consist of about 16,000 spots. With modern equipment the resolution limit can be extended to better than 0.15 nm (or more than 100,000 unique spots).

Consider the fact that with a conventional light microscope we cannot distinguish two small objects that are much closer together than the distance represented by the wavelength of the light with which we observe them. This is about 460 nm for blue light. By comparison the wavelength of the copper K_α radiation, which is used in protein crystallography, is 0.1542 nm entirely appropriate for seeing the individual atoms of which matter is composed. Recognizing this, W. L. Bragg in Cambridge, England, in 1913 used X-ray diffraction to establish the structures of NaCl, KCl, and KBr in the crystalline state. The science of X-ray crystallography had been founded. In 1926 James Summer crystalled the enzyme urease crystallization of other enzymes soon followed.

In 1934, J. E. Bernal brought back to Cambridge from Uppsala, Sweden, some crystals of pepsin almost 2 mm long that had been grown in T. Svedberg's laboratory. Bernal and Crowfoot showed that these delicate crystals, which contained almost 50% water, gave a sharp diffraction pattern when they were protected by enclosure in a narrow capillary tube containing some of the mother liquor from which the crystals had been grown. After this, diffraction patterns were obtained for many protein crystals, and in 1937 Max Perutz chose for his thesis work at Cambridge the X-ray crystallography of hemoglobin. The project seemed hopeless at times, but in 1968, 31 years later, Perutz had determined the structure. If X-rays could be focused easily, could one build an X-ray microscope that would permit the immediate viewing of molecular structures? X-ray holograms at the molecular level have been obtained.

However, currently the only practical X-ray microscope for protein structures involves the measurement of diffraction patterns created by the scattering of X-rays (or of neutrons) from the crystalline lattice. The details of this procedure can be found in other sources. It is sufficient to point out here that a pattern of many spots such as that in Fig. 1.19 is, obtained. In a typical determination of protein structure from 10,000 to several times that number of spots must be measured. The needed information is contained in the coordinates of the spots (*i.e.*, in the angles through which the scattering occurs) and in the intensities of the spots. The structure is obtained by a Fourier synthesis in which a large number of mathematical functions in the form of three-dimensional sinusoidal waves are summed. The wavelengths of these functions are determined by the positions of the spots in the diffraction pattern.

The amplitudes of the waves are related to the intensities of the spots. However, the waves are not all in phase and the phases must be learned in some other way. This presents a difficult problem. Mathematical methods have been devised that automatically determine the phases for small molecules and usually allow the structures to be established quickly. For proteins it is more difficult. In 1954, Perutz introduced the *isomorphous replacement method* for determining phases. In this procedure a heavy metal, such as mercury or platinum, is introduced at one or more locations in the protein molecule. A favourite procedure is to use mercury derivatives that combine with SH groups. The resulting heavy metal-containing crystals must be isomorphous with the native, *i.e.*, the molecules must be packed the same and the dimensions of the crystal lattice must be the same.

However, the presence of the heavy metal alters the intensities of the spots in the diffraction pattern and from these changes in intensity the phases can be determined. Besides the solution to the phase problem, another development that was absolutely essential was the construction of large and fast computers. It would have been impossible for Perutz to determine the structure of hemoglobin in 1937, even if he had already known how to use heavy metals to determine phases. The first protein structure to be learned was that of myoglobin, which was established by Kendrew *et al.* in 1960. That of the enzyme lysozyme was deduced by Blake *et al.* in 1965.

Since then, new structures have appeared at an accelerating rate so that today we know the detailed architecture of over 6,000 different proteins with about 300 distinctly different folding patterns. New structures are being determined at the rate of about one per day. X-ray diffraction has also been very important to the study of naturally or artifically oriented fibrous proteins and provided the first experimental indications of the β structure of proteins. Suppose that you have isolated a new protein. How can you learn its three-dimensional structure? The first step is crystallization of the protein, something that a biochemist may be able to do.

Crystallization is done in many ways, often by the slow diffusion of one solution into another or by the slow removal of solvent through controlled evaporation. Ammonium sulfate and polyethylene glycol are two commonly used precipitants. The presence of the neutral detergent β-octyl glucoside improves some crystals. Droplets of protein solution mixed with the precipitant are often suspended on microscope cover glasses in small transparent wells or are placed in depression plates within closed plastic boxes. In either case, a reservoir of a solution with a higher concentration of precipitant is present in the same compartment.

Water evaporates from the samples into the larger reservoir, concentrating the protein and causing its crystallization. Crystals grow slightly better in a spacecraft than on Earth. Some proteins, notably myosin from muscle, crystallize well only after reductive methylation of all lysine side chains to dimethyllysine with formaldehyde and sodium borohydride (Eq. 1.34). The next step is for a protein crystallographer to mount a small perfect crystal in a closed silica capillary tube and to use an X-ray camera to record diffraction patterns such as that in Fig. 1.19. These patterns indicate how perfectly the crystal is formed and how well it diffracts X-rays.

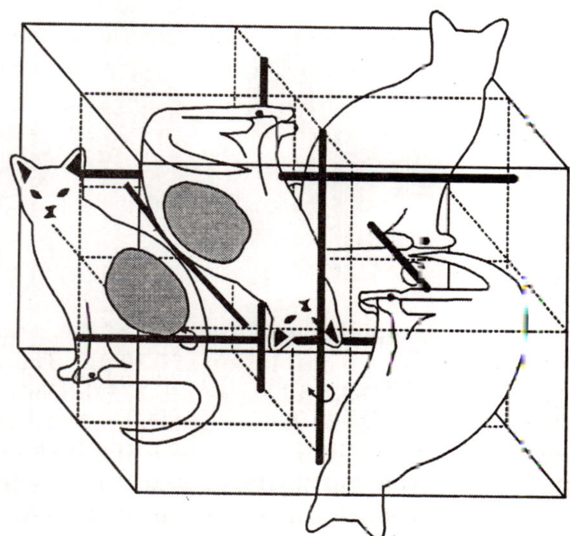

Fig. 1.20. Diagram showing an asymmetric unit as it might appear in a unit cell of space group $P2_12_12_1$. This unit cell has three pairs of nonintersecting twofold *screw axes* which are marked by the shaded rods. These are designated by arrows and the symbol. Two asymmetric units are related one to another by rotation around a twofold axis together with translation by one-half the dimension of the unit cell.

The patterns are also used to calculate the dimensions of the unit cell and to assign the crystal to one of the seven *crystal systems* and one of the 65 enantiomorphic *space groups*. This provides important information about the relationship of one molecule to another within the unit cell of the crystal. The unit cell (Fig. 1.20) is a parallelopiped whose sides are parallel to crystallographic axes and which, by translation in three dimensions, gives rise to the entire crystal. The unit cell must contain at least one *asymmetric unit,* the smallest unit of structure that lacks any element of crystal symmetry. The asymmetric unit may consist of one or a small number of protein subunits. In crystals of the triclinic system all of the asymmetric units are aligned in the same manner and there are no axes of symmetry.

Monoclinic crystals have a single *twofold or dyad axis.* The unit cell might contain two molecules related one to another by the dyad or one dimeric molecule with the dyad located as in Fig. 1.11c. Orthorhombic crystals have three mutually perpendicular twofold axes and the unit cell is a rectangular solid. Trigonal, tetragonal, and hexagonal crystal systems have three, four and sixfold axes of symmetry, respectively, while the cubic crystal contains four threefold axes along with diagonals of the cube as well as two-fold axes passing through the faces.

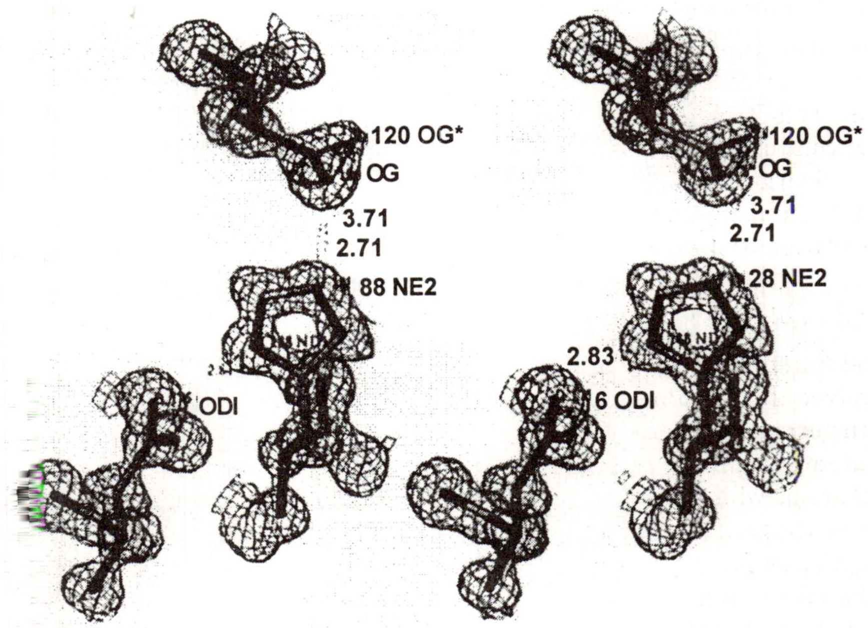

Fig. 1.21. Stereoscopic view of a section of the structure of cutinase from the fungus *Fusarium solani* determined to a resolution of 0.10 nm. The three amino acid residues shown are serine 120 (top), histidine 188, and aspartate 175 (lower left). The structure is presented as a contour map with a "wire mesh" drawn at a "cutoff" level of density equal to 1σ above the average, where σ is the root mean square density of the entire map. The side chains of these three residues constitute the "catalytic triad" in the active site of this enzyme. At this resolution more than one conformation of a group may often be seen. For example, the gamma oxygen (OG) of S120 is seen in two positions, the major one being toward His 188. When the map is drawn with a lower contour level the N–H proton on His 188 that is hydrogen bonded to Asp 175 can also be seen.

Within each crystal system there are several space groups. An example is the orthorhombic space group $P2_12_12_1$, which is often met with small organic molecules and proteins. In this space group the unit cell contains three mutually perpendicular but nonintersecting twofold screw axes (Fig. 1.20). The position of one molecule in the unit cell is related to the next by both a 180° degree rotation about the twofold axes and a translation of one-half the length of the unit cell. There is such a screw axis for each of the three directions. The third step in the structure determination is collection of the X-ray diffraction data. This may be done with a *diffractometer* in which a narrow collimated pencil source of X-rays is aimed at the crystal and the intensities and positions of the diffracted beams are measured automatically.

The computer controlled diffractometer is able to measure the anlges to within less than one-hundredth of a degree. If sufficient time is allowed, very weak spots can be counted. Today, diffractometers are more likely to be used for preliminary measurements, while the major data collection is done with an *area detector,* an instrument that collects many reflections simultaneously. There are some difficulties. For example, during the long periods of irradiation needed to measure the weak spots with a diffractometer the protein crystals decompose and must be replaced frequently. Data collection may last for months.

The newest methods utilize more powerful X-ray sources and often *synchrotron radiation,* which delivers very short and extremely intense pulses of X-rays and allows data to be collected on very small well-formed crystals. The fourthstep is the preparation of isomorphous crystals of heavy metal-containing derivatives. The heavy metal may be allowed to react with the protein before crystallization or may be diffused into preformed crystals. A variety of both cationic and anionic metal complexes, even large $Ta_6Br_{12}{}^{2+}$ tantalum clusters, have been used. Two or more different heavy metal derivatives are often required for calculation of the phases.

The heavy metal atoms must be present at very small number of locations in the unit cell. An entire data set must be collected for each of these derivatives.

The evaluation of the phases from these data is a complex mathematical process which usually involves the calculation first of a "difference Patterson projection. This is derived by Fourier transformation of the differences between the scattering intensities from the native and heavy atom-containing crystals. The Patterson map is used to locate the coordinates of the heavy metal atoms which are then refined and used to compute the phases for the native protein. Alternative methods of solving the phase problem are also used now. When a transition metal such as Fe, Co, or Ni is present in the protein, anomolous scattering of X-rays at several wavelengths (from synchrotron radiation) can be used to obtain phases. Many protein structures have been obtained using this multiple wavelength anomalous diffraction (MAD phasing) method.

Selenocysteine is often incorporated into a protein that may be produced in bacteria using recombinant DNA procedures. Crystals are prepared both with protein enriched in Se and without enrichment. In some ways it is better to incorporate tellurium (^{127}Te) in telluromethionine. In the fifth step of an X-ray structure determination the *electron density map* is calculated using the intensities and phase information. This map can be thought of as a true three-dimensional image of the molecule revealed by the X-ray microscope.

It is usually displayed as a stereoscopic view on a computer graphics system (Fig. 1.21). It is also often prepared in the form of a series of transparencies mounted on plastic sheets. Each sheet represents a layer, perhaps 0.1 nm thick, with contour lines representing different levels of electron density. Using the electron density map a three-dimensional model of the protein can be built. For years, the customary procedure was to construct a model at a scale of 2 cm = 0.1 nm using an *optical comparator,* but crystallographers now use computer graphics systems. The three-dimensional image of the electron density map and a computer-generated atomic model are superimposed on the computer screen. An example is shown in Fig. 1.21.

When the superposition has been completed, the coordinates of all atoms are present in the memory of the computer. The final step in the structure determination is *refinement* using various mathematical methods. From the coordinates of the atoms in the model the expected diffraction pattern is computed and is compared with that actually observed. The differences between predicted and observed density are squared and summed. The sum of the squares constitutes an error function which is then minimized by moving the various atoms in the model short distances while keeping bond lengths, angles, and van der Waals distances within acceptable limits and recalculating the error function. This complex "refinement" pro-re must be repeated literally hundreds of thou-sandsof times with every part of the structure being varied.

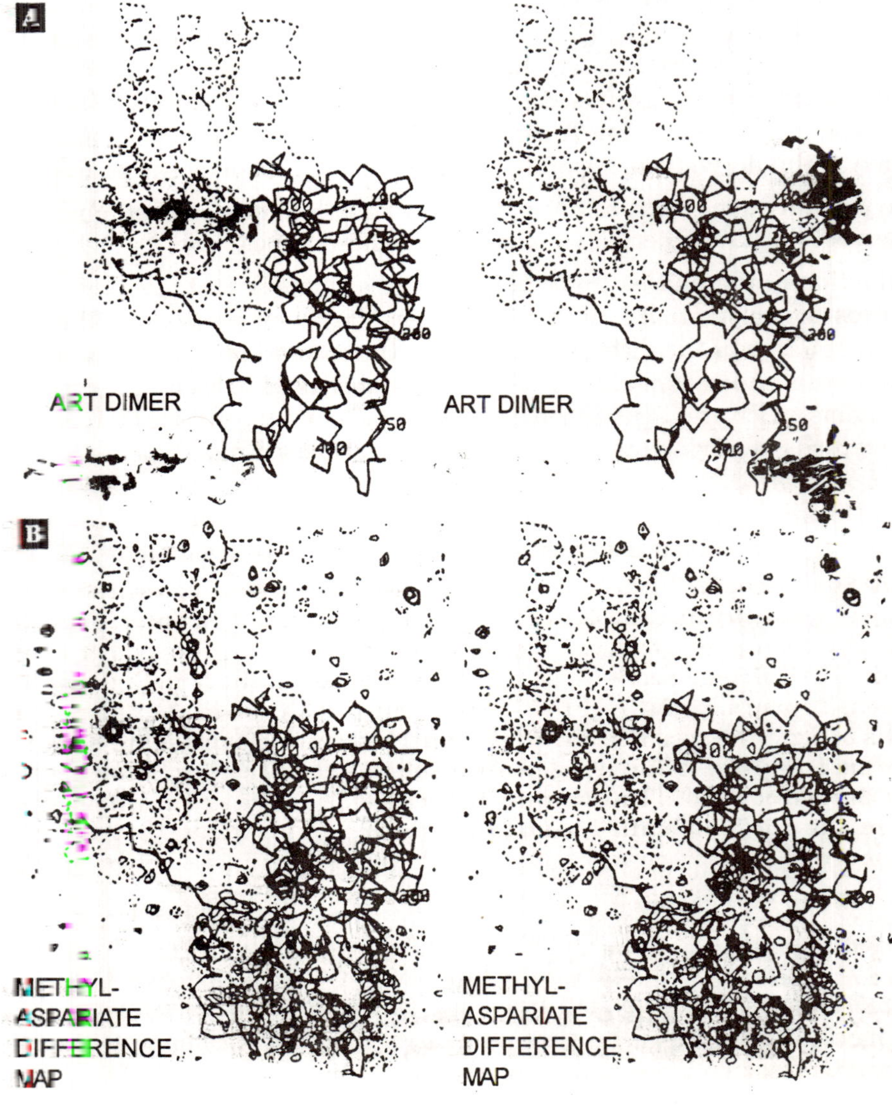

Fig. 1.22 (A) Stereoscopic a-carbon plot of the cystolic aspartate aminotransferase dimer viewed down its dyad symmetry axis. Bold lines are used for one subunit (subunit 1) and dashed lines for subunit 2. The coenzyme pyridoxal 5'-phosphate (Fig. 1.24) is seen most clearly in subunit 2 (center left). (B) Thirteen sections, spaced 0.1 nm apart, of the 2-methylaspartate difference electron density map superimposed on the a-carbon plot shown in (A). The map is contoured in increments of ± 2σ (the zero level omitted), where σ = root mean square density of the entire difference map. Positive difference density is shown as solid contours and negative difference density as dashed contours. The alternating series of negative and positive difference density features in the small domain of subunit 1 (lower right) show that the binding of L-2-methylaspartate between the two domains of this subunit induces a right-to-left movement of the small domain. (C) Nine sections, spaced 0.1 run apart, of a part of the 2-methylaspartate difference map superimposed on the atomic model shown in (A) and (D). The coenzyme is shown as the internal aldimine with Lys 258. The positive and negative contours on the two sides of the coenzyme ring indicate that the coenzyme tilts over to form the external aldimine when substrates react. (D) Superimposed structure of the active site of the enzyme in its free form as in (A) (bold lines) and the refined structure of the α-methylaspartate complex, (dashed lines). This illustrates the tilting of the coenzyme ring.

Structures at very high resolution (0.07-0.1 nm) may reveal multiple positions for hydrogen-bonded side chains as well as hydrogen atoms and even bonding electrons. Once the three-dimensional structure of the protein is known, further experiments are usually done using the X-ray diffraction technique. Since protein crystals contain channels of solvent between the packed molecules, it is usually possible to diffuse small molecules into the crystal. These may be substrates, inhibitors, or allosteric effectors. Diffraction data are collected after diffusion of the small molecules into the crystal and a *difference electron density map* may be calculated. This may show exactly where those molecules were bound.

An example is shown in Fig. 1.22. Sometimes difference maps not only show the binding to an enzyme of a substrate or other small molecule but also reveal conformational changes in proteins. Such is the case for Fig. 1.22, which shows the binding of α-methylaspartate, an inhibitor that behaves initially like a substrate and goes part way through the reaction sequence for aspartate amino-transferase until further reaction is blocked by the methyl group. This difference map also shows that the part of the protein to the left of the binding site in the figure has moved.

Using the X-ray data it was possible subsequently to "deduce the nature of the conformational change. Examination of the effect of temperature on the diffraction pattern of a protein can give direct information about the mobility of different parts of the mole cules. This complements information obtainable in solution from NMR spectra. Because hydrogen atoms contain only one electron, and therefore scatter X-rays very weakly, they are usually not seen at all in X-ray structures of proteins.

"However, neutrons are scattered strongly by hydrogen atoms and *neutron diffraction* is a useful tool in protein structure determination. It has been used to locate tightly bonded protons that do not exchange with ^2H_2O as well as bound water (^2H_2O). The development of synchrotron radiation as an X-ray source has permitted accumulation of data for electron density difference maps in less than 1 *s* and it is expected that such data can eventually be acquired in ~ 1 ps. If a suitable photochemical reaction can be initiated by a picosecond laser flash, a substrate within a crystalline enzyme can be watched as it goes through its catalytic cycle:

An example is the release of inorganic phosphate ions from a "caged phosphate" (Eq. 1.49) and study of the reaction of the released phosphate with glycogen phosphorylase.

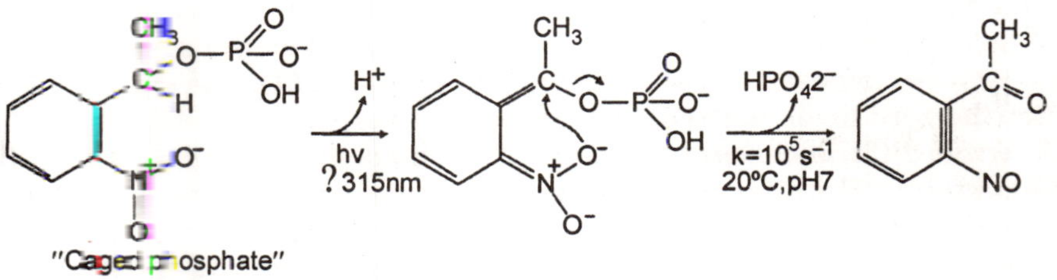

...(1.49)

However, caged substrates usually must diffuse some distance before reacting, so very rapid events cannot be studied. An alternative approach is to diffuse substrates into crystals at a low temperature at which reaction is extremely slow but a substrate may become seated in an active site ready to react. In favorable cases such "frozen" Michaelis complexes may be heated by a short laser pulse to a temperature at which the reaction is faster and the steps in the reaction may be observed by X-ray diffraction.

NUCLEAR MAGNETIC RESONANCE (NMR)

Organic chemists and biochemists alike have long relied on NMR spectroscopy to assist in identification and determination of structures of small compounds. Most students have some familiarity with this technique and practical information is available in many places. Measurements can be done on solids, liquids, or gases but are most often done on solutions held in special narrow NMR tubes. Volumes of samples are typically 0.5 ml (for a 1–5 mM solution containing 1 – 5 μmol of protein) but less for small molecules and higher concentrations. Newer techniques allow use of a volume/as small as 5 nl and containing <0.1 nmol of sample

For many years progress in biochemical application of NMR was slow, but a dramatic increase in the power of the spectrometers, driven in great measure by the revolution in computer technology, has made NMR spectroscopy a major force in the determination of structures and functions of proteins and nucleic acids.

Basic Principles of NMR Spectroscopy

The basis of NMR spectroscopy lies in the absorption of electromagnetic radiation at radiofrequencies by atomic nuclei. All nuclei with odd mass numbers (e.g., ^{1}H, ^{13}C, ^{15}N, ^{17}O, ^{19}F, and ^{31}P), as well as those with an even mass number but an odd atomic number, have magnetic properties. Absorption of a quantum of energy E = *hv* occurs only when the nuclei are in the strong magnetic field of the NMR spectrometer and when the frequency *v* of the applied electromagnetic radiation is appropriate for "resonance" with the nucleus being observed.

In the widely used "500-megahertz" NMR spectrometers the liquid helium-cooled superconducting electromagnet has a field strength of 11.75 tesla (T). In this field a proton resonates at ~ 500 megahertz (MHz) and nuclei of ^{31}P, ^{13}C, and ^{15}N at ~202, 125, and 50 MHz, respectively. At 500 MHz the energy of a quantum is only $E = 3.3 \times 10{\sim}^{3?} \times 10^{8}$ J = 0.2 J moH, more than four orders of magnitude less than the average energy of thermal motion of molecules (3.7 kj moH). Thus, the spin transitions induced in the NMR spectrometer have no significant effect on the chemical properties of molecules. The resonance frequency *v* at which absorption occurs in the spectrometer is given by where H_0 is the strength of the external magnetic field, *p.* is the magnetic moment of the nucleus being investigated, and *h is* Planck's constant.

The basis for NMR spectroscopy lies in the fact that nuclei in different positions in a molecule resonates at slightly different frequencies. In a protein each one of the hundreds or thousands of protons resonate at its own frequency. With older NMR instruments a spectrum at the constant magnetic field H_0 can be obtained by varying the frequency and observing the values at which absorption occurs, much as is done for ultraviolet, visible, and infrared spectra With newer pulsed NMR spectrometers the measurement is done differently but the spectra look the same.

$$v = \mu H_0 / h \qquad \text{...(1.50)}$$

The higher the magnetic field, the greater the variation in resonance frequency and the higher the sensitivity. The most powerful commercial NMR spectrometers currently available operate at about 750 MHz for ^{1}H and a few higher frequency instruments have been built. The *proton NMR* spectrum of the coenzyme pyridoxal phosphate in $^{2}H_2O$ is shown in Fig. 1.23 as obtained with a 60-MHz spectrometer. Four things can be measured from such a spectrum: (1) the *intensity* (area under the band). In a proton NMR spectrum, areas are usually proportional to the numbers of equivalent protons giving rise to absorption bands; (2) the *chemical shift,* the difference in frequency between the peak observed for a given proton and a peak of some standard reference compound. In Fig. 1.24 the reference peak is at the right edge; (3) the *width* at half-height (in hertz), a quantity that can provide information about molecular motion and about chemical exchange; and (4) *coupling constants* which measure interactions between nearby magnetic nuclei.

These are extremely important to the determination of structures of both small and large molecules,. "With a magnetic field of H_0 = 11.75" I and a 500–MHz oscillator, the positions of proton resonances in organic compounds are spread over a range of –10,000 Hz. This is 20 parts per million (ppm) relative to 500 MHz. Positions of individual resonances are usually given in ppm and are always measured in terms of a shift from the resonance position of some standard substance. For protons this is most often *tetramethyl-silane* (TMS), an inert substance that can be added directly to the sample in its glass tube. Biochemists often use ^2H_2O as solvent and the water-soluble sodium 3-trimethylsilyl 1-propane sulfonate (*DSS* or Tier's salt) as a standard. Its position is insignificantly different from that of TMS. For NMR spectra measured in ^2H_2O, the "pD" of the medium is sometimes indicated. It has often been taken as the pH meter reading plus 0.4.

However, because of uncertainty about the meaning of pD, most workers cite the apparent pH measured with a glass electrode and standardized against aqueous buffers. It is important to describe how the measurement was made when publishing results.

Band Widths

The narrowness of a band in an NMR spectrum is limited by the *Heisenberg uncertainty principle,* which states that $\Delta E \times \Delta f = h/2n$, where h is Planck's constant, ΔE is the uncertainty in the energy, and Af is the lifetime of the magnetically excited state. Since $E = hv$ for electromagnetic radiation, E is directly proportional to the width of the absorption band (customarily measured at one-half its full height).

The magnetic nucleus is well shielded from external influences and the lifetime of its excited state tends to be long. Hence, Δv is small, often amounting to less than 0.2 Hz. This fact is very favourable for the success of high-resolution proton magnetic resonance. However, bands are often much broader for large macromolecules.

The Chemical Shift

In a molecule such as TMS, the electrons surrounding the nuclei "shield" the nucleus so that it does not experience the full external magnetic field. For this reason, absorption occurs at a high frequency (high energy). Protons that are bound to an atom deficient in electrons (because of attachment to electron withdrawing atoms or groups) are *deshielded.* The greater the deshielding, the further *downfield* from the TMS position is the NMR peak. "The magnitude of this chemical shift may be stated in hertz, but it is most often expressed in ppm as δ (Eq. 1.51).

The value of δ is the shift in frequency relative to frequency of the oscillator in parts per million and is independent of the field strength. It still depends upon use of a particular reference standard which must be stated when a δ value is given. In the spectrum shown in Fig. 1.23, the methyl protons appear 2.45 ppm below the DSS peak but still at a relatively high field. Characteristic chemical shift ranges for other protons extend to ~20 ppm. Aromatic rings lead to

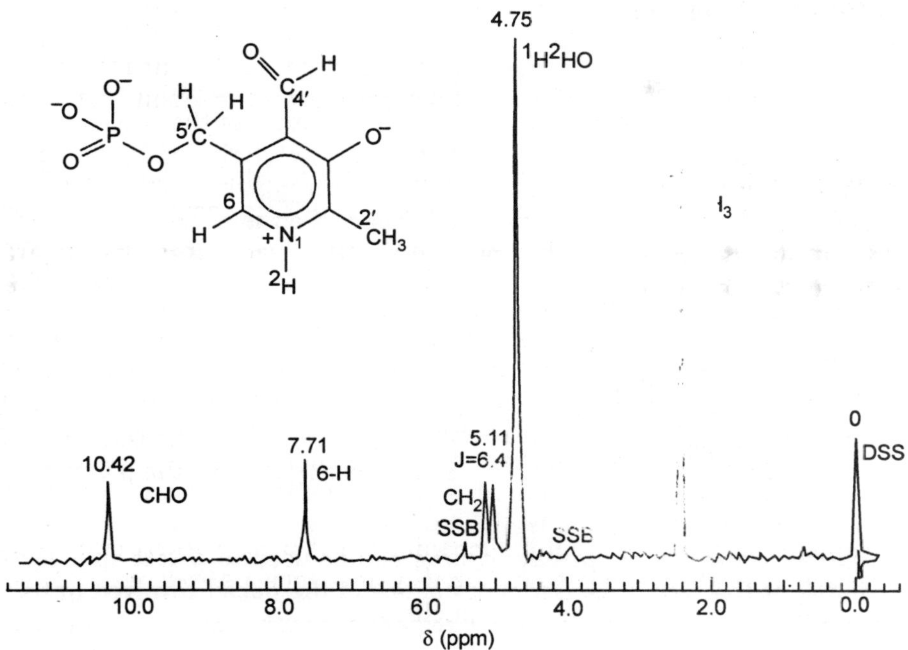

Fig. 1.23. The 60-MHz proton magnetic resonance spectrum of pyridoxal 5¢-phosphate at neutral pH (apparent pH = 6.65). The internal standard is DSS. Chemical shifts in parts per million are indicated beside the peaks.

$$\delta(\text{ppm}) = \frac{\Delta\upsilon(\text{Hz}) \times 10^6}{\upsilon(\text{Hz}) \text{ of oscillator}} \qquad ...(1.51)$$

strong deshielding of attached protons because of a *ring current* induced in the circulating π electrons. Thus, in Fig. 1.23 the peaks of the methylene protons which are adjacent to the aromatic ring occur at 5.10 and 5.12 ppm. The 6-H, which is bound directly to the ring, is more strongly deshielded and appears at 7.71 ppm. The hydrogen of the aldehyde groups is deshielded as a result of a similar "diamagnetic electronic circulation" in the carbonyl group. Its peak is even further downfield at 10.4 ppm. Ring current and other effects on chemical shifts are important in NMR spectroscopy of proteins.

Aromatic proton resonances sometimes stand out because they have been shifted far downfield. Ring current effects on chemical shifts can be predicted quite accurately if three-dimensional structures are known. Additionally, computer programs are available for predicting them. Hydrogen bonding has a very large effect on the chemical shift of protons. The resonance of a strongly hydrogen-bonded proton is usually shifted downfield from its position in non-hydrogen-bonding media. This is especially true for hydrogen bonds to charged groups, e.g., the NH of a histidine or tryptophan side chain hydrogen bonded to a carboxylate group, a situation often met in the active sites of proteins.

The chemical shift of ^{13}C in the carboxyl group is also affected. While *ring current shifts* can be predicted quite well, it is much more difficult to predict the total chemical shift.

Scalar Coupling (J Coupling)

The energy of the spin transition of a hydrogen nucleus is strongly influenced by the local presence of other magnetic nuclei, e.g., other protons that are covalently attached to the same or an adjacent atom. These neighboring protons can be in either of the two spin states, a fact that results in easily measured differences in the energy of the NMR transition under consideration. This spin-spin interaction (coupling) leads to a splitting of NMR bands of protons into two or more closely spaced bands. The ethyl group often appears in NMR spectra as a "quartet" of four evenly spaced peaks that arise from the CH_2 group and a triplet of peaks arising from the CH_3 protons.

Protons attached to the same carbon (geminal protons), and in similar environments, do not ordinarily split each other's peaks, while the protons on the neighboring carbon do. The *coupling constant J* is the difference in hertz between the successive peaks in a multiple. It is a field-independent quantity and the same no matter what the frequency of the spectrometer. In Fig. 1.24 the peak of the methylene protons is split by ^{1}H–^{31}P coupling, with a value of $J \sim 6.4$ Hz. While spin-spin coupling is most pronounced when magnetic nuclei are close together in a structure, the effect can sometimes be transmitted through up to five covalent bonds. The technique of *double irradiation or spin decoupling* can be used to detect spin coupling. The sample is irradiated at the resonance frequency of one of the nuclei involved in the coupling, while the spectrum is observed in the frequency region of the other nucleus of the coupled pair.

Under these conditions the multiplet collapses into a singlet and the mutual coupling of the two nuclei is established. The coupling can be seen directly in appropriate two-dimensional NMR spectra. The coupling constant between two *vicinal protons* which are attached to adjacent carbon atoms (or other atoms) depends upon the torsion angle.

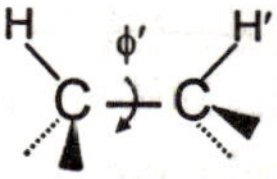

The *Karplus equation* (Eq. 1.52) relates J to the torsion angle ϕ' (so labeled to distinguish it from peptide torsion angle ϕ;

$$J_{H,H'} A \cos^2 \phi' + B \cos \phi' + C \qquad ...(1.52)$$

This equation was predicted on theoretical grounds, but the constants, *A, B,* and C are empirical. Other forms of the equation, some of them simplified, have also been proposed. The Karplus relationship is often used to estimate time-averaged torsion angles in peptides. For a C_αH-NH torsion angle the parameters *A, B* and C of Eq. 1.52 are ~6.4, 1.4, and 1.9, respectively. For a C_αH - C_βH they are ~9.5, 1.6, and 1.8, respectively. Coupling constants between ^{J}H and ^{13}C or ^{15}N are of importance in determination of three-dimensional structure of proteins. *The nuclear Overhauser effect* (NOE)lsthe result of transfer of magnetization from one nucleus to a nearby nucleus directly through space rather than via *J*-coupling. This was observed first as a result of irradiation of a resonance in a one-dimensional spectrum resulting in an increased intensity of the resonance of the nearby nucleus.

Magnetization transfer can occur from a given nucleus to one or more nearby nuclei. Each such transfer that is detected is usually referred to simply as an NOE. For an NOE to be observable the two nuclei must be very close together, <0.5 run (5 Å). The strength of the magnetization transfer falls off approximately as the sixth power of the interatomic distance. Consider two hydrogen atoms, both tightly hydrogen-bonded to an intervening oxygen atom, e.g., of a carboxylate or phosphate group. The expected H–H distance would be ~0.3 nm and a strong NOE between them would be anticipated.

Two nonbonded hydrogen atoms (e.g., on methyl groups of amino acid side chains) can be as close together as 0.24 nm at van der Waals contact and could show a very strong NOE. However, contact is rarely this close in proteins unless in a hydrogen bond.

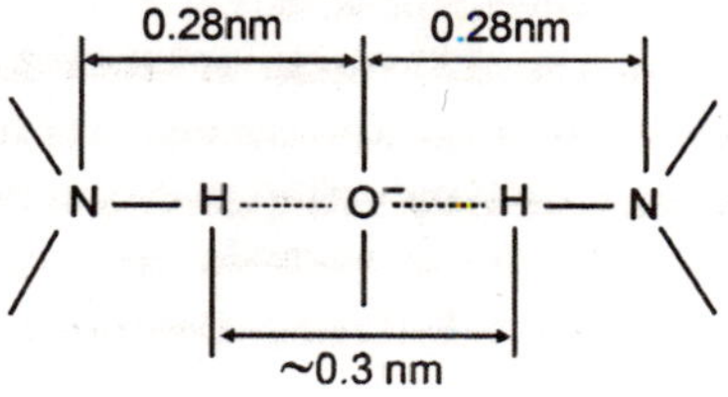

0.28nm 0.28nm

N — H ········ O ········ H — N

~0.3 nm

Nuclei Other than Hydrogen

Denterium (^{2}H)

The natural abundance is very low so that use of ^{2}H-labeled compounds is practical for study of metabolism, e.g., for following an ^{2}H label in glucose into products of fermentation or in mammalian blood flow. Deuterium NMR has been used extensively to study lipid bilayers.

Carbon-13

Use of ^{13}C in NMR developed slowly because of the low natural abundance of this isotope. Another complication was the occurrence of ^{13}C-H coupling involving the many protons normally present in organic compounds. The latter problem was solved by the development of *wide-band proton decoupling* (noise decoupling). With a natural abundance of only 1.1%, ^{13}C is rarely present in a molecule at adjacent positions. Thus, ^{13}C–^{13}C coupling does not introduce complexities and in a noise-decoupled natural abundance spectrum each carbon atom gives rise to a single peak. Even so, ^{13}C NMR spectroscopy was not practical until pulsed Fourier transform (FT) spectrometers were developed. Chemical shifts in ^{13}C spectra are often 200 ppm or more downfield relative to TMS.

The effects of substituents attached to a carbon atom are often additive when two or more substituents are attached to the same atom. It is often necessary (but costly) to prepare compounds enriched in ^{13}C beyond the natural abundance. For proteins this may be done by growing an organism on a medium containing [^{13}C]glucose or a single amino acid enriched in ^{13}C. Using metabolites enriched in ^{13}C, it is also possible to observe metabolism of living tissues directly. For example, glycogen synthesis from ^{13}C-containing glucose has been observed in a human leg muscle using a wide-bore magnet and surface coils for transmitting and receiving.

Nitrogen – 15

Despite difficulties associated with low natural abundance (0.37%) and low sensitivity, ^{15}N NMR is practical and with isotopically enriched samples has become very important. Proteins with a high content of ^{15}N can be produced easily and inexpensively from cloned genes in bacterial plasmids. For example, cells of *E. coli* can be grown on a minimal medium containing [^{15}N] NH_4Cl. Since ^{13}C can also be introduced in a similar way it is possible to incorporate both isomers simultaneously. Production of uniformly labeled protein containing ^{15}N and/or ^{13}C provides the basis for multidimensional isotope-edited spectra necessary for protein structure determination (next section) and for study of tautomerization of histidine rings. ^{15}N chemical shifts of groups in proteins are spread over a broad range table 1.3.

Phosphorus –31

NMR spectroscopy using ^{31}P, the ordinary isotope of phosphorus, also has many uses. Application of ^{31}P NMR to living tissues has been extraordinarily informative. The many phosphorus nuclei in nucleotides, coenzymes, and phosphorylated metabolites and proteins are all suitable objects of investigation by NMR techniques.

Fluorine -19

Although not abundant in nature, ^{19}F gives an easily detected NMR signal and can be incorporated in place of hydrogen atoms into many biochemical compounds including proteins. In one study genetic methods were used to place 3-fluoro-tyrosine separately into eight positions in the lac repressor protein Measurements of ^{19}F NMR spectra were used to study domain movement. Active site groups of enzymes can be modified to incorporate ^{19}F. Binding of fluorinated substrates can be studied. Nontoxic ^{19}F-containing compounds are useful as intracellular pH indicators, the NMR spectrometer serving as the pH meter. An atom of fluorine attached to an aromatic ring is highly and predictibly sensitive to inductive effects of substituents in the para position to the fluorine. For example, fluorine at the 6-position in pyridoxal phosphate can be observed inenzymes and reports changes in coenzyme structure.

Some other Nuclei

Here are a few reported uses of NMR on other nuclei. ^{3}He, binding into little cavities in fullerenes; ^{11}B, binding of boronic acids to active sites, Na, measurement of intracellular [Na^+]; ^{5}Cl and ^{37}Cl, binding to serum albumin; Cd, reporter that can replace Zn^{2+} (no magnetic moment) in active sites of many enzymes and in nonenzymatic systems as well; Tl, replacing K^+ in enzyme binding sites, ^{17}O, study of dynamics of protein hydration, and ^{77}Se, observation of acetylchymotrypsin intermediate.

Fourier Transform Spectrometers and Two-Dimensional NMR

Although NMR spectroscopy was widely used by the 1950s it was revolutionized by two developments, pioneered by Richard Ernst in the mid 1960s and 1970s. The first of these was pulsed Fourier transform (FT) spectroscopy, which permits rapid accumulation of high-resolution spectra. In an FT NMR spectrometer a strong pulse of radio frequency (RF) radiation is delivered to the sample over a period of a few microseconds and its effects are observed at all frequencies simultaneously. Although 1-2 or more seconds must be allowed before the next pulse is delivered, one complete NMR spectrum is obtained with each pulse.

Often the results of hundreds, thousands, or even hundreds of thousands of pulses are added to provide greater sensitivity. With good temperature control this may be accomplished over periods of minutes to days.

Table 1.3. Approximate Chemical Shift Ranges in ^{1}H– and in ^{15}N-NMR Spectra

Group	*^{1}H chemical shift (ppm from TMS)*	*^{15}N chemical shift (ppm from liquid NH_3)*
$—CH_3$	0 – 4.0	
$—CH_2—$	1.1 — 4.4	
—CH	2.4 – 5	
Peptide αH		
random coil[a]	3.9–5.0	
–OH[b]	~5 – 6	
>C=C–H	5–8	
$–NH_2$	31–37	
NH		
Peptide	7–12	103–142
Aromatic H	7–9	
Imidazole		
$C^{\varepsilon 1}$–H	~7.7	
$C^{\delta 2}$–H	~7.0	
N–H	~10	165–180
Imidazolium		
$C^{\varepsilon 1}$–H	~8.7	
$C^{\delta 2}$–H	~7.4	
N–H	~10–18	
–CHO	9.4–10.4	
Aldehyde		
–COOH	11.3–12.2	
Indole NH	~10	130–145
Guanidinium		
$–N^{\eta}H_2$	69–77	
$–N^{\varepsilon}H$	31–37	

Free Induction Decay

The strong exciting RF pulse is delivered with an orientation at right angles to that of the static field H_0 of the magnet and whose direction defines the *z* axis. As a result of this pulse, the magnetization of a nucleus is tilted away from the *z* axis and *precesses around the z axis* at its resonance (Larmor) frequency, which is ~500 MHz for ^{1}H in a 500-MHz spectrometer. The frequency that is measured is actually a difference from the "carrier frequency" of the RF pulse. The precessing magnetization of the nuclei has a component in the *xy* plane which induces an electrical signal in the coil of the NMR probe which defines the *y* axis. This signal, which contains the Larmor frequencies of all of the nuclei of a given element, is recorded as a function of time over a period of a few seconds.

The signal decays away exponentially. However, this curve of *free induction decay* (FID) is not smooth but contains within it all of the Larmor frequencies. If enough points (perhaps 500 in a 2-s acquisition) are recorded and stored in the computer's memory, Fourier transformation of the data will produce the frequency-dependent NMR spectrum.

Relaxation Times T_1 and T_2

When very strong pulse of elecctromagnetic radiation is applied in the NMR spectrometer, virtually all of the nuclei are placed in the magneticall excited state. If another pulse were applied immediately, little energy would be absorbed because the system is *saturated.* In the older "continuous-wave" NMR spectrometers, the older "continuous-wave" NMR spectrometers, the energy is always kept small so that little saturation occurs. However, in FT NMR instruments, the strong pulses lead to a high degree of saturation. Application of repeated pulses would produce no useful information were it not for the fact that the excited nuclei soon relax back to their equilibrium energy distribution.

Relaxation occurs through interactions of the nuclei with fluctuating magnetic fields in the environment. For organic molecules in solution the fluctuations that are most often effective in bringing about relaxation are the result of moving electrical dipoles in the immediate vicinity. Even so, relaxation of protons in water requires seconds. Relaxation ot nuclear magnetic states is characterized by two relaxation times. The longitudinal or *spin-lattice relaxation time* T_2 measures the rate of relaxation of the net magnetic vector of the nuclei in the direction of H_0. The transverse or *spin-spin relaxation time* T_2 measures the relaxation in the *xy* plane perpendicular to the direction of H_0.

The two relaxation times can be measufed independently. In general, $T_2 < T_1$. For solids, T_2 is quite short ($\sim 10^{-5}$ s) whereas relaxation times of seconds are observed in solutions. This lengthening of the lifetime of the excited state in going from solid to liquid leads to a narrowing of absorption lines and explains why NMR bands in liquids are often narrow. However, an increase in viscosity or a loss of fluidity in a membrane leads to broadening.

How can T_2 and T_1 be measured? T_2 for fluids can often be estimated from the width of the band Δv at half-height. However, pulsed NMR methods are usually employed, the measurement of T_1 being especially easy.

$$T_2 \approx 1/\pi \, \Delta v \qquad \text{...(1.53)}$$

An attempt is often made to relate T_1 and T_2 to the molecular dynamics of a system. For this purpose a relationship is sought between Tj or T_2 and the *correlation time* T_c of the nuclei under investigation. The correlation time is the time constant for exponential decay of the fluctuations in the medium that are responsible for relaxation of the magnetism of the nuclei. In general, $1/\tau_c$ can be thought of as a rate constant made up of the sum of all the rate constants for various independent processes that lead to relaxation. One of the most important of these $(1/\tau_1)$ is for molecular tumbling.

$$1/\tau_1 \approx (3k_BT)/4\pi\eta r^3 \qquad \text{...(1.54)}$$

This equation is closely related to that of rotational diffusion. Another term is the reciprocal of the residence time τ_m, the mean time that a pair of dipoles are close enough together to lead to relaxation. In the usual solvents at room temperature, τ_c is of the order of 10^{-12} to 10^{-10} s.

Thus, relaxation rates in solutions are considerably faster than the frequencies of radiation absorbed in the NMR spectrometer (~10^8 s^{-1}). Relaxation is relatively ineffective and T_1 and T_2 are usually large and equal.

Bands remain sharp. As the correlation time increases (as happens, for example, if the viscosity is increased), T_1 and T_2 decrease with T_1 reaching a minimum when $\tau_c \sim \upsilon$, the frequency of the absorbed radiation. Lines are broadened and hyperfine lines (from coupling between nuclei) cannot be resolved. As τ_c is increased further, T_2 reaches a constant low value, while T_1 rises again. NMR measurements can be made in the region where τ_c exceeds υ, a circumstance that is favored by the use of high-frequency spectrometers.

On the other hand, in fluids it is more customary to work in the range of "extreme motional narrowing" at low values of τ_c. Both T_1 and T_2 rise as the mobility of the molecules increases. A Limitation of use of NMR measurements of proteins comes from the increase in tumbling time with increasing size of the molecules. Since $1/\tau_r$ is often the most important term in the relaxation rate constant, only small proteins of mass <20 kDa give very sharp bands.

Nevertheless, usable spectra are often obtainable on proteins ten times this size. A practical problem in ^{13}C NMR arises from slow relaxation (long T_1). Partial saturation is attained and signal intensities are reduced for those carbon atoms for which relaxation is especially ineffective. Relaxation times can be measured separately for each carbon atom in a molecule and can yield a wealth of information about the *segmental motion* of groups within a molecule.

Although the relationships between relaxation times and molecular motion are complex, they are often relatively simple for ^{13}C. Carbon atoms are usually surrounded by attached hydrogen atoms, and dipole-dipole interactions with these hydrogen atoms cause most of the nuclear relaxation. For a carbon atom attached to N equivalent protons in a molecule undergoing rapid tumbling, holds. Where $h = h/2n$ and γ_c and γ_H are the magnetogyric ratios of carbon and hydrogen nuclei.

$$1/T_1 = \frac{Nh^2\gamma_c^2\gamma_H^{\ 2}\tau_{eff}}{r^6} \qquad \text{...(1.55)}$$

This equation permits a calculation of an effective correlation time τ_{eff} for each carbon atom. T_1 and T_2 can also be evaluated for individual ^{15}N or ^{13}C nuclei in labeled proteins.

Two-dimensional and Multidimensional NMR Spectra

Proteins have such complex NMR spectra that, the mean time that a pair of except for small regions at the upfield and downfield ends (Fig. 1.26), it is impossible to interpret one-dimensional spectra. A solution to this problem came from the development by Jeener, Ernst, and Freeman of methods of displaying NMR spectra in two dimensions. All two dimensional and multidimensional NMR methods make use of one basic procedure.

After the initial RF pulse, a second or a series of subsequent RF pulses are introduced. This is done before the nuclei have had time to relax completely. The time t_1 from the initial pulse to the second pulse is called the *evolution period* and allows accumulation of information about NOEs or *J*-coupling, whether homonuclear (e.g. ^{1}H –^{1}H or ^{13}C–^{13}C) or heteronuclear (e.g., ^{1}H–^{13}C–^{13}C or ^{1}H–^{15}N). Two-dimensional spectra usually require hours or days of acquisition because separate FIDs are collected for a series of many different values of t_1. This provides a second timescale for the experiment. The second pulse is often used to rotate the directions of magnetization of the nuclei that are being observed from the *xy* plane into the *yz* plane but with the z component at 180° to the H° vector. One very important type of two-dimensional plot is the *NOESY* (NOE spectroscopy) spectrum, which detects NOEs between all excited nuclei that are close enough together (Fig. 1.24B).

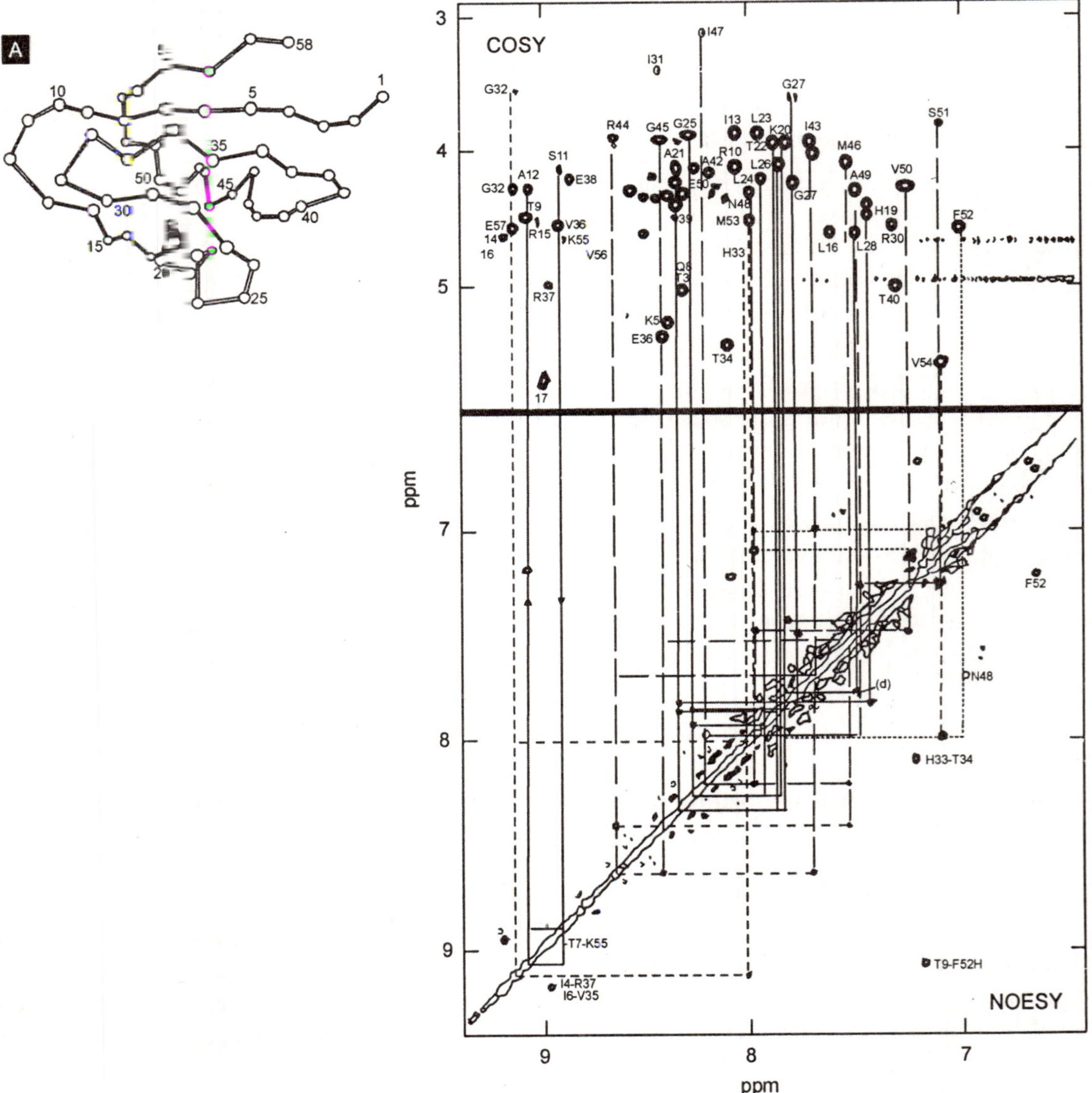

Fig. 1.24 (A) Alpha-carbon plot of the structure of ribosomal protein L30 from *E. coli* as deduced by NMR spectroscopy and model building. (B) Combined COSY-NOESY diagram for ribosomal protein L30 used for elucidation of d_{NK} connectivities. The upper part of the diagram represents the fingerprint region of a COSY spectrum recorded for the protein dissolved in H_2O. The sequential assignments of the crosspeaks is indicated. The lower part of the diagram is part of a NOESY spectrum in H_2O. The d_{NN} "walks" are indicated by : (→—) S11–A12; (—) H19 to L26; (- - - -) I43 to S51; (- - - -) G32 to H33; (——) F52 to V54.

The transfer of magnetization occurs during a *mixing time* Δ, which follows the second pulse. For a protein Δ may be 25–300 ms. A third 90° pulse returns the z components of the magnetization to be parallel with the y axis and the FID is collected over the period t_2. The amplitude of the resonances detected is modulated by the frequencies that existed during the evolution period. After Fourier analysis a two-dimensional plot with two frequency axes is generated and is usually displayed as a contour plot. Along the diagonal of Fig. 1.24B are peaks representing the one-dimensional spectrum.

All of the peaks off of the diagonal are NOEs which can be related back to the peaks on the diagonal as shown by the horizontal and vertical lines. A second important two-dimensional method is *correlation spec-troscopy* (*COSY*), in which the pairs of off -diagonal peaks result from spin-spin coupling. A related method called *TOCSY* provides correlations that extend through more than three bonds. The COSY plot in Fig. 1.25B is for the synthetic cyclic decapeptide cyclo-(Δ^3–Pro–D–*p*–Cl–Phe–D–Trp–Ser–Tyr–D–Trp–N–Me–Leu–Arg–Pro–β–Ala). It was obtained in *six* hours on a 500-MHZ instrument.

The region marked I reveals couplings of protons on α-carbons to those on adjacent β carbons within the same residue ($J_{\alpha\beta}$; Fig. 1.26) and other couplings within the side chain. Each amino acid has a characteristic pattern. From careful study of this region it is possible to correlate each α-H resonance with a particular amino acid side chain. However, some residues are difficult to distinguish, e.g., His, Trp, Phe, and Tyr have similar $J_{\alpha\beta}$ values. Region II of Fig. 1.25B reveals connectivities of α and β hydrogens to N-H protons in the 7– 9 ppm region. Each α-H is coupled to the N-H of the same residue ($J_{\alpha H}$, Fig. 1.26).

A section of a COSY plot is also shown in Fig. 1.24B and indicates how resonances can be related to those in the NOESY plot made on the same sample. Of great importance in the determination of protein structures is the use of ^{15}N- or ^{13}C-enriched samples to obtain isotope-edited spectra. For example in *HSQC* or in *^{15}N-multiple quantum cohenence (HMQC)* spectra we see only NH protons in a plot of ^{1}H chemical shift in one dimension versus the ^{15}N chemical shift of the attached nitrogen atom in the other (Fig. 1.25). Furthermore, as shown in this figure, particular types of NH bonds (peptide, $-NH_2$, imidazole, indole, amide, and guanidinium) appear in different regions. There is only one peptide NH per residue and, for a small protein, each may be separately visible.

Three-Dimensional Structures and Dynamics of Proteins

The first three-dimensional structure of a small protein was determined solely from NMR measurements in 1984. To date hundreds of "NMR structures" have been deduced. For small proteins of M_r < 10,000 two-dimensional COSY and NOESY spectra can suffice. For larger proteins use of ^{15}N- and/or ^{13}C-enriched proteins is essential. "This permits generation of a third and even a fourth frequency *axis* and three-and four dimensional NMR. For example, using various complex pulse sequences an ^{15}N-correlated ^{1}H–^{2}H NOESY spectrum can be generated. This shows directly which NOEs arise from NH protons with resonances in the amide region of an HSQC or HMQC spectrum.

With both ^{15}N and ^{13}C present, many additional coupling patterns and values can be observed. It is also possible to measure NOEs from atoms in a protein to those of a relatively weakly bound ligand such as a coenzyme or substrate analog and to determine the conformation of the bound ligand from this *transferred NOE*.

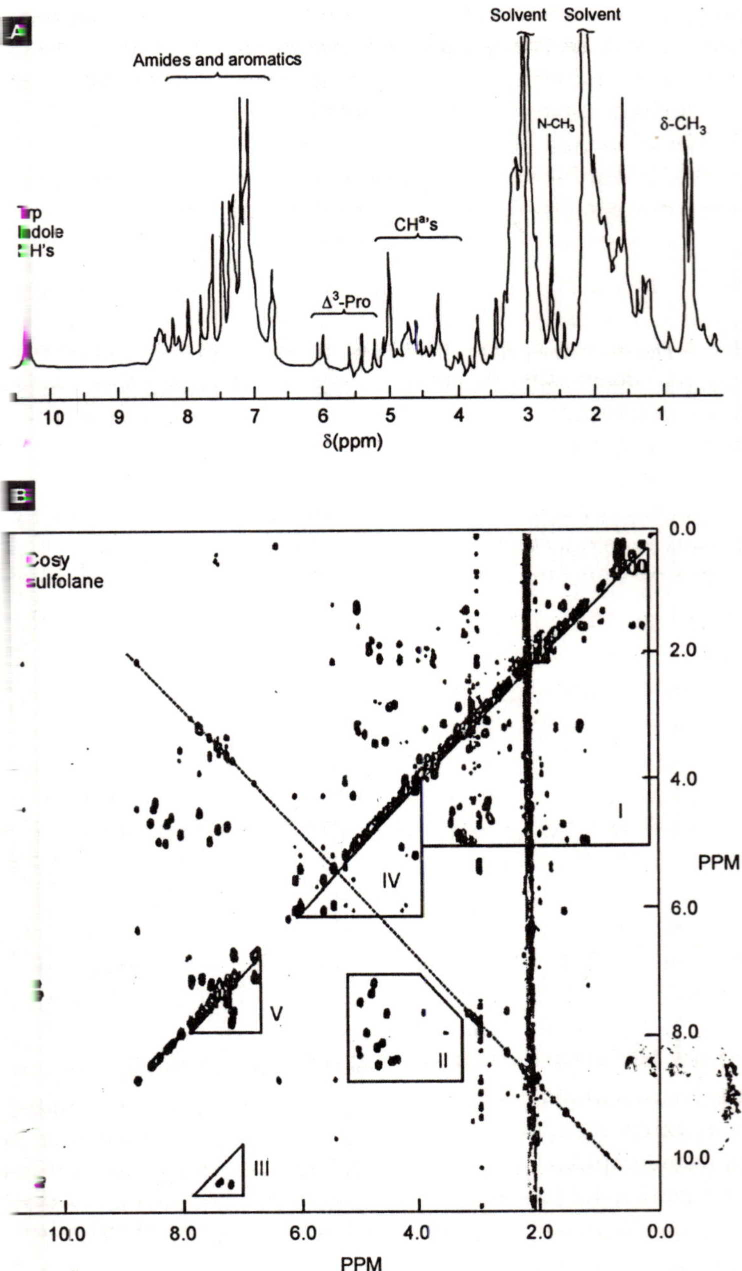

Fig. 1.25. Proton NMR spectra of a cyclic decapeptide analog of the gona-dotropin-releasing hormone in the solvent sulfolane at 500 MHz. (A) One-dimensional spectrum. This figure also illustrates upfield methyl group regions, α-hydrogen, amide, aromatic, and downfield (10–20 ppm) regions. The indole NH resonances are shifted downfield by the ring current of the indole. (B) COSY spectrum plotted as a contour map. The outlined areas represent five unique J coupled regions; area I, the C^{α} to C^{β} to C^{γ} etc., side chain connectivities; area II, the NH to C^{α} connectivities; area III, the indole NH of tryptophan; area IV, the connectivities of the Δ^3-Prol residue; and area V, the aromatic resonances. Peaks that appear on the solid vertical line and the dashed diagonal lines are artifacts.

Assignment of Resonances

After acquisition of the necessary data, which may require 10–15 mg of protein, the observed resonances must be assigned to specific amino acid residues in the peptide chain. The connectivities of the individual CH and NH groups that have been identified in the COSY spectrum and information about the relationship of one atom to another atom nearby in space are required. The closest neighbours to either αH or peptide NH protons are often protons in a neighbouring residue ($d_{\alpha N}$, d_{NN}, Fig. 1.26). 'As was pointed out in the preceding section, NOE correlations obtained from plots such as that in provide much of the information needed to establish which resonances belong to each residue in a known sequence.

The three-dimensional structure of the ribosomal protein L30 of *E. coli.* (Fig. 1.24A) was deduced entirely by NMR spectroscopy. A downfield part of the NOESY and COSY plots used is shown in Fig. 1.25B. This figure also shows how cross-peaks in the NOESY spectrum were correlated with identified COSY peaks. It is helpful initially to hunt for unique dipeptides that can be identified in the NOESY spectrum. Ambiguities that arise can be resolved by use of various additional techniques.

NOEs and Distance Constraints

NOESY plots also contain the essential information needed to determine which side chains *distant* in the sequence are close together in space. A NOE observed for a pair of nuclei falls off as the inverse sixth power of the distance between them. For this reason, NOEs are observed only for pairs of atoms closer than about 0.4 nm. It is possible, in principle, to calculate distances between nuclei from the NOE intensity, but this is not accurate. Often, the NOE cross-peaks are grouped into three categories that correspond to maximum possible distances of 0.25, 0.30, and 0.40 nm. These can be related for the most part to intraresidue (e.g., $d_{N\alpha'}$< of Fig. 1.26) sequential (e.g., $d_{\alpha N}$, and $d_{NN'}$ Fig. 1.26) and long range backbone-backbone distances.

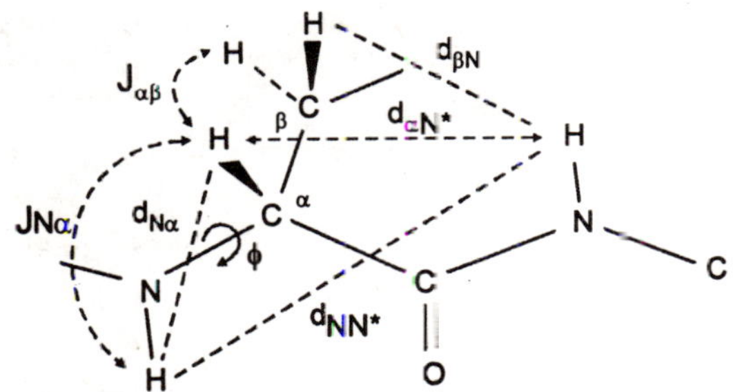

Fig. 1.26. Illustration of some distances ($d_{NN'}$ d_{HK} and $d_{\beta N}$ obtained from NOESY spectra and some coupling constants ($J_{\alpha\beta}$ and $J_{N\alpha}$ obtained from ^1H COSY spectra or *J*-resolved spectra. The coupling constants and proton chemical shifts provide a "fingerprint" for each residue and $J_{N\alpha}$ may also provide an estimated value for torsion angle f. The distances establish residue-to-residue connectivities as well as distance constraints that may permit a calculation of three-dimensional structure. Additional coupling constants can be measured for ^{15}N- or ^{13}C- enriched proteins. Coupling from β-hydrogens to other side chain hydrogens provides "fingerprint" information about individual residues.

These values constitute a series of *distance constraints* which are applied while making an automated computer search for a folding pattern that will meet these constraints and at the same time have acceptable torsion angles and good side chain packing throughout. This process makes use of distance geometry algorithms and other methods. early success was the solution of a 75-residue amylase inhibitor independently by cry stall ographers[9es]"and NMR spectrosco-pists.^^The NMR structure was based on 401 NOE distance constraints, 168 distance constraints imposed by hydrogen bonds and 50 torsion angles deduced from J values. Recently, refinement of NMR structures has been done as in X-ray crystallography. Some structures have been refined using both NMR and X-ray data. The spectra in Fig. 1.24, 1.25 and 1.27 are for relatively small proteins. Spectra of larger proteins are more complex and lines are broader.

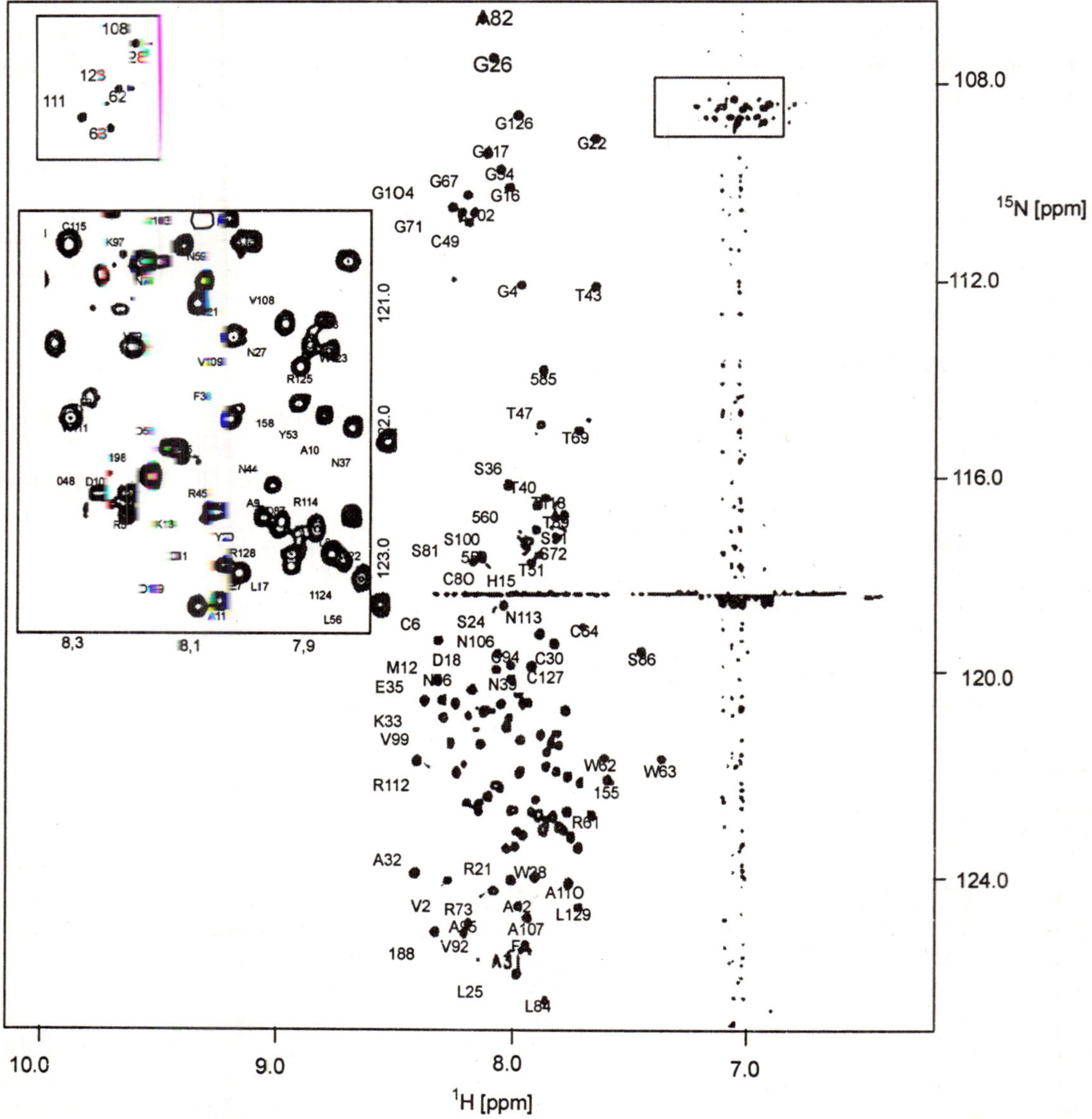

Fig. 1.27. A $^{15}N - {}^{1}H$ HSQC spectrum of partially denatured 129-residue hen lysozyme. Boxes enclose the tryptophan indole region (upper left), the arginine side chain N$^{\varepsilon}$ region (upper left), and a portion of the amide NH region (lower center and enlarged in the insert). Resonances of pairs of hydrogen atoms in side chain (Asn and Gin) amide groups are indicated by horizontal lines.

Many techniques are used to simplify spectra. The NMR spectrum of a protein is simplified considerably if the protein is denatured by heating, and ^1H NMR spectra of "random coil" proteins can be predicted well from tables of standard chemical shifts for the individual amino acids. Many amide NH protons exchange with solvent rapidly, making it easier to assign the remaining peaks. However, nearly all NH peaks will be seen in an HMQC or HSQC (Fig. 1.27) spectrum. Partial, or even complete, substitution of deuterium for hydrogen will also simplify spectra.

Microorganisms that will grow in a medium rich in D_2O can be used as sources of partially deuterated proteins. Because the remaining protons usually have ^2H rather than ^1H as a neighbour, dipolar line broadening is reduced and sharper resonances are observed. Substitution of ^{15}N for ^{14}N in the backbone amide groups can also yield spectra with narrower lines. *Isotope-edited* NMR spectra allow simplification of complex two-dimensional spectra by observation of only those protons attached to an isotopically labeled nucleus, e.g. ^{13}C or ^{15}N. Measurement of ^{15}NH-$C_{\alpha H}J$ couplings facilitates structure determinations.

Many aspects of the dynamics of the action of enzymes of this size can be studied by NMR spectroscopy. New approaches allow study of proteins up to -30 kDa in size. Computer programs can analyze data and make automatic assignments of resonances. Also useful are techniques for measuring NMR spectra on solids, including microcrystalline proteins. In some cases, e.g., for heme proteins, ffie crystals can be oriented in a magnetic field permitting measurements of NMR spectra with more than one orientation of the crystals. This potentially affords more information than the usual techniques.

For example, growth of bacteria in ^2H_2O media containing ^1H-containing pyruvate yields proteins with almost complete deuteration in the C-α an C-β positions but with highly protonated methyl groups. This gives rise to good – CH_3 to – CH_3 NOEs and provides other advantages in both NMR spectroscopy and mass spectrometry. Isotopically enriched amino acids, e.g., ^{13}C-enriched leucine, or isoleucine containing ^{15}N, ^{13}C, and ^2H as well can be fed to growing bacteria. The use of paramagnetic shifts by ions such as Gd^{3+} may be helpful. Very large shifts are sometimes induced in heme proteins by the Fe^{3+} of the heme that is embedded within the protein.

Other Information from NMR Spectra

NMR Titrations

How do pH changes affect NMR resonances? Resonances of ^1H nuclei close enough to a proton with a pK_a in the pH region under study will experience a shift, which may be either upfield or down-field when the proton dissociates. The resonance of ^{13}C in a carboxyl group will shift downfield when the proton on the carboxyl group dissociates. If the proton that dissociates is tightly hydrogen bonded in a protein, its rate of dissociation may be *slow* compared to the NMR frequency used. If so, the original peak will decrease in value as the pH is raised and a new peak willjppear at a position characteristic of the dissociated form.

However, protons attached to N or O are usually in *rapid* exchange with the solvent. In this case, the NMR resonance of the nucleus being observed will move continuously from one chemical shift value at low pH to a different one at high pH. In an intermediate case the resonance will shift and broaden. Both the $C^{\delta 2}$–H and $C^{\varepsilon 1}$–H protons of histidine can often be seen in proteins (Fig. 1.28 A). As is shown, in Fig. 1.28 A and B, their chemical shifts are strongly dependent upon the state of protonation of the ring nitrogen atoms.

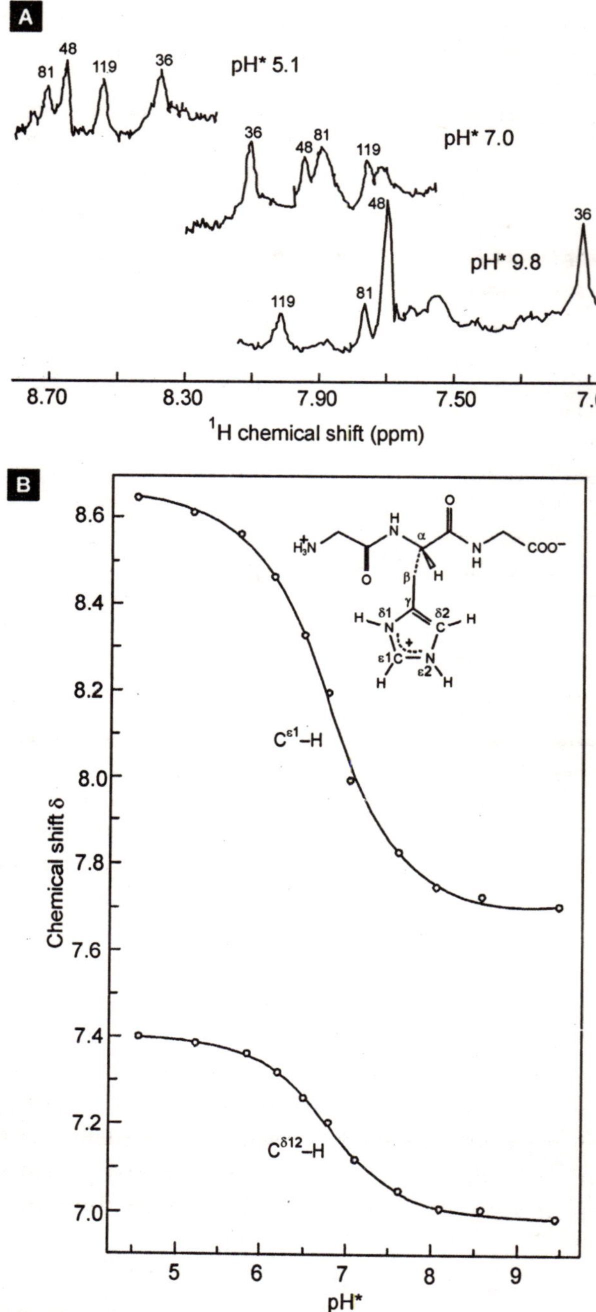

Fig. 1.28. (A) ^{1}H NMR spectra of human myoglobin in D_2O showing the C–H$^{\varepsilon 1}$ resonances of the imidazole rings of histidines 36,48, 81, and 119 at three values of the apparent pH(pH*). (B) ^{1}H NMR titration curves for the histidine C$^{\varepsilon 1}$ and C–$^{\delta 2}$ ring protons of glycyl-L-histidylglycine. Data obtained at 100 MHz using 0.1 M tripeptide in D_2O containing 0.3 M NaCl. The pH values (labeled pH*) are uncorrected glass electrode pH meter readings of D_2O solutions using an electrode standardized with normal H_2O buffers. Chemical shifts are downfield from TMS.

At low pH, the positive charge that is shared by the two NH groups attracts electrons away from both CH positions, causing deshielding of the CH protons. The effect is greater for the $C^{\varepsilon 2}$–H than for $C^{\varepsilon 2}$-H protons (Table 1.3). If the groups being titrated do not interact strongly with other nearby basic or *acidic* groups, pX_a values can be estimated for individual histidines. With ^{15}N-containing proteins the tautomeric states ($NH^{\varepsilon 2}$ vs $NH^{\delta 1}$) of the imidazole rings can also be deduced. Observation of imidazole rings shows that in proteins pK_a values of imidazolium groups are sometimes less than five and sometimes greater than ten.

Observing Exchangeable Protons

^1H spectra of proteins are often recorded in D_2O because of interference from the very strong absorption of H_2O at approximately 4.8 ppm. However, resonances of some rapidly exchanging protons are lost. Special pulse sequences, as well as improvements in spectrometer design, allow many of these resonances to be seen in H_2O. At the far downfield end ($\delta > 10$) of ^1H NMR spectra there are often weak peaks arising from NH protons of imidazole or indole side chains that can be observed in H_2O.

These resonances are often shifted 2–5 ppm downfield from the positions given in Table 1.3. The ^1H resonance for a carboxyl (–COOH) proton is shifted to 20 ppm or more in spectra of very strongly hydrogen-bonded anionic complexes such as the malonate dianion· The strongest hydrogen bonds cause the greatest downfield shifts because the negative charge pulls the ^1H proton away from the electrons of the atom to which it is attached, deshielding the proton. A good example is provided by the imidazole NH proton of the active site of trypsin and related serine proteases, which is seen at ~16 ppm.

Another example is provided by aspartate aminotrans ferase, whose ^1H NMR spectrum in a dilute aqueous phosphate buffer is shown in Fig. 1.29. The peak labeled A is the resonance of the NH proton on the ring of the pyridoxal phosphate coenzyme and peak B belongs to an adjacent imidazole group of histidine 143. Both of these protons move with pH changes around a pK_a of ~ 6.2 which is associated with the Schiff base proton 6–8 run away. Peak A moves upfield 2.0 ppm and peak B downfield 1.0 ppm

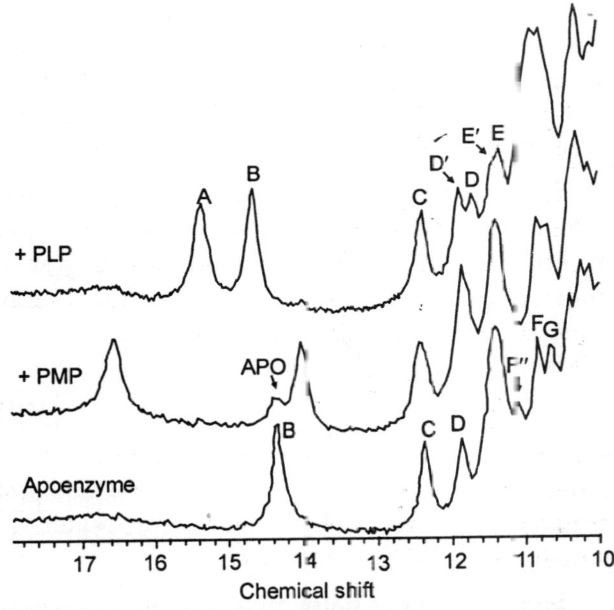

Fig. 1.29. Spectra of the pyridoxal phosphate (PLP), pyridoxamine phosphate (PMP) and apoenzyme forms of pig cytosolic aspartate aminotransferase at pH 8.3, 21°C. Some excess apoenzyme is present in the sample of the PMP form. Spectra were recorded at 500 MH_2. Chemical shift values are in parts per million relative to that of H_2O taken as 4.80 ppm at 22°C. Peak A is from a proton on the ring nitrogen of PLP or PMP, peaks B and D are from imidazole NH groups of histidines 143 and 189, and peaks C and D' are from amide NH groups hydrogen bonded to carboxyl groups.

when the pH is raised around this pK_a. These hydrogen-bonded protojis act as sensitive "reporters" of the electronic environment of the active site. Many proteins contain carboxylate or phosphate groups in their active sites and observation of NMR resonances of protons hydrogen bonded to them or to groups in substrates or inhibitors may be a useful technique for study of many enzymes and other proteins.

Exchange Rates of Amide Protons

The NH protons of the peptide backbone can be observed in H_2O in the 6–11 ppm region. If the spectrum is recorded for a sample in D_2O, many of these resonances disappear gradually as the protons of the peptide units exchange with the deuterium ions of the medium. A study of the observed exchange rates can shed light on the dynamics of protein molecules in solution. Exchange of amide protons in proteins with ^2H or ^3H has been studied on a relatively slow timescale by such techniques as observation of infrared vibrations of the amide group.

However, the development of two-dimensional NMR dramatically improved the ability to study proton exchange. The measurements may involve use of quenching by rapid solvent change, special pulse sequences, and study of T_1 relaxation rates (seconds timescale). Recently, electrospray mass spectrometry has also been exploited. Exchange patterns for large proteins may be followed, and proteolytic fragmentation into short peptides may be used after various lengths of exchange time to investigate the exchange in specific regions of a protein.

Unfolding of a protein under denaturing conditions can be studied, as can refolding of a completely denatured protein. Amide hydrogen exchange is usually discussed in terms of a model proposed by Linderstroϕm-Lang in 1955. He suggested that portions of protein molecules unfold sporadically to allow rapid exchange which can be catalyzed by H^+, HO^-, or other acids or bases. The pH dependence of exchange for a given amide NH can be described as follows:

$$k_{ex} = k_H[H^+] + k_{OH}[OH^-] + k_w \quad ...(1.56)$$

where k_H, k_{OH} and k_w are rate constants for acid-catalyzed, base-catalyzed, and the very slow water-catalyzed exchange. While many amide protons exchange rapidly, hydrogen-bonded NH protons in well-packed hydrophobic core regions exchange slowly, some times remaining in the protein for years in a D_2O solution.

Some unusually stable small proteins such as the seed protein *crambin* show little exchange. The $C^{\varepsilon 1}$ protons of histidine imidazoles also exchange slowly with D_2O from the medium. The exchange rates are rapid at higher temperatures, with an average half-life of about 11 min at 65°C. Different residues may exchange at different rates. Binding of substrates or inhibitors can stabilize the protein, slowing all exchange rates of both amide and imidazole groups.

Solid-state NMR and Other Topics

Little can be said about these topics, but NMR measurement on solid crystalline materials is now practiced and is providing a wealth of information. It is being applied more often to biochemically related problems. An important medical application of NMR is in *imaging,* a topic dealt already any where in this chapter.

Table 1.4. Selected World Wide Web Servers Related to Protein Structures and Sequences[a]

Question/area	*Tool*	*Access/URL*
Database search by comparison of 3D structures	Dali Server	http://www.embl-heidelberg.de/dali/dali.html
Structural classification of proteins	SCOP	http://www.bio.cam.ac.uk/scop/
Summary and analysis of PDP structures	PDBSum	http://www.biochem.ucl.ac.uk/bsm/pdbsum
Retrieve 3D coordinates	Protein Data Bank	http://www.rcsb.org/pdb/
SWISS-PROT sequence database, Swissmodel homology modeling,etc.	ExPASy	http://expasy.hcuge,ch/
Molecular graphics viewer for PCs and workstations	RasMol	http://www.bernstein-plus-sons.com/software/rasmol
World Wide Web-Entrez and Molecular Modeling Database access		http://www.ncbi.nlm.nih.gov
Protein Science Kinemages	MAGE and PREKIN	http://www.prosci.uci.edu/kinemages/ Kinemagelndex.html
Pedro's Biomolecular Research Tools		http://www.fmi.ch/biology/research_tools.html
Predict secondary structure from sequence	Predict Protein server	http://www.sander.embl-heidelberg.de
Browse databanks in molecular biology	SRS server	http://www.embl-heidelberg/src/srsc
The Human Genome Database		http://gdbwww.gdb.org/
Image Library of Biological Macromolecules		http://www.imb-jena.de/IMAGE.html
Bacterial Nomenclature		http://www.gbf-braunschweig.de/DSMZ/bactnom/ bactnam.htm

THE PROTEIN DATA BANK, THREE-DIMENSIONAL STRUCTURES AND COMPUTATION

The x,y,z coordinates of all atoms in published, refined three-dimensional structures have been deposited in the Protein Data Bank (Table 1.4). Many other related databases are available, eg., covering molecular modeling, gene sequences, proteome data, and much, much more. A good way to keep up to date is to read the "computer corner" in *Trends in Biochemical Sciences (TIBS).* Most databases can be reached on the World Wide Web. A selected list is given in Table 1.4. The widely used viewer called RasMol can be used with your PC or MacIntosh computer or UNIX workstation. Another way to view macromolecules is to use the *Protein Science* Kinemages ("kinetic images") using the program MAGE. Although it is mentioned in a few places, this book does not begin to describe the rapid growth of computation in biochemistry, biophysics, and biology in general.

Very fast methods of protein structure determination are being developed. One of the major goals in current computation is to predict folding patterns of proteins from their sequences. By comparing sequences we can often guess an approximate structure but accurate predictions are still not possible.

Having a structure, we would like to predict properties and reactivities and to be able to guess how two more macromolecules interact to form macromolecular complexes. We would like to understand the complex chain of nonpolar and electrostatic interactions that underlie the fundamental properties of catalysis, movement, and responsiveness of organisms. Many computers and ingeneous minds are working to help us match theory with reality in these areas. Read the current journals!

CHAPTER

2 Metabolism

CARBOHYDRATE METABOLISM

Carbohydrate metabolism in the organism tissues encompasses enzymic processes leading either to the breakdown of carbohydrates (catabolic pathways), or to the synthesis thereof (anabolic pathways). Carbohydrate breakdown leads to energy release or intermediary products that are necessary for other biochemical processes. The carbohydrate synthesis serves for replenishment of polysaccharide reserve or for renewal of structural carbohydrates. The effectiveness of various routes of carbohydrate metabolism in tissues and organs is defined by the availability of appropriate enzymes in them;

Carbohydrate Catabolism in Tissues

A number of routes for carbohydrate catabolism in tissues are known. They include *glycolysis* and its variant, *glycogenolysis,* which are auxiliary pathways to energy production, respectively, by breakdown of glucose (or other monosaccharides) and glycogen to lactate (under anaerobic conditions) or to CO_2 and H_2O (under aerobic conditions).

There is known one more catabolic route for carbohydrates commonly referred to the *pentose phosphate cycle* (also called *hexose monophosphate shunt,* or *phosphogluconate pathway*). As a tribute to the biochemists who have played a decisive role in its investigation, the pentose phosphate cycle is also referred to as the Warburg-Dickens-Horecker pathway. The pentose phosphate cycle represents a multienzyme system in which the important intermediates are, as the name implies, pentose phosphates. This cycle may be regarded as a branching, or shunt, at the glucose 6-phosphate step in the overall glycolysis.

Penotose Phosphate Cycle

To provide for all steps of the pentose phosphate cycle, ht least three glucose 6-phosphate molecules are required. Let us consider separate reactions of this cycle.

1. Dehydrogenation of glucose 6-phosphate is the reaction that directs glucose 6-phosphate via the pentose phosphate pathway; it is catalyzed by glucose-6-phosphate dehydrogenase (in the schemes below, for a fuller description of the cyclic process, three glucose 6-phosphate molecules are used)

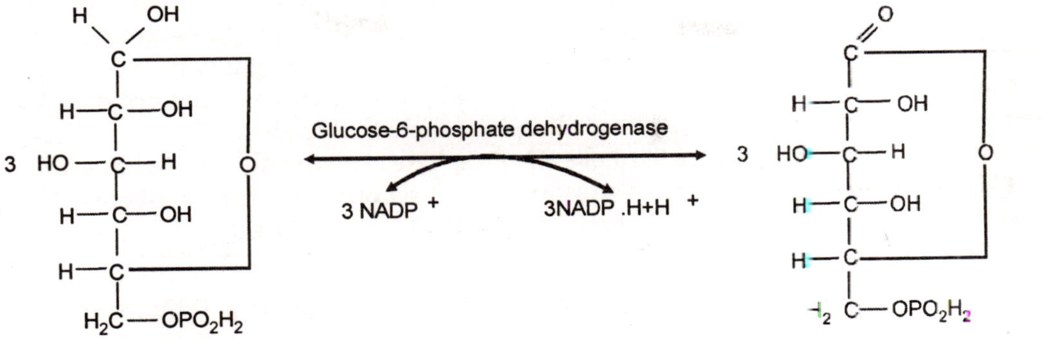

Glucose-6-phosphate dehydrogenase

3 glucose 6-phosphate + 3 NADP$^+$ ⟶ 3 6-phosphogluconate lactone + 3NADP.H+H$^+$

Glucose-6-phosphate dehydrogenase is a dimer with a molecular mass of about 1,35,000. Up to eight electrophoretically separable isoenzymes for this enzyme are known. A specific feature of the above reaction is the formation of NADP–H_2. The reaction equilibrium is strongly shifted to the right, since the lactone formed is liable to hydrolysis, which is spontaneous or *lactonase-assisted.*

2. Hydrolysis of 6-phosphogluconate lactone to 6-phosphogluconate:

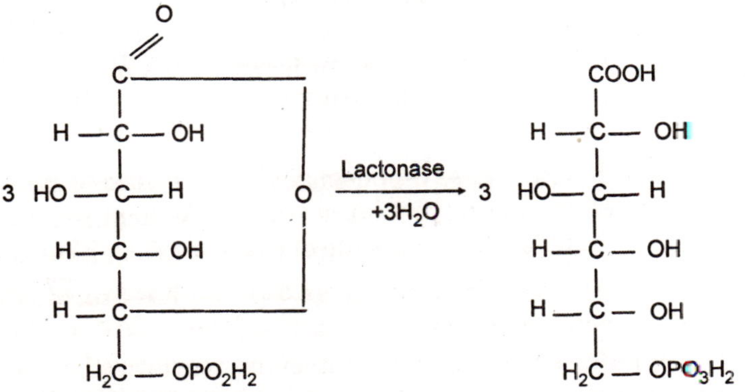

3 6-phosphogluconate lactone —Lactonase, $+3H_2O$⟶ 3 6-phosphogluconate

3. Dehydrogenation of 6-phosphogluconate to ribulose 5-phosphate. This reaction is catalyzed by *Q-phosphogluconate dehydrogenase* according to the scheme:

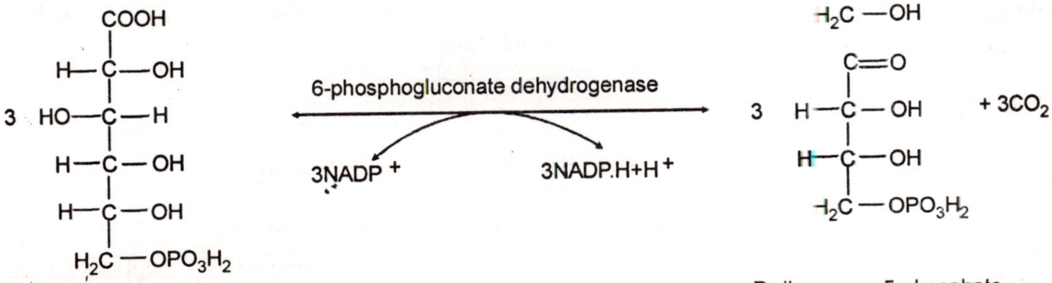

6-phosphogluconate dehydrogenase

3 6-phosphogluconate + 3NADP$^+$ ⟶ 3 D-ribulose 5-phosphate + 3NADP.H+H$^+$ + $3CO_2$

The reaction equilibrium is shifted to the right. 6-Phosphogluconate dehydrogenase is a dimer with a molecular mass of about 100 000. Several isoenzymes are known for this dehydrogenase. A specific feature of this reaction is that dehydrogenation leads to an unstable intermediate which is immediately decarboxylated on the surface of the enzyme. This is the second oxidation reaction in the pentose phosphate cycle that leads to NADP–H_2; therefore, the conversion of glucose 6-phosphate to ribulose 5-phosphate is commonly referred to as the *oxidative phase of the pentose phosphate cycle.* The sequence of reactions starting from ribulose 5-phosphate to the formation of initial glucose 6-phosphate is called the *nonoxidative, or anaerobic, phase* of this cycle. ;

4. Interconversion or isomerization of pentose phosphates. Ribulose 5-phosphate is capable of a reversible isomerization to other pentose phosphates—xylulose 5-phosphate and ribose 5-phosphate. These reactions are catalyzed by two respective enzymes, viz., pentose-phosphate epimerase and pentose-phosphate isomerase, according to the scheme below :

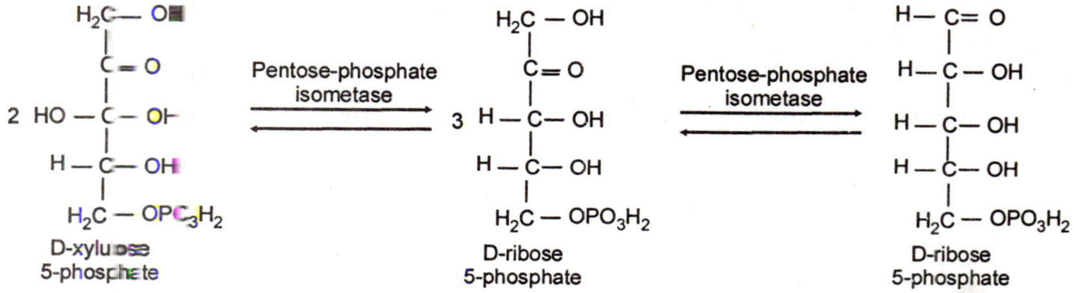

2 D-xylulose 5-phosphate (H_2C—OH, C=O, HO—C—OH, H—C—OH, H_2C—OPO_3H_2) $\underset{\text{isometase}}{\overset{\text{Pentose-phosphate}}{\rightleftharpoons}}$ 3 D-ribose 5-phosphate (H_2C—OH, C=O, H—C—OH, H—C—OH, H_2C—OPO_3H_2) $\underset{\text{isometase}}{\overset{\text{Pentose-phosphate}}{\rightleftharpoons}}$ D-ribose 5-phosphate (H—C=O, H—C—OH, H—C—OH, H—C—OH, H_2C—OPO_3H_2)

Two other pentose phosphates (ribose 5-phosphate and xylulose 5-phosphate), which are derived from ribulose 5-phosphate, are important for the subsequent reaction of the cycle. Two molecules of xylulose 5-phosphate and one molecule of ribose 5-phosphate are required for this.

5. Transfer of glycolic aldehyde from xylulose 5-phosphate onto ribose 5-phosphate or the first transketolase reaction. The next reaction, which is catalyzed by *transketolase,* involves the pentose phosphates produced by the foregoing reaction (the transferable moiety is Shown in the box).

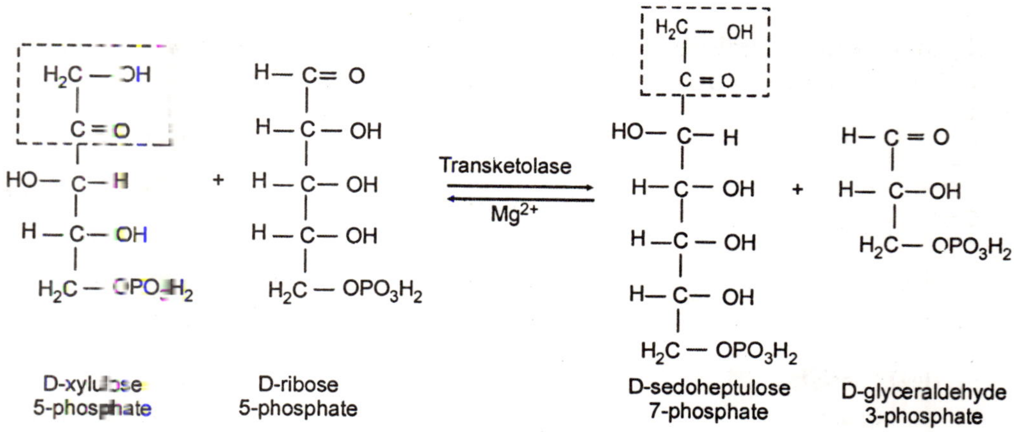

D-xylulose 5-phosphate ([H_2C—OH, C=O], HO—C—H, H—C—OH, H_2C—OPO_3H_2) + D-ribose 5-phosphate (H—C=O, H—C—OH, H—C—OH, H—C—OH, H_2C—OPO_3H_2) $\underset{Mg^{2+}}{\overset{\text{Transketolase}}{\rightleftharpoons}}$ D-sedoheptulose 7-phosphate ([H_2C—OH, C=O], HO—C—H, H—C—OH, H—C—OH, H—C—OH, H_2C—OPO_3H_2) + D-glyceraldehyde 3-phosphate (H—C=O, H—C—OH, H_2C—OPO_3H_2)

A ribose 5-phosphate molecule and one of the two xylulose 5-phosphate molecules are used during the first transketolase reaction. The other xylulose 5-phosphate molecule is consumed later, in the second transketolase reaction.

Transketolase is a dimer with a molecular mass of 1,40,000. Its coenzyme is *thiamine bisphosphate.* Mg^{2+} ions are required for the reaction. Both transketolase reaction products are used as substrates at the next step of the cycle.

6. Transfer of dihydroxyacetone moiety from sedoheptulose 7-phosphate onto glyceraldehyde 3-phosphate. This reaction is reversible and is catalyzed by *transaldolase* according to the scheme :

$$\begin{array}{c} H_2C-OH \\ | \\ C=O \\ | \\ HO-C-H \\ | \\ H-C-OH \\ | \\ H-C-OH \\ | \\ H-C-OH \\ | \\ H_2C-OPO_3H_2 \end{array} + \begin{array}{c} H-C-OH \\ | \\ H-C-OH \\ | \\ H_2C-OPO_3H_2 \end{array} \xrightleftharpoons{\text{Transketolase}} \begin{array}{c} H_2C-OH \\ | \\ C=O \\ | \\ HO-C-H \\ | \\ H-C-OH \\ | \\ H-C-OH \\ | \\ H_2C-OPO_3H_2 \end{array} + \begin{array}{c} H-C=O \\ | \\ H-C-OH \\ | \\ H-C-OH \\ | \\ H_2C-OPO_3H_2 \end{array}$$

D-sedoheptulose 7-phosphate; D-glyceraldehyde 3-phosphate; D-fructose 5-phosphate; D-erythrose 4-phosphate

Transaldolase is a dimer with a molecular mass of about 70 000. The fructose 6-phosphate molecule produced by this reaction enters the glycolysis, while erythrose 4-phosphate is used as a substrate at the subsequent steps of the cycle.

7. Transfer of glycolic aldehyde from xylulose 5-phosphate onto erythrose 4-phosphate or the second transketloase reaction. This reaction is related to the first transketolase reaction and is catalyzed by the same enzyme. The only distinction is that erythrose 4-phosphate acts as an acceptor for glycolic aldehyde:

$$\begin{array}{c} H_2C-OH \\ | \\ C=O \\ | \\ HO-C-H \\ | \\ H-C-OH \\ | \\ H_2C-OPO_3H_2 \end{array} + \begin{array}{c} H-C=O \\ | \\ H-C-OH \\ | \\ H-C-OH \\ | \\ H_2C-OPO_3H_2 \end{array} \xrightleftharpoons{\text{Transketolase}} \begin{array}{c} H_2C-OH \\ | \\ C=O \\ | \\ HO-C-H \\ | \\ H-C-OH \\ | \\ H-C-OH \\ | \\ H_2C-OPO_3H_2 \end{array} + \begin{array}{c} H-C-CH \\ | \\ H-C-CH \\ | \\ H_2C-OPO_3H_2 \end{array}$$

D-xylulose 5-phosphate; D-erythrose 4-phosphate; D-fructose 5-phosphate; D-glyceraldehyde 3-phosphate

Fructose 6-phosphate and glyceraldehyde 3-phosphate also enter the glycolysis.

Thus, in the course of reactions Catalyzed by the intrinsic enzymes of the pentose phosphate cycle, two fructose 6-phosphate molecules, one glyceraldehyde 3-phosphate molecule, and three carbon dioxide molecules are produced from three glucose 6-phosphate molecules. In addition, six NADP • H_2 molecules are formed. The overall scheme for the pentose phosphate cycle is:

3 Glucose 6-phosphate— 6NADP$^+$ → 2 Fructose 6-phosphate
+ Glyceraldehyde 3–phosphate + 6NADP.H_2 + 3CO_2

Interrelation of the Pentose Phosphate Cycle and Glycolysis

These two pathways for carbohydrate conversion are closely related (Fig. 2.1). The products of the pentose phosphate route— fructose 6-phosphate and glyceral dehyde 3-phosphate — are

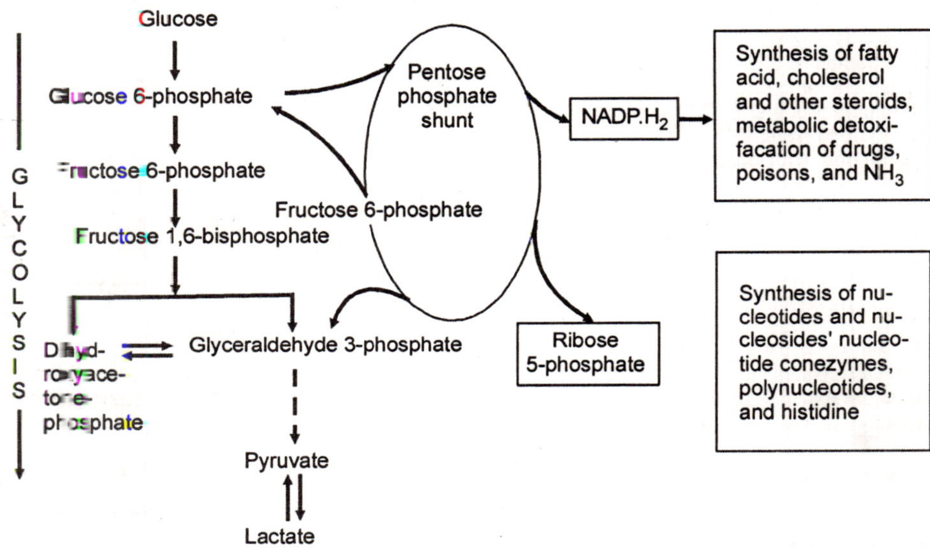

Fig. 2.1 Scheme for integration of pentose phosphate shunt and glycolysis

likewise glycolysis metabolites; for this reason, they are involved in glycolysis and undergo conversion by glycolytic enzymes. Two molecules of fructose 6-phosphate can regenerate to two glucose 6-phosphate molecules through the agency of the glycolytic enzyme glucose-phosphate isomerase. Here the pentose phosphate pathway functions as a cycle. The other product, glyceral-dehyde 3-phosphate, enters the glycolysis to be either converted to lactate (under anaerobic conditions) or oxidized to CO_2 and H_2O (under aerobic conditions).

As can easily be estimated, the conversion of glyceraldehyde 3-phosphate to lactate leads to the formation of two ATP molecules, while the combustion to CO_2 and H_2O produces 20 ATP molecules. It follows therefore that under physiological conditions, when the pentose phosphate pathway for carbohydrate conversion is included in the glycolysis, the overall process of glucose 1 6-phosphate conversion may be expressed via the pentose phosphate cycle. Under anaerobic conditions:

3 Glucose 6-phosphate + 6NADP$^+$ + 2P_1 → 2 Glucos 6-phosphate + Lactate
+ 2ATP + 6NADP.H_2 + 3CO_2

Under aerobic conditions:

3 Glucose 6-phosphate + 6NADP$^+$ + 20ADP + 20P_1
→ 2 Glucose 6-phosphate + 6NADP.H_2 + 6CO_2 + 6H_2O + 20ATP

At first glance, the energetic value of this conversion of glucose 6-phosphate via the pentose phosphate cycle appears to be inferior to that of the aerobic glycolysis pathway, the latter providing a maximum of 38 ATP molecules. However, it should be borne in mind that a major portion of energy is stored in NADP $\bullet H_2$, and 6 NADP $\bullet H_2$ molecules are energetically equivalent to 18 ATP molecules. Consequently, the energetic effect remains the same.

The Biological Function of the Pentose Phosphate Cycle

The biological function of the pentose phosphate cycle involves the production of two compounds: NADP $\bullet H_2$, which is a "reductive force" in the synthesis of various materials, and the metabolite ribose 5-phosphate, which is used as a building material in the synthesis of various species (see Fig. 2.1). The major functions of the pentose phosphate cycle are:

1. amphibolic function: the cycle is a route to degradation of carbohydrates and, simultaneously,, to the supply of materials used in synthetic reactions (NADP $\bullet H_2$ and ribose 5-phosphate);
2. energetic function, since the involvement of pentose phosphate cycle products (glyceraldehyde 3-phosphate) in the glycolysis produces energy;
3. synthetic function, as a major function associated with the use of NADP $\bullet H_2$ and ribose 5-phosphate.

NADP-H_2 is used:

1. in the detoxification of drugs and poison's in the monooxygenase oxidation chain of the endoplasmic reticulum of the liver;
2. in the synthesis of fatty acids and other structural and reserve lipids;
3. in the synthesis of cholesterol and its derivatives—bile acids, steroid hormones (corticosteroids, female and male sex hormones), and vitamins D; and
4. in the neutralization of ammonia under reductive animation.

Ribose 5-phosphate is used in the synthesis of histidine, nucleosides and nucleotides (nucleotide mono-, di-, and triphosphates), nucleotide coenzymes (NAD, NADP, FAD, and CoA), and polymeric nucleotide derivatives (DNA, RNA, and short-chain oligonucleotides). The pentose phosphate pathway for carbohydrate conversion is primarily operative in the organs and tissues in which an intensive utilization of NADP $\bullet H_2$ is needed for reactions of reductive synthesis and for reactions involving ribose 5-phosphate in the synthesis of nucleotides and nucleic acids. For this reason, a high activity of this pathway is observed in fat tissue, liver, mammary gland tissue (especially during lactation, since the milk fat synthesis is essential in this case), adrenal glands, gonadal glands, marrow, and lymphoid tissue.

Relatively high is the activity of pentose phosphate shunt dehydrogenases in the erythrocytes. A low activity of the pentose phosphate pathway is observed in muscular tissue (heart and skeletal muscle).

BIOSYNTHESIS OF CARBOHYDRATES IN TISSUES

In the human and animal tissues and organs, synthesis of carbohydrates occurs. Since glucose is the starting structural unit for producing other monosaccharides and for assembling polysaccharides, it is expedient to consider potential routes for the glucose synthesis in tissues and organs. Formation of glucose from noncarbohy-drate materials is attested by the fact that, under prolonged starvation (in an extreme contingency or as applied in therapy), the polysaccharide carbohydrate reserves are rapidly consumed, while the glucose level in the circulating blood is maintained to supply tissues, especially brain, with energy.

Gluconeogenesis

The synthesis of glucose from noncarbohydrate sources is referred to as the *gluconeogenesis.* It is feasible only in certain organism tissues. The major site for gluconeogenesis is the liver. To a lesser extent, the kidneys and intestinal mucosa are involved in this process.

Mechanism for Gluconeogenesis. Since the glycolysis involves three energetically irreversible steps at the pyruvate kinase, phosphofructokinase, and hexokinase levels, the production of glucose from simple noncarbohydrate materials, for example, pyruvate or lactate, by a reversal of glycolysis ("from bottom upwards") is impossible. Therefore, indirect reaction routes are to be sought for.

The first indirect route in glucose synthesis involves the formation of phosphoenol-pyruvate from pyruvate without the intervention of pyruvate kinase. This route is catalyzed by two enzymes. At first, pyruvate is converted into oxaloacetate. This reaction occurs in the mitochondria as the pyruvate molecules enter them, and is catalyzed by *pyruvate carboxylase* according to the scheme

$$\underset{\text{pyruvate}}{CH_3-\underset{\substack{\| \\ O}}{C}-COOH} + HCO_3^- + ATP \xrightarrow{\text{Pyruvate carboxylase}} \underset{\text{oxaloacetate}}{HOOC-CH_2-\underset{\substack{\| \\ O}}{C}-COOH} + ADP + H_3PO_3$$

This enzyme, similar to all CO_2-assimilating enzymes, contains *biotin* for a cofactor. Oxaloacetate is released from the mitochondria into the cytoplasm to enter gluconeogenesis. In the cytoplasm, oxaloacetate converts to phosphoenolpyruvate via a reaction catalyzed by phosphoenolpyruvate carboxylase:

$$\underset{\text{oxaloacetate}}{HOOC-CH_2-\underset{\substack{| \\ O}}{C}-COOH} + GTP(ATP) \xrightarrow{\text{Phosphoenolpyruvate carboxylase}}$$

$$\longrightarrow \underset{\text{phosphoenolpyruvate}}{CH_2=\underset{\substack{| \\ O\sim PO_3H_2}}{C}-COOH} + GDP(ADP) + CO_2$$

The reaction equilibrium is shifted to the right. The major supplier of phosphate groups is GTP, but for this purpose, ATP may also be available.

All of the glycolysis reactions ranging from phosphoenolpyruvate to fructose 1, 6-bisphosphate are reversible, and the phosphoenolpyruvate molecules formed are consumed for producing fructose 1,6-bisphosphate by making use of the same glycolysis enzymes.

The second indirect route involves the formation of fructose 6-phosphate from fructose 1,6-bisphosphate without the intervention of phosphofructokinase reaction. This route is catalyzed by *fructose bisphosphatase:*

Fructose 1,6-bisphosphate + H_2O $\xrightarrow{\text{Fructose bisphosphatase}}$ Fructose 6-phosphate + H_3PO_4

The reaction is irreversibly shifted to the right. Fructose 6-phosphate is isomerized to glucose 6-phosphate by glucose-phosphate isomerase.

The third indirect route involves the formation of free glucose from glucose 6-phosphate by circumventing the hexokinase reaction. This route is catalyzed by *glucose 6-phosphatase:*

Glucose 6-phosphatase + H_2O $\xrightarrow{\text{Fructose bisphosphatase}}$ Glucose+ H_3PO_4

The free glucose produced by this reaction is supplied to the blood from the tissues. The general scheme for pyruvate gluconeogenesis is presented in Fig. 2.2. As exemplified by gluconeogenesis, one may easily envision the economical organization of these metabolic routes, since, apart from four special gluconeogenesis enzymes pyruvate carboxylase,, phosphopyruvate carboxylase, fructose bisphosphatase, and glucose 6-phosphatase—individual glycolytic enzymes are also used in the gluconeogenesis.

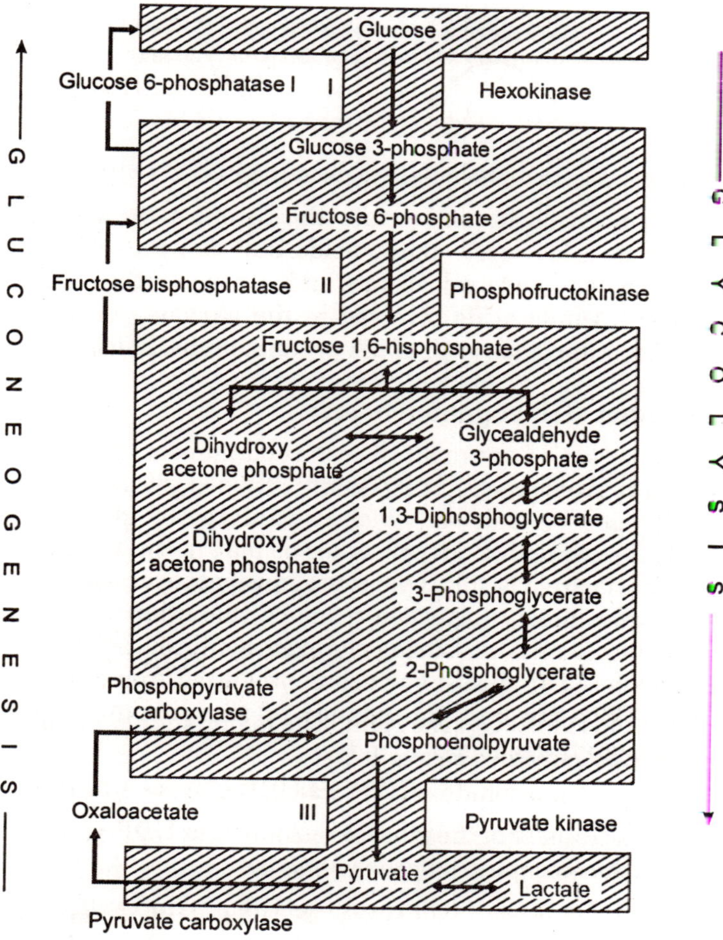

Fig. 2.2. Schematic representation of gluconeogenesis

Noncarbohydrate Sources for Gluconeogenesis: In addition to pyruvate and lactate, which are delivered to the liver and kidneys, other noncarbohydrate compounds serve as substrates for glucose synthesis. In accordance with the gluconeogenesis scheme (see Fig. 2.2) it may be anticipated that all materials of non-carbohydrate nature that are amenable to conversion to a glucolysis metabolite (first group of materials), to pyruvate (second group), or to oxaloacetate (third group), can serve as potential sources for glucose synthesis. For example, glycerol may be included in the first group of materials, since this triol, which can convert, to dihydroxyacetone phosphate, can further take up, depending on the reaction conditions, either gluconeogenesis route, or glycolysis route. The involvement of glycerol in gluconeogenesis proceeds according to the scheme:

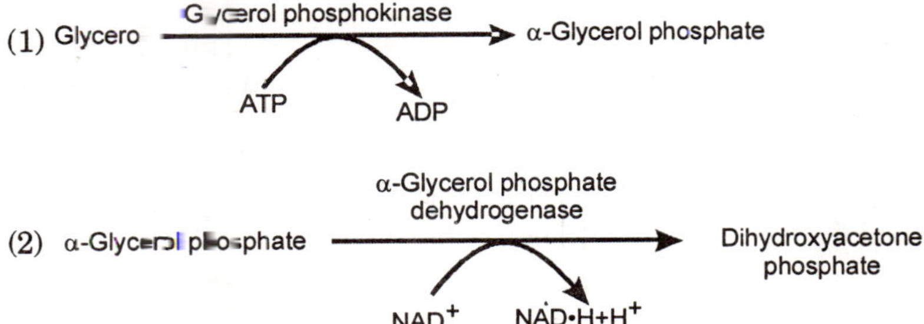

Subsequently, dihydroxyacetone phosphate is used in glucose synthesis.

The Krebs cycle acids convertible to oxaloacetate (third group materials) may also be assigned to gluconeogenesis substrates. However, amino acids that can convert the major source for gluconeogenesis to both pyruvate and oxaloacetate and, consequently, to glucose are the major source for gluconeogenesis. Amino acids involved in the gluconeogenesis are referred to as *glycogenic amino acids*. They encompass all of the protein amino acids, barring leucine.

Biosynthesis of Glycogen (Glycogenogenesis)

Synthesis of glycogen is carried out in all the cells of organism (the erythrocytes, perhaps, being the only exception), but this process is especially active in the skeletal muscles and in the liver. The reaction of glycogen breakdown, which is catalyzed by glycogen phosphorylase, is nearly irreversible; for this reason, this enzyme takes no part in the synthesis of glycogen. Two Routes for glycogen synthesis are possible. One route involves a successive addition of glucose units to the extant glycogen moiety (glycogen primer); the other one originates in glucose molecules. The source of glucose residues during glycogen synthesis is the active form of glucose, uridine diphosphate glucose (UDP-glucose), which is produced from glucose 1-phosphate and uridine triphosphate (UTP) through the agency of the enzyme *glucose–1–phosphate uridyltransferase* according to the scheme:

$$\text{Glucose 1-phosphate} + \text{UTP} \rightleftarrows \text{UDP-glucose} + H_4P_2O_7.$$

The next step involves a transfer of the glucose residue from UDP-glucose onto the glycogen primer through the aid of the enzyme *glycogen synthetase:*

$$\text{UDP-glucose} + (\text{Glucose})_n \rightarrow \text{UDP} + (\text{Glucose})_{n+1}$$

To be noted, the glycogen synthetase catalyzes the formation of α-1 → 4-glycoside bonds only. The "branching" enzyme amylo-(α-1,4 → α-1,6)-transglycolysase, transfers short fragments (two or three glucose residues) from one portion of the glycogen molecule onto another and forms α-1 → 6-glycosidic bonds (branch points). The alternating action of these two enzymes results in the lengthening of the glycogen molecule.

If the synthesis starts from glucose molecules, then the initial step is the transfer of glucose residues from UDP-glucose onto an intermediary acceptor—*dolichol phosphate* (membrane-bound polyprenol phosphate). Dolichol phosphate assists in the synthesis of an oligosaccharide which is then transferred onto the protein. The successive addition of oligosaccharide chains to the glycogen molecule proceeds in the same mariner as in the former route. The feasibility of the latter route is substantiated by the fact that glycogen is always bound with some protein.

Interrelation of Glycogen Synthesis and Degradation. Glycogen synthetase exists in two interconvertible forms. The *phosphorylated,* or *inactive,* form is called *glycogen synthetase* D; the *nonphosphorylated,* or *active,* form is referred to as *glycogen synthetase* I. The transition from one form to the other is accomplished by two enzymes, glycogen synthetase kinase (1) and glycogen synthetase phosphatase (2) according to the scheme:

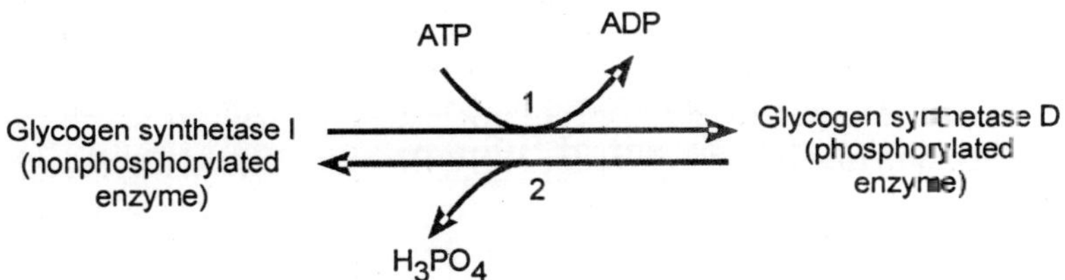

The processes of glycogen synthesis and degradation in the cells are controlled via phosphorylation mechanisms involving the key enzymes of glycogen metabolism — glycogen synthetase and glycogen phosphorylase. The, activation of adenylate cyclase (for example, with adrenalin or glucagon) leads to the production of cAMP which triggers the "cascade" mechanism of phosphorylation of glycogen synthetase and glycogen phosphorylase, with the resultant formation of the inactive (plosphorylated) glycogen synthetase D and active glycogen synthetase I. This favours glycogen synthesis.

Biosynthesis of Other Oligosaccharides and Polysaccharides

Homo- and heteropolysaccharides are carbohydrate components of plasmic and structural glycoproteins in the mammals. These carbohydrates are made up of a small set of monosaccharides: galactose, mannose, N-acetylglycosamine, N-acetyl-galactosamine, fucose, and sialic acid. The connective tissue polysaccharides contain glucuronic acid, iduronic acid, xylose, and sulphated derivatives of N-acetyl-glucosamine or N-acetylgalactosamine. These homo- and heteroglycans can be assembled by activated monosaccharide forms, such as nucleoside phosphate derivatives of monosaccharides. UDP-derivatives of glucose, galactosp, N-acetyl-glucosamine, N-acetylgalactosamine, glucuronic acid and xylose; GDP-derivatives of mannose and fucose; and CMP-derivatives of sialic acid are used in the synthesis. Most nucleoside phosphate saccharides are produced through reaction of monosaccharides or their derivatives with the corresponding nucleoside triphosphates. Occasionally, certain monosaccharides convert to other monosaccharides as constitutive components of nucleoside diphosphate saccharides. For example, UDP-glucuronic acid and UDP-xylose are formed from UDP-glucose:

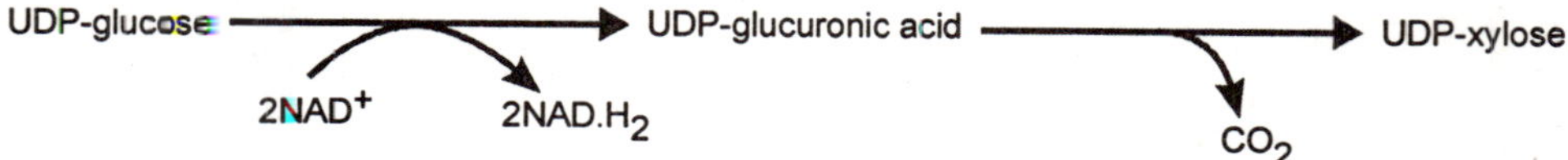

GDP-fucose is formed from GDP-mannose:

For example, it is to be noted that UDP-glucuronic acid which is formed in the tissues, is used not only for polysaccharide synthesis, but also for neutralization and removal of toxic and useless materials or foreign compounds from the organism. Synthesis of oligosaccharides is carried out with the participation of specific *glycosyltransferases* whose diversity and functional specialization in each cell provide for the production of various types of homo- and heteroglycans.

In the cell, glycosyltransferases are bound with the membranes of endoplasmic reticulum or Golgi apparatus i.e. the organelles primarily responsible for assembling oligosaccharides. It may be presumed that initially the assembly is made on the poly-prenol phosphate molecule onto which the monosaccharide residues from nucleoside phosphate saccharides are successively transferred by glycosyltransferases. Subsequently, the synthetized oligosaccharide is transferred onto proteins to form glycoproteins. Simultaneous transfer of glycosyl residues and assembly of protein-based oligo-and polysaccharides are also possible.

CARBOHYDRATE METABOLISM CONTROL IN THE ORGANISM

The carbohydrate metabolic routes in various tissues of the organism discussed above differ in intensity, which is denned by metabolic features specific of each tissue and organ. However, from the standpoint of activity of the whole organism, certain specializations of the carbohydrate metabolic routes in individual tissues are profitably complementary. For example, strenuous muscular exertion requires energy which is initially supplied by the breakdown of glycogen to lactic acid. The latter compound is excreted into the blood to be supplied to the hepatic tissue, where it is used for the synthesis of glucose during gluconeogenesis. From the liver, glucose is delivered in the blood to the skeletal muscles to be consumed for energy generation and to be deposited as glycogen. This intertissue (or interorgan) cycle in the carbohydrate metabolism is referred to as the Cori cycle (called also glucoselactate cycle):

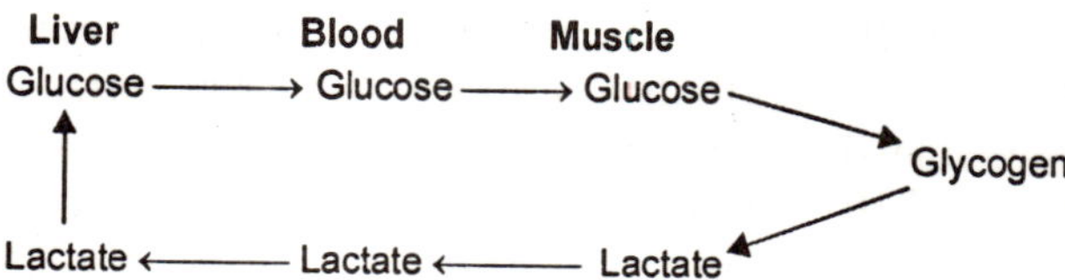

The maintenance of a constant glucose level in the blood is of primary importance for the organism, since glucose is the major energy substrate for the nervous tissue. The normal glucose content in the blood is 3.3 tp 4.0 mmol/litre. An increased concentration of glucose in blood is known as *hyperglycemia*. If hyperglycemia reaches as high as 9 to 10 mmol/litre, the glucose excess is released into the urine, i.e. *glucosuria* sets in. On the contrary; a decreased glucose percentage in the blood is known as *hypoglycemia*. Hypoglycemia as low as about 1.5 mmol/litre leads to the syncopal state, while a still lower glucose concentration results in high excitability of the nervous system and ultimately leads to convulsions and coma.

To gain a better understanding of the mechanism that controls the glucose level in the blood, it is important to examine processes that contribute to an increased or lowered glucose concentration (Fig. 2.3).

Processes leading to hyperglycemia:

(1) absorption of glucose from the intestine (alimentary hyperglycemia);

(2) breakdown of glycogen to glucose (commonly, in liver);

(3) gluconeogenesis (in liver and kidney).

Processes leading to hypoglycemia:

(1) transport of glucose from the blood to tissues followed by glucose oxidation to end products;

(2) synthesis of glycogen from glucose in liver and skeletal muscles;

(3) production of triacylglycerol from glucose in fat tissue.

The dietary intake 6f carbohydrates leads to a short-term (within 1 or 2 hours) hyperglycemia and, occasionally, glucosuria.

Starvation stimulates the consumption of glycogen reserves in liver and skeletal muscles, which prevents the development of hypoglycemia, but within the space of a few hours only. Then, under lasting starvation, the glucose level is maintained solely owing to the gluconeogenesis, primarily at the expense of proteinic amino acids which suffer degradation in the tissues.

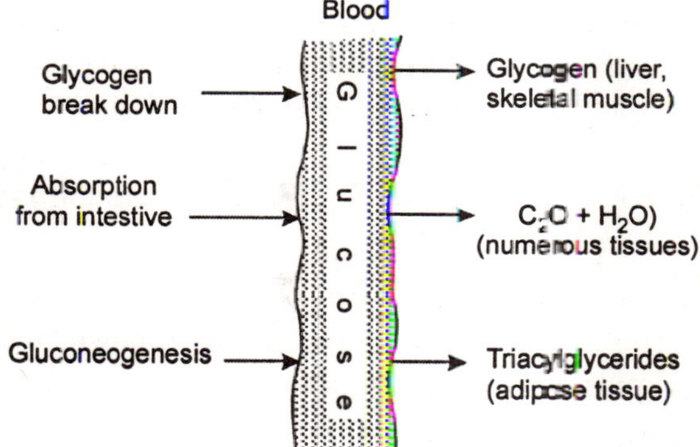

Fig. 2.3. Scheme for the processes conducive to an increase (on the left) and to a decrease (on the right) of glucose level in blood

In point of fact, the potential ability to sustain starvation is determined by the protein reserves available for glucose production. The glucose level in the blood is monitored by neurohormonal mechanisms. Excitation of the sympathetic portion of the autonomic nervous system increases the glucose level in the blood, while excitation of the parasympathetic portion produces a reverse effect. The only hormone capable of reducing the glucose content is *insulin*. It stimulates all of the three processes of glucose assimilation (intracellular transport and degradation of glucose, synthesis of glycogen, and synthesis of triglyceride from glucose in fat tissue). All other hormones make the glucose level increase; for this reason, they are occasionally referred to as *contrainsular* hormones. These include adrenalin, glucagon, thyroxin and triidothyronin, somatotropin (which stimulate glycogen degradation), and glucocorticoids (which stimulate gluconeogenesis).

LIPID METABOLISM

Lipids are continually renewed in the organism tissues. The major part of lipids in the human body is represented by triacylglycerides which occur as inclusions in most tissues; the fat tissue, which consists nearly totally of triacylglycerides, is especially lipid-rich. Since triacylglycerides play an important role in the energetics of the organism, the processes of their renewal (the conversion half-time for triacylglycerides in different organs varies from 2 to 18 days) involve both mobilization and deposition during energy production. Compound lipids (phospholipids, sphingolipids, glycolipids, and cholesterol and its esters) that make part of the biomembrane are subject to a less active renewal as compared with triacylglycerides. Their renewal is associated either with the restoration of an impaired portion of the membrane, or with the replacement of a "defective" molecule by a new one. The renewal of tissue lipids involves their preliminary intracellular hydrolysis by enzymes.

DEGRADATION OF LIPIDS IN TISSUES

Intracellular Hydrolysis of Lipids

Hydrolysis of triacylglycerides in tissues is effected by a tissue enzyme, *tri-acylglyceride lipase,* which hydrolyzes triacylglycerides to glycerol and free fatty acids. There are a variety of tissue lipases that differ primarily in their optimum pH and their location in the cell. The acidic lipase is contained in lysosomes; the basic lipase, in microsomes; and the neutral lipase, in cytoplasm. A specific feature of the tissue lipase is its sensitivity to hormones which, by activating adenylate cyclase, elicit the transition of the inactive tissue lipase to its active form via phosphorylation with protein kinase. This mechanism bears resemblance to the activation of phosphorylase B. Lipases mobilize triacylglycerides. This process is also known as the tissue *lipolysis.* The cell membrane phosphoglycerides are hyclrolyzed with *phospholipases* A_1 A_2, C, and D, which are located chiefly in lysosomes. However, certain phospholipases also occur in other cell organelles.

Hydrolysis of phosphoglycerides yields glycerol, fatty acids, nitrogenous alcohols, and inorganic phosphate. There are also known specific enzymes for hydrolysis of sphingolipids and glycolipids; the enzymes are involved in the renewal of these lipids. As is known, hydrolysis of intracellular lipids does not lead to a storage of glycerol and fatty acids. This indicates that the hydrolysis rate for the lipids is balanced against the rate of their intracellular oxidation. In the adipose tissue, glycerol and fatty acids as produced by triacylglyceride hydrolysis are not subject to oxidation and are released into the blood to be supplied to other organs.

Oxidation of Glycerol

The glycerol metabolism is closely related to the glycolysis involving glycerol metabolites according to the following scheme:

$$\begin{matrix} H_2C-OH \\ | \\ HC-OH \\ | \\ H_2C-OH \end{matrix} \xrightarrow[\text{ATP} \rightarrow \text{ADP}]{\text{Glycerol phosphokinase}} \underset{\alpha\text{-Glycerol phosphate}}{\begin{matrix} H_2C-OH \\ | \\ HC-OH \\ | \\ H_2C-OPO_3H_2 \end{matrix}} \xrightarrow[\text{NAD}^+ \rightarrow \text{NAD}\cdot\text{H} + \text{H}^+]{\alpha\text{-Glycerol-phosphate dehydrogenase}} \underset{\text{Dihydroxyacetone phosphate}}{\begin{matrix} H_2C-OH \\ | \\ C=O \\ | \\ H_2C-OPO_3H_2 \end{matrix}}$$

At first, glycerol is converted to α-glycerol phosphate through the agency of *glycerol phosphokinase.* α-Glycerol phosphate, by the action of NAD-dependent *α-glycerol-phosphate dehydrogenase,* is converted to dihydroxyacetone phosphate, which, as a common glycolysis metabolite, enters glycolysis to be reduced by enzymes to Jactate under anaerobic conditions, or to CO_2 and H_2O under aerobic conditions. Conversion of one glycerol molecule yields one ATP molecule under anaerobic, and 19 ATP molecules, under aerobic conditions. Glycerol is a profitable energy substrate and is used as an energy source practically by all origans and tissues.

Oxidation of Fatty Acids

Oxidation of higher fatty acids was first studied in 1904 by Knoop who fed animals with phenyl-substituted fatty acids and analyzed the products in the urine. He showed that the fatty acid oxidation results in the successive cleavage of two carbon moieties from the carboxyl end. Knoop coined the fatty acid oxidation mechanism as β-*oxidation.* As has been established by Kennedy and Lehninger in 1948-1949, oxidation of fatty acids occurs in the mitochondria only. Lynen and coworkers (1954-1958) have outlined major enzymic processes in fatty acid oxidation.

At the present time, the β-oxidation of fatty acids is referred to as the Knoop Lynen cycle. The fatty acids, as produced by intracellular hydrolysis of triacylglycerides or supplied to the cell from the blood, must be brought into a state of activation. Their activation is effected in the cytoplasm with the participation of *acyl-CoA synthetase* according to the scheme:

$$CH_3-(CH_2)_n-CH_2-CH_2-COOH + ATP + CoASH \xrightarrow[\text{acyl-CoA synthetase}]{Mg^{2+}}$$

$$\longrightarrow CH_3-(CH_2)_n-CH_2-CH_2-\overset{\overset{\displaystyle O}{\|}}{C}\sim SCoA + AMP + H_4P_2O_7$$

Since the activation process is effected extramitochondrially, transport of acyls across the membran6 into the mitochondria is necessary. The transport is accomplished with the participation of *carnitine,* which takes up the acyl from acyl-CoA on the outer membrane side. Acylcarnitine assisted by carnitine translocase diffuses to the inner side of the membrane to give its acyl to the Co A located in the matrix.

The process of reversible acyl transfer between Co A and carnitine on the outer and inner sides of trie membrane is effected by the enzyme *acyl-Co A -carnitine transf erase 2.4.* The oxidation of fatty acids within the Knoop-Lynen cycle occurs in the matrix. The Knoop-Lynen cycle includes four enzymes that act successively on acetyl-CoA. These are: *acyl-CoA dehydrogenase* (FAD-dependent enzyme), *enoyl-CoA hydratase, 3-hydroxyacyl-CoA dehydrogenase* (NAD-dependent enzyme), and *acetyl-CoA acyltransf erase.* Each turn, or revolution, of the "fatty acid spiral" produces an acetic acid residue split from the fatty acid as acetyl-CoA to yield one FAD $\bullet H_2$ molecule and one NAD $\bullet H_2$ molecule (Fig. 2.3).

The cycle turns are then repeated until the fatty acid chain becomes shortened to a four-carbon fragment, i.e. butyryl-CoA. In the last turn, butyryl-CoA splits apart, and two, rather than one, acetyl-CoA molecules are formed. The oxidation products of an even-numbered fatty acid are acetyl-CoA, FAD $\bullet H_2$ and NAD $\bullet H_2$. Subsequently, acetyl-CoA enters the Krebs cycle, and FAD $\bullet H_2$ and NAD $\bullet H_2$ are directly supplied to the respiratory chain.

The specific behaviour of odd-numbered fatty acids under oxidation is that one propionyl-CoA molecule ($CH_3-CH_2-CO \sim SCoA$) per oxidized fatty acid molecule, alongside the products acetyl-CoA, FAD $\bullet H_2$ and NAD $\bullet H_2$ (common to even-numbered fatty acids), is formed. Propionyl-CoA converts to succinyl-CoA:

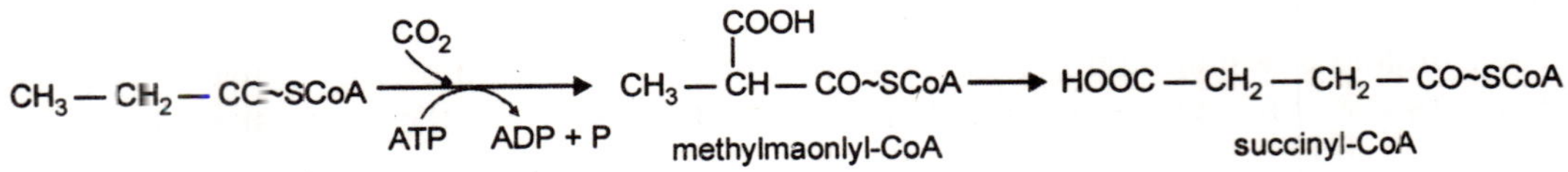

$$CH_3-CH_2-CO{\sim}SCoA \xrightarrow[ATP \quad ADP + P]{CO_2} CH_3-\underset{}{\overset{COOH}{\overset{|}{C}H}}-CO{\sim}SCoA \longrightarrow HOOC-CH_2-CH_2-CO{\sim}SCoA$$

methylmaonlyl-CoA succinyl-CoA

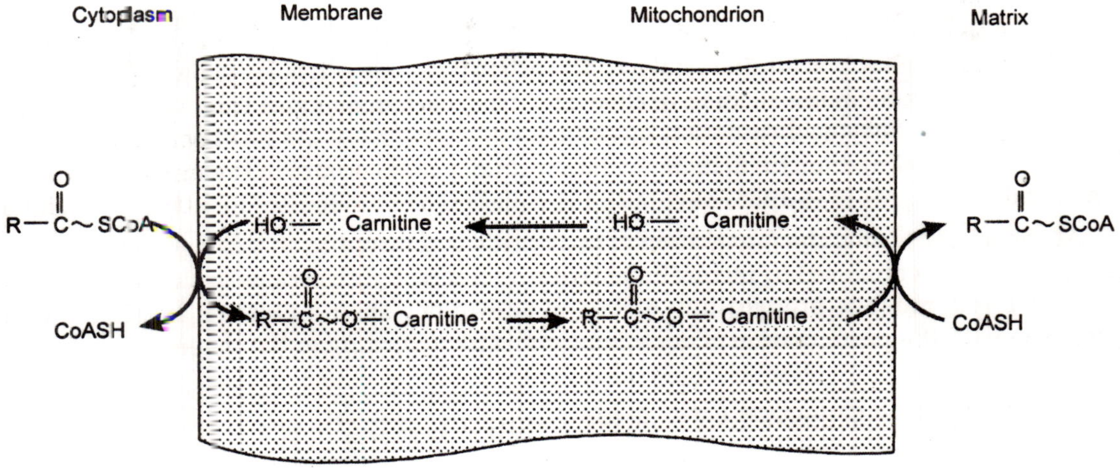

Cytoplasm Membrane Mitochondrion Matrix

$R-\overset{O}{\overset{\|}{C}}{\sim}SCoA$; CoASH ; HO — Carnitine ; $R-\overset{O}{\overset{\|}{C}}{\sim}O-$ Carnitine ; HO — Carnitine ; $R-\overset{O}{\overset{\|}{C}}{\sim}O-$ Carnitine ; $R-\overset{O}{\overset{\|}{C}}{\sim}SCoA$; CoASH

Fig. 2.4. Transport of fatty acids across the mitochondrial membrane.

Carboxylation of propionyl-CoA is accomplished by *propionyl-CoA carboxylase* (biotin, which is the carboxyl group carrier, serves as a coenzyme for this enzyme); the presence of ATP is also required. The methylmalonyl-CoA formed is converted by *methylmalonyl-CoA mutase* (whose coenzyme, deoxyadenosylcobalamin, is a derivative of vitamin B_{12}) to succinyl-CoA; the latter enters the Krebs cycle.

The specific behaviour of unsaturated fatty acids under oxidation is determined by the position and the number of double bonds in the fatty acid molecule. The stepwise oxidation of an unsaturated acid to the position ,of a double bond in it proceeds in a mariner similar to that of saturated acid oxidation. If th'e double bond retains the same configuration (*trans*-configuration) and position ($\Delta^{2,3}$) as those of the enoyl-CoA, which is produced during the oxidation of saturated fatty acids, the subsequent oxidation proceeds via conventional route. Otherwise, the oxidation reaction proceeds with the involvement of an accessory enzyme, $\Delta^{3,4}$-cis-$\Delta^{2,3}$-*trans-enoyl-CoA isomerase;* this facilitates the translocation of the double bond to an appropriate position and alters the double-bond configuration from *cis* to *trans*. The unsaturated fatty acid oxidation proceeds at a rate higher over that for saturated acids.

For example, if the oxidation rate for saturated stearic acid is taken as a reference value, the oxidation rate for oleic acid is 11 times, linolic acid, 114 times, linolenic acid, 170 times, and arachidonic acid, nearly 200 times as high as that for stearic acid. In addition to β-oxidation, two other oxidation routes are known for fatty acids, referred to as α-and ω-oxidation. However, they exhibit a lower activity and initially involve the formation of α- and ω-hydroxy acids, with subsequent conversions thereof. These oxidation routes are of inferior energetic value as compared with β-oxidation; presumably, they are implicated in special functions of the cell.

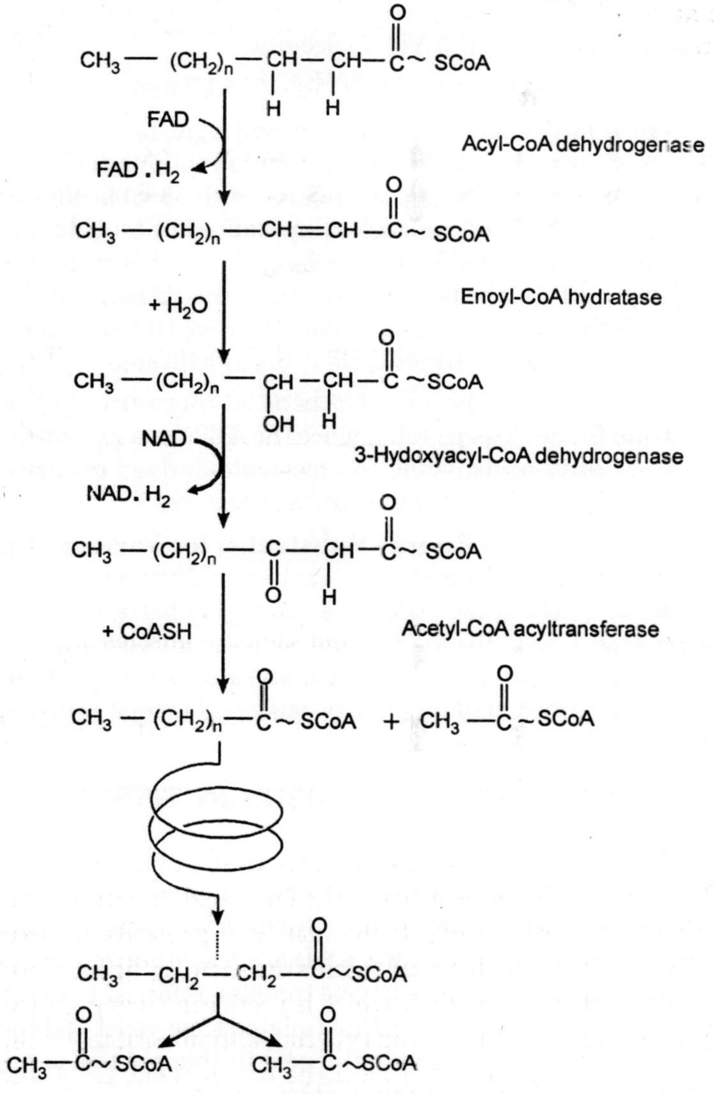

$$CH_3-(CH_2)_n-CH_2-CH_2-C(=O)\sim SCoA$$

FAD → $FAD.H_2$ — Acyl-CoA dehydrogenase

$$CH_3-(CH_2)_n-CH=CH-C(=O)\sim SCoA$$

$+ H_2O$ — Enoyl-CoA hydratase

$$CH_3-(CH_2)_n-CH(OH)-CH_2-C(=O)\sim SCoA$$

NAD → $NAD.H_2$ — 3-Hydoxyacyl-CoA dehydrogenase

$$CH_3-(CH_2)_n-C(=O)-CH_2-C(=O)\sim SCoA$$

+ CoASH — Acetyl-CoA acyltransferase

$$CH_3-(CH_2)_n-C(=O)\sim SCoA + CH_3-C(=O)\sim SCoA$$

$$\vdots$$

$$CH_3-CH_2-CH_2-C(=O)\sim SCoA$$

$$CH_3-C(=O)\sim SCoA \qquad CH_3-C(=O)\sim SCoA$$

Fig. 2.5. Scheme for oxidation of fatty acids in the Knoop-Lynen cycle.

Energy Balance of Fatty Acid Oxidation. The energetic value of an even-numbered fatty acid is estimated in the following manner. Complete oxidation of a fatty acid composed of $2n$ carbon atoms yields n acetyl-CoA molecules each acetyl containing two carbon atoms) and $n—1$ $FAD \bullet H_2$ and $NAD \bullet H_2$ molecules (since the last turn of the "fatty acid spiral" yields two acetyl-CoA molecules, one $FAD \bullet H_2$ molecule and one $NAD \bullet H_2$ molecule). Oxidation of $FAD \bullet H_2$ gives two ATP molecules, and oxidation of $NAD \bullet H_2$, three ATP molecules, i.e. a total of 5 ATP molecules, or, in the general case, $5(n — 1)$ ATP molecules. As has been noted above, the complete oxidation of one acetyl-CoA molecule results in the formation of 12 ATP molecules, while n

acetyl-GoA molecules provide $12n$ ATP molecules. One ATP molecule being used for the fatty acid activation, $12n$ — 1 ATP molecules remain. Now, the ATP balance for the complete oxidation of an even-numbered fatty acid may be expressed by the formula where n is equal to half the number of carbon atoms in a given fatty acid. For example, the palmitic acid molecule, which contains 16 carbon atoms, produces 130 ATP molecules.

$$5(n - 1) + 12n - 1 = (17n - 6) \text{ ATP molecules}$$

The energetic value of fatty acids is superior, for example, to that of glucose. For example, the complete oxidation of capronic acid (whose molecule contains the same number of carbon atoms as glucose) yields 45 ATP molecules as compared with 38 molecules which can be derived from glucose. However, the acetyl-CoA molecules as produced by β-oxidation require a sufficient amount of oxaloacetate to be degraded by the Krebs cycle. In this respect, carbohydrates have an advantage over fatty acids, since the breakdown of the former species leads to pyruvate serving as a source for both acetyl-CoA and oxaloacetate (pyruvate-carboxylase reaction), i.e. the acetyl-CoA conversion within the Krebs cycle is thus facilitated.

It is not without reason that in the older biochemical literature the notion that "fats burn down in the carbohydrate flame" was popular, since the ATP from glycolysis can be used for the cytoplasmic activation of fatty acids, while the pyruvate-derived oxaloacetate facilitates the insertion of fatty acid acetyl residues into the Krebs cycle.

Importance of Fatty Acids as Energy Substrates for Various Organs and Tissues. The organism tissues differ in the extent of utilization of fatty acids and their intermediary oxidation products, the so-called ketone bodies, as energy substrates. Fatty acids are actively consumed in the heart as well as in the kidneys and skeletal muscles (under prolonged physical exertion). In these organs, the ketone bodies undergo oxidation to yield additional energy. In the nervous tissue, the consumed amount of fatty acids and ketone bodies as sources of energy is insignificant.

BIOSYNTHESIS OF LIPIDS IN TISSUES

Biosynthesis of Fatty Acids

In the organism tissues, fatty acids are continually renewed in order to provide not only for the energy requirements, but also for the synthesis of multicomponent lipids (triacylglycerides, phospholipids, etc.). In the organism cells, fatty acids are resynthetized from simpler compounds through the aid of a supramolecular multienzyme complex referred to as *fatty, acid synthetase*. At the Lynen laboratory, this synthetase was first isolated from yeast and then frohn the liver of birds and mammals. Since in mammals palmitic acid in this process is a major product, this multienzyme complex is also called *palmitate synthetase.*

Biosynthesis of fatty acids exhibits a number of specific features:

1. fatty acid biosynthesis, as distinct from oxidation, is localized in the endo-plasmic reticulum;
2. the source for the synthesis is malonyl-CoA, which is produced from acetyl-GoA;
3. acetyl-CoA is involved in the synthetic reactions as a primer only;
4. NADP-H_2 is used to reduce fatty acid biosynthesis intermediates;
5. all the steps of malonyl-CoA fatty acid biosynthesis are cyclic processes that occur on the surface of palmitate synthetase.

Production of Malonyl-CoA for the Fatty Acid Biosynthesis. Acetyl-CoA serves as a substrate in the production of malonyl-CoA. There are several routes by which acetyl-CoA is supplied to the cytoplasm. One route is the transfer of acetyl residues from the mitochondrial matrix across the mitochondrial membrane into the cytoplasm. This process resembles a fatty acid transport and is likewise effected with the participation of carnitine and the enzyme *acetyl-CoA-carnitine transferase.* Another route is the production of acetyl-CoA from citrate. Citrate is delivered from the mitochondria and undergoes cleavage in the cytoplasm by the action of the enzyme *ATP-citrate lyase:*

$$\text{Citrate} + \text{ATP} + \text{CoA} \rightarrow \text{Acetyl-CoA} + \text{Oxaloacetate} + \text{ADP} + P_1$$

The reaction is practically irreversible, and is shifted to the right.

The acetyl-CoA supplied to the cytoplasm via the above routes is used for the synthesis of malonyl-CoA:

$$\underset{\text{acetyl-CoA}}{CH_3-C(=O)\sim SCoA} + HCO_3 + ATP \xrightarrow[\text{(E-biotin)}]{Mg^{2+}} \underset{\text{malonyl-CoA}}{HOOC-CH_2-C(=O)\sim SCoA} + ATP + H_3PO_4$$

The reaction is catalyzed by the biotin. enzyme *acetyl-CoA carboxylase* (E-biotin) assisted by Mg^{2+} ions. This enzyme is a tetramer with a molecular mass of 4,00,000-5,00,000.

Steps of Fatty Acid Biosynthesis Assisted by Palmitate Synthetase. Palmitate synthetase is composed of seven enzymes; of these, each is assigned a definite function. The *acyl carrier protein* (AGP) is located at the centre of the multienzyme complex; the other six enzymes occupy peripheral positions. AGP acts both as an acceptor and a distributor of acyl groups. AGP contains a covalently bound *4-phosphopantethein* bearing a free SH group for accepting an acyl moiety. In addition-to this central SH group, palmitate synthetase has a peripheral SH group. Both SH groups participate as acyl acceptors in the synthesis of fatty acids on the surface of the multienzyme complex.

The cyclic process of fatty acid synthesis may be represented by a series of consecutive reactions $\left(\text{hereafter palmitate synthetase is designated by the symbol } <^{SH}_{SH}\right)$

1. Transfer of acetyl moiety from acetyl-CoA onto the synthetase:

$$E<^{SH}_{SH} + CH_3-C(=O)\sim SCoA \rightarrow E<^{S\sim C(=O)-CH_3}_{SH} + CoASH$$

This reaction is carried out by the first enzyme of palmitate synthetase—*acetyltransacylase*, which possesses an SH group. At this stage of the synthesis, the acetyl acts as a primer.

2. Transfer of malonyl moiety from malonyl-CoA onto the synthetase:

$$E(S\sim \overset{O}{\overset{\|}{C}}-CH_3)(SH) + HOOC-CH_2-\overset{}{C}(=O)\sim SCoA \longrightarrow E(S\sim \overset{O}{\overset{\|}{C}}-CH_3)(S\sim \overset{O}{\overset{\|}{C}}-CH_2-COOH) + CoASH$$

The reaction is effected by the second synthetase enzyme—*malonyltransacylase.*

3. Acetyl-malonyl condensation and decarboxylation of the product formed:

$$E(S\sim \overset{O}{\overset{\|}{C}}-CH_3)(S\sim \overset{O}{\overset{\|}{C}}-CH_2-COOH) \longrightarrow E(SH)(S\sim \overset{O}{\overset{\|}{C}}-CH_2-\overset{O}{\overset{\|}{C}}-CH_3) + CO_2$$

The reaction is catalyzed by the third synthetase enzyme—β-*ketoacyl synthetase.* An acetoacetyl, which is bound to synthetase, is formed at this stage.

4. The first reduction of the intermediate with the involvement of NADP $\bullet H_2$

$$E(SH)(S\sim \overset{O}{\overset{\|}{C}}-CH{=}CH-CH_3) \xrightarrow[NADP.H+H^+ \;\rightarrow\; NADP^+]{} E(SH)(S\sim \overset{O}{\overset{\|}{C}}-CH_2-CH_2-CH_3)$$

The reaction is catalyzed by the fourth synthetase enzyme—β-*ketoacyl reductase,* to yield intermediary hydroxybutyryl.

5. Dehydration of the intermediate:

$$E(SH)(S\sim \overset{O}{\overset{\|}{C}}-CH_2-\overset{O}{\overset{\|}{CH}}-CH_3) \longrightarrow E(SH)(S\sim \overset{O}{\overset{\|}{C}}-CH{=}CH-CH_3) + H_2O$$

The reaction is catalyzed by the fifth synthetase *enzyme—hydroxyacyl hydratase,* to produce crotonyl.

6. The second reduction of the intermediary product with the involvement of NADP $\bullet H_2$:

$$E(SH)(S\sim \overset{O}{\overset{\|}{C}}-CH_2-\overset{O}{\overset{\|}{C}}-CH_3) \xrightarrow[NADP\text{-}H+H^+ \;\rightarrow\; NADP^+]{} E(SH)(S\sim \overset{O}{\overset{\|}{C}}-CH_2-\overset{OH}{\overset{|}{CH}}-CH_3)$$

The reaction is catalyzed by the sixth synthetase enzyme—*enoylreductase,* to form an enzyme-bound butyryl. The butyryl thus synthetized is transferred, through the mediacy of the first synthetase enzyme, acetyltransacylase, onto the SH group (the upper one in the Scheme) initially bound to the acetyl primer. The SH group (the lower one in the Scheme), thus freed, accepts a new malonyl residue:

$$\text{E}\begin{matrix}\diagup \text{SH} \\ \diagdown \text{S} \sim \overset{\overset{\displaystyle O}{\|}}{C} - CH_2 - CH_2 - CH_3\end{matrix} \longrightarrow \text{E}\begin{matrix}\diagup \text{S} \sim \overset{\overset{\displaystyle O}{\|}}{C} - CH_2 - CH_2 - CH_3 \\ \diagdown \text{SH}\end{matrix}$$

The synthetic cycle is thus repeated.

Seven cycles are implicated in palmitic acid biosynthesis and, accordingly, seven malonyl residues and one acetyl are required. Acetyl is the end moiety in fatty acid biosynthesis. The palmitic acid thus synthetized is either transferred onto the outer CoA to produce acyl-CoA, or, more commonly, is hydrolyzed by the specific *palmitate deacylase* to yield a free fatty acid.

Fatty Acid Chain Elongation. The mitochondria and endoplasmic reticulum provide the conditions for an eventual chain elongation of the cell-synthetized or dietary fatty acids. This process is different from the fatty acid biosynthesis in the proper sense of the term. In the mitochondria, the chain elongation is achieved through the aid of an enzyme complex by adding acetyl residues from acetyl-CoA. In the endoplasmic reticulum, the chain elongation is accomplished by an enzyme complex through making use of malonyl-CoA.

Biosynthesis of Unsaturated Fatty Acids. In the mammalian tissues, the formation of monoene fatty acids is only possible. Oleic acid is derived from stearic acid, and palmitooleic acid, from palmitic acid. This synthesis is carried out in the endoplasmic reticulum of the liver cells via the monooxigenase oxidation chain. Any other unsaturated fatty acids are not produced in the human organism and must be supplied in vegetable food (plants are capable of generating polyene fatty acids). Polyene fatty acids are essential food factors for mammals.

BIOSYNTHESIS OF TRIGLYCERIDES

Triglyceride biosynthesis proceeds with the involvement of the lipids deposited in fat tissue or in other tissues of the organism. This process is localized in the hyaloplasm of cells.

α-Glycerol phosphate and acyl-CoA, rather than corresponding free glycerol and free fatty acid, are utilized in the direct synthesis of triglycerides. α-Glycerol phosphate is produced either by phosphorylating the glycerol supplied to the tissue, or by reducing dihydroxyacetone phosphate as an intermediary product of glycolysis.

The first step of triglyceride biosynthesis is the formation of phosphatidic acid with the involvement of *glycerophosphate acyltransferase:*

$$\begin{array}{l} H_2C-OH \\ \quad | \\ H-C-OH \\ \quad | \\ H_2C-OPO_3H_2 \end{array} \xrightarrow[\text{R—CO}\sim\text{SCoA, R'—CO}\sim\text{SCoA} \;\to\; 2\text{ CoA SH}]{} \begin{array}{l} H_2C-O-\overset{\overset{\displaystyle O}{\|}}{C}-R \\ \quad | \\ H-C-O-\overset{\overset{\displaystyle O}{\|}}{C}-R' \\ \quad | \\ H_2C-OPO_3H_2 \end{array}$$

α- glycerol phosphate → phosphatidic acid

Further, the phosphatidic acid is subject to an attack by phosphatidate phosphatase to yield diglyceride:

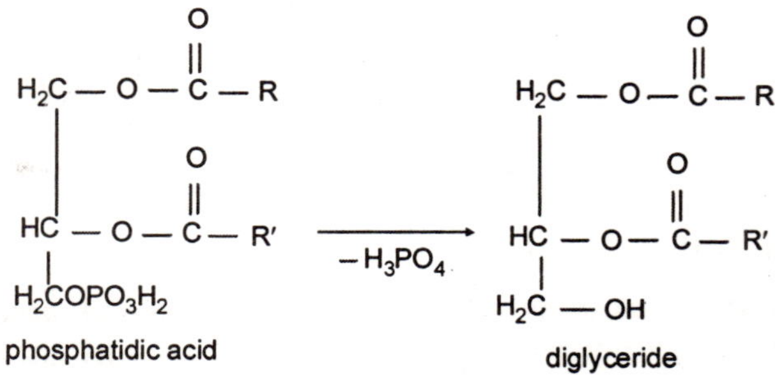

$$\begin{array}{l} H_2C-O-\overset{\overset{O}{\|}}{C}-R \\ \;| \\ HC-O-\overset{\overset{O}{\|}}{C}-R' \\ \;| \\ H_2COPO_3H_2 \end{array} \xrightarrow[-H_3PO_4]{} \begin{array}{l} H_2C-O-\overset{\overset{O}{\|}}{C}-R \\ \;| \\ HC-O-\overset{\overset{O}{\|}}{C}-R' \\ \;| \\ H_2C-OH \end{array}$$

phosphatidic acid — diglyceride

The third acyl residue is transferred onto diglyceride by means of *diglyceride acyltransferase:*

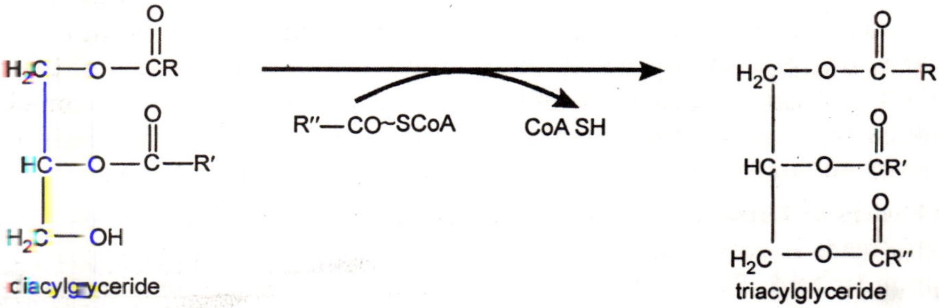

$$\begin{array}{l} H_2C-O-\overset{\overset{O}{\|}}{C}R \\ \;| \\ HC-O-\overset{\overset{O}{\|}}{C}-R' \\ \;| \\ H_2C-OH \end{array} \xrightarrow{R''-CO\sim SCoA \;\; \to \;\; CoA\,SH} \begin{array}{l} H_2C-O-\overset{\overset{O}{\|}}{C}-R \\ \;| \\ HC-O-\overset{\overset{O}{\|}}{C}R' \\ \;| \\ H_2C-O-\overset{\overset{O}{\|}}{C}R'' \end{array}$$

diacylglyceride — triacylglyceride

The triacylglyceride thus synthetized is stored as fat inclusions in the cell cytoplasm.

Phospholipid Biosynthesis

Biosynthesis of phospholipids is associated with the renewal of membranes. This process is accomplished in the tissue hyaloplasm. The first steps of phospholipid and triglyceride biosyntheses coincide; subsequently, these routes diverge at the level of phosphatidic acid and diglyceride. Two routes to phospholipid biosynthesis are known; in either, the participation of CTP is necessary. The first route involves phosphatidic acid in phosphoglyceride biosynthesis. Phosphatidic acid reacts with CTP to yield CDP-diglyceride which, as a coenzyme, can participate in the transfer of diglyceride onto serine (or inositol) to produce phosphatidylserine (or phosphatidylinositol).

Serine phosphatides are liable to decarboxylation (pyridoxal phosphate acting as a coenzyme) to yield ethanolamine phosphatides. The latter species are subject to methylation by S-adenosylmethionine (which donates three methyl groups), tetrahydrofolic acid and methylcobalamin acting as methyl group carriers. The second synthetic route involves activation of an alcohol (for example, choline) to produce CDP-choline. The latter participates in the transfer of choline onto diglyceride to form phosphatidylcholine. The phospholipids thus obtained are transported by lipid-carrior cytoplasmic proteins to the membranes (cellular or intracellular) to replace the used or impaired phospholipid molecules.

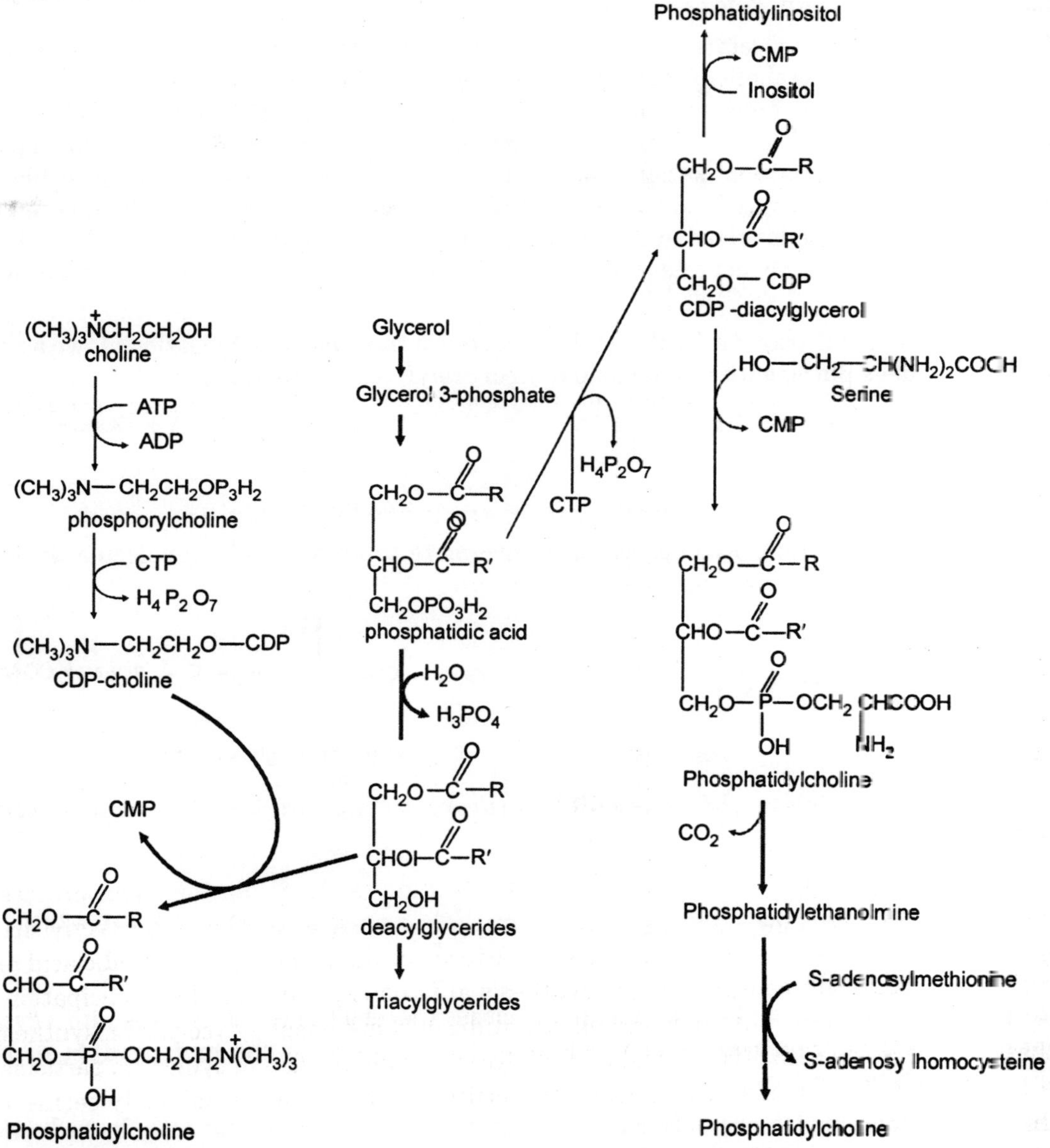

Phosphatidylinositol

CMP

Inositol

$CH_2O-\overset{O}{\overset{\|}{C}}-R$

$CHO-\overset{O}{\overset{\|}{C}}-R'$

CH_2O-CDP

CDP-diacylglycerol

$HO-CH_2-CH(NH_2)_2COOH$

Serine

CMP

$(CH_3)_3\overset{+}{N}CH_2CH_2OH$

choline

ATP

ADP

$(CH_3)_3N-CH_2CH_2OP_3H_2$

phosphorylcholine

CTP

$H_4P_2O_7$

$(CH_3)_3N-CH_2CH_2O-CDP$

CDP-choline

Glycerol

Glycerol 3-phosphate

$H_4P_2O_7$

CTP

$CH_2O-\overset{O}{\overset{\|}{C}}-R$

$CHO-\overset{O}{\overset{\|}{C}}-R'$

$CH_2OPO_3H_2$

phosphatidic acid

H_2O

H_3PO_4

$CH_2O-\overset{O}{\overset{\|}{C}}-R$

$CHOH-\overset{O}{\overset{\|}{C}}-R'$

CH_2OH

deacylglycerides

Triacylglycerides

$CH_2O-\overset{O}{\overset{\|}{C}}-R$

$CHO-\overset{O}{\overset{\|}{C}}-R'$

$CH_2O-\overset{O}{\overset{\|}{P}}(OH)-OCH_2CHCOOH(NH_2)$

Phosphatidylcholine

CO_2

Phosphatidylethanolamine

S-adenosylmethionine

S-adenosylhomocysteine

Phosphatidylcholine

CMP

$CH_2O-\overset{O}{\overset{\|}{C}}-R$

$CHO-\overset{O}{\overset{\|}{C}}-R'$

$CH_2O-\overset{O}{\overset{\|}{P}}(OH)-OCH_2CH_2\overset{+}{N}(CH_3)_3$

Phosphatidylcholine

Fig. 2.6. Two pathways for the synthesis of certain phospholipids (after Berezov and Korovkin)

Because of the competition between the phospholipid and triglyceride synthetic routes for common substrates, all substances that favour the former route impede the tissue deposition of triglycerides. Such substances are referred to as *lipotropic factors*. They include: choline, inositol, and serine, as structural components of phospholipids; pyridoxal phosphate, as an agent facilitating the decarboxylation of serine phosphatides; methionine, as a donor of methyl groups; and folic acid and cyanocobalamin, involved in the formation of methyl group transfer coenzymes (tetrahydrofolic acid and methylcobalamin). They may be used as drugs preventing excessive deposition of triglycerides in tissues (the so-called fatty infiltration).

Biosynthesis of Ketone Bodies

Three compounds: acetoacetate, β-hydroxybutyrate, and acetone, are known as ketone bodies. They are suboxidized metabolic intermediates, chielfy those of fatty acids and of the carbon skeletons of the so called ketogenic amino acids (leucine, isoleucine, lysine, phenylalanine, tyrosine, and tryptophan). The keton body production, or ketogenesis, is effected in the hepatic mitochondria (in other tissues, ketogenesis is inoperative). Two pathways are possible for ketogenesis. The more active of the two is the *hydroxymethyl glutarate cycle* which is named after the key intermediate involved in this cycle. The other one is the *deacylase* cycle. In activity, this cycle is inferior to the former one. Acetyl-CoA is the starting compound for the biosynthesis of ketone bodies.

Hydroxymethyl Glutarate Cycle. At the first step of this cycle, condensation of two acetyl-CoA molecules takes place, with the participation of *acetyl CoA acetyltransferase:*

$$\underset{\text{acetyl-CoA}}{CH_3-\underset{\|}{\overset{}{C}}{\sim}SCoA} + \underset{\text{acetyl-CoA}}{CH_3-C{\sim}SCoA} \rightarrow \underset{\text{acetoacetyl-CoA}}{CH_3-\overset{O}{\overset{\|}{C}}-CH_2-\overset{O}{\overset{\|}{C}}{\sim}SCoA} + COASH$$

Further, acetoacetyl-CoA becomes coupled once more to an acetyl-CoA molecule through the assistance of *hydroxymethylglutaryl-CoA synthase:*

$$\underset{\text{acetoacetyl-CoA}}{CH_3-\overset{O}{\overset{\|}{C}}-CH_2-\overset{O}{\overset{\|}{C}}{\sim}SCoA} + \underset{\text{acetyl-CoA}}{CH_3-\overset{O}{\overset{\|}{C}}{\sim}SCoA} \rightarrow \underset{\beta\text{- hydroxy-}\beta\text{-methylglutaryl-CoA}}{HOOC-CH_2-\overset{CH_3}{\underset{OH}{C}}-CH_2-C{\sim}SCoA} + COASH$$

β-Hydroxy-β-methylglutaryl-CoA is split by *hydroxymethylglutaryl-CoA lyase* into acetyl-CoA and acetoacetate:

$$HOOC-CH_2-\overset{CH_3}{\underset{OH}{C}}-CH_2-\overset{O}{\overset{\|}{C}}{\sim}SCoA \rightarrow CH_3-\overset{O}{\overset{\|}{C}}{\sim}SCoA + HOOC-CH_2-\overset{O}{\overset{\|}{C}}-CH_3$$

Acetyl-CoA is again used at the first step and closes thereby the whole process into a cycle. Acetoacetate, as a representative of the ketone body family, is the end product of the hydroxymethyl glutarate cycle.

The other ketone bodies are derived from acetoacetate: β-hydroxybutyrate, by reduction with the involvement of NAD-dependent *hydroxybutyrate dehydrogenase,* and acetone, by decarboxylation of acetoacetate with the participation of *aceto-acetate decarboxylase:*

$$HOOC-CH_2-\overset{O}{\overset{\|}{C}}-CH_3 \xrightarrow[NAD.H+H^+ \;\rightarrow\; NAD^+]{} \underset{\beta\text{ - hydroxybutyrate}}{HOOC-CH_2-\overset{OH}{\overset{|}{CH}}-CH_3}$$

$$HOOC-CH_2-\overset{O}{\overset{\|}{C}}-CH_3 \xrightarrow{-CO_2} \underset{\text{acetone}}{H_3C-\underset{O}{\underset{\|}{C}}-CH_3}$$

The Deacylase Pathway for Ketogenesis is feasible after the formation of acetoacetyl-CoA which is subject to hydrolysis to acetoacetate in the liver with the involvement of *acetoacetyl-CoA hydrolase,* or *deacylase.*

In the liver, the ketone bodies suffer no transformation, and are excreted into the blood. The normal contents of ketone bodies (as acetoacetate or β-hydroxybutyrate) amount to mere 0.1–0.6 mmol/litre). Other tissues and organs (heart, lung, kidney, muscle, and nervous tissue), as distinct from the liver, utilize the ketone bodies as energy substrates. In the cells of these tissues, acetoacetate and β-hydroxybutyrate enter ultimately the Krebs cycle and "burn down" to CO_2 and H_2O to release energy.

Biosynthesis of Cholestrol

In the experiments with acetic acid labelled radioisotopically and fed to animals, it has been established that the cholesterol carbon framework is made up entirely of the acetic acid carbon.

Biosynthesis of cholesterol from acetyl-CoA proceeds, assisted by the enzymes of endoplasmic reticulum and hyaloplasm, in many tissues and organs. This process is especially active in the liver of adult humans.

Cholesterol biosynthesis is a multistage process; in general, it may be divided, into three steps:

1. production of mevalonic acid from acetyl-CoA;
2. synthesis of an "active isoprene" from mevalonic acid followed by the condensation of the former to squalene;
3. conversion of squalene to cholesterol.

The initial reactions in the first step, prior to the formation of β-hydroxy-β-methylglutaryl-CoA from acetyl-CoA, resemble those involved in ketogenesis with the only distinction that ketogenesis occurs in the mitochondria, while cholesterol biosynthesis is carried out extramitochondrially:

2 Acetyl-CoA → Acetoacetyl-CoA + Acetyl-CoA → β-Hydroxy-β-methylglutaryl-CoA

Further, β-hydroxy-β-methylglutaryl-CoA is converted with *hydroxymethylglutaryl-CoA reductase* to mevalonic acid :

$$HOOC{-}CH_2{-}\underset{OH}{\overset{CH_3}{C}}{-}CH_2{-}\underset{O}{\overset{}{\underset{\|}{C}}}{\sim}SCoA \xrightarrow[2NADP.H+H^+ \;\curvearrowright\; 2NADP^+]{} HOOC\ CH_2{-}\underset{OH}{\overset{CH_3}{C}}{-}CH_2{-}CH_2OH + CoASH$$

This reaction is irreversible and is a rate-limiting stage of the overall cholesterol biosynthesis.

An alternative route to mevalonic acid is also possible, which differs from the former one in that the formation of β-hydroxy-β-methylglutaryl residue occurs on the surface of an acyl carrier protein (like in fatty acid biosynthesis). The intermediary product in this route, β-hydroxy-β-methylglutaryl-S-ACP, is reduced by another enzyme to mevalonic acid.

During the second step, mevalonic acid is implicated in a number of enzymic reactions involving ATP, and is converted to isopentyl pyrophosphate and to its isomer 3,3-dimethylallyl pyrophosphate. Actually, the two compounds constitute the "active isoprene", which is consumed in the production of squalene.

During the third step, cholesterol is generated from squalene:

$$\text{Squalene} \rightarrow \text{Lanosterol} \rightarrow \text{Cholesterol}$$

The steroid ring hydroxylation proceeds with the involvement of the monooxygenase chain of endoplasmic reticulum membranes.

Cholesterol esters are produced by transferring an acyl moiety from acyl-CoA or from phosphatidylcholine onto the cholesterol hydroxyl group. The latter process is catalyzed by *phosphatidylcholine cholesterol acyltransferase:*

$$\text{Phosphatidylcholine + Cholesterol} \rightarrow \text{Lysophosphatidylcholine + Cholesterol ester}$$

Cholesterol esters are produced especially actively in the intestinal mucosa and in the liver.

Thus, the tissue cholesterol can be synthetized from any materials whose breakdown leads to acetyl-CoA. These include carbohydrates, amino acids, fatty acids, and glycerol. The liver plays a decisive role in the cholesterol metabolism. The liver accounts for 90% of the overall endogenic cholesterol and its esters; the liver is also implicated in the biliary secretion of cholesterol and in the distribution of cholesterol among other organs, since the liver is responsible for the synthesis of apoproteins for pre-β-lipoproteins, α-lipoproteins, and β-lipoproteins which transport the secreted cholesterol in the blood.

In part, cholesterol is decomposed by intestinal microflora; however, its major part is reduced to coprostanol and cholestanol which, together with a small amount of nonconverted cholesterol, are excreted in the feces. Cholesterol, mostly esterified, is utilized in the buildup of cell biomembranes. Besides, cholesterol is a precursor to biologically important steroid compounds: bile acids (in liver), steroid hormones (in adrenal cortex, male and female sexual glands, and placenta), and vitamin D_3, or cholecalciferol (in skin).

REGULATION OF LIPID METABOLISM IN THE ORGANISM

The rate of lipid metabolism in the organism tissues is dependent on the dietary supply of lipids and on the neurohormonal regulation. An excessive intake of high-calory food (carbohydrates and triglycerides) impedes the consumption of endogenic triglyceride reserves stored in the fat tissues. Moreover, carbohydrates provide a very favourable basis for the neogenesis of various lipids: for this reason, a large dietary intake of carbohydrate-rich food exerts a significant influence on the production of triglycerides and cholesterol in the organism. Synthesis of endogenic cholesterol is also controlled by exogenous cholesterol supplied in food: the more dietary cholesterol is digested, the less endogenic cholesterol is produced in the liver.

Exogenous cholesterol inhibits the activity of hydroxymethylglutaryl-CoA reductase and the cyclization of squalene to lanosterol. The dietary ratio of various lipids plays an important role in the lipid metabolism in the organism. The available amounts of polyene fatty acids and phospholipids acting as solvents for fat-soluble vitamins affect not only the absorption of the latter species, but also the solubility and stability of cholesterol in the organism fluids (blood plasma and lymph) and biliary ducts.

Vegetable oils with a high percentage of phospholipids and polyene fatty acids impede an excessive accumulation of cholesterol and its deposition in blood vessels and other tissues, and facilitate the removal of cholesterol excess from the organism. These processes are most markedly affected by corn oil, safflower oil, cottonseed oil, and sunflower oil. The consumption of unsaturated fatty acids contained in vegetable oils produces a favourable effect on the synthesis of endogenic phospholipids (for which these acids are substrates); polyene fatty acids are also needed in the production of other materials, for example, prostaglandins.

Unsaturated fatty acids act as uncouplers for the oxidative phosphorylation and thus accelerate oxidation processes in the mitochondria and control thereby an excessive triglyceride deposition in the tissues. The lipotropic factors exercise a marked effect on the biosynthesis of phospholipids and triglycerides. As has been mentioned above, they facilitate the phospholipid synthesis. The dietary deficiency of lipotropic factors favours the triglyceride production in the organism. Starvation elicits mobilization of triglycerides from the adipose tissue and inhibits the endogenic cholesterol synthesis owing to the low activity of hydroxy-methylglutaryl-CoA reductase. The latter process provides the possibility for the active production of ketone bodies in the liver. The neurohormonal control of lipid metabolism chiefly affects the mobilization and synthesis of triglycerides in the fat tissue.

The lipolysis in tissues is dependent upon the activity of triglyceride lipase. All the regulators that favour the conversion of the inactive (nonphosphory-lated) lipase to the active (phosphorylated) one, stimulate the lipolysis and the release of fatty acids into the blood. Adrenalin and noradrenalin (secreted in the sympathetic nerve endings), hormones (glucagon, adrenalin, thyroxin, triiodothyronine, somatotropin, β-lipotropin, corticotropin, etc.), tissue hormones, including biogenic amines (histamine, serotonin, etc.) act as stimulators for this process.

Insulin, on the contrary, inhibits the adenylate cyclase activity, preventing thereby the formation of active lipase in the fat tissue, *i.e.* retards the lipolysis. In addition, insulin favours the neogenesis of triacyl-glycerides from carbohydrates, which, on the whole, provides for lipid deposition in the fat tissues as well as for the cholesterol production in other tissues. The thyroid hormones thyroxin and triiodotfiyronine assist in the oxidation of the cholesterol side chain and in the biliary excretion of cholesterol in the intestine.

PATHOLOGY OF LIPID METABOLISM

Most commonly, the lipid metabolism pathology is manifest as *hyperlipemia* (elevated concentration of lipids in blood) and *tissue lipidoses* (excessive lipid deposition in tissues). Normally, the lipid contents in the blood plasma are: total lipids, 4-8 g/litre; triglycerides, 0.5-2.1 mmol/litre; total phospholipids, 2.0–3.5 mmol/litre; total cholesterol, 4.0–10.0 mmol/litre (esterified cholesterol accounts for 2/3 of total cholesterol).

Hyperlipemia may manifest itself by an increased concentration of lipids, or certain groups thereof. For example, hypercholesterolemia and hypertriglyceridemia may be mentioned in this connection. Since practically all the blood plasma lipids make part of lipoproteins, hyperlipemias may be reduced to one of the hyper-lipoproteinemia forms which differ in the varied ratios of plasma lipoproteins of different groups. Distinguished are exogenous, or alimentary, hyperlipemias which are actually associated with a normally increased blood lipid concentration after the intake of a food high in fat, and endogenic hyperlipemias caused by impaired lipid metabolism.

The endogenic hyperlipemia may be due to a primary hereditary defect in apoproteins or in a lipid metabolism enzyme. However, of more frequent occurrence are hyperlipemias attributable to secondary causes, for example, to regulatory disturbances of the lipid metabolism or to unfavourable environmental factors. Five types of primary hyperlipoproteinemias are distinguished.

Hyperlipoproteinemia, Type I, is characterized by the enhanced content of chylomicrons in the blood plasma; simultaneously, the percentage of α- and β-lipoproteins may be lowered. The triglyceride content is 8-10 times above the norm, while cholesterol does not exceed the normal level. Presumably, Type I is associated with a defective lipoprotein lipase that destroys chylomicrons.

Hyperlipoproteinemia, Type II, is characterized by an increased β-lipoprotein content in the blood plasma and, respectively, by a 1.5-2-fold higher, against the norm, cholesterol concentration. The familial form of hyperlipoproteinemia. Type IIa, is also known, which manifests itself in the occurrence of a defective apoprotein for β-lipoproteins, and in a slower breakdown of these materials in the tissues.

Hyperlipoproteinemia, Type III, is a rare hereditary disease (also called familial dysbetalipoproteinemia) manifested by the occurrence of an uncommon β-lipoprotein form. Cholesterol and triglyceride contents in the patients may occasionally be 2-5 times superior to the norm.

Hyperlipoproteinemia, Type IV, is characterized by increased contents of pre-β-lipoproteins and triglycerides (2-5 fold) in the blood plasma. Its incidence rate is higher in aged patients. Hereditary forms of this disease (called also familial hyperprebetalipoproteinemia) have been described.

Hyperlipoproteinemia, Type V. This pathology is manifested by increased contents of "chylomicrons, pre-β-lipoproteins, triglycerides, and cholesterol in the patients blood plasma.

Secondary hyperlipoproteinemias, which arise from a disordered lipid tissue metabolism or its impaired control, are observed in diabetes mellitus, thyroid gland hypofunction, alcoholism, etc.

Tissue Lipidoses. Hyperlipoproteinemias may lead to tissue lipidoses. Lipidoses can also arise from hereditary defects of the enzymes involved in the synthesis and breakdown of lipids in the tissues. We now discuss certain instances of tissue lipidoses. *Atherosclerosis* is a widespread pathology, manifested chiefly by the deposition of cholesterol in arterial walls, which results in the formation of lipid plaques (atheromas).

Lipid plaques are specific foreign bodies around which the connective tissue develops abnormally (this process is called sclerosis). This leads to the calcification of the impaired site of a blood vessel. The blood vessels become inelastic and compact, the blood supply through the vessels is impeded, and the plaques may develop into thrombi. Atherosclerosis results from hyperlipoproteinemia. All of the lipoproteins, excepting chylomicrons, are capable of penetrating the vessel wall. However, α-lipoproteins, which are rich in proteins and phospholipids, are liable to an easy breakdown within the vessel wall, or are apt to leave it because of their small size. β-Lipoproteins and, partly, pre-β-lipoproteins containing much cholesterol exhibit atherogenic properties.

Elevated concentrations of lipids of these groups and an increased vessel wall permeability are conducive to deposition of atherogenic lipoproteins within the walls, with the subsequent development of atherosclerosis. *Fatty infiltration of the liver.* In this pathology, the triglyceride concentration in the liver is 10-fold superior to the norm. The accumulation of fat in the cytoplasm of hepatic cells leads to an impaired liver function. The causes of this pathology are numerous; one of these may be a deficiency in lipotropic factors and the associated therewith synthesis of excess triglycerides.

Ketosis is a pathologic state produced by an excess of ketone bodies in the organism. However, ketosis may be regarded as a lipid metabolism pathology with a certain reserve, since excessive biosynthesis of ketone bodies in the liver is sequent upon an intensive hepatic oxidation not only of fatty acids, but also of ketogenic amino acids. The breakdown of the carbon frameworks of these amino acids leads to the formation of acetyl-CoA and acetoacetyl-CoA, which are used in ketogenesis. The ketosis is accompanied by *ketonemia* and *ketonuria,* which is manifested by the increased concentration of ketone bodies in blood and their excretion in the urine.

In an aggravated form of ketosis, the ketone body concentration in blood may be as high as 10-20 mmol/litre. The ketone bodies are normally present in the daily urine in trace amounts, while in pathology, 1 to 10 g (or even more) of ketone bodies per day is excreted in the urine. Most commonly, ketonemia and ketonuria are observed in diabetes mellitus (the manifest ketosis symptoms are dependent on the extent of diabetes mellitus), as well as in prolonged starvation or in "steroid" diabetes.

APPLICATIONS OF LIPIDS AND THEIR COMPONENTS IN PHARMACOTHERAPY

Fat-emulgated preparations for parenteral administration have been elaborated for clinical applications. Since these are administered to the patients intravenously, the size of fat emulsion particles should not exceed the size of the largest naturally occurring lipoproteins—chylomicrons, i.e. about 1 μm. Fat emulsions on the basis of corn oil (preparation lipomaize), cottonseed oil (lipofundin, lipomol), soybean oil (intralipid) have been proposed. These preparations are composed of lipids (10 to 20%), emulsifying agents (phosphatides and other materials) and, occasionally, glycerol. They are prescribed to asthenic patients for increasing energy resources of the organism. In addition, lipotropic preparations (methionine, choline, and inositol), which make part of natural phospholipids, are used in the prophylaxis of fatty infiltration of the liver.

AMINO ACID AND PROTEIN METABOLISM

Amino acids constitute the major source of nitrogen for the mammalian organism; for this reason, they are a linking chain in catabolism and anabolism of nitrogenous materials. In the adult human organism, about 400 g of protein per day is renewed. The renewal rate for different proteins varies from a few minutes to 10 and even more days. To be noted, certain proteins, such as collagen, are practically not subject to renewal. On the whole, the half-life for all the proteins in the human organism is about 80 days.

One fourth of the protein amino acids (about 100 g) is degraded irreversibly and must therefore be replenished at the expense of the dietary amino acids and, in part, of endogenically synthetized amino acids. The intracellular concentrations of free amino acids are small and, under normal biological conditions, relatively constant. This is indicative of the fact that in the cells, a certain stationary amino acid level, or a *free amino acid fund,* is maintained, which reflects a dynamic balance of the amino acids that are supplied and consumed.

There are known several *routes for the free amino acid supply,* contributing to the amino acid fund in the cell.

1. Transport of amino acids from the extracellular fluid (commonly observed when alimentary amino acids are absorbed). Amino acids penetrate the cell by secondary active transport

at the expense of the membrane sodium ion gradient energy. The transport is effected through the aid of the five transport systems (carrier proteins) that have been mentioned above in discussing amino acid absorption in the intestine.

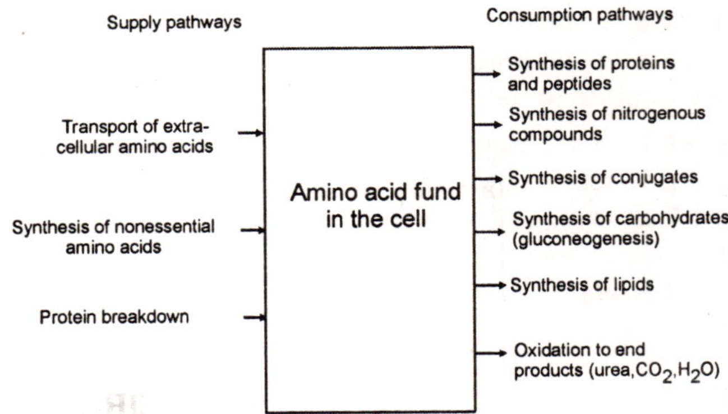

Fig. 2.7. Process affecting the amino acid fund in the cells

2. Production of nonessential amino acids.
3. Intracellular hydrolysis of proteins. This is the major route to amino acid supply.

Consumption routes for amino acids, leading to a reduced intracellular amino acid level, are as follows.

1. Synthesis of proteins and peptides as a major pathway of amino acid consumption.
2. Synthesis of nonproteinic nitrogenous compounds—purines, pyrimidines, porphyrins, choline, creatine, melanin, certain vitamins, coenzymes (nicotinamide, folic acid, and coenzyme A), tissue regulators (histamine and serotonine), and mediators (adrenalin, noradrenalin, and acetylcholine).
3. Synthesis of carbohydrates (gluconeogenesis) using the amino acid carbon frameworks.
4. Synthesis of lipids using acetyl residues of the amino acid carbon frameworks.
5. Oxidation of amino acids to the end metabolites. This is a route to energy generation via amino acid degradation.

Breakdown of Proteins to Amino Acids in Tissues

The first step of protein renewal is proteolysis as effected by *tissue proteinases,* or *cathepsins.* Cathepsins are located chiefly in the lysosomes, as many other hydrolytic enzymes. However, cathepsins also occur in other ceil organeiles: mitochondria, endoplasmic reticulum, and hyaloplasm. The lysosomal cathepsins are most active in an acidic medium, therefore they are referred to as acidic cathepsins.

In the cell cytoplasm and other organelles, cathepsins are optimally active in neutral and weakly basic media. Proteins liable to proteolysis interact initially with the Golgi apparatus and with the endoplasmic reticulum of the cell to produce the so-called *autophagosomes* (intracytoplasmic vacuoles). The autophagosomes are attacked by the primary lysosomes to be fused with the latter to form *autolysosomes* (or secondary lysosomes). The lysosomal cathepsins accomplish fast hydrolysis of the proteins that have been taken up by the organelles.

The proteolysis involving lysosomal cathepsins is assisted by the proteinases ot cytoplasm sap. Cathepsins differ both in optimal pH and in specificity towards the protein substrates and peptide bonds they hydrolyze. Cathepsins are classified into *exopepti-dases,* which catalyze the hydrolytic scission of terminal peptide bonds from the N- or C-end of the polypeptide chain, and *endopeptidases,* which catalyze the hydrolysis of peptide bonds in the interior of the peptide chain.

Depending on the specific features of the catalytic groups of the active centre, distinguished are: *thiol cathepsins* (which contain cysteine as the -catalytic centre), *asparagine cathepsins,* or *carboxycathepsins* (which contain aspartic acid as the catalytic centre), and *ferine cathepsins* (with the catalytic centre represented by serine). Different cathepsins are denoted by Latin characters. Currently, in mammals, the following cathepsins are known.

Tissue Thiol Proteinases. *Cathepsin* B has an optimum pH at 5.5-6.0. Cathepsin B is an endopeptidase, but it can also split dipeptides from the C-end, i.e. act as a dipeptidyldipeptidase. Cathepsin B hydrolyzes various intracellular proteins (glycolytic enzymes, immunoglobulins, myofibrillar proteins, collagen, and hemoglobin); it is also capable of converting proinsulin to insulin in the pancreas. Cathepsin B has been identified in many tissues of the human organism.

Cathepsin N, or collagenolytic enzyme, is an endopeptidase, and acts on collagen only. The optimum pH for proteolysis of native collagen is 3.6, and of soluble collagen, 6.0. Cathepsin N occurs in the lysosomes or in the cytoplasm of human spleen and placenta. Its presence in other organs and tissues has not as yet been verified.

Cathepsin H is both an endopeptidase and an amino peptidase. As an amino peptidase, it splits the N-terminal amino acids from the polypeptide chain. It can hydrolyze numerous soluble cytoplasmic proteins. Its optimal pH is 6.0–7.0. Cathepsin H is highly active in the human liver.

Cathepsin L is an endopeptidase exhibiting a high hydrolytic activity towards cytoplasmic proteins with a short renewal period. Its optimum pH is 5.0. Cathepsin L has been found in all of the organism tissues.

Cathepsin C, or *dipeptidyldipeptidase* I, is an exopeptidase. It is capable of cleaving N-terminal dipeptides from a polypeptide chain. It cannot hydrolyze proline-containing bonds. Its optimum pH is 5.0 to 6.0. The specific feature of this enzyme is its polymerase activity (the ability to catalyze the formation of macromolecules from simple components) at pH 7.0 to 8.0.

Cathepsin S is an endopeptidase, with optimum pH 3.0 to 4.0. It occurs in the spleen and in lymph nodes.

Tissue Asparagine Proteinases. Cathepsin D is an endopeptidase capable of splitting peptide bonds formed by aromatic amino acids. Its optimum pH is 3.5 to 4.0; enzymatically, it is similar to pepsin. Cathepsin D is believed to be a primitive representative of digestive enzymes of one-cell organisms. Presumably, in the course of evolution, its specialization has led to the emergence of pepsin. Cathepsin D is present in the lysosomes of nearly all organs and tissues. Its highest activity has been observed in the spleen, kidney, and lung. It hydrolyzes numerous cytoplasmic proteins, myosin (major muscle protein), major myelin protein, and hemoglobin. When present in the cartilage, it affects the hydrolysis of proteoglycans, but at pH 5.0 only.

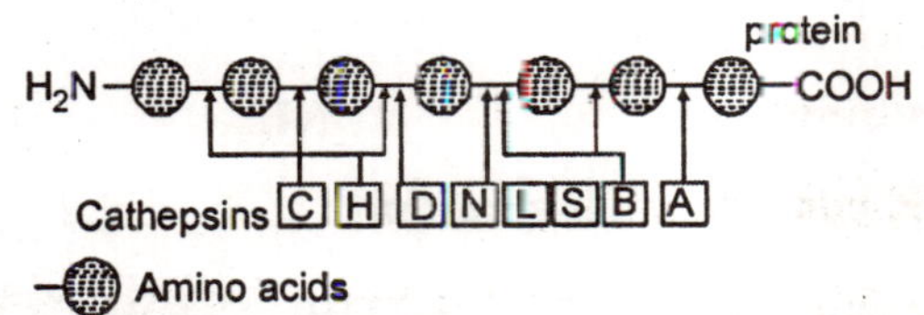

Fig. 2.8. Scheme for hydrolysis of intracellular proteins by cathepsins

Tissue Serine Proteinases. *Cathepsin* A, or *carboxypeptidase* A, is an exopeptidase whose enzymic activity is similar to that of pancreatic carboxypeptidase A.

It splits C-terminal amino acids of a polypeptide chain and exhibits a slight activity as an endopeptidase (hydrolyzes bonds involving a tyrosine carboxyl group). Its optimum pH is 5.0 to 5.5. Schematically, the action of various cathepsins on the intracellular hydrolysis of proteins is presented. Here the resultant products are free amino acids and dipeptides. The latter species are hydrolyzed by cellular dipeptidases to amino acids.

Biological Role of Cathepsins

Proteolysis, as it is effected in tissues, is important for the renewal of proteins, for the disposal of defective protein molecules, and for the energetic mobilization of endogenic proteins (especially in starvation). Therefore, cathepsins perform not only a destructive function, but are also involved in reconstructive processes. Deficiency in cathepsins reduces the possibility for renewal of tissue proteins, which leads to the accumulation of impaired and functionally inhibited proteins.

Cathepsins are capable of a *limited proteolysis,* i.e. they are apt to split off a particular fragment from a polypeptide chain. This function as performed by cathepsins may be regarded as a regulatory one—since the proteins resynthetized after the cathepsin intervention regain their activity (this phenomenon is reminescent of the production of enzymes from digestive proenzymes). The limited proteolysis (carried out in specialized neurosecretory cells) liberates neuropeptides that perform mediatory and hormonal functions. Pro-hormones produced by the endocrine glands are converted to active protein hormones by the same mechanism.

AMINO ACID CATABOLIC PATHWAYS

The catabolic pathways by which amino acids are degraded to the end products may arbitrarily be divided into three groups: (1) catabolic pathways involving the conversion of NH_S groups (any amino acid has at least one α-NH_2 group); (2) pathways for the breakdown of amino acid carbon frameworks; (3) decarboxylation of amino acid α-COOH groups. The third pathway is a version for the amino acid carbon framework conversion. It is employed in the production of biogenic amines and will be dealt with later in discussing the processes involving formation and degradation of mediators.

Conversion of α-amino Groups of Amino Acids

In the organism tissues, the cleavage of an amino group from an amino acid leads to the production of ammonia. This process is called deamination. Four types of deamination are known:

(1) reductive deamination

$$\underset{\displaystyle NH_2}{R—\overset{}{C}H—COOH} \xrightarrow{+2H} R—CH_2—COOH + NH_2$$

2. hydrolytic deamination

$$\underset{\displaystyle NH_2}{R—CH—COOH} \xrightarrow{+H_2O} \underset{\displaystyle OH}{R—CH—COOH} + NH_3$$

3. intramolecular deamination

$$R{-}CH_2{-}\underset{\displaystyle NH_2}{\underset{|}{CH}} \longrightarrow R{-}CH{=}\,CH{-}COOH + NH_3$$

4. oxidative demination

$$R{-}\underset{\displaystyle NH_2}{\underset{|}{CH}}{-}COOH \xrightarrow{+\frac{1}{2}O_2} R{-}\underset{\displaystyle O}{\underset{\|}{C}}{-}COOH + NH_3$$

Ammonia is the product common to all the deamination types. In addition to ammonia, other species—fatty acids, hydroxy acids, unsaturated acids, and keto acids—are also formed. For most organisms, including human and animal, the oxidative deamination is a typical process; however, in certain amino acids, for example, histidine, intramolecular deamination occurs.

Oxidative deamination can proceed via a direct route or an indirect route (transdeamination).

Direct oxidative deamination is effected by L- and *D-amino acid oxidases* which are found in peroxisomes. The L-amino acid oxidase contains **FMN**, and the D-amino acid oxidase, **FAD**, as a coenzyme. The reaction proceeds according to the scheme below

$$R{-}\underset{\displaystyle NH_2}{\underset{|}{CH}}{-}COOH \xrightarrow[\text{E-FMN, E-FAD} \;\to\; \text{E-FMN.}H_2\text{, E-FAD.}H_2]{} R{-}\underset{\displaystyle NH}{\underset{\|}{C}}{-}COOH \xrightarrow{+H_2O} R{-}\underset{\displaystyle O}{\underset{\|}{C}}{-}COOH + NH_3$$

E-FMN / E-FAD ⇄ E-FMN.H_2 / E-FAD.H_2; O_2 → H_2O_2; H_2O_2 —catalase→ H_2O

Keto acids, ammonia, and H_2O_2 are the reaction products. In peroxisomes, hydrogen peroxide undergoes an immediate catalase-assisted decomposition to water and oxygen. Under physiological pH, the L-amino acid oxidase exhibits a low activity; the D-amino acid oxidase is more active. However, the physiological importance of this oxidase remains as yet unclear, since the dietary proteins contain L-amino acids only. Presumably, L-amino acids are in part converted by intestinal bacterial isomerases to D-amino acids to be deaminated in the tissues by D-amino acid oxidases. On the whole, however, direct oxidative deamination plays but a minor role in the conversion of amino acidic NH_2-groups.

Transdeamination is the major route to amino acid deamination U is accomplished in two steps. The first step is *transamination,* i.e. the transfer of an amino group from an amino acid onto an α-keto acid without the intermediary formation of ammonia; the second step is the oxidative deamination of an amino acid in the proper sense of the term. Since the amino groups produced at the first step are "collected" together as constituents of glutamic acid, the second step involves the oxidative deamination of this acid. We now discuss each of the transdeamination steps in greater detail.

Transamination of amino acids has been discovered by the Soviet biochemists A.E. Braunshtein and M.G. Kritsman in 1937. This process may schematically be presented as:

$$\underset{\alpha\text{-amino acid-1}}{R\text{—}CH(NH_2)\text{—}COOH} + \underset{\alpha\text{-keto acid-2}}{R'\text{—}C(=O)\text{—}COOH} \rightleftharpoons \underset{\text{keto analogue of }\alpha\text{-amino acid-1}}{R\text{—}C(=O)\text{—}COOH} + \underset{\alpha\text{-amino acid-2}}{R'\text{—}CH(NH_2)\text{—}COOH}$$

The transamination reaction is a reversible one; it is catalyzed by enzymes *amino-transferases*, or *transaminases*. Aminotransferases are ubiquitous and occur in all animal and plant cells, as well as in microorganisms. At present, over 50 amino-transferases have been reported. The majority of them act on L-amino acids only; however, in the microorganisms, aminotransferases responsive to D-amino acids only have also been identified.

Natural α-amino acids, numerous β–, γ–, δ—, and ε-amino acids as well as amino acid amides (glutamine and asparagine) serve as amino group sources in the transamination reaction.

Most of the known aminotransferases exhibit a group specificity, using a small number of amino acids as substrates. Three α-keto acids—pyruvate, oxaloacetate, and 2-oxoglutarate—are engaged as amino-group acceptors in transamination reactions. Most commonly, 2-oxoglutarate serves as an NH_2-group acceptor to produce glutamic acid. The transfer of an amino group onto pyruvate or oxaloacetate produces, respectively, alanine or aspartic acid according to the scheme:

Starting amino acid + Pyruvate (or oxaloacetate) → Keto analogue of starting amino acid + Alanine (or aspartic acid)

Further, the NH_2 groups from alanine and aspartic acid are transferred onto 2-oxoglutarate. This reaction is catalyzed by highly active aminotransferases: *alanine aminotransferase* (ALT) and *aspartate aminotransferase* (AST), both exhibiting substrate specificity:

$$\underset{\text{alanine}}{CH_3\text{—}CH(NH_2)\text{—}COOH} + \underset{\text{2-oxoglutaric acid}}{HOOC\text{—}CH_2\text{—}CH_2\text{—}C(=O)\text{—}COOH} \xrightarrow{ALT} \underset{\text{pyruvic acid}}{CH_3\text{—}C(=O)\text{—}COOH} + \underset{\text{glutamic acid}}{COO\text{—}CH_2\text{—}CH_2\text{—}CH(NH_2)\text{—}COO}$$

$$\underset{\text{aspartic acid}}{HOOC\text{—}CH_2\text{—}CH(NH_2)\text{—}COOH} + \underset{\text{2-oxoglutaric acid}}{HOOC\text{—}CH_2\text{—}CH_2\text{—}C(=O)\text{—}COOH} \xrightarrow{AST} \underset{\text{oxaloacetate}}{HOOC\text{—}CH_2\text{—}C(=O)\text{—}COOH} + \underset{\text{glutamic acid}}{HOOC\text{—}CH_2\text{—}CH_2\text{—}CH(NH_2)\text{—}COOH}$$

An aminotransferase is composed of an apoenzyme and a coenzyme. Derivatives of pyridoxine (vitamin B_6)— *pyridoxal 5-phosphate* (PALP) and *pyridoxamine 5-phosphate* (PAMP)—are coenzymes for aminotrasferases. The two coenzymes are reversibly interconvertible during transamination. It should be noted that aminotransferases require both coenzymes for the reactions they catalyze, as distinct from other enzymes that require only one of the two and are, therefore, either pyridoxal-phosphate-dependent, or pyridoxamine-phosphate-dependent.

The mechanism for enzymic amino acid transamination has been studied by a number of authors (Braunshtein, Shemyakin; Metzler, Ikawa, Snell). According to this mechanism at the first stage, the amino acid NHj group reacts with the aldehyde group of pyridoxal phosphate, O—CH—PALP, to produce an intermediate Schiff's base of *aldimine* type, isomerizable to its *ketimine* form $H_2N—CH_2—$ PAMP (Schiff's base of pyridoxamine phosphate):

$$\mathrm{H{-}C(R)(COOH){-}NH_2 + O{=}CH{-}PALP} \underset{+H_2O}{\overset{-H_2O}{\rightleftharpoons}} \underset{\text{aldimine}}{\mathrm{H{-}C(R)(COOH){-}N{=}CH{-}PALP}}$$

$$\rightleftharpoons \underset{\text{ketimine}}{\mathrm{C(R)(COOH){=}N{-}CH_2{-}PAMP}} \underset{-H_2O}{\overset{+H_2O}{\rightleftharpoons}} \mathrm{C(R)(COOH){=}O + H_2N{-}CH_2{-}PAMP}$$

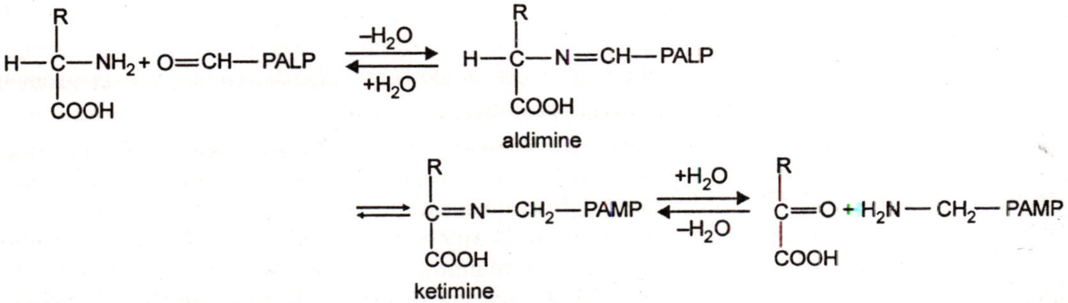

Further, ketimine is hydrolyzed to form a keto analogue of the initial amino acid and PAMP. At the second step, PAMP reacts with α-keto acid (an amino group acceptor), and the whole process is thus duplicated in the reversed order, i.e. ketimine is the first to be formed, and then aldimine.

The latter species is liable to hydrolysis to yield a new amino acid and PALP. Thus, aminotransferase coenzymes perform the function of an amino group carrier via conversion from the aldehyde to the aminated form, and vice versa. Later, more detailed studies on the transamination mechanism have provided evidence that initially the pyridoxal-phosphate aldehyde group forms a Schiff's base (aldimine) with the ε-amino group of lysine residue at the catalytic site of transaminase apoenzyme. Therefore, in the course of the reaction, the amino group of the substrate (i.e. amino acid) replaces theε-amino group of the apoen-zymic lysine residue to form a complex "PALP—substrate amino acid".

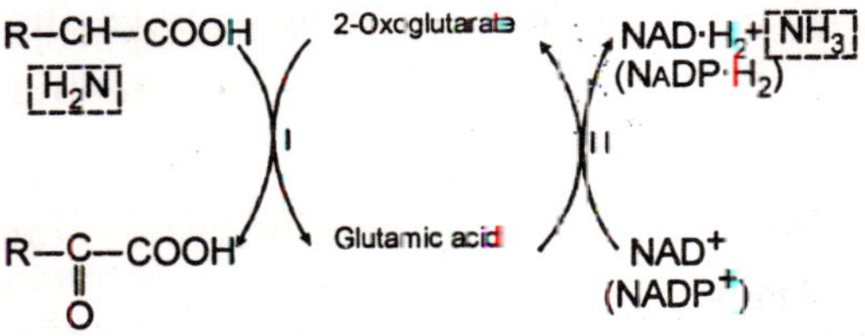

R–CH–COOH [H_2N] → (I; 2-Oxoglutarate → Glutamic acid) → R–C(=O)–COOH; Glutamic acid → (II; NAD⁺ (NADP⁺) → NAD·H_2 (NADP·H_2) + [NH_3]) → 2-Oxoglutarate

Fig. 2.9. General scheme for transdeamination reactions (according to Braunshtein).

Oxidative deamination of glutamic acid. The biological significance of transamination reactions resides in combining the "amino groups derived from all degraded amino acids to build up a molecule of a single species only—in this particular case, glutamic acid. Glutamic acid is supplied to the cells of mitochondria, within which the second transdeamination stage, the deamination

***I*—transaminatkw; II—deamination**

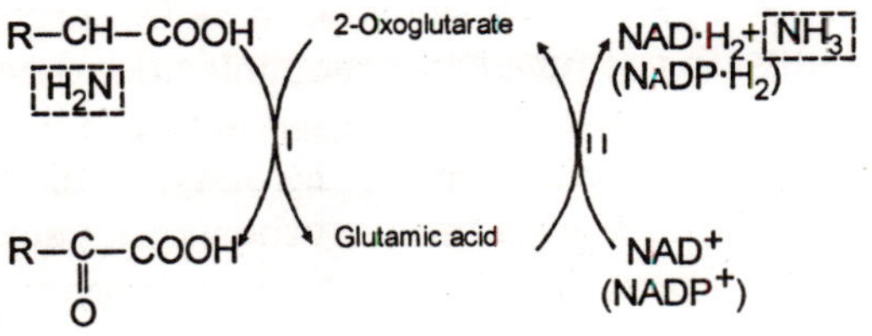

R–CH–COOH [H_2N] → (I; 2-Oxoglutarate → Glutamic acid) → R–C(=O)–COOH; Glutamic acid → (II; NAD⁺ (NADP⁺) → NAD·H_2 (NADP·H_2) + [NH_3]) → 2-Oxoglutarate

proper of glutamic acid, takes place. This reaction is catalyzed by *glutamate dehydrogenase,* which can use both NAD⁺ and NADP⁺ as a coenzyme

$$\underset{\text{glutamic acid}}{HOOC\text{–}CH_2\text{–}CH_2\text{–}CH(NH_2)\text{–}COOH} \xrightarrow[NAD^+\ (NADP^+)\ \to\ NAD.H+H^+\ (NAD.H+H^+)]{\text{Glutamate dehydrogenase}} \underset{\text{2-oxoglutaric acid}}{HOOC\text{–}CH_2\text{–}CH_2\text{–}C(=O)\text{–}COOH} + NH_3$$

Schematically, the transdeamination process is sketched. Both stages of this process are reversible. The reverse sequence of reactions leading to ammo acid synthesis from α-keto acids and ammonia was named *transreamination* by Braunshtein. The initial stage of transreamination involves reductive animation of 2-oxoglutarate involving ammonia and a reduced NADP. The process is catalyzed by NADP-dependent glutamate dehydrogenase to produce glutamic acid. Then the glutamate NH_2 group is transferred onto any α-keto acid to produce the corresponding amino acid. If a tissue contains appropriate α-keto acids, these are apt to be converted to amino acids by the transreamination pathway. In the animal tissues, α-keto acids are produced from nonessential amino acids, while in plants and bacteria, α-keto acids are derived from all available amino acids.

Routes to Metabolic Detoxificatioh of Ammonia

In the organism, ammonia is produced by the following processes:

(1) deamination of amino acids;

(2) deamination of biogenic amines (histamine, serotonin, cysteine, etc.);

(3) deamination of purine bases (guanine and adenine);

(4) deamination of amino acid amides (asparagine and glutamine);

(5) breakdown of pyrimidine bases (uracil, thymine, and cytosine).

Ammonia is a very toxic compound, especially for nervous cells. Its accumulation in the organism elicits excitation of the nervous system. For this reason, to counteract the adverse effects, mechanisms for neutralization of ammonia are operative in the organism tissues. These mechanisms include: (1) production of urea; (2) reductive amination, or transreamination; (3) production of amino acid amides (asparagine and glutamine); and (4) production of ammonium salts. The major route to detoxification of ammonia is urea synthesis. In the 19th centurey the Russian biochemists M.V. Nentsky and S.S. Salazkin showed that in the liver, urea was formed from ammonia and carbonic acid. Krebs and Heusen leit established that urea synthesis was a cyclic process involving ornithine as a catalyst.

Cohen and Ratner demonstrated that the synthesis of carbamoyl phosphate was the initial reaction of this cycle. The urea cycle (also referred to as the arginine-urea cycle, or ornithine cycle) is shown schematically. To produce one urea molecule, three ATP molecules are needed to be consumed. Urea is a material which is inert towards the living organism. The major site for its formation in the organism is the liver which contains all the enzymes involved in urea production. In the brain, of the total set of urea synthesis enzymes, only carbamoyl phosphate synthetase is missing, which precludes the cerebral formation of urea. Disturbances in the hepatic function lead to a decreased urea production, with the resultant diminution of urea blood concentration and urea excretion in the urine.

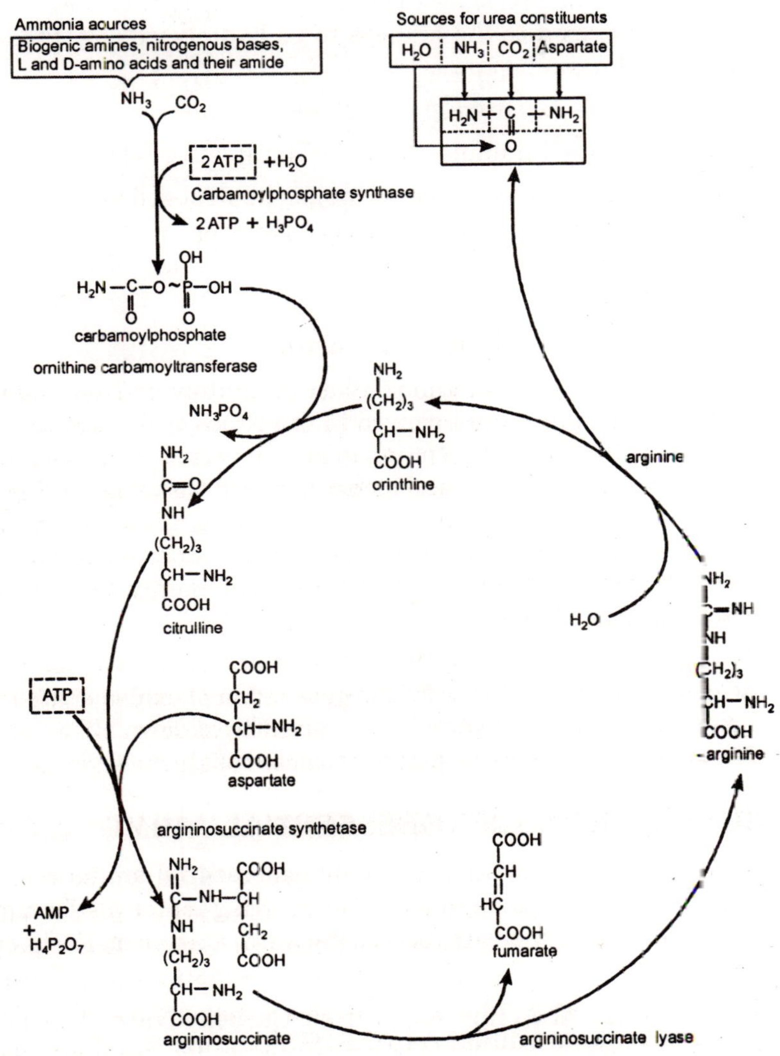

Fig. 5.10. Urea production cycle

Reductive Animation. Viewed from the standpoint of an ammonia-binding process, it is little effective, since it requires a significant amount of 2-oxoglutarate for its effectuation.

Production of Asparagine and Glutamine. It is an important subsidiary route to binding ammonia. It proceeds with the involvement of *asparagine synthetasr* and *glutamine synthetase* according to the schemes :

$$\text{L-Aspartate} + \text{ATP} + NH_3 \rightarrow \text{Asparagine} + \text{AMP} + H_4P_2O_7$$

$$\text{L-Glutamate} + \text{ATP} + NH_3 \rightarrow \text{Glutamine} + \text{ADP} + H_3PO_4$$

This process is active in nervous and muscle tissues and in kidney.

Ammonium Salt Production. Glutamine and, to a lesser extent, asparagine are believed to be a kind of vehicles for ammonia, since the two species, after their formation in the tissues, are delivered in the bloodstream to the kidneys to be hydrolyzed by the action of specific enzymes, *glutaminase* and *aspartaginase:*

$$\text{L-Asparagine} + H_2O \rightarrow \text{Aspartate} + NH_3$$

$$\text{L-Glutamine} + H_2O \rightarrow \text{Glutamate} + NH_3$$

Ammonia, liberated in the tubules of the kidney, is neutralized *to ammonium* salts:

$$NH_3 + H^+ + Cl^- \rightarrow NH_4Cl$$

which are excreted in the urine.

Conversion of Amino Acid Carbon Frameworks

The carbon frameworks of protein amino acids ultimately convert, after the scission of amino groups, to five products that are involved in the Krebs cycle. Glycine, alanine, leucine, cysteine, serine, threonine, lysine, and tryptophan are converted to acetyl-CoA; phenylalanine and tyrosine, to acetyl-CoA and fumarate; isoleucine, to acetyl-CoA and succinyl-CoA; valine and methionine, to succinyl-CoA; arginine, histidine, glutamine, glutamic acid, and proline, to 2-oxoglutarate; and asparagine and aspartic acid, to oxaloacetate. When an amino acid "burns down" to CO_2 and H_2O, it releases a significant amount of energy, nearly equivalent to that produced by aerobic oxidation of glucose.

An active breakdown of amino acids leads to the formation of acetyl-CoA and, subsequently, to the synthesis of ketone bodies in the liver. The generation of oxaloacetate and other acids of the Krebs cycle during the amino acid hydrocarbon radical breakdown allows utilization of amino acids, contained in the liver and kidneys, for the purposes of gluconeogenesis.

BIOSYNTHESIS *OF* NONESSENTIAL AMIMO ACIDS

In the mammalian tissues, biosynthesis of only nonessential amino acids is possible; the essential amino acids must be supplied in food. The starting materials for the biosynthesis of nonessential amino acids are intermediates of carbohydrate catabolism, Krebs cycle metabolites, and essential amino acids.

Production of Nonessential Amino Acids from Carbohydrate Catabolites, Pyruvate, 3-phosphoglycerate, and ribose 5-phosphate serve as sources for the synthesis of amino acids. The former two are glycolysis metabolites, and the latter one is a pentose phosphate cycle metabolite.

Alanine is produced from pyruvate via two routes: (1) transamination with the participation of alanine aminotransferase (ALT):

$$\text{Pyruvate} + \text{Glutamate} \xrightleftharpoons{\text{ALT}} \text{Alanine} + \text{2-Oxoglutarate}$$

2. reductive amination involving alanine dehydrogenase (ALDH):

$$\text{Pyruvate} + NH_3 + \text{NAD.H} + H^+ + H_2O \xrightleftharpoons{\text{ALDH}} \text{Alanine} + NAD^+ ...$$

3-Phosphoglycerate leads to serine, and this in turn to glycine. Therefore, 3-phosphoglycerate may be regarded as a common precursor to serine and glycine:

$$3-\text{Phosphoglycerate} \rightarrow \text{3-Phosphoserine} \xrightarrow[-H_3PO_4]{} \text{Serine}$$

$$\text{Serine} + \text{THFA} \rightarrow \text{Glycine} + N^5, N^{10}\text{-Methylene-THFA}$$

Histidine is produced from ribose 5-phosphate. Initially, ribose 5-phosphate is converted to α-5-phosphoribosyl 1-pyrophosphate (PRPP), which participates in the subsequent biosynthesis of histidine as well as purines:

Ribose 5-phosphate → Phosphoribosyl pyrophosphate → Histidine

However, the possibility for biosynthesis of histidine via this route is rather limited, and for this reason histidine is included among half-essential amino acids.

Synthesis of Nonessential Amino Acids from the Krebs Cycle Metabolites. Oxaloacetate and 2-oxoglutarate are sources for amino acid synthesis. Aspartic acid and asparagine are produced, respectively, from oxaloacetate via transamination with the participation" of aspartate aminotransferase, and via amination involving asparagine synthetase:

$$\text{Oxaloacetate} + \text{Glutamate} \rightleftharpoons \text{Aspartic acid} + \text{2-Oxoglutarate}$$

$$\text{Aspartic acid} + NH_3 + \text{ATP} \rightleftharpoons \text{Asparagine} + \text{ADP} + P_1$$

Glutamic acid, glutamine, proline, and, finally, hydroxyproline are produced from 2-oxoglutarate. Synthesis of glutamic acid and glutamine proceeds, respectively, with the involvement of glutamate dehydrogenase and glutamine synthetase. Glutamic acid is converted to proline during metabolism; hydroxyproline is formed only when proline becomes inserted into the polypeptide chain to be hydroxylized by proline hydroxylase.

Synthesis of Nonessential Amino Acids from Essential Amino Acids. From the essential amino acid phenylalanine, the nonessential amino acid tyrosine is formed. This process is catalyzed by phenylalanine hydroxylase for which dihydrobiopterin acts as a cofactor, and NADP $\bullet H_2$ as a reductant:

$$\text{Phenylalanine} + \text{NADP} \bullet H + H^+ + O_2 \rightarrow \text{Tyrosine} + \text{NADP}^+ + H_2O$$

Since phenylalanine is an essential amino acid, the possibility for endogenic synthesis of tyrosine is somewhat limited.

Methionine is capable of converting to cysteine. Sulphur is supplied by methionine itself, and the remaining portion of the molecule is a serine residue. The active form of methionine, S-*adenosylmethionine*, is involved in the synthesis. Demethylation of this species leads to S-adenosylhomocysteine and, subsequently, to homocysteine:

$$\text{Methionine} \rightarrow \text{S-Adenosylmethionine} \rightarrow \text{S-Adenosylhomocysteine} \xrightarrow[+H_2O]{}$$

$$\rightarrow \text{Homocysteine} + \text{Adenosine}$$

The interaction of homocysteine with cysteine proceeds with the participation of *cystathionine synthetase* to produce *cystathionine*. At the ultimate stage, cystathionine is split by *cystathionase* to yield free cysteine :

$$\text{Cystathionine} \rightarrow \text{2-Oxobutyrate} + NH_3 + \text{Cysteine}$$

Pyridoxal phosphate acts as a cofactor for cystathionase.

Ornithine is converted to *arginine* in the mammalian tissues. However, arginine is produced on a smaller scale, since it serves as a substrate in the synthesis of urea. Therefore, arginine is a half-essential amino acid.

AMINO ACIDS AS USED FOR PRODUCTION OF NONPROTEINIC NITROGENOUS COMPOUNDS

Amino acids are widely used for the biosynthesis of a large variety of nonproteinic nitrogenous materials: choline, phosphatides, creatine, mediators (including biogenic amines), pigments, vitamins, coenzymes, porphyrins, purine and pyrimidine bases. In plants, amino acids are starting materials for the synthesis of alkaloids and various products of secondary metabolism; and in microorganisms, for the synthesis of antibiotics.

Biosynthesis and Breakdown of Creatine. In the human and animal tissues, creatine synthesis proceeds with the involvement of three amino acids: arginine, glycine, and methionine. Creatine synthesis proceeds in two steps. The first step involves the formation of guanidinoacetic acid from arginine and glycine with the participation of the enzyme *glycine-amidine transferase* (in the Scheme below, the transferable moiety is shown included within the dashed box):

$$\underset{\text{L-arginine}}{H_2N{-}C({=}NH){-}NH{-}(CH_2)_3{-}CH(NH_2){-}COOH} + \underset{\text{glycine}}{NH_2{-}CH_2{-}COOH} \longrightarrow \underset{\text{guanidinoacetic acid}}{H_2N{-}C({=}NH){-}NH{-}CH_2{-}COOH} + \underset{\text{L-ornithine}}{H_2N{-}(CH_2)_3{-}CH(NH_2){-}COOH}$$

The first step proceeds actively in the kidney and pancreas.

At the second step, methylation of guanidinoacetic acid through the assistance of *guanidinoacetate methyltransferase* takes place. The active form of methionine, S-adenosyl-methionine, acts as a methyl group donor:

$$\underset{\text{guanidinoacetic acid}}{H_2N{-}C({=}NH){-}NH{-}CH_2{-}COOH} \xrightarrow[\text{s-adenosylmethionine} \;\to\; \text{s-adenosylhomocysteine}]{} \underset{\text{creatine}}{H_2N{-}C({=}NH){-}N(CH_3){-}CH_2{-}COOH}$$

This step takes place in the liver and "the pancreas which provide for the conditions and for the appropriate enzyme required for the synthesis of creatine from guanidinoacetic acid. It is commonly believed that the creatine that has been synthetized in the liver and in the pancreas is supplied via blood circulation to other organs and tissues: brain, skeletal muscles, heart, etc.

However, many details of creatine metabolism remain unclear. In particular, most creatine has been found to be present in the human heart and other muscle tissues, although creatine practically does not enter them. In the cells, creatine is involved in the energy transfer via reversible transphosphorylation with ATP. A product of creatine breakdown is creatinine produced via a nonenzymic pathway:

$$\underset{\text{creatine}}{H_2N-\underset{\underset{NH}{\|}}{C}-\underset{\underset{CH_3}{|}}{N}-CH_2-COOH} \xrightarrow[-H_2O]{} \underset{\text{creatinine}}{HN-\underset{\underset{NH}{\|}}{C}-\underset{\underset{CH_3}{|}}{N}-CH_2-CO}$$

About 2% of creatine contained in the organism is liable to conversion to creatinine. In small amounts, creatine and creatinine are present in the blood plasma. In the creatine-synthetizing organs (kidney, liver, pancreas), creatine concentrations are small (0.1 – 0.4 g/kg). In highest amounts, creatine is found in skeletal muscles (25–55 g/kg), heart (15–30 g/kg), and brain tissue (10–15 g/kg).

Urinary creatine is found only in children. In adult humans, creatinine is excreted in the urine (4.4 to 17.6 mmole per day); creatinine excretion has been found to be directly related to the muscular development of the human body. Creatine found in the urine of an adult human is indicative of a pathology.

Production and Breakdown of Mediators. Mediators are produced in the nervous tissue and certain other cells. Neuromediators, which are generated in the nerve endings, are transmitter substances that participate in the transmission of the nervous impulse to other nervous cells or peripheral organs and tissues. Tissue mediators take part in the metabolic tissue regulation.

In the production of a number of mediators, of special importance is the reaction of amino acid decarboxylation specifically catalyzable by decarboxylases (enzymes that contain pyridoxal phosphate as a coenzyme);

$$\underset{\text{amino acid}}{R-\underset{\underset{NH_2}{|}}{CH}-COOH} \xrightarrow[\text{(PALP)}]{\text{Decarboxylase}} \underset{\text{biogenic amine}}{R-CH_2-NH_3}+CO_2$$

The decarboxylation products are amines exhibiting a high biological activity; for this reason, they are referred to as *biogenic amines.* Most mediators belong to this group of compounds.

A route to detoxification of biogenic amines is the oxidative deamination with the involvement of *aminooxidases;*

$$R-CH_2-NH_2 + O_2 + H_2O \xrightarrow{\text{Aminooxidase}} R-C\overset{\displaystyle O}{\underset{\displaystyle H}{\lessgtr}} + NH_3 + H_2O_2$$

Aminooxidases are differentiated into a monoaminooxidase (MAO) type and a diaminooxidase (DAO) type. FAD serves as a coenzyme for MAO, and pyridoxal phosphate, for DAO (the reaction requires the presence of Cu^{2+} ions). MAO is bound with the cell mitochondria, arid DAO is bound in the cytoplasm. In small amounts, these enzymes are present in the blood. Primary, secondary, and tertiary amines are inactivated by MAO, while histamine, putrescine, cadaverine, and to a lesser extent, aliphatic amines are mostly inactivated by DAO. Aldehydes produced by deamination of biogenic amines are oxidized to organic acids by *aldehyde dehydrogenases:*

$$R{-}C(=O){-}H + NAD{\cdot}H + H^+ + H_2O \rightarrow R{-}COOH + NAD^+$$

Production of Histamine. Histamine is synthetized from histidine by the action of *histidine decarboxylase:*

$$\text{Histidine (imidazole–}CH_2{-}CH(NH_2){-}COOH) \xrightarrow{-CO_2} \text{Histamine (imidazole–}CH_2{-}CH_2{-}NH_2)$$

Nearly all the tissues and organs contain histamine. Especially abundant is histamine in the lung tissue and in the skin; it also occurs in the marrow and in the subcortex. A large quantity of histamine is produced and deposited in the mast cells of connective tissue, where it is stored in a bound state as a protein-heparin complex. Histamine is released from the mast cells by the agents called histamine liberators. In large amounts histamine is produced in the gastric mucosa, where it affects the secretion of pepsin and hydrochloric acid. In the blood, histamine is bound with basophil and eosinophil granules. Histamine in small amounts is always present in The blood plasma and other biological fluids.

Histamine serves .as a mediator of allergic reactions and is released in large amounts in pathologic processes.

Production of Serotonin: Serotonin is derived from tryptophan:

Tryptophan (indole–$CH_2{-}CH(NH_2){-}COOH$) —Triptophan monooxigenase (O_2, tetrahydropteridine → H_2O, dihydropteridine)→ 5-hydroxytryptophan (HO–indole–$CH_2{-}CH(NH_2){-}COOH$) —decarboxylase (PALP), $-CO_2$→ 5-hydroxytryptamine (serotonin) (HO–indole–$CH_2{-}CH_2{-}NH_2$)

In adult humans, about 90% of serotonin is contained in the enterochromafnn cells of the intestine. The rest is found in the mast cells of the skin, in the spleen, liver, kidneys, and lungs. In relatively large amounts serotonin occurs in the blood thrombocytes and in the central nervous system. It is also produced.in the cortical grey substance of the brain and in the hypothalamus. Serotonin acts as a mediator in the nervous system and as a local functional regulator for the peripheral organs and tissues.

Production of γ-Aminobutyric acid (GABA). GABA is produced from glutamic acid by the action of *glutamate decarboxylase:*

$$\underset{\text{glutamate}}{HOOC{-}CH_2{-}CH_2{-}CH(NH_2){-}COOH} \xrightarrow{-CO_2} \underset{\text{GABA}}{HOOC{-}CH_2{-}CH_2{-}CH_2{-}NH_2}$$

The synthesis proceeds in the inhibitory synapses of the nervous system; for these, GABA acts as a mediator. The largest amounts of GABA are found in the sub-cortical formations of the brain (black substance, pale globe, and hypothalamus). In the peripheral organs, it occurs in trace amounts only.

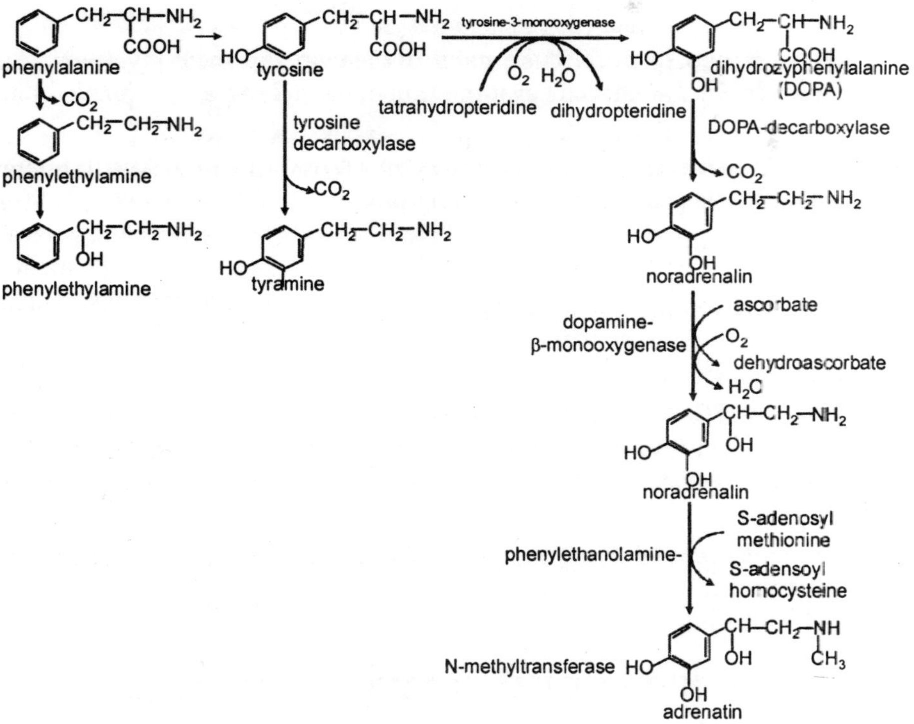

Fig. 2.11. Scheme for formation of catecholamines

Production of Catecholamines: Catecholamines constitute a group of biogenic amines with mediatory and hormonal functions. They are produced from phenylalanine and tyrosine (Fig. 2.11). The following biogenic amines are derived from phenylalanine and tyrosine: phenylethylamine, phenylethanolamine, 3,4-dihydro-xyphenylalanine (DOPA), dopamine (3,4-dihydroxyphenethylamine), noradrenalin, and adrenalin. The catecholamines that exhibit mediatory and hormonal functions to a great extent are adrenalin and noradrenalin.

Dopamine has been found to exhibit mediatory properties; presumably, the same holds true for tyramine. Noradrenalin and adrenalin are synthetized in the chromaffin cells of *subslanlia medullaris* of the adrenal glands (for performing a hormonal function), in the adrenergic synapses of the brain, and in the sympathetic nerve endings of the vegetative nervous system. Dopamine is formed in the synapses of the central nervous system (hypothalamus, limbic system, etc.), in the retina and sympathetic ganglia, where it acts as a mediator in the dopaminergic synapses. Small amounts of dopamine are found in the medulla of the adrenal glands.

In larger amounts, as compared with other Catecholamines, dopamine occurs in the liver, lungs, and intestine. Adrenalin and noradrenalin are inactivated via two pathways: with the involvement of *monoaminooxidase* or *catechol O-methyltransferase,* i.e. via deamination and methylation.

Production of Taurine. Taurine is a biogenic amine; it is produced from cysteine:

$$\underset{\text{cysteine}}{\begin{array}{c} H_2C-SH \\ | \\ H-C-NH_2 \\ | \\ COOH \end{array}} \longrightarrow \underset{\text{cysteine sulfinic acid}}{\begin{array}{c} H_2C-SO_2H \\ | \\ H-C-NH_2 \\ | \\ COOH \end{array}} \longrightarrow \underset{\text{cysteic acid}}{\begin{array}{c} H_2C-SO_3H \\ | \\ H-C-NH_2 \\ | \\ COOH \end{array}} \xrightarrow[CO_2]{} \underset{\text{taurine}}{\begin{array}{c} H_2C-SO_3H \\ | \\ CH_2 \\ | \\ NH_2 \end{array}}$$

This compound is synthetized in various organs. In the liver, taurine is involved in conjugation reactions with bile acids. In the nervous system, taurine is presumably capable of acting as a synaptic mediator.

Production of Certain Vitamins and Coenzymes from Amino Acids. In human tissues, *nicotinamide* (vitamin PP) is produced from tryptophan:

$$\text{Tryptophan} \xrightarrow{n \text{ steps}} \text{Nicotinic acid . Nicotinamide}$$

Therefore, tryptophan supplied in food can in part compensate for a deficit of dietary nicotinamide.

Cysteine is used for the biosynthesis of coenzymes of pantothenic acid (vitamin B_5), 4-phosphapantetheine, and CoA. Glutathione, which acts as a coenzyme for certain oxireductases, is produced from cysteine, glutamic acid, and glycine by a two-step reaction. At the first step of this synthesis glutaminylcysteine syn-thetase is operative, and at the second step, glutathione synthetase.

It appears more expedient to discuss syntheses of porphyrins and nitrogenous nucleotide bases with the involvement of amino acids in the subsequent chapters concerned with the metabolism of these materials.

AMINO ACID METABOLISM CONTROL IN THE ORGANISM

Amino acid balance in the human organism is dependent on the nutritive value of dietary proteins. Lack of any of the essential amino acids in the alimentary diet or a prolonged deficit of half-essential amino acids entails a disorder in the utilization of other amino acids (both essential and nonessential) in protein biosynthesis. An amino acid imbalance sets in. In norm, consumption of amino acids during biosynthetic processes is at balance with their supply (hydrolysis of tissue proteins or synthesis of nonessential amino acids). Amino acids are readily excreted from the cells into the pericellular fluid and the blood by simple diffusion. In the bloodstream, they are delivered to the tissues and organs. A fact to be noted is that amino acids are supplied to various tissues selectively, especially to the brain tissue. The reason for this remains largely unclear.

Amino acid transport to the cells of different organs is apparently determined by the efficiency of the carrier systems involved with respect to-different groups of amino acids. In the kidneys the reabsorption of amino acids from the urine into the blood occurs; owing to this recovery mechanism, only a small amount of amino acids is lost into the urine. The dietary supply of amino acids is one of the factors that regulate amino acid metabolism. An excessive consumption of food high in proteins results in an increased delivery of amino acids to the liver. The amino acids supplied favour an increased activity of the liver enzymes (enzymes for deamination of amino acids and for urea synthesis) in the catabolic conversion of amino acids to the end products. This mechanism provides for the removal of an amino acid excess supplied to the organism.

During starvation, on the contrary, an active breakdown of tissue proteins to free amino acids takes place. The amino acid balance control at various stages of amino acid transport can be effected by means of the intestinal epithelium (in absorption), peripheral tissues (in intracellular penetration), and the tubules of the kidney (in reabsorption). Transmembrane amino acid transport is stimulated chiefly by insulin.

In addition, insulin and other hormones can vary the amino acid consumption rate in biosynthetic reactions, or the protein degradation rate in tissues. Under physiological conditions, insulin, somatotropin, thyroid hormones, and male and female sex hormones facilitate the utilization of amino acids in protein biosynthesis and other synthetic processes. Glucocorticoids are capable of accelerating, depending on the tissue, both processes: protein degradatron.and amino acid production in certain tissues, and consumption of amino acids in protein biosynthesis, in others. Since amino acids constitute the major source of nitrogen for all the nitrogenous materials, they are responsible for the nitrogen balance in the organism.

Nitrogen balance is the difference between the amount of nitrogen supplied to the organism in nitrogen-containing nutrients (chiefly proteins and amino acids) and that of nitrogen-containing materials discharged from the organism in the urine, feces, and sweat. If the amount of nitrogen supplied is equal to that lost from the organism, a *nitrogen equilibrium* sets in. If the amount of nitrogen ingested is larger than that excreted from the organism, the *nitrogen balance is said to be positive.* With the nitrogen eliminated superior to that supplied, the *nitrogen balance is negative.* The positive nitrogen balance is observed during the active synthesis of proteins from amino acids (during prenatal development of the fetus, during juvenile growth, in the diet rich in meat, in the administration of anabolic hormones, etc.). The negative nitrogen balance develops under complete protein deprivation, under prolonged reduction of motor activity, in grave chronic diseases, burns, and other states adverse to the normal protein digestion and amino acid absorption, or conducive to the degradation of tissue protein.

AMINO ACIDS AS MEDICINAL PREPARATIONS

In the practical medicine, preparations on the basis of protein hydrolysates and individual amino acids are used.

Protein Hydrolysate Preparations (preparations for parenteral–nutrition). These are mixtures of amino acids produced by acidic or enzymic hydrolysis of proteins. They encompass hydrolysine, casein hydrolysate, aminopeptide, cerebrolysine, aminocrovin, and fibrinosol. These preparations are used for compensation of protein deprivation of the organism and for maintaining the nitrogen equilibrium; they can also provide for a positive nitrogen balance in patients with surgical gastroenteric intervention, with protein digestion disturbances, or with deep burns.

Preparations from Individual Amino Acids. Methionine and other methionine-rich hydrolysates are used as lipotropic factors, and in treating protein deficiency in chronic diseases.

Promising appears to be the use of cysteine (as well as methionine, which converts to cysteine in the organism) in the impaired metabolism of sulphur-containing proteins (collagen; proteins constitutive of the lens, cornea, etc.) and in the poisoning with heavy metal salts towards which the amino acids act as binding agents.

For clinical purposes, glutamic and aspartic acids are widely used (the latter mostly as potassium and magnesium salts of aspartic acid), which play a decisive role in the detoxification of ammonia and in other synthetic reactions of amino acid metabolism. At the present time, preparations on the basis of multicomponent mixtures of crystalline amino acids (especially essential ones) are extensively studied for potential applications in pure form or as additives, dosed at defined ratios, to other drugs of natural origin.

HEMOPROTEIN METABOLISM

Conjugated macromolecules (glycoproteins and lipoproteins) are renewed via the pathways that have been discussed above. The protein moiety is hydrolytically degraded to amino acids, and the nonprotein moiety, via conversion routes characteristic of carbohydrates and lipid breakdown, respectively. A specific feature of hemoproteins is the metabolic involvement of the nonprotein moiety of these conjugated proteins.

The hemoglobin of blood erythrocytes and of marrow cells accounts for a major portion (about 83%) of hemoproteins in the human organism. The remainder are myoglobin of skeletal muscles and heart (about 17%) and. cellular hemoproteins— cytochromes, catalase, peroxidase, etc. (not exceeding 1%).

Synthesis of Hemoproteins

Glycine and succinyl-CoA are the starting compounds in heme synthesis (Fig. 2.12). The reaction involving the pyridoxal-assisted enzyme δ-*aminolevulinate synthetase* yields δ-*aminolevulinic acid*. Then two molecules of δ-aminolevulenic acid combine, with the participation of *porptiobilinogen synthetase*, to form *por-phobilinogen*, a direct precursor to porphyrins. One of these is coproporphyrin III which is directly converted to protoporphyrin IX. The insertion of iron ions into the protoporphyrin IX ring is effected with the assistance of *ferrochelatase*. At 'the ultimate step, heme becomes complexed with globin to form hemoglobin or myoglobin.

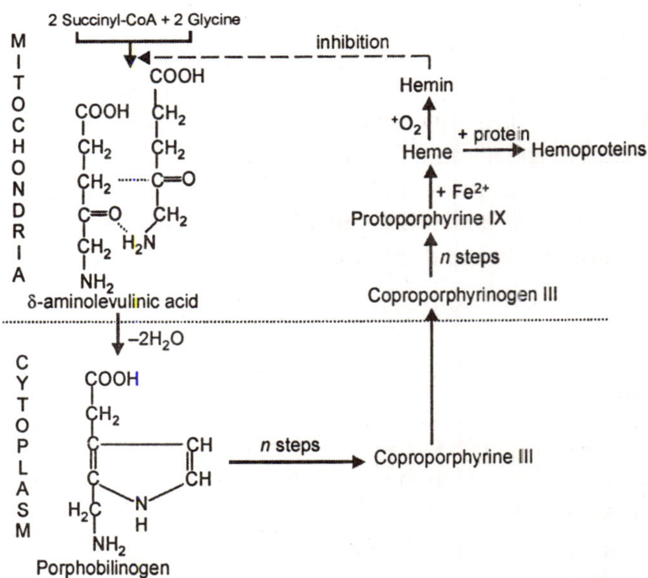

Fig. 2.12. Scheme for the synthesis of heme in the cells

In the synthesis of other hemoproteins, heme adds to the specific protein moiety of cytochromes or other heme-containing enzymes. If there is not a sufficient quantity of protein to bind the heme formed, the latter is oxidized by oxygen to hemin which inhibits the first reaction step of porphyrin synthesis (Fig. 2.12). The enzymes involved in heme biosynthesis are found in the marrow, nucleated erythrocytes, liver, kidneys, and intestinal mucosa. The reactions leading to δ-aminolevulinic acid proceed in the mitochondria; the production of porphobilinogen and the subsequent synthesis of coproporphyrinogen III occur in the cytoplasm, and the synthesis of heme from coproporp-hyrinogen III, in the mitochondria.

Breakdown of Hemoproteins

In the human organism, about 9 g of hemoproteins per day is degraded, mainly owing to the breakdown of erythrocytic hemoglobin. Erythrocytes undergo degradation (lysis) in the bloodstream or in the spleen in the range of about 120 days. Hemoglobin, released from the erythrocytes in the blood, becomes bound with haptoglobin (a group of glycoproteins in the α_2-globulin fraction of blood plasma) and is supplied, as a hemoglobin-haptoglobin complex, to the cells of the reticuloendothelial system (RES), chiefly to those of the spleen. Hemoglobin is oxidized to methemoglobin (Fe^{3+}) to be subsequently degraded. Haptoglobin is split off to pass into the blood. In RES cells, the first stage of hemoglobin degradation takes place, leading to bilirubin:

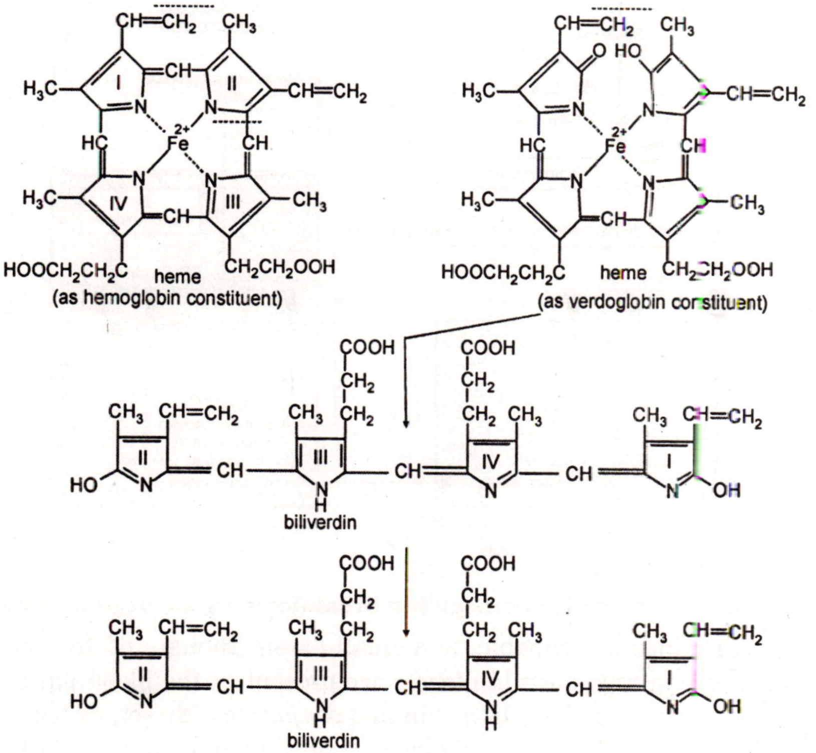

heme (as hemoglobin constituent)

heme (as verdoglobin constituent)

biliverdin

biliverdin

Initially, the heme α-methyne bridge linking two neighbouring pyrrole rings is oxidatively cleaved by *hemoxigenase*. The heme ring is ruptured to form *verdoglobin*. The iron, released from verdoglobin, becomes bound to a carrier protein to be supplied in the bloodstream to the marrow and to globin. Giobin is hydrolyzed by the spleen cathepsins to amino acids. The opened tetrapyrrole chain of verdoglobin is called *biliverdin* (a bile pigment of green colour). Biliverdin can be reduced to bilirubin, a yellow-red pigment.

Bilirubin is sparingly soluble in water and, when released into the blood, it becomes bound with the plasma protein albumin. The albumin-bilirubin complex is the most important normal transport form for bile pigments. In the bloodstream, bilirubin is carried over to the liver cells. Since bilirubin is a lipophil, it passes easily across the liver cell membrane and, in doing so, liberates itself from albumin.

The second stage of hemoprotein conversion takes place inside the liver cells (Fig. 2.13). Conjugated forms of bilirubin, *bilirubin glucuronides,* are produced in the liver. UDP-glucuronic acid acts as a donor of glucuronic acid. The reaction is catalyzed by *VDP-glucuronosyltransferase* to yield bilirubin monoglucuronide (20%) and bilirubin diglucuronide (80%). Bilirubin glucuronides are compounds readily-soluble in water.

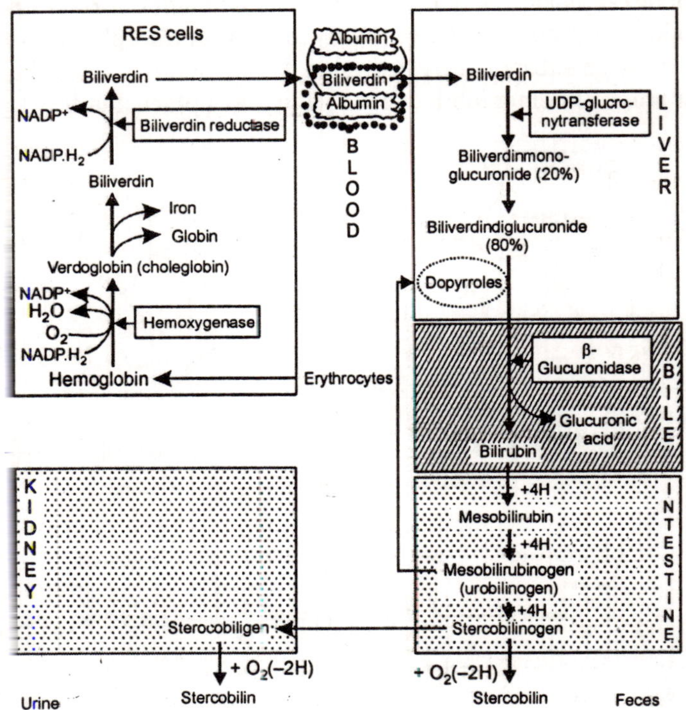

Fig. 2.13. Scheme for hemoglobin breakdown in the organism cells

Bilirubin glucuronides are capable, to a small extent though, of diffusing into the blood capillary. For this reason, two bilirubin forms are present in the blood plasma: *unconjugated* (also referred to as indirect, or free) bilirubin and *conjugated* (direct, or bound) bilirubin. The former accounts for about 75% of the total blood plasma bilirubin, and the latter, for about 25%. Bilirubin glucuronides are excreted in the intestinal bile where the third, final, stage of hemoprotein degradation occurs.

In the bile tracts, glucuronic acid is split off from bilirubin glucuronides, with the recovery of unconjugated bilirubin. In the small intestine, a small portion of bilirubin can be absorbed and, through the portal vein, be supplied to the liver to be again excreted in the bile from the intestine. This hepatoenteric circulation, bearing resemblance to bile acid circulation, is presumably of biological importance. The major portion of bilirubin undergoes multiple reductions by the intestinal bacteria or by the reductases of intestinal mucosa: In the small intestine, bilirubin is converted to *mesobilirubin* and then to *mesobilirubinogen* (or urobilinogen). The latter is absorbed in the small intestine, and through the portal vein is delivered to the liver in which urobilinogen is irreversibly degraded to mono- and dipyrroles. The intact urobilinogen is returned in the bile to the intestine'. In the large intestine, mesobilirubinogen (urobilinogen) is reduced by anaerobic bacteria to stercobilinogen which, like urobilinogen, is a colourless substance. Most stercobilinogen is excreted in the feces and undergoes rapid oxidation with atmospheric oxygen to Stercobilin, an orange-yellow pigment, chiefly responsible for the colour of the feces. A small amount of stercobilinogen is absorbed in the rectum, and through the hemorrhoidal vein system, in bypass to the liver, is delivered in the bloodstream to the kidneys to be excreted in the urine.

Urinary stercobilinogen, like fecal stercobilinogen, is oxidized to Stercobilin which is in part responsible for the light-yellow colour of the normal urine. It was formerly believed that urobilinogen was excreted in the urine (hence the name coined for this material) to be oxidized to urobilin exhibiting the same colour as Stercobilin. Presumably, the resemblance in colour mightjiave led to erroneous conclusions, since urobilin and Stercobilin structurally are different species. Normally, urobilinogen is excreted neither in urine, nor in feces. Thus, the intermediate product of heme breakdown, bilirubin, is normally not accumulated in the blood, but is rapidly entrapped by the liver cells.

The concentrations (absolute and relative) of different forms of bilirubin in the blood allow concluding, on the one hand, on the hemoprotein degradation rate in RES cells, and, on the other hand, on the bilirubin conversion in the liver. Normally, the total bilirubin concentration in the blood serum amounts to 8–20 µmol/litre (here unconjugated bilirubra accounts for 75%). In humans, the end product of bilirubin conversion, stercobilinogen, is chiefly excreted in the feces (about 300 mg per day) and, to a lesser extent, in the urine (about 1–4 mg per day). The composition of the pigments excreted in the feces and urine is influenced by the intestinal microflora.

In neonates, whose intestine is devoid of microflora, bilirubin being excreted is oxidized to biliverdin, which is green in colour; for this reason, the meconium (newborns' feces) retains a dark green grass-like colour. In infants, within the first year of life, as the intestine becomes populated with microflora, the composition of excreted pigments and the colour of feces gradually change. The greenish fecal colour, produced by the presence of biliverdin, disappears, since the intestinal bacteria become increasingly active in the reduction of bilirubin to stercobilinogen.

Pathology of Bile Pigment Metabolism

The production, conversion, and excretion of bilirubin in the organism may become disturbed owing to a variety of factors. An increased concentration of bilirubin in the blood leads to its deposition in tissues, including skin and mucosa, and imparts to these a yellow-brownish colour (the colour of bilirubin). Such states of the organism are referred to as *jaundices*. A variety of jaundices are known, of which typical are *hemolytic* jaundice, *parenchymatous* (or hepatocellular) jaundice, and *obstructive* jaundice.

Hemolytic jaundice may arise from a number of causes conducive to an extensive intravascular (or tissue) breakdown of erythrocytes, or to their tissue breakdown (in RES cells). A large amount of unconjugated bilirubin, excreted from RES cells in the bloodstream, fail to become conjugated in the liver, which results in a high bilirubin level in the blood. Owing to an excess of excreted stercobilin and urobilin, the feces acquire an intense, nearly dark coloration, and the abundantly excreted urobilin colours—the urine intensely yellow-orange.

Parenchymatous jaundice arises because of impaired liver cells (as afflicted with viruses or toxic hepatotropic compounds); the liver cells become more permeable to invading agents, including bilirubin glucuronides normally excreted from hepatic cells *An* the blood in small amounts. The impaired hepatic cells exhibit a reduced activity towards the uptake of bilirubin from the blood and fail to provide for the normal production of bilirubin glucuronides in them. For this reason, despite the normal hemolysis, the concentrations of unconjugated and conjugated bilirubin increase, albeit to a lesser extent as in hemolytic jaundice. Because of a diminished excretion of stercobilin and urobilin, the feces and the urine are but slightly coloured. However, a small quantity of urinary unconjugated bilirubin, normally absent in the urine, is observed.

Obstructive jaundice as its name implies, arises due to an impeded outflow of bile in the intestine. As a consequence, conjugated bilirubin is reexcreted from the liver cells in the blood. The blood contains an increased amount of conjugated bilirubin which, being a readily soluble compound, is excreted at a higher rate in the urine. Because of this, the urine acquires a beer-like colour, and is discharged with a yellow bright foam. The feces, devoid of bile pigments, are imparted a grey-white, clay-like colour.

Inogenous jaundice (jaundice in newborns) *is* believed to be of physiological origin. This disease arises from the developmental insufficiency of glucuronosyltransferase, the bilirubin conjugation enzyme. Therefore, an increased degradation of erythrocytes, irrespective of its origin, leads to an enhanced level of unconjugated bilirubin in the blood and, consequently, to an icteric indisposition. Commonly, physiological jaundice in newborns is transient within two weeks as the hepatic secretion of glucuronosyltransferase increases. In premature infants, physiological jaundice may be of longer duration. However, a prolonged increase of unconjugated bilirubin concentration in the blood may constitute a health hazard due to the toxic effect of bilirubin on the developing brain.

Occasionally, convulsions and irreversible nervous system disturbances are observed in infants. In adults, the cerebral cells are little permeable to bilirubin and as a rule, hyperbilirubinemia evokes no complications. In certain instances, in order to stimulate the activity of glucuronosyltransferase, the administration of phenobarbital is recommendable, which increases the enzyme secretion and alleviates the icteric processes. The impaired pigment metabolism is also observed in intestinal dysbacteriosis produced by suppression of normal microflora in the intestine (for example, as a result of prolonged therapy with tetracycline antibiotics). In dysbacteriosis intermediary products of bilirubin reduction are excreted in the feces; under deep suppression of the intestinal microflora, bilirubin itself, when oxidized to biliverdin, colours the feces greenish.

METABOLISM OF PURINE AND PYRIMIDINE MONONUCLEOTIDES

Mammals are not in need of dietary supply of nitrogenous bases or nucleotides despite the ability of the mammalian organism to assimilate directly these nutrients with food rich in nucleic acids. In the mammalian tissues, the consumed purine and pyrimidine nucleotides are continually renewed via their synthesis from simple compounds. Nucleotide phosphates, which are present in normal cells in small and relatively constant amounts, act as a link in the pathways of supply and consumption of mononucleotides:

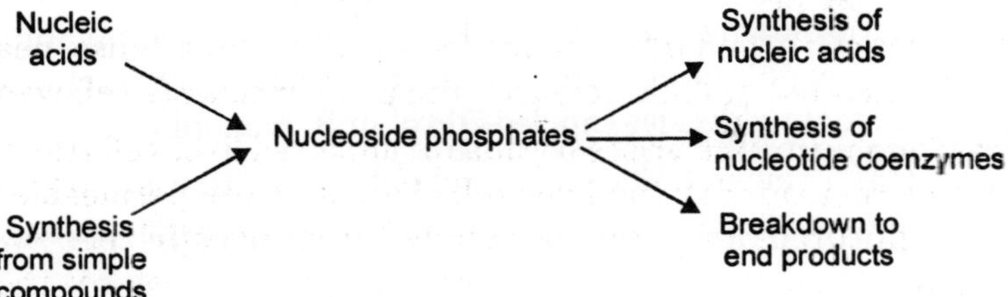

Biosynthesis of Mononucleotides

Biosynthesis of Purine Mononucleotides. The starting compound in the synthesis of purine nucleotides is phosphoribosyl pyrophosphate (PRPP), which can be derived from ribose 5-phosphate and ATP. The prime purine mononucleotide that crowns a long chain of synthetic reactions is inosine monophosphate (IMP), which is a parent species for generating other purine nucleoside phosphates via AMP and xanthosine monophosphate (XMP). The genesis of the purine ring is shown in fig. 2.14b.

Biosynthesis of Pyrimidine Mononucleotides. The starting compounds m the synthesis of pyrimidine mononucleotides are carbamoyl phosphate and aspartic acid They lead, through a long chain of reactions, to uridine 5′-monophosphate (UMP) and other pyrimidine mononucleotides (Fig. 2.15). The constituent atoms of the pyrimidine ring are supplied by carbomoyl phosphate and aspartic acid.

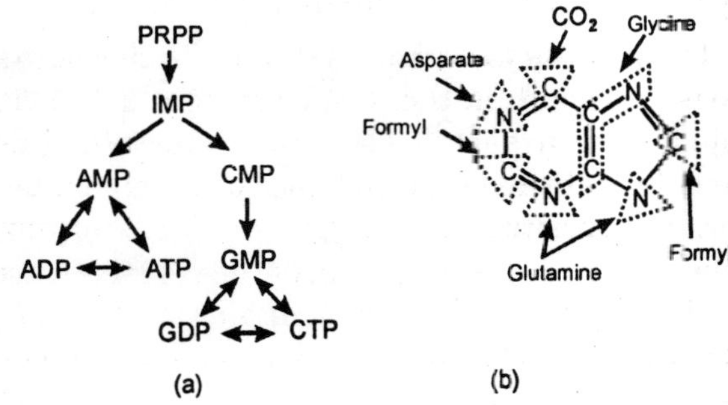

Fig. 2.14. Scheme for the synthesis of purine mononucleotides (a) and genesis of nucleotide purine ring (b)

Biosynthesis of Deoxyribonucleotides. precursors to DNA, is effected by reducing ribose to 2′-deoxyribose within the preformed nucleotide. A special protein, thioredoxin, bearing two SH groups in its molecule, acts as a reductant in this process.

Breakdown of Nucleic Acids and Nucleotides

Nucleic acids undergo hydrolysis in the organism tissues through the assistance of nucleases—deoxyribonucleases (DNAses) and ribonucleases (RNAses). These enzymes catalyze the cleavage of endo- or terminal internucleotide bonds (3′,5′-phosphodiester bonds) in DNA and RNA. The nucleases that split bonds in the

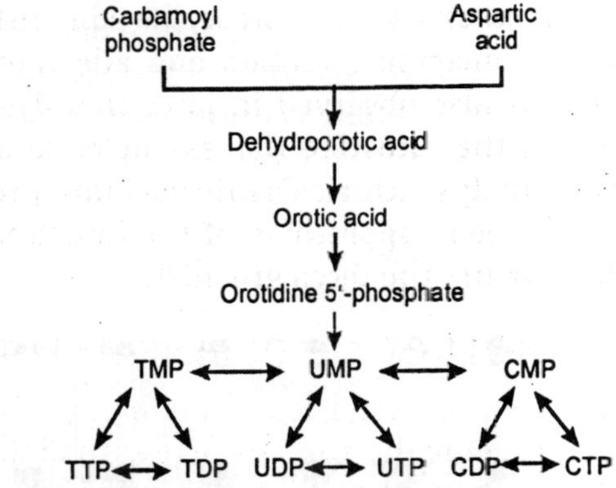

Fig. 2.15. Scheme for the synthesis of pyramidine nucleotides

interior of a polynucleotide chain are referred to as *endonucleases,* and those that split off terminal nucleotides, as *exonucleases.* Some of the exonucleases are capable of hydrolyzing both DNA and RNA. When a polynucleotide chain is attacked by nucleases, it degrades to-oligonucleotides and mononucleotides (nucleoside 5′-or 3′-monophosphates).

Mononucleotides break down to the end metabolites via either hydrolytic, or phosphorolytic pathways (in the former instance water, and in the latter, phosphoric acid are used as bond-cleaving agents).

Mononucleotides break down to free bases, pentose, and phosphate through the agency of *nucleotidases* (5′-nucleotidase and 3′-nucleofidase) and *nucleoside hydrolases.*

Purine bases are oxidized to uric acid:

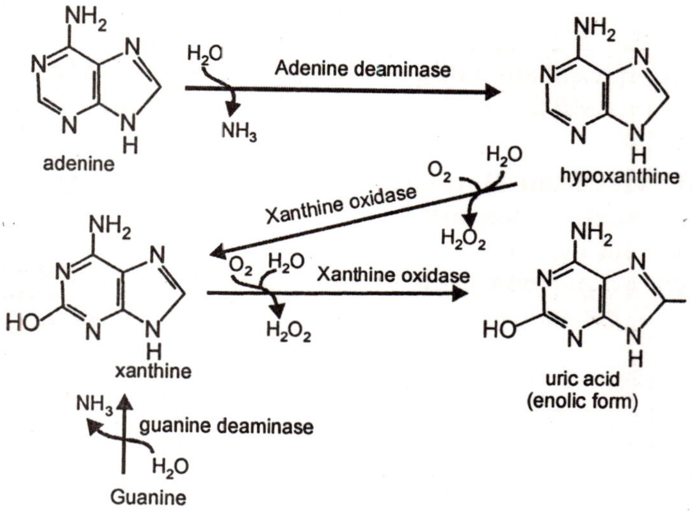

Uric acid is the end product of purine nucleotide metabolism in the human organism, and is excreted in the urine. Conversion of pyrimidine bases proceeds to β-alanine, CO_2, and ammonia. β-Alanine is used for the synthesis of muscle di-peptides carnosine and anserine; or is excreted in the urine. Thus, uric acid concentrations in the blood and the urine reflect the extent of nucleic acid cleavage in the organism. When adhering to a special purine-free diet, one can, by urinary uric acid analysis, estimate the endogenic purine metabolism in one's organism. Since uric acid is sparingly soluble in water, its normal concentrations in the fluid media of the organism are close to the solubility limit.

An increased level of uric acid in the blood, known as *hyperuricemia,* leads to the deposition of sodium urate crystals in tissues, especially in the regions where a local drop in the medium pH occurs. The disease attendant to the deposition of urates in the joints or the kidneys is called *podagra.* Its occurrence is wide enough, especially in human males. Occasionally, podagra may be caused by the congenital (partial or total) deficiency of the enzyme involved in the reutilization of free bases in the nucleotide synthesis. The podagral processes can be alleviated by administering preparations capable of reducing the production of uric acid during purine oxidation (for example, allopurinol), or stimulating the renal excretion of uric acid (for example, anturane and cinchophen).

CHAPTER

3 Animal Polysaccharides

MICROBIAL GLYCOPROTEINS

Bacterial and Fungal Glycoproteins

No evidence has been found to substantiate the claim for the presence of covalently-bound carbohydrate in energy-transducing adenosine triphosphatases of *Escherichia coli*. Incomplete removal of sodium dodecyl sulphate (SDS) after SDS-electrophoresis of the protein results in a positive reaction with the periodate-Schiff staining method. Tunicamycin and 2-deoxy-D-*arabinc*-hexose have been shown to inhibit the *in vivo* glycosylation of polypeptides in polysomes from *Aspergillus niger,* but have no effect on the *in vitro* glycosylation.

Although no new oligosaccharide chains were synthesized, pre-existing chains were extended. The applicability of natural abundance ^{13}C n.m.r. spectroscopy to studies on the structure and mobility of the carbohydrate residues of glycoproteins has been evaluated by examining the spectra at 63.4 kG of native and denatured glucoamylase from *A. niger* Integrated intensities reveal an average of about one hundred and twenty carbohydrate residues per molecule, composed of about eight α-D-mannopyranosyl, ten α-D-glucopyranosyl, and a maximum of fifteen β-D-mannopyranosyl residues. All of the observed carbohydrate residues are involved in O-glycosidic linkages to L-serine, L-threonine, or other carbohydrate residues. Although the neutral- and amino-sugar composition of the plasma membrane of *Dictyostelium discoideum* has been observed to remain unchanged during the early stages of differentiation, the amounts of glycoproteins increase and that of one particular component (mol. wt. 1.25×10^5) decreased dramatically. More than half of the spherule-wall fragments of the slime mould *Physarum polycephalum* are composed of a glycoprotein fraction. In addition, proteoglycans having a high carbohydrate to protein ratio were identified.

A family of glycopeptides having disperse molecular weights, charge, and sugar composition have been isolated from the culture filtrates of the tomato-leaf fungal pathogen *Fulvia fulvia.* D-Galactose, D-glucose, and D-mannose are the principal sugars with smaller amounts of 2-amino-2-deoxy-D-glucose, 2-amino-2-deoxy-D-galactose, and D-glucuronic acid.

The proposed structure, similar to that suggested for the peptidophospho-D-galacto-D-mannan secreted from *Penicillium charlesii.* The extension of the oligosaccharide chains of the glycoprotein of *Neurospora crassa* does not require the presence of a lipid intermediate. After the synthesis of the simple D-mannosyl-protein core, a second D-mannosyl unit is transferred directly from GDP-D-mannose. Different patterns of glycosylation have been identified for the *exo*-(1 → 3)-β-D-glucanases and *endo*-(1 → 3)-β-D-glucanases of *Pichia polymorpha.* Contrary to other reports, the *endo*-(1 → 3)-β-D-glucanases like the *exo*-(1 → 3)-β-D-glucanases are glycoproteins, containing residues of D-glucose and D-mannose.

Studies on the excretion of these enzymes by both protoplasts and whole cells appear to support the hypothesis that glycosylation takes place before or during secretion of the enzymes into the periplasmic space and no further glycosylation occurs during their secretion into the supernatant culture. The properties of β-D-fructofuranosidase (invertase) of *Saccharomyces cerevisiae,* and of the modified enzyme after removal of about 90% of the carbohydrate residues, have been found to differ considerably, not only in stability but also in their capacity to be renatured following inactivation by guanidinium chloride.

The covalently bound carbohydrate of yeast β-D-fructofuranosidase has been shown to be involved in the promotion of folding of the protein to its most stable conformation. After removal by *endo*-β-2-acetamido-2-deoxy-D-glucanase H of about 90% of the carbohydrate associated with this enzyme, no significant differences were detected in the catalytic and physical properties of the carbohydrate-depleted and the unmodified enzyme, although the former was much less stable to multiple freeze-thaw treatment, and to digestion by proteolytic enzymes. Many proteins and glycoproteins secreted by mutants of *S. cerevisiae* deficient in killer toxin expression show altered electrophoretic mobility and isoelectric points when compared with those of the parent strain.

The observed differences indicate that secretion in these mutants is abnormal, probably at the level of protein processing. An unusual glycoprotein, isolated from the spores of *Ustilago tritia,* contains residues of L-fucose, 2-acetamido-2-deoxy-D-glucose, 2-acetamido-2-deoxy-D-galactose, D-galactose, D-glucose, and D-mannose.

VIRAL GLYCOPROTEINS

The following aspects of viral glycoproteins have been reviewed; glycoprotein and protein precursors to plasma membranes in vesicular stomatitis virus infected He La cells, membrane glycoproteins of enveloped viruses, membrane assembly, including the synthesis and intracellular processing of vesicular stomatitis virus, and the structure and replication of α-viruses. The adherence of streptococci to canine kidney cells has been shown to involve the recognition of viral haemagglutinin, a glycoprotein whose synthesis is blocked by tunicamycin and streptovirudin.

Antibody in the sera of chickens has been measured for reactivity to determinants of the envelope glycoprotein of endogenous and exogenous avian tumour viruses. The pattern of reactivity of animals infected with exogenous sarcoma virus served to define a class of determinants present on the envelope glycoprotein of exogenous viruses and absent from that of the endogenous virus Rous-associated virus-O. Significant changes in the transport of D-glucose and in glycolysis have been observed after treatment of normal chicken fibroblasts with the major glycoprotein from avian myoblastosis virus.

An avian sarcoma virus envelope glycoprotein (mol. wt. 8.5×10^4), which specifically binds to chick embryo fibroblasts, has been purified with retention of its biological binding activity. A glycoprotein of this type should be useful for the analysis of early events of avian RNA tumour virus infection as well as for the purification and analysis of cellular receptor sites for avian type C viral envelope glycoproteins. Glycopeptides isolated from avian RNA tumour viral glycoproteins have been characterized by their broad molecular weight distribution. The oligo-saccharide chains of these glycopeptides are made up of both D-mannose-rich neutral chains and complex chains bearing terminal *N*-acetylneuraminic acid residues.

A photoreactive glycolipid probe, 2-[12-(4-azido-2-nitrophenoxy) stearamido]-2-deoxy-D-[1-C]glucose has been used to map the envelope proteins and glycoproteins of Newcastle disease virus. The Bunyaviridae is a family of arthropod-borne viruses characterized by possession of genomes consisting of three distinct size classes of RNA, designated L (large), M (medium), and S (small). M (medium) RNA has been shown to code for two viral glycoproteins (G1 and G2) of snowshoe hare (SSH) virus, LaCrosse (LAC) virus, and an LAC/SSH recombinant virus. A correlation between Epstein-Barr virus membrane antigen and three surface glycoproteins (mol. wts. 2.36, 2.12, and 1.41 $\times$ 10^5) has been established on the basis of radioimmunoprecipitation and immunoadsorption experiments. The physical location in viral DNA of the templates specifying the major viral glycoproteins of Herpes simplex virus and of mutations causing alterations in the behaviour of infected cells has been reported.

In Herpes simplex virus type 1 a glycoprotein (mol. wt. 5.2×10^4) containing an oligo-D-mannosyl core (mol. wt. 1.8×10^3) is processed by further glycosylation and sialylation to produce a more acidic and heterogeneous molecule, which was identified as the type-common envelope glycoprotein D (CP-1 antigen). Three glycoprotein antigens of Herpes simplex virus type 1 react as targets for immunocytolysis reactions in the presence of either complement or mononuclear leukocytes from human peripheral blood. A glycoprotein with affinity for the Fc region of immunoglobulin G has been isolated from extracts of cultured cells infected with Herpes simplex virus type 1. Evidence is provided that it could account for at least some, and possibly all, of the Fc-binding activity detected on the surfaces of infected cells.

The glycoprotein accumulates in two electrophoretically distinct forms (mol. wts. 8.0 and 6.5×10^4) and is distinct from glycoproteins characterized previously from this virus. Herpes simplex virus cells labelled with D-galactose and 2-amino-2-deoxy-D-glucose have been subjected to proteolytic digestion. High molecular weight fractions containing the amino-sugar, which were present in the digests of uninfected cells, were absent. Changes were also observed in the relative binding of the glycopeptides from the two types of cells to concanavalin A. Two glycoproteins (GP1 and GP2, mol. wts. 1.05×10^5 and 8.2×10^4, respectively) isolated from influenza C virus have been found to share common amino-acid sequences.

The larger molecular weight glycoprotein is considered to be the primary viral gene product, which may be converted into the lower molecular weight glycoprotein by proteolytic cleavage a mechanism that appears to be essential for maximal viral infectivity. Four different glycopeptides, each one only containing *N*-glycosidic linkages between 2-acetamido-2-deoxy-D-glucose and L-asparagine, have been isolated after proteolytic digestion of influenza A viral glycoproteins. Two of the glycopeptides contained residues of 2-acetamido-2-deoxy-D-glucose, D-mannose, D-galactose, and L-fucose, whilst the other two contained residues of D-mannose and 2-acetamido-2-deoxy-D-glucose.

The types of carbohydrate side chains associated with haemagglutinin, one of the two distinct glycoproteins on the surface of influenza A virions, has been found to vary depending on host cell type as well as virus strain. Further, the distribution of oligosaccharide chains associated with the haemagglutinin varied among influenza virus strains of the same as well as different antigenic subtypes. Two possible mechanisms by which changes in amino-acid sequence can cause variation in the relative amounts of the two different types of glycopeptides have been considered. The variable genes of influenza A viruses coding for haemagglutinin and neuraminidase have been found to consist of a relatively small highly conserved region, which is presumably involved in the functional integrity of the gene products and a relatively large highly variable region, presumably involved in the immunological properties. Evidence for the extrusion of haemagglutinin into membrane vesicles and its glycosylation during synthesis has been presented. Glycosylation of influenza viral glycoproteins has been investigated by pulse-labelling of infected baby hamster kidney cells with radioactive sugar precursors.

On the basis of this work and earlier work, steps involved in the biosynthesis of these glycoproteins have been postulated. The polypeptide portion of the viral glycoprotein is synthesized in association with rough membranes. During or immediately after its synthesis, oligosaccharide cores that contain D-mannose and 2-amino-2-deoxy-D-glucose are transferred from lipid-linked intermediates to glycosylation sites on the polypeptides. The glycoproteins then migrate from the rough to smooth membranes, where some D-mannosyl residues are removed from a population of oligosaccharide cores and subsequently branching of the chains occurs by addition of D-galactosyl, L-fucosyl-, and 2-acetamido-2-deoxy-D-glucosyl residues in a stepwise manner. The remaining oligosaccharide cores do not undergo either this trimming process or the addition of sugars at branch points.

After further migration from smooth membranes to plasma membranes, the two types of glycoproteins are incorporated into mature virions. A particular monoclonal antibody specific for the influenza virus haemagglutinin molecule is capable of blocking virus-specific cytotoxic T cells in an *in vitro* assay. The binding appears to be in the context of the histocompatibility antigen H-2D^d but not of H-2K^d. Another monoclonal antibody that has been isolated is capable of blocking both sets of HA-specific effector T cells. Particles of hepatitis B surface antigen have been labelled after treatment with neuraminidase and D-galactose oxidase, followed by reduction with sodium borotritide. The incorporated label was detected chiefly in a glycoprotein (mol. wt. 2.8×10^4) that retained its serological activity after treatment with neuraminidase. A radioimmunoassay procedure has demonstrated that a glycoprotein (mol. wt. 3.6×10^4) from a murine mammary tumour virus has both type-specific and group-specific antigenic determinants.

A glycoprotein antigen (mol. wt. 7.5×10^4), which competes with murine C-type viral glycoprotein in an inter-species radioimmunoassay, has been purified from the surface of EL-4 tumour cells. Serological and structural studies show this glycoprotein to be distinct from any of the viral glycoproteins of similar molecular weight so far examined. The glycosylated *env* gene precursor of Moloney murine leukaemia virus has been isolated by selective immunoprecipitation. By use of tunicamycin to inhibit nascent glycosylation, it was demonstrated that the precursor contained an apoprotein (mol. wt. 6×10^4) as compared with the complete virion glycoprotein (mol. wt. 7×10^4). The precursor appears to contain two distinct major sites of glycosylation which are within a 3×10^4 dalton region of the apoprotein sequence. Other workers have observed that the molecular weight of the precursor of the polypeptide encoded by the *env* gene of Rauscher murine leukaemia virus is reduced from 8.5×10^4 to 6.8×10^4 when synthesized in the presence of tunicamycin.

Most, if not all, of the carbohydrate content of the larger molecular weight glycoprotein is linked to L-asparaginyl residues, and glycosylation appears to be necessary for normal processing of precursor proteins into viral particles. Immunoprecipitation has been used in the isolation of a major *gag*-related internal structural glycoprotein (mol. wt. 8×10^4) from Moloney murine leukaemia virus. In cells, this glycoprotein appears to be further glycosylated to a glycoprotein (mol. wt. 9.5×10^4), which is rapidly cleaved and released from the cell as two soluble *gag*-related glycoproteins (mol. wts. 5.5×10^4 and 4.5×10^4, respectively).

Analysis of the carbohydrate components of two ecotropic murine leukaemia viruses have indicated that these viruses possess complex glycopeptides, containing residues of D-mannose, D-galactose, L-fucose, and 2-amino-2-deoxy-D-glucose, as well as glycopeptides with high levels of D-mannosyl residues, but lacking D-galactosyl or L-fucosyl residues. Glycosylation of the viral glycoproteins appears to be controlled by features of the viral polypeptide sequence as well as the host ceil of origin. The products of the major histocompatibility locus have been shown to interact with viral antigens on the cell surface of thymocytes. A number of polypeptides bind to the major envelope glycoprotein of Rauscher murine leukaemia virus.

The receptors for this glycoprotein, which probably involve lipoproteins, have been prepared from plasma membrane preparations from mouse cells. Evidence for a cell-associated precursor of two virion glycopeptides (GP1 and GP2) of lymphocytic choriomeningitis virus has been reported. Proteolytic cleavage of the precursor molecule occurs at a single site or at spatially-close multiple sites releasing GP1, a glycopeptide containing residues of L-fucose, D-galactose, and 2-amino-2-deoxy-D-glucose, and GP2, a glycopeptide containing only residues of L-fucose of 2-amino-2-deoxy-D-glucose. The Friend spleen focus-forming virus encodes a glycoprotein (mol. wt. 5.5×10^4).

Although it appears to be structurally related to the lymphatic leukaemia viral glycoprotein (mol. wt. 7.5×10^4), the former glycoprotein has distinctive structural features in the polypeptide portion of the molecule. Cells infected with Rous sarcoma virus, and then treated with tunicamycin, synthesize a polypeptide (mol. wt. 6.2 × 10^4) containing antigenic determinants of both the envelope glycoproteins (mol. wts. 8.5 X 10^4 and 3.5 × 10^4) of the virus. In addition, the polypeptide contained all the tryptic peptides of the precursor glycoprotein (mol. wt. 9.2 × 10^4) of the two envelope glycoproteins. The binding of Semliki Forest virus and its spike glycoproteins to baby hamster kidney cells has been followed by electron microscopy and quantitative binding assays. Binding of the virus to the cell involves two different modes; near neutral pH the virus binds to specific glycoproteins and at low pH it binds non-specifically to the lipids of the plasma membrane.

Separation of the membrane glycoproteins El and E2 of this virus has been achieved by affinity chromatography using immobilized concanavalin A. Glycopeptide El, which binds to the lectin, contains two peripheral sugar branches on the polypeptide chain containing D-mannosyl-, D-galactosyl-, 2-acetamido-2-deoxy-D-glucosyl-and neuraminosyl-residues, whereas the glycopeptide E2 contains single chains containing D-mannosyl- and 2-acetamido-2-deoxy-D-glucosyl residues. Metabolic labelling of the precursor of the Semliki Forest viral envelope glycoproteins E2 and E3 has revealed that the precursor glycoprotein contains mainly glycan chains with a high level of D-mannosyl residues, whereas its products, the viral glycoproteins E2 and E3, contain an additional lactosamine-type glycan. Semliki Forest virus membrane glycoproteins, after removal of their L-fucosyl, D-galactosyl, neuraminosyl, and distal 2-acetamido-2-deoxy-D-glucosyl residues, have been digested with an *endo*-β-D-2-acetamido-2-deoxy-glucanase (EC 3.2.1.96) to release a tetrasaccharide.

The absolute structure of this tetra-saccharide is not yet known, although it contains two α-D-mannosyl, one β-D-mannosyl, and one 2-acetamido-2-deoxy-D-glucosyl residues. Semliki Forest virus glycoproteins have been desialylized and then subjected to hydrazinolysis to release glycans whose molecular sizes range from 1.85 – 2.15 × 10^3. Sendai virus, one of the paramyxovirus group, has two glycoproteins that constitute the spikes on the virus envelope. The HN glycoprotein is responsible for the attachment of the virus to the cell membrane, resulting in the agglutination of cells, and is associated with neuraminidase activity. The F glycoprotein is required for envelope fusion and is also required for biological activities such as infectivity, haemolysis, and cell fusion. Disulphide bonds in the HN glycoprotein may be split using mild conditions with a concomitant loss of biological activity of the virus.

More drastic conditions are necessary for the cleavage of interpeptide disulphide bonds between sub-units on the F glycoprotein, when both fragments remain on the virion. This is in contrast to the release of sub-units of influenza virus, and a model for the arrangement of these sub-units on viral membranes is proposed. A method has been developed for the reconstitution of biologically active membranes with the individual HN and F glycoproteins of Sendai virus. Two mouse L-cell variant lines (CL3 and CL6), selected for their resistance to ricin, are restricted in their ability to replicate Sindbis and Semliki Forest viruses. The CL3 cells exhibit an increase in CMP-neuraminic acid : glycoprotein neuraminosyltransferase and GM_3-synthase activities, whereas the CL6 cells contain decreased UDP-D-galactose: glycoprotein galactosyltransferase and UDP-2-acetamido-2-deoxy-D-glucose:glycoprotein 2-acetamido-2-deoxy-D-gluco-syltransferase activities.

The adsorption of Sindbis virus to CL6 cells was considerably reduced, suggesting a loss or inaccessibility of the receptors for the virus accounted for a major defect in the viral protein. In contrast, CL3 cell line synthesized Sindbis viral RNA and proteins, but was unable to convert the precursor glycoprotein, PE2, into the structural protein E2. The cleavage of PE2 to E2 was blocked in both CL3 and CL6 cells infected with Semliki virus. A study of the tryptic peptides released from glycoproteins (mol. wt. 7×10^4) of ecotropic and amphotropic murine leukaemia viruses suggests that the viruses are structurally divergent. The glycoproteins isolated from complete and defective rabies viruses contain the same monosaccharides, *viz.* neuraminic acid, D-galactose, D-mannose, L-fucose, and 2-acetamido-2-deoxy-D-glucose.

The isoelectric points, molecular sizes, and glycosylation patterns of the two groups of glycoproteins were found to differ. The carbohydrate moieties of the glycoproteins from the paramyxovirus SV5 contain residues of L-fucose, D-mannose, D-galactose, and 2-amino-2-deoxy-D-glucose (in the molar ratio 1 : 3 : 4 : 2). Glycopeptides isolated from the HN_1, F_1, and F_2 glycoproteins contain three, three, and one carbohydrate chains, respectively. The structure (1) is unusual for one isolated from a eukaryotic source because of the presence of acetaldehyde in an acetal linkage. Ten percent of the carbohydrate chains are esterified with sulphate, located at O-6 of the amino-sugar residues. One or two moles of covalently bound fatty acid per mole of Sindbis virus glycoprotein E1 have been reported.

The covalent attachment of the fatty acids to the polypeptide is alkali-labile and is thought to occur during maturation of the viral glycoprotein. The structure (2) of the oligosaccharide chains of the glycoprotein E2, isolated from growth of this virus in hamster and chicken cells, is based on sugar composition, linkage analysis, and sequential degradation with enzymes.

Glycoproteins isolated from the virus grown separately in the two hosts show no variation in the structure of the oligosaccharide chains. It is noteworthy that the neuraminic acid residues in the oligosaccharide chains of the Sindbis viral glycoprotein are linked exclusively to O-3 of the D-galactosyl residues, whereas in mammalian glycoproteins it is to O-6 or to a mixture of O-3 and O-6 D-galactosyl residues. Glycoprotein-glycoprotein interactions have been proposed to account for the observations of hexagonal arrays of glycoproteins on the surface of Sindbis virus. A variant of Sindbis virus that contains a smaller molecular weight form of the viral glycoprotein E2 has been isolated.

The molecular weight of the precursor glycoprotein and the glycosylation pattern of the smaller E2 glycoprotein are normal, thus indicating that the latter is formed by an aberrant proteolytic cleavage. The envelope glycoprotein of Venezuelan equine encephalomyelitis virus is derived by proteolytic cleavage of a precursor glycoprotein, as has been shown with many other viruses. A comparison of the envelope glycoproteins of representative strains of this virus, using polyacrylamide gel electrophoresis and isoelectric focusing has been presented, together with a description of a method for serological sub-typing of the virus strains using antisera to these glycoproteins.

A procedure has been described for the selective isolation of mutants of vesicular stomatitis virus, defective in the production of the viral glycoprotein.

The glycoprotein of vesicular stomatitis virus has been selectively liberated from the virion membrane by the dialysable non-ionic detergent octyl β-D-glucopyranoside. When mixtures

```
   Me
   |
   CH
  /  \
 O    O
  \  /
β-D-Galp-(1 → 3)-β-D-Galp-(1 → 2)-α-D-Manp
                                      1
                                      ↓
  Me                                  3
  |                          β-D-Manp-(1 → 4)-β-D-Glcp-NAc-(1 → 4)-β-D-GlcpNAc-L-Asn
  CH                                  6                6
 /  \                                 ↑                ↑
O    O                                1                1
 \  /
β-D-Galp-(1 → 3)-β-D-Galp-(1 → 2)-α-D-Manp           L-Fucp

                                   (1)
```

```
Neup5Ac-(2 → 3)-β-D-Galp-(1 → 4)-β-D-GlcpNAc-(1 → 2)-α-D-Manp-2)-α-D-Manp
                                                           1
                                                           ↓
                                                           3
                                                  β-D-Manp-(1 → 4)-β-D-Glcp-NAc-(1 → 4)-β-D-GlcpNAc-L-Asn
                                                           6                                  6
                                                           ↑                                  ↑
                                                           1                                  1
Neup5 Ac-(2 → 3)-β-D-Galp-(1 → 4)-β-D-GlcpNAc-(1 → 2)-α-D-Manp                             (L-Fuc)n
                          n = 0 or 1
                                              (2)
```

of the viral glycoprotein and egg lecithin were dialysed until free from the detergent, glycoprotein vesicles formed spontaneously with spikes protruding in the same external orientation as the original virion membrane.

Pulse-labelling of HeLa cells infected with vesicular stomatitis virus has resulted in the isolation from the bound ribosomes of a peptidyl-transfer RNA species specifically labelled in the carbohydrate moiety. The size distribution of the nascent polypeptide chains that serve as a substrate for the cellular glycosyltransferases has been determined by polyacrylamide gel electrophoresis. Attachment of the carbohydrate chains starts when approximately half of the amino-acid sequence of the viral glycoprotein has been synthesized. The non-glycosylated proteins of the virus, which are produced by growing the virus in the presence of tunicamycin, are temperature-sensitive and undergo aggregation at elevated temperatures.

Proteins with different sequences of amino-acids appear to have differing requirements for glycosylation, and the glycosylation probably influences the folding and ultimate conformation of the glycoprotein. The general effects of glycosylation and restriction on different bacteriophage T4 gene expression and replication have been examined.

PLANT GLYCOPROTEINS

Review articles dealing with chemical and biochemical aspects of glycoproteins, and the involvement of the Golgi apparatus in the biosynthesis and secretion of plant glycoproteins and polysaccharides, have been published. The inhibition of the *in vitro* transfer of 2-acetamido-2-deoxy-D-glucose and D-mannose from UDP-2-acetamido-2-deoxy-D-glucose and GDP-D-mannose, respectively, into lipid-linked intermediates has been demonstrated in mung beans and carrots in the presence of amphomycin.

The antibiotic prevents the D-mannosylation of protein, but does not block the *in vitro* transfer of D-glucose from UDP-D-glucose to β-D-glucans. Infection of muskmelon *(Cucumis melo)* seedlings by the fungus *Colleto-trichum lagenarium* causes a ten-fold increase in the level of glycoproteins containing L-hydroxyproline. Observed increases in diseased plants accounted for a cell-wall enrichment of glycosylated L-hydroxyprolyl and L-seryl residues. The extent of accumulation of these glycoproteins was inhibited by treating the plants with ethylene, or growing them in the presence of *L-trans*-hydroxy-proline.

The increase in cell wall L-hydroxyproline-rich glycoproteins mediated by treatment with ethylene was paralleled by an increasing resistance of the host to the pathogen. Inversely, inhibition of the synthesis of these glycoproteins in diseased plants was correlated to an accelerated and more intense colonization of the host by the pathogen. An L-arabino-D-galactan-protein complex has been isolated from the style canal of *Gladiolus* by affinity chromatography on the D-galactose-binding lectin from the giant clam *Tridacna maxima*.

The procedure should prove useful in the isolation of polysaccharides and glycoproteins containing β-D-galactopyranosyl residues. Three globulins isolated from the seeds of *Lupinus augustifolius* have been shown to contain six, four, and two major sub-units, which are all glycosylated to varying degrees with residues of D-mannose, D-galactose, and 2-amino-2-deoxy-D-glucose. Proteolytic digestion of two of the glycoproteins has yielded glycopeptides (mol. wt. 5×10^3). In addition to hemicellulose, pectic substances, and cellulose, high levels of L-hydroxyproline have been detected in the cellulose fraction of the endosperm wall of rice *(Oryza sativa)*. Glycoproteins containing this amino-acid may be bound to the cellulose microfibrils. A series of oligosaccharides, containing up to four L-arabinosyl residues linked to L-hydroxyproline, has been isolated from the cell walls of *Phaseolus vulgaris.*

In cell suspensions of cultures of *P. vulgaris,* variations in the patterns of L-arabinosylation were observed only during the lag phase and early log phase of the cultures. A glycoprotein, containing only small amounts of L-hydroxy-proline and consisting of a single polypeptide chain, has been isolated from *P. vulgaris.* Treatment with alkali released two polypeptide chains, with most of the carbohydrate residues linked to L-serine on the polypeptide of lower molecular weight. Methylation analysis indicated the presence of 1,2-, 1,3-, or 1,5-substituted L-arabinofuranosyl residues, non-reducing terminal D-galactosyl residues, and a chain of (1 → 4)-β-D-glucosyl residues. Soybean 7S protein contains residues of D-mannose and 2-acetamido-2-deoxy-D-glucose.

After proteolytic digestion, three related glycopeptides were isolated and structural analysis has allowed a possible structure (3) to be deduced. The stem and root-infecting pathogen of soybean, *Phytophthora megasperma,* excretes glycoproteins. They have proved to be poor, non-specific elicitors of phytoalexin accumulation. Free and membrane-bound ribosomes prepared from wheat germ have binding sites for concanavalin A. Four glycoproteins (mol. wts. 1.4 – 2.4 × 10) were identified by sodium dodecyl sulphate-polyacrylamide gel electrophoresis in both classes of ribosomes. Membrane preparations from the alga *Prototheca zopfli* incorporate D-mannose and 2-acetamido-2-deoxy-D-glucose from their respective sugar nucleotides into acid labile glycolipids whose properties are consistent with those of dolichyl D-mannopyranosyl phosphate.

The intermediate appears to be involved in the formation of glycoproteins containing high proportions of D-mannosyl residues of the type present in more-evolved eukaryotes. Fluorescein-conjugated lectins and lactoperoxidase-catalysed iodination reactions have been used to locate the cell surface glycoproteins of the multi-cellular alga *Volvox carteri.*

CELL AND TISSUE GLYCOPROTEINS

Various glycopeptides containing L-fucose and 2-acetamido-2-deoxy-D-glucose, but no neuraminic acid, have been isolated from garfish olfactory nerve. A light-microscopical, immunocytochemical method for detecting myelin-associated glycoproteins in myelin and myelin-forming cells of developing rat nervous systems has been reported. When horse-radish peroxidase is activated with glutaraldehyde and then covalently coupled to wheat germ agglutinin, the conjugate is forty times more sensitive than the free horse-radish peroxidase in tracing the retrograde connections from the submandibular gland to the superior cervical ganglion of the rat.

Solubilized glycoproteins of purified rat brain myelin bind to immobilized concanavalin A and other lectins. Examination of proteins for D-glucosyl-L-lysine in normal human lens, senile and diabetic cataracts shows that D-glucosylation does not appear to be a primary factor in cataract formation. A small molecular weight oligosaccharide (16) attached separately at two sites near the *N*-terminus of bovine rhodopsin has been examined by a combination of methylation analysis, acetolysis, and enzymic studies. This structure is identical to a proposed intermediate in the biosynthesis of complex L-asparagine-linked oligosaccharides. Three oligosaccharides have been released from bovine rhodopsin by hydrazinolysis.

One of the components has an identical structure to (16) except for the absence of the L-asparagine residue, and structures [(17) and (18)] were established for the other two oligosaccharides.

β-D-Glc*p*NAc-(1 → 2)-α-D-Man*p*-(1 → 3)-β-D-Man*p*-(1 → 4)-β-D-Glc*p*NAc-(1 → 4)-β-D-Glc*p*NAc-LAsn
6
↑
1
α-D-Man*p*

(16)

β-D-Glc*p*NAc-(1 → 2)-α-D-Man*p*-(1 → 3)-β-D-Man*p*-(1 → 4)-β-D-Glc*p*NAc-(1 → 4)-D-Glc*p*NAc
6
↑
1
α-D-Man*p*
6/3
↑
1
α-D-Man*p*

(17)

β-D-Glc*p*NAc-(1 → 2)-α-D-Man*p*-(1 → 3)-β-D-Man*p*-(1 → 4)-β-D-Glc*p*NAc-(1 → 4)-D-Glc*p*NAc
6
↑
1
α-D-Man*p*-(1 → 3)-α-D-Man*p*
6
↑
1
α-D-Man*p*

(18)

Bovine D-galactosyltransferase (lactose synthase) has been used to label glycoproteins of mouse retina by transferring D-[^{14}C]galactose to non-reducing 2-acetamido-2-deoxy-D-glucosyl residues. Information was obtained concerning changes in the patterns of glycoproteins during the development of mouse neuronal retina. Photoreceptor loss in mice with retinal degeneration was paralleled by the disappearance of certain glycoproteins.

Sixteen glycoproteins have been separated from rat brain by a combination of hydroxyapatite and lectin affinity chromatography. A simple reproducible procedure for the fractionation of brain glycoproteins using hydroxyapatite-gel column chromatography has been developed. Whole rat brain on digestion with proteolytic enzymes releases an *N*-glycosidic glycopeptide fraction which has been purified using immobilized concanavalin A.

One of the glycopeptides has an unusual sugar sequence (19) that has been shown to be part of the determinant of the so-called X-antigen.

$$\begin{array}{c} \alpha\text{-Neu}p\text{5Ac-}(2 \rightarrow 3)\text{-}\beta\text{-D-Gal}p\text{-}(1 \rightarrow 4)\text{-}\beta\text{-D-Glc}p\text{NAc-}(1 \rightarrow \\ \qquad\qquad\qquad\qquad\qquad\qquad 3 \\ \qquad\qquad\qquad\qquad\qquad\qquad \uparrow \\ \qquad\qquad\qquad\qquad\qquad\qquad 1 \\ \qquad\qquad\qquad\qquad\qquad\qquad \alpha\text{-L-Fuc}p \end{array}$$

(19)

A glycoprotein antigen (mol. wt. 7.6 × 10^4) comprising three sub-units of identical size has been isolated from a cell-free extract of a human lung adeno-carcinoma. All the properties of this glycoprotein are very similar to those of another antigen purified from a separate human lung tumour. A patient with an inherited deficiency of lysosomal N^4-(2-acetamido-2-deoxy-β-D-glucopyranosyl) hydrogen L-asparaginate amidohydrolase accumulated, in the liver, oligosaccharides whose structures are related to the inner-core region of a number of different glycoproteins. They have been characterized by g.l.c.–m.s. of the appropriate derivatives as N^4-(acetamido-2-deoxy-(3-D-glucopyranosyl) hydrogen L-asparaginate and N^4-[O^2-α-D-mannopyranosyl-(1 → 6)-O-β-D-mannopyranosyl-(1 → 4)-O-2-acetamido-2-deoxy-D-glucopyranosyl-(1 → 4)-2-acetamido-2-deoxy-β-D-glucopyranosyl] hydrogen L-asparaginate. The hepatic storage of glycopeptides and glycolipids in a cat affected by G_{M_1} -gangliosidosis has been reported.

Two major oligosaccharide chains, derived from the glycopeptides, had molecular weights of 1.8 × 10^3 and 1.35 × 10^3 and contained residues of D-galactose, D-mannose, and 2-acetamido-2-deoxy-D-glucose in the molar proportions 2.0:3.1:4.1 and 1.0:2.2:2.7, respectively. Mechanisms by which newly synthesized glycoproteins are transferred from the hepatocytes to the bile have been reported. The effects of the bacterial tripeptide, leupeptin, on the lysosomal degradation of glycoproteins in perfused rat liver have been examined. Leupeptin inhibits digestion of endocytosed proteins such as asialofetuin within the lysosomes. A hydrophilic glycoprotein with insulin-binding activity has been isolated from rat liver by affinity chromatography on columns of immobilized insulin.

Treatment of the glycoprotein with either neuraminidase or trypsin caused a decrease in the insulin-binding activity. The glycoprotein and glycosaminoglycan content of liver in which hepatic cancer has been induced with 3′-methyl-4-dimethylaminoazobenzene show a gradual increase in quantity with time. Various structural features of proteins, such as polypeptide size and net charge, are known to influence their *in vivo* degradative rates.

The presence of covalently bound carbohydrate has been shown to be a third feature of protein structure that correlates with protein half-lives within cells. Glycoproteins have been found to be degraded more rapidly than unfractionated proteins or fractions enriched for non-glycoproteins in rat liver, muscle, brain, and kidney, and also in human fibroblasts grown in culture. A DNA double-helix unwinding glycoprotein, isolated from roe-deer liver, contains residues of L-fucose, D-xylose, and D-mannose together with two unidentified sugars.

The glycoprotein is involved in the unwinding of adenine-thymine deoxyribose base pairs in DNA. From results obtained using a sandwich technique with immunofluorescence, it appears that Tamm-Horsfall glycoprotein is present in hamster kidney slices only in the ascending limb of the loop of Henle and the distal convoluted tubule. The glycoprotein may operate as a barrier to decrease the passage of water molecules by trapping them at the membrane of the cells.

The structure of the oligosaccharide chains in Tamm-Horsfall glycoprotein has been studied using the native glycoprotein as an acceptor for calf kidney sialyltransferase (CMP *N*-acetylneuraminic acid:glycoprotein sialyltransferase) in competition studies with desialylized α_1-acid glycoprotein and fetuin. The results indicate that the carbohydrate structures of the Tamm-Horsfall glycoprotein are most probably not identical with the *N*-glycosidically-bound oligosaccharide chains present in the other two glycoproteins. Localization of glycoproteins in lysosomes and prelysosomes of a kidney epithelial cell line from monkey has been established by use of high resolution autoradiography after incorporation of L-[^{3}H]fucose and by a semicarbazide-silver proteinate staining method after oxidation with periodate.

The glycoproteins appear to be transported through different components of the vacuolar system of the cell in a sequential fashion rather than *via* independent pathways. The majority of the glycoproteins in the lysosomal membranes do not have acid hydrolase activity. A homogeneous sulphated glycoprotein composed of protein (40%), carbohydrate (53%), and sulphate (1.6%) and containing D-galactose, 2-amino-2-deoxy-D-galactose, 2-amino-2-deoxy-D-glucose, and neuraminic acid as the principal sugars, has been isolated from the small intestine of the rabbit. Resonance Raman scattering spectroscopy within the visible absorption band of uteroferrin, a glycoprotein of the porcine uterus, results in an intense Raman spectrum that is similar to that of Fe^{3+} transferrin, indicating similarities in the metal binding sites of the two macromolecules. Sulphated glycoproteins (mol. wts. > 2.4×10^5) are biosynthesized, at least in the uterine luminal epithelial cells of the rat, under the stimulation of oestrogen but not progesterone, which inhibits their biosynthesis. Glycoproteins having chemical and antigenic properties similar to those of carcinoembryonic antigen have been isolated from human prostate.

Carcinoembryonic antigen prepared by identical procedures from three different liver metastases has been shown to have different degrees of homogeneity on electrophoresis and crossed immunoelectrophoresis, and also has different antigenic activities. A minor isoenzyme of human prostatic acid phosphatase (pI 5.5) has been separated from the major isoenzyme (pI 4.9). Immunological similarities between the two enzymes were demonstrated. Ribonuclease isolated from HeLa cell lysosomes has been characterized as a glycoprotein on the basis of its binding to immobilized concanavalin A. Glycoproteins associated with two rat ascites mammary adenocarcinoma cell lines differ in their behaviour towards lectins and their xenotransplantability.

Differences were observed in the complexity and size of their oligosaccharide chains. The maximum release of a sialoglycoprotein, but not a sialoglycolipid, from a murine tumour cell by neuraminidase from *Clostridium perfringens* was obtained only in the presence of physiological concentration of Ca^{2+}. Glycoproteins, isolated from granulation tissue and fractionated by gel filtration and ion-exchange chromatography, inhibit collagen synthesis in rat embryo tendon cells. The glycoproteins may function in the feedback regulation of granuloma development. A glycopeptide, composed of thirty-nine ammo-acid residues isolated from pig, ox, and sheep pituitary, exhibits a high degree of sequence conservation.

The glycopeptide arising from each source contains one oligosaccharide chain attached to L-asparagine at position 6 in the sequence and contains five residues of 2-amino-2-deoxy-D-glucose. Part of the microheterogeneity shown by isoelectric focusing of glycoproteins from the electric organ of the eel *Electrophorus electricus* is attributable to variations in the content of neuraminic acid. Treatment of the glycoproteins with neuraminidase reduces the number from ten to four.

The microheterogeneity is due to post-translational modification of oligosaccharides on a common polypeptide backbone. A glycoprotein (mol. wt. 1.5×10^5) has been isolated from whole human platelets by affinity chromatography on immobilized wheat-germ agglutinin. Extraction of the whole platelets with lithium di-iodosalicylate released three major glycoprotein fractions. A glycoprotein, designated thrombospondin, containing three disulphide-linked chains, each having a molecular weight of 1.4×10^5, is released from human blood platelets in response to thrombin.

Isolation of this glycoprotein was achieved by affinity chromatography on immobilized heparin. Its position in the cytoplasmic granule fraction of the platelets, from which it is released with known coagulation factors, suggests a physiological role related to the coagulation scheme. A novel glycoprotein containing *N*-glycolylneuraminic acid has been isolated from the eggs of rainbow trout *(Salmo irideus)*. All of the carbohydrate chains, which contain residues of D-galactose and 2-acetamido-2-deoxy-D-galactose, are terminated by the disaccharide *O*-α-*N*-glycolylneuraminyl-(2 → 8)-*N*-glycolylneuraminic acid. The identity of this disaccharide has been confirmed by methylation and g.l.c.–m.s. analysis.

The embryos of the sea urchin *(Strongylocentrotus purpuratus)* synthesize several classes of sulphated glycoproteins during gastrulation. During development of the embryos, glycoproteins containing both unsulphated and sulphated *N*-glycosidically-linked oligosaccharides are synthesized *via* lipid-linked intermediates. The biosynthesis of these molecules appears to be a prerequisite to the differentiation and morphogenesis that occurs during gastrulation. The structures [(20)-(22)] of three glycopeptides, each having a high level of D-mannosyl residues and present in Chinese hamster ovary cells, have been determined. Oligosaccharides with identical structures have been found in other soluble glycoproteins such as Unit A glycopeptide of thyroglobulin, Taka-amylase (a microbial α-amylase), and ovalbumin.

The structure (23) of the oligosaccharide portion of a low molecular weight lipid-linked intermediate, which is primarily involved in the assembly of a major lipid-linked oligosaccharide of Chinese hamster ovary cells, has been established.

R^1-α-D-Man*p*-(1 → 6)-α-D-Man*p*-(1 → 6)–β-D-Man*p*-(1 → 4)-β-D-Glc*p*NAc-(1 → 4)-D-Glc*p*NAc-L-Asn

3 ↑ α-D-Man*p* | R^2 (branch on the first α-D-Man*p*)

3 ↑ α-D-Man*p* 2 ↑ 1 α-D-Man*p* ↑ R^3 (branch on the β-D-Man*p*)

(20) $R^1 = R^2 = R^3$ = α-D-Man*p*-(1 → 2)

(21) $R^1 = R^3$ = α-D-Man*p*-(1 → 2), R^2 = H

(22) $R^1 = R^2 = R^3$ = H

Antigens isolated from the eggs of *Schistosoma mansoni* contain at least six glycoproteins capable of binding to immobilized concanavalin A. These glycoproteins are able to elicit lymphocyte blastogenic responses. The same glycoproteins can also be isolated by affinity chromatography on immobilized anti-(soluble egg antigen) antibody. An oligosaccharide, containing residues of D-glucose, D-mannose, and 2-acetamido-2-deoxy-D-glucose in the molar ratio 3 : 9 : 2, is the major or exclusive species transferred *en bloc* from the lipid-linked intermediate to proteins in intact, uninfected chick embryo fibroblasts.

The initial stages of processing of this oligosaccharide species after transfer involve the sequential removal of the three D-glucosyl residues. The three D-glucosyl residues are attached as a trisaccharide to a D-mannosyl residue at one of the non-reducing termini of the oligosaccharide, and the 2-acetamido-2-deoxy-D-glucosyl residues are located at the reducing terminus. Treatment with tunicamycin affects the binding and degradation of low-density lipoproteins by cultured human fibroblasts, probably by inhibition of glycosylation of the glycoprotein receptor for the lipoproteins.

The large-scale purification of neutral and acidic glycopeptides from human diploid fibroblasts has been reported. D-Mannose 6-phosphate has been identified as a constituent of glycoproteins that inhibit the assimilation of bovine testicular β-D-galactosidase by human skin fibroblasts. Phosphorylated oligosaccharides have been isolated from the inhibitor and D-mannose 6-phosphate appears to be associated with (1 → 2)-linked D-mannosyl residues in the glycoprotein.

The recognition marker encompasses a structure larger than a single D-mannosyl 6-phosphate residue. D-Mannose 6-phosphate is present as a recognition marker in human spleen β-D-glucuronidase.

α-D-Man*p*-(1 → 2)-α-D-Man*p*-(1 → 2)–α-D-Man*p*-(1 → 3)-β-D-Man*p*-(1 → 4/3)-β-D-Glc*p*NAc-(1 → 4)-D-Glc*p*NAc

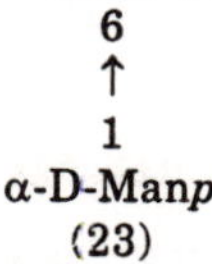

(23)

Differentiation antigens and glycoproteins of lymphocytes have been the subject of a review.

Human peripheral blood lymphocytes have been fractionated into sub-classes by an affinity chromatographic technique involving immobilized lentil lectin. Two groups of cells were isolated, differing both in their surface properties and in certain *in vitro* functional capabilities. Glycoproteins have been purified from rat thymocytes and spleen. Monoclonal antibodies to these glycoproteins have been prepared in order to study the relationship between the thymocyte glycoproteins and rat I_a-antigens.

Metabolic labelling of the carbohydrate residues of glycoproteins from mitogen-stimulated mouse lymphocytes has demonstrated that by using a combination of different labelled sugars, such as L-fucose, D-mannose, and D-galactose, it is possible not only to differentiate between activated T- and B-cells, but also between certain sub-populations. The method has been used to discriminate early- and late-responding concanavalin A-reactive thymocytes.

The interactions of peanut agglutinin with human thymocytes, peripheral blood lymphocytes, and peripheral blood cells of various types of leukaemia have been investigated.

The majority of human thymocytes bind to the lectin. The validity of using the lectin as a marker of immature blood cells and its potential clinical application have been discussed. Human fibroblast and leukocyte interferons show a strong affinity for the copper chelate of bis-carboxymethylaminoagarose.

Human leukocyte interferon adsorbs non-specifically to immobilized hyperimmune serum immuno- globulin G. The interferon can be recovered by washing the adsorbent with 1,2-dihydroxyethane. An improved procedure for the isolation of interferons produced by mouse Ehrlich ascites tumour cells infected with Newcastle disease virus provides interferons of three size classes (mol. wts. 3.3×10^4, 2.6×10^4, and 2.0×10^4). The differences in size may be due, in part, to unequal glycosylation. Glycosylation of interferon, produced by mouse L cells after induction by Newcastle disease virus, has been inhibited by tunicamycin with no loss of antiviral activity. Human lymphoid cells, when stimulated *in vitro* with concanavalin A or phytohaemagglutinin, release lymphotoxins, a group of cytologically active molecules. They have been separated into various classes and sub-classes by gel and ion-exchange chromatography, and also by affinity chromatography on immobilized lectins, thus indicating their glycoprotein nature.

Treatment of the glycoproteins with glycosidases has no effect on the antigenicity of these molecules and the carbohydrate moieties do not appear to be necessary for their *in vitro* lytic activity. D-[^{14}C]Glucose has been incorporated into lipid-linked oligosaccharides containing up to eleven monosaccharide units by explants of lactating rabbit mammary glands. Considerable redistribution of the label occurred into D-mannosyl and 2-acetamido-2-deoxy-D-glucosyl residues. A hypothesis that has been presented may give some meaning to the various changes in carbohydrate attached to proteins and lipids in malignant cells. Extensive changes in the populations of glycolipids and glycoproteins as seen in malignant cells may result in persistent cell division, decreased intercellular adhesiveness, altered transport, altered immunogenicity, and loss of specialized function.

CELL-SURFACE AND MEMBRANE GLYCOPROTEINS

The continued interest in this field of study is reflected in the large number of review articles which have been published. The proceedings of an ICN-UCLA Symposium, in Colorado in 1977, have been published under the title 'Cell Surface Carbohydrates and Biological Recognition', and the proceedings of another symposium on cell membranes of normal and neoplastic cells has appeared as 'Cell Surface Carbohydrate Chemistry'. Cell-surface antigens, with special reference to the histocompatibility antigens, have been reviewed.

A comprehensive survey of the metabolism of cell-surface components in both eukaryotic and prokaryotic cells has been published. Other reviews have dealt with the structure, biosynthesis, and biological functions of cell-surface glycoproteins, the purification of cell-membrane glycoproteins, the role of fibronectin in malignant transformation, fibronectin-adhesive glycoproteins of cell-surface blood, and protein-bound oligosaccharides of cell membranes. The organization of the glycoproteins and glycolipids on the surface of plasma membranes and their participation in cell contacts, transformation, and regulation of the cell cycle have been discussed.

A model of intracellular recognition is presented and the role of the surface matrix of the cell is discussed. Neuraminic acid residues of glycoproteins have been spin-labelled in order to study the effects of various binding processes upon the mobility and environment of the sugar in tissue sections from human colon and erythrocyte components. Quantitative data obtained with fluorescent-substituted lectins is in agreement with data obtained for radioactively labelled lectins for the determination of cell-surface glycoconjugates.

A survey of the application of a lectin-based method to the light-microscopical localization of D-mannose-rich glycoproteins and a range of methods for the ultrastructural characterization of glycoconjugates on cell surfaces has been reported. Several differences, including the presence of an additional glycoprotein, have been noted on the surface labelling of hamster variant cell lines when compared to wild cells. The accompanied resistance to the cytotoxic effect of con-canavalin A provides support for the view that modified cell-surface glycoproteins are directly involved in the complex concanavalin A-resistant phenotype, which includes changes in some fundamental biological properties. Hamster-cell fibronectin contains two sub-units (mol. wt. 2.3×10^5 each) in a disulphide-bonded dimer.

The disulphide bonds, the single free thiol group and the carbohydrate units, are distributed randomly along domains that can be separated by proteolytic cleavage. A fragment (mol. wt. 2.0×10^5) containing all or most of the carbohydrate side chains, but relatively little L-cystine, contains the gelatin-binding site. Intact disulphide bonds are required for the binding of fibronectin to cells and to gelatin, and blockage of the free thiol groups prevents the binding to cells. Treatment of fibronectin with cyanogen bromide and subtilisin releases two fragments (mol. wts. 5.0×10^4 and 3.0×10^4), which have been isolated by affinity chromatography on immobilized gelatin.

Since the higher molecular weight fragment, which was derived from the *C*-terminus of the protein chain of the glycoprotein, also mediated the adhesion of fibroblasts to collagen, it contains both the collagen- and cell-binding sites in the fibronectin molecule. Similar results have been independently published. Fibronectin is a sensitive target for leukocyte proteases under physiological conditions. Because of the sensitivity of fibronectin to neutral proteases from inflammatory cells and its apparent role in tissue organization, it is possible that the morphological and functional changes accompanying inflammation may reflect destruction of fibronectin in addition to other components of the connective tissue matrix.

Plasma transglutaminase (Factor XIII*a*) catalyses the cross-linking of fibronectin and collagen. The reaction may be important in wound healing and tissue repair. Heparin induces the transition of fibronectin from a globular to an elongated form, capable of forming filamentous precipitates which adsorb native collagen. This reaction is inhibited by hyaluronic acid and putrescine. Peritoneal macrophages are capable of the *in vitro* bio-synthesis and secretion of a protein that by immunological, biochemical, and functional criteria is identical to fibronectin. A change in glycoprotein components of the plasma membrane in liver cells is part of the complex adaptation to diet containing high fat and high D-glucose levels. Membranes from rats fed on a high fat diet bind significantly less [^{125}I]concanavalin A than do those from rats fed on a fat-free diet.

Analysis of delipidated membranes show significantly lower values for neuraminic acid, 2-amino-2-deoxy-D-hexose, L-fucose, and D-mannose, but not D-galactose. Differences between glycoproteins from diabetic and normal rat-liver membranes have been described. The hepatic-binding protein, which binds to terminal D-galactosyl residues of asialoglycoproteins, and the receptors for *Ricinus communis* lectin have been localized simultaneously on the plasma membrane of isolated rat hepatocytes. The specificity of the combining site of the rabbit hepatic lectin has been explored by a competitive binding assay using a [^{3}H]-labelled precursor blood-group I substance with terminal non-reducing β-D-galactosyl residues and the lectin immobilized on macroporous agarose.

Data on the relative inhibiting power of various polysaccharides, glycoproteins, and oligosaccharides, and also monosaccharides and their glycosides have been recorded. A recycling mechanism may exist wherein the receptor is not destroyed in the lysosomes and remains available for subsequent reinsertion into the hepatocyte plasma membrane as part of a continuing process for the clearance of D-galactosyl-terminated ligands in the circulation.

The existence and distribution on parenchymal and non-parenchymal hepatic cells of separate and distinct receptors for selectively modified proteins has been reported. Whereas D-galactosyl-terminated glycoproteins were found preferentially in the hepatocytes, glycoproteins terminated in D-mannosyl or 2-acetamido-2-deoxy-D-glucosyl residues were recovered largely in the sinusoidal cells of the liver. These findings have been substantiated by the observation that the selective hepatic uptake of the lysosomal glycosidase β-D-2-acetamido-2-deoxyhexosidase, which is a glycoprotein with terminal D-mannosyl residues, is by the sinusoidal cells. Isolated specific cell-surface components of rat hepatocytes involved in intracellular adhesion do not appear to be receptors for either fibronectin or heparin.

The endocytosis of human urine α-D-2-acetamido-2-deoxy-glucosidase is mediated by at least three different cell-surface receptors on rat hepatocytes, involving the recognition of D-galactosyl, D-mannosyl 6-phosphate, and 2-acetamido-2- deoxy-D-glucosyl residues. A binding protein exists in rat-liver microsomes with specificity for β-D-glucuronidase. Once removed from the membrane, the protein undergoes self-aggregation forming large macromolecular complexes. The binding protein shows organ localization, being present in liver and kidney but not spleen.

Lysolecithin has been used to solubilize the D-galactosyl- and 2-acetamido-2-deoxy-D-glucosyl-transferases from rat-liver microsomal membranes. A method, which is not dependent on cell fractionation, has been developed to analyse the sub-cellular location of membrane glycoproteins. The method uses the conversion of glycoproteins containing neuraminic acid residues into forms having a differing mobility in a two-dimensional electrophoretic system when intact cells are treated with neuraminidase. Using this method, it is possible to determine whether or not a glycoprotein is present only on the plasma membrane or whether it is shared by membrane systems.

Most of the glycoproteins of rat hepatoma tissue culture cells are shared by the plasma membrane and by an intracellular membrane system. Treatment of hepatoma tissue cells with trypsin at a concentration that results in cell aggregation releases a glycopeptide (mol. wt. 5.5×10^4) from a membrane glycoprotein (mol. wt. 8.5×10^4). The glycopeptide is much more resistant to further degradation by proteases and also accumulates in the medium during normal cell growth. Plasma membranes of beef kidney glomeruli have been prepared and found to contain 9% carbohydrate. Laminin, a high molecular weight non-collagenous glycoprotein consisting of at least two polypeptide chains (mol. wt. 2.2×10^5 and 4.4×10^5) and joined by disulphide bonds, has been isolated from a mouse tumour.

Immunological studies have indicated that this glycoprotein is distinct from fibronectin and is produced by various cultured cells. A glycopeptide consisting of fifteen amino-acids and two oligosaccharide units, one linked to L-hydroxylysine and the other to L-asparagine, has been isolated from bovine glomerular basement membrane. The oligosaccharide side chains appear to be attached to the polypeptide backbone within a space of not more than twelve amino-acid residues. A low apparent turnover in organ culture for most of the high molecular weight human intestinal brush border membrane glycoproteins has been reported.

Some of the glycoproteins co-migrate on polyacrylamide gel electro-phoresis with enzymic activities of known intestinal glycosidases. Rabbit intestinal brush borders incorporate [^{35}S]sulphate into glycoprotein and glycolipid as sulphate ester, which is the major contributor to the anionic character of the membrane. D-Glucose is the predominant sugar component of the glycoprotein fraction, which also contains smaller amounts of D-mannose, D-galactose, L-fucose, 2-amino-2-deoxy-D-glucose, and 2-amino-2-deoxy-D-galactose. Biosynthetic studies have indicated that the glycoproteins and glycolipids are actively synthesized and degraded within the mature small intestinal enterocyte. Colchicine, given intraperitoneally to rats before and after injection of L-[^{3}H]fucose, markedly inhibits the incorporation of labelled glycoprotein into microvillus membrane with a corresponding increase in incorporation into the lateral basal membrane.

By using specific immunochemical techniques in combination with two-dimensional electrophoresis, it is possible to identify a set of four glycoprotein sub-units present only in placental brush border membranes. A group of seven placental glycoproteins was found to be common to liver and kidney membranes. A two-dimensional mapping procedure, involving isoelectric focusing and sodium dodecylsulphate electrophoresis, has been used in a comparative study of the surface anatomy of baby hamster kidney cells and three clones resistant to ricin.

Abnormal patterns were observed in all three mutant clones indicating different mechanisms of ricin resistance; glycoproteins, which may be involved in cellular reactions, were identified. Tunicamycin causes selective inhibition of protein glycosylation in baby hamster kidney fibroblasts. Membrane glycoproteins continue to be transported to the cell surface despite the absence of oligosaccharides linked to L-asparagine. The localization of brain glycoprotein NSA3 has been studied by indirect immunofluorescence of sections of rat brain. Grey matter membranes from calf brain catalyse the transfer of *NN′*-diacetyl-chitobiose from dolichyl diphosphate to at least one membrane-associated polypeptide acceptor, *via* an *N*-glycosidic linkage. Axolemma-enriched membranes from bovine myelinated axons catalyse the synthesis of doliehyl-linked derivatives of D-mannose and 2-acetamido-2-deoxy-D-glucose.

The lipid-mediated glycosylation involves P′-dolichyl phosphate D-mannose and a dolichyl phosphoryl oligosaccharide consisting of eight hexose units terminated at the reducing end with *NN′*-diacetylchitobiose. The dolichyl-bound oligosaccharide is transferred to an L-asparagine residue in endogenous membrane polypeptide acceptors to give a glycoprotein (mol. wt. 2.4×10^4). Evidence for the transfer of an oligosaccharide containing D-glucosyl, D-mannosyl, and two 2-acetamido-2-deoxy-D-glucosyl residues from exogenous glycosylated oligosaccharide lipid to endogenous membrane glycoproteins by a membrane preparation from calf brain has been reported. D-[^{14}C]Glucose is incorporated into protein acceptors of bovine retinal membranes. Treatment of the resulting glycoprotein with α-amylase released maltose, indicating the presence of (1 → 4)-linked α-D-glucosyl residues. The externally disposed polypeptides in chick synaptic plasma membranes are glycoproteins, containing non-reducing terminal D-galactosyl residues.

A myelin-associated glycoprotein on the surface of rat, bovine, and human myelin membranes is one of the principal receptors for concanavalin A. A complex glycoprotein I, consisting of two polypeptides (mol. wt. 2.1×10^5 and 1.5×10^5) has been isolated from human platelet membranes. Glycocalicin, a soluble loosely bound membrane glycoprotein related to the glyco-protein I system, was also purified. The two polypeptides of glycoprotein I produce essentially the same peptides on tryptic hydrolysis, but these are different from those of glycocalicin.

Human platelet membranes contain sixteen glycoproteins having molecular weights ranging from 3.5×10^4 to 3.0×10^5, which have been resolved by polyacrylamide gel electrophoresis. Essentially all of the carbohydrate is exposed to the exterior surface of the membrane. A glycoprotein, with binding properties for the Fc fragment of immunoglobulin G, has been isolated from human platelets. The F_c fragment binding-glycoprotein, which has a molecular weight of 2.55×10^5, and forms *in vitro* complexes with immunoglobulin G, dissociated into sub-units of molecular weight 5.0×10^4 upon reduction. Wheat-germ agglutinin binds to the surface of human platelets and leads to their agglutination.

Unlike other aggregating agents, the clumping ability of platelets by the lectin is markedly reduced on removal of cell-surface neuraminic acid. Since the terminal neuraminic acid residues have been implicated in the clearance of platelets from the circulation, the lectin may prove to be a useful tool in the exploration of those clinical conditions in which platelet survival is shortened. Wheat-germ agglutinin also stimulates platelets by a mechanism which closely mimics the action of thrombin, and is independent of intracellular cross-linking. The lectin, which binds mainly to glycoprotein I on the platelet membrane, increases the subsequent binding of thrombin to fixed platelets. The platelets of patients with T_n syndrome contain abnormalities in the glycoproteins of the membranes. These abnormalities may arise from the incomplete synthesis of oligosaccharide chains since the platelets also exhibit low T transferase (UDP-D-galactose: 2-acetamido-2-deoxy-D-galactose 3-β-D-galactosyltransferase) activity. A complete absence of glycocaliacin and a significant reduction of membrane-bound glycoprotein I in the platelets of patients with Bernard-Soulier syndrome has been reported.

In the case of platelets from patients with Glanzmann's thrombasthenia, a significant reduction of glycoprotein II (mol. wt. 1.2×10^5) and glycoprotein III (mol. wt. 1.0×10^5) were observed, whereas patients with storage pool disease had increased levels of glycoprotein IV (mol. wt. 8.5×10^4). Glycoproteins isolated from membrane-enriched fractions of platelets from dogs with thrombobasthenic thrombopathia and von Willebrand's disease and from normal controls give patterns on polyacrylamide electrophoresis similar to those for the analogous human conditions. The major surface glycoprotein of normal human blood granulocytes is markedly reduced in granulocytes from three patients with a chromosomal abnormality in all or most bone marrow mitoses. Solubilized surface proteins from normal human lymphocytes have been obtained by mild digestion with trypsin, and the binding of these components to labelled *Ricinus sanguineus* has been studied.

A method for the direct visualization of lymphocyte membrane receptors has made use of fluoresceinyl derivatives of glycosylated cytochemical markers such as bovine serum albumin covalently attached to either D-mannose, 2-acetamido-2-deoxy-D-galactose, L-fucose, or lactose. Glycoproteins from human or mouse lymphocytes and labelled either by the D-galactose oxidase or lactoperoxidase techniques have been isolated by affinity chromatography on the immobilized haemagglutinin from *Helix pomatia*. The major cell surface glycoprotein (mol. wt. 1.5×10^5) from both normal and thymus-derived human lymphocytes was responsible for almost all the binding to the haemagglutinin. This was not observed with various B cells at different steps of differentiation.

However, two of four B lymphoma lines and a myeloma line contained instead a glycoprotein (mol. wt. 2.0×10^5) capable of binding to the haemagglutinin. Studies on the mouse lymphocyte lines showed that the T-derived lymphocytes from both normal and malignant cells, but not from adult B-cells expressed a major glycoprotein (mol. wt. 1.3×10^5) capable of binding to *H. pomatia* haemagglutinin. An affinity chromatographic method based on immobilized *Helix pomatia* haemagglutinin has been used to differentiate mouse natural killer cells from other lymphoid sub-populations.

A major membrane glycoprotein (mol. wt. 5.4×10^4) has been isolated from membrane preparations of human B-type lymphoid lines. Its expression on cultured human cell lines and normal lymphocytes, as well as various types of leukaemic cells, has been determined. Mouse B- and T-cells differ in the relative amounts of different types of carbohydrate units present in their cell-surface glycoproteins. The major oligosaccharide component in both cell types, which was labile to alkali, was identified as 2-acetamido-2-deoxy-3-O-β-D-galactosyl-D-galactose. The surface glycoproteins of rat lymphoid cells have been labelled after periodate oxidation followed by reduction with sodium borotritide.

The labelled glycoproteins were studied for binding to lentil lectin, and for immunoprecipitation or affinity chromatography/immunoadsorption with antibodies to rat cell-surface antigens. All of the labelled glycoproteins have been identified as previously characterized differentiation antigens of rat thymocytes or of B- and T-lymphocytes. Direct evidence has been obtained for the binding of diphtheria toxin to specific glycoproteins from the plasma membranes of guinea-pig lymph node cells. These glycoproteins also bind to immobilized lentil lectin and are not detectable in cell membranes of mouse L cells that are resistant to the toxin. A glycoprotein (mol. wt. 1.0×10^5), which binds to diphtheria toxin, has been isolated from the surface of lymph nodes and thymus cells of hamsters. Lectins such as concanavalin A do not appear to block the binding of toxin to receptors on mammalian cells, although they appear to prevent the internalization of the toxin.

In studies on the effects of diphtheria toxin receptor on Chinese hamster cells, concanavalin A, wheat-germ agglutinin, and a glycopeptide from ovalbumin all acted as inhibitors of the cytotoxic effect. A decreased affinity of toxin for a cell-surface receptor has been suggested, since a mutant was isolated having ten to fifteen-fold more resistance to the toxin than wild-type cells. Different types of leukaemic cells have been surface-labelled by sequential treatment with neuraminidase, D-galactose oxidase, and sodium borotritide. Analysis of the surface glycoprotein profile by polyacrylamide gel electrophoresis has provided a method for the classification of leukaemic cells according to cell type and the stage of differentiation. Specific changes in the pattern of surface glycoproteins of a human promyelocytic leukaemia cell line during morphological and functional differentiation have been observed.

The major surface glycoprotein (mol. wt. 1.6×10^5) after induced differentiation is replaced by another glycoprotein (mol. wt. 1.3×10^5) having properties similar to those of a major surface glycoprotein from human blood granulocytes. Surface glycoproteins, including the receptor for immunoglobulin E, from rat basophilic leukaemia cells, bind to immobilized lentil lectin. The immobilization of *N*-acetylneuraminidase and its potential application in the modification of cell surfaces of leukaemia, mouse thymus, and spleen lymphocytes has been described. Structural and functional aspects of differentiation antigens and glycoproteins of lymphocytes have been reviewed. Serologically and biochemically pure preparations of HLA-A and -B antigens have been isolated from a series of immunoadsorption columns constructed from monoclonal antibodies having specificity for these antigens.

Similar methodology has been used in the preparation of the H-2K^k antigen, which is expressed at levels of 2×10^6 molecules/cell by a murine lymphoma. Purified preparations of the histocompatibility-Z antigen from mouse liver contain two moles of alloantigenic heavy chain per mole of β_2-microglobulin. Denaturing conditions were required to remove the β_2-microglobulin component. Peptides have been isolated and the amino-acid composition recorded of peptides arising from tryptic and chymotryptic hydrolysis of the histocompatibility antigen HLA-B7. The complete amino-acid sequence of the papain-solubilized heavy chain of the antigen consists of 271 amino-acid residues with a single glycan moiety linked at position 86 by an N-glycosidic bond. Significant homology was observed between certain regions and immunoglobulin G-constant domains and β_2-microglobulin.

Major differences between amino-acid sequences of murine H-2K^b antigen and HLA glycoprotein occur within delineated regions. Alveolar macrophages bind glycoconjugates having D-mannosyl, D-glucosyl, or 2-acetamido-2-deoxy-D-glucosyl residues in a non-reducing terminal position. The interaction between α-macroglobulin-trypsin complexes and receptors on the surface of rabbit alveolar macrophages, and the dependency of the interaction on divalent cations has been described. Proteases that bind to α-macroglobulin can be rapidly internalized *via* specific cell-surface receptors and then degraded, thus preventing potentially injurious proteolytic activity. A relationship between rabbit macrophage plasma-membrane glycoproteins and rabbit lung lavages, which contain a glycoprotein inhibitor of phagocytosis, has been demonstrated immunoelectrophoretically. Five glycoproteins of adrenal chromaffin granules have been isolated by lectin affinity chromatography.

The D-galactose oxidase labelling technique has revealed that at least the carbohydrate portion of the major glycoprotein is present on the inner side of the granule membranes. Conjugates of horse-radish peroxidase with lectins or with cholera toxin have been used for the cytochemical detection of carbohydrates on the cell surfaces of cultured neurons and neuroblastoma cells. The lectins and the toxin undergo endocytosis in the cisternae and vesicles of the Golgi-endoplasmic reticulum-lysosome complex. Treatment of neuroblastoma N2*a* cells with N^6O^2-dibutyryl adenosine 3′, 5′-cyclic monophosphate and theophylline results in changes in the overall degree of glycoproteins and changes in the turnover of glycoproteins (mol. wt. 1.0×10^5). The organization of binding sites on mouse neuroblastoma cells for fluorescein-labelled lectins has been reported.

Receptors for concanavalin A, wheat-germ agglutinin, and *Ricinus communis* lectin redistribute and become internalized, although the mode of redistribution differs from concanavalin A to the other two lectins. Solubilized receptors from mouse submaxillary glands have been adsorbed on immobilized concanavalin A and a ligand-binding assay performed on the immobilized receptor. Of three different groups of glycoproteins which are released from the surface of TA3-H*a* murine adenocarcinoma ceils, two groups contain epiglycanin with receptors for *Vicia graminea* lectin and are believed to be responsible for the ability of this line to kill allogeneic mice.

The cellular origin of these glycoproteins has been studied. Unequivocal evidence for the presence in epiglycanin of 2-acetamido-2-deoxy-3-*O*-*β*-D-galactopyranosyl-D-galactosyl units has been presented. Using an indirect immunoperoxidase technique an antigen, immunologically related to a glycoprotein of mouse mammary tumour virus, has been identified in sections of human breast cancer. Cross-reactivity between the human and murine antigens is due to the polypeptide rather than the oligosaccharide component. The plasma membranes of AH66 hepatoma ascites cells contain a glycoprotein resembling glycophorin A from human erythrocyte membranes in having a high content of 2-acetamido-2-deoxy-D-galactose, 2-acetamido-2-deoxy-D-glucose, D-galactose, and neuraminic acid.

In the TA-3 tumour system, the proportion of Af-glycolylneuraminic acid in the neuraminic acid of the cell-surface glycoproteins of an ascites cell line correlates directly with the proportion of N-glycosyl-linked carbohydrate chains in the glycoprotein. The role of cell-surface glycoproteins in malignant transformation has been discussed. An outline of the teratoma system in relation to cell-surface antigens of mouse embryonal carcinoma cells has been published. A membrane-associated glycoprotein has been solubilized by proteolytic digestion of membranes of rat mammary carcinoma cells.

The material was not further characterized other than by its affinity for *Lens culinaris* lectin. A class of glycoproteins recognized as cell-surface markers by lectins and antibody appears to be expressed on the surface of embryonal carcinoma cells. Because of their unusual structural characteristics and abundance, the carbohydrate chains may play a role in the early stages of embryogenesis. High molecular weight D-galactosyl glycoproteins that are characteristic of early embryonic carcinoma cells are at least partly exposed to the cell surface and they contain binding sites for peanut agglutinin. Several ^{12S}I-labelled plasma-membrane glycoproteins of HeLa cells have been isolated by affinity chromatography on wheat-germ agglutinin immobilized on agarose. An abnormal membrane glycoprotein that is associated with malignancy in a wide range of different tumours binds to both concanavalin A and to wheat-germ agglutinin. Several physical parameters of this glycoprotein resemble those of the insulin receptor. A major group of glycoproteins, which have a high molecular weight and stain only faintly with Coomassie Blue, but bind to concanavalin A, have been identified in the plasma membranes of Chinese hamster fibroblasts. Their presence is only revealed by two-dimensional electrophoresis and not by conventional electrophoretic techniques.

The alteration of membrane components of hamster-kidney fibroblasts transformed by Rous sarcoma virus is expressed in an increase in the levels of three glycoproteins, when compared with.those of control cells. The structure (24) of the major cell-surface glycoprotein from hamster-embryo fibroblasts has been established. The structure of a surface glycoprotein from virus-transformed hamster-kidney cells has only been partially characterized and may well prove to have the same structure as that of the embryo fibroblast glycoprotein.

Normal human fibroblasts contain two classes of binding sites on the surface membrane for concanavalin A. The membrane receptor on human fibroblasts for epidermal growth factor

```
Neup5Ac-(2 → 3)-β-D-Galp-(1 → 4)-β-D-GlcpNAc-(1 → 2)-α-D-Manp
                                                           1
                                                           ↓
                                                           6
                                                      β-D-Manp-(1 → 4)-β-D-Glcp-NAc-(1 → 4)-D-GlcpNAc-L-Asx
                                                          3                                  6
                                                          ↑                                  ↑
                                                          1                                  1
Neup5 Ac-(2 → 3)-β-D-Galp-(1 → 4)-β-D-GlcpNAc-(1 → 2)-α-D-Manp                            α-L-Fucp
```

(24)

```
(α-Neup5Ac)m,-(2 → 3)-β-D-Galp-(1 → 4)-β-D-GlcpNAc-(1 → 2)-α-D-Manp
                                                              1
                                                              ↓
                                                              6
                                                      β-D-Manp-(1 → 4)-β-D-Glcp-NAc-(1 → 4)-D-GlcpNAc-L-Asn
                                                          3                                  6
                                                          ↑                                  ↑
                                                          1                                  1
(α-Neup5Ac)m,-(2 → 3)-β-D-Galp-(1 → 4)-β-D-GlcpNAc-(1 → 2)-α-D-Manp                  (α-L-Fucp)n
```

(25) m = 0 on α-sub-units; m = 1 on β-sub-units; n = 0 or 1

is associated with a glycoprotein or glycolipid containing D-mannosyl, D-galactosyl, and 2-acetamido-2-deoxy-D-glucosyl residues. At least half of the membrane glycoproteins of the rough and smooth microsomes and Golgi vesicles from rat liver are associated with the cytoplasmic surface of the membranes. The cell-surface receptor of cultured rat basophilic leukaemia cells for immunoglobulin E has been purified by affinity chromatography on immobilized immunoglobulin E. The purified receptor exists in monomeric (mol. wt. 4.0×10^4) and polymeric (mol. wt. 1.7×10^5) forms, both of which are capable of binding to immunoglobulin E. Chinese hamster ovary plasma cell membranes contain a glycoprotein (mol. wt. 1.0×10^5), which precipitates with wheat-germ agglutinin, and a glycoprotein (mol. wt. 1.3×10^5) which precipitates with concanavalin A.

The advantages of such a lectin immunoprecipitation technique over affinity chromatography for the isolation of glycoproteins is discussed. The biosynthesis of a secretory protein and a transmembrane viral glycoprotein, when compared by different experimental approaches, appear to involve an initially common pathway.

HORMONAL GLYCOPROTEINS

Studies on the sub-units of human glycoprotein hormones in relation to re-production have been reviewed. One volume of a treatise on hormonal proteins and peptides has been devoted to thyroid hormones. The ectopic production of human chorionic gonadotrophin (hCG) and its α- and β-sub-units has been discussed. Like thyroid-stimulating hormone (TSH) human chorionic gonadotrophin interacts with a receptor that has a ganglioside or ganglioside-like structure, and undergoes a change in conformation upon interacting with its receptor.

The α- and β-sub-units of human chorionic gonadotrophin each contain two *N*-glycosidically linked oligosaccharide chains with structures (25) similar to those of many other *N*-glycosidic carbohydrate units of animal glycoproteins. The precise location and structure (26) of the four *O*-glycosidically-linked oligosaccharide units in the *C*-terminal peptide of β-sub-unit chains of human chorionic gonadoptrophin and a revised amino-acid sequence for this peptide have been reported.

$$\begin{array}{l} \alpha\text{-Neu}p\text{5Ac-}(2 \rightarrow 3)\text{-}\beta\text{-D-Gal}p\text{-}(1 \rightarrow 3)\text{-D-Gal}cp\text{NAc-}(1 \rightarrow \text{L-Ser} \\ \qquad\qquad\qquad\qquad\qquad\qquad\qquad\qquad\quad 6 \\ \qquad\qquad\qquad\qquad\qquad\qquad\qquad\qquad\quad \uparrow \\ \qquad\qquad\qquad\qquad\qquad\qquad\qquad\qquad\quad 2 \\ \qquad\qquad\qquad\qquad\qquad\qquad\qquad (\alpha\text{-Neu}p\text{5Ac})_n \end{array}$$

(26)

The chorionic gonadotrophin excreted by patients with hydatidform mole resembles that from urine of normal pregnant women in amino-acid composition, but differs in having a lower content of D-mannosyl and 2-acetamido-2-deoxy-D-glucosyl residues. Any biological role for the carbohydrate moiety of human chorionic gonadotrophin must also involve the protein moiety since a glycopeptide fraction prepared from human chorionic gonadotrophin exhibited no cross-reactivity with anti-(human chorionic gonadotrophin) antiserum.

Likewise there was no competition of this glycopeptide with human chorionic gonadotrophin for testicular receptors. First trimester placental mRNA directs the synthesis of a pre-protein form of the α-sub-unit of human chorionic gonadotrophin, containing a segment of twenty-four amino-acid residues at its N-terminus. In the presence of microsomal membranes, the pre-protein is cleaved into products with a homogeneous N-terminus.

Evidence is submitted to suggest that this processing step is unlikely to account for the heterogeneity that has been observed previously in the structure of this region of the sub-unit. An improved method for the purification of radiolabelled glycoprotein hormones is based on their competitive elution from immobilized concanavalin A with methyl α-D-glucopyranoside. The method separates free iodine, damaged hormone, and radiolabelled bovine serum albumin. After removal of five amino-acid residues from the α-sub-unit of bovine luteinizing hormone (LH) with carboxypeptidase, followed by covalent cross-linking of the product, no significant changes in conformation have been detected by c.d. measurements although the biological activity is almost completely abolished.

The cell-free synthesis of bovine pituitary proteins from total cellular pituitary RNA has been studied. The products include the α-sub-unit of bovine luteinizing hormone, which was identified by immunoprecipitation and sodium dodecyl sulphate-gel electrophoresis. Increases in the extent of covalent cross-linking between α- and β-sub-units of bovine luteinizing hormone with carbodiimide results in a concomitant decrease in the receptor binding activity of the non-dissociable product. Perturbations in sub-unit interactions presumably due to the presence of the cross-links have been examined.

MILK GLYCOPROTEINS

Approximately half of the casein micelle from bovine milk binds to wheat-germ agglutinin *via* neuraminic acid moieties of K-casein. The structures of a tri-saccharide (27) and a tetrasaccharide (28) linked O-glycosidically to L-threonine residues in bovine K-casein have been reported.

$$\begin{array}{c} \alpha\text{-Neu}p\text{5Ac-}(2 \rightarrow 3)\text{-}\beta\text{-D-Gal}p\text{-}(1 \rightarrow 3)\text{-D-Galc}p\text{NAc-}(1 \rightarrow \text{L-Thr} \\ \qquad\qquad\qquad\qquad\qquad 6 \\ \qquad\qquad\qquad\qquad\qquad \uparrow \\ \qquad\qquad\qquad\qquad\qquad 2 \\ \qquad\qquad\qquad\qquad\qquad (\alpha\text{-Neu}p\text{5Ac})_n \end{array}$$

$$(27)\ n = 0$$

$$(28)\ n = 1$$

α-Lactalbumin and casein have been measured by radioimmunoassay in rat mammary gland at various stages of gestation and lactation. Rabbit α-lactalbumin contains five residues of 2-amino-2-deoxy-D-glucose per mole of protein, as well as other sugars. The amino-acid sequence has been determined for the 122 amino-acids and includes a single carbohydrate moiety, probably at L-asparagine-45. A protein that is antigenically identical to the soluble glycoprotein of the milk fat globule membrane has been identified in whey.

By indirect elimination, the major protein components of whey (β-lactoglobulin, α-lactalbumin, and immunoglobulin G) were shown not to be the anti-(soluble glycoprotein) antibody. The effects of tunicamycin on the glycosylation of lactating rabbit-mammary glycoproteins has been reported. Although lipid-linked sugars are not involved in the glycosylation of casein, which is probably glycosylated by direct sequential transfer of monosaccharides to protein, they serve as intermediates in the glycosylation of a group of mammary polypeptides. The carbohydrate chains of the free secretory component from human milk have the general composition and structural features (29) of *N*-glycosidically linked oligosaccharides of numerous other mammalian glycoproteins. Two unusual features are the di-L-fucosyl disaccharide unit and the single 2-acetamido-2-deoxy-D-glucosyl residue instead of the usual two ammo-sugar residues at the reducing end of the molecule.

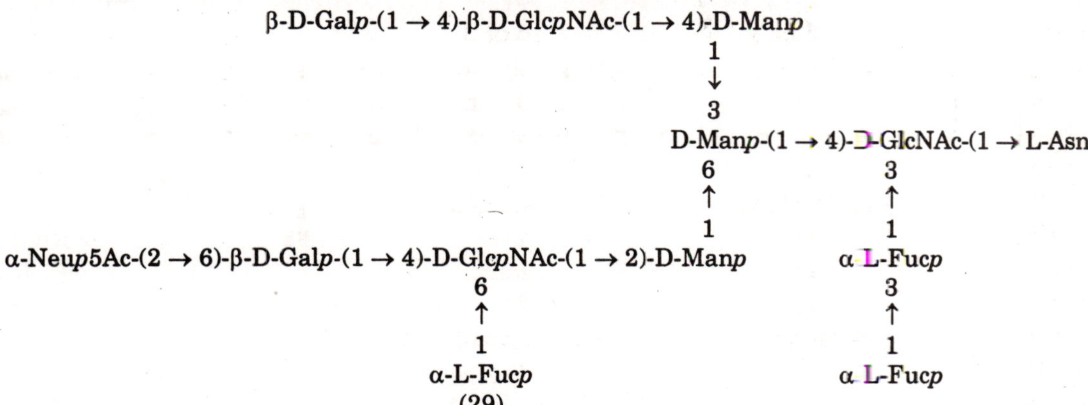

(29)

Multiple forms of β_2-microglobulin have been detected in human colostrum. One form (mol. wt. 4.8×10^4), identical to lactollin, is a tetrameric form of β_2-microglobulin, and three others (mol. wt. 1.2×10^4) are separable by electrophoresis.

PLASMA AND SERUM GLYCOPROTEINS

The sera of certain animal species contain substances which are not antibodies, although they bind to hepatitis B surface antigen. They are probably glycoproteins and can be selectively removed from sera by reaction with purified hepatitis B surface antigen. A heterogeneous population of serum glycoproteins, which bind to immobilized *Ricinus communis* lectin, have been isolated and shown to account for the inhibition of the binding of desialylized glycoprotein to the hepatocyte membrane. These glycoproteins are present in small amounts in normal human serum, but in increased levels in serum from patients with cirrhosis. [^{3}H]-Labelled raffinose has been covalently attached to a variety of serum glycoproteins and used in detecting the tissue and cellular sites of catabolism of long-lived circulating glycoproteins. A method using metal chelate chromatography has been developed for the large-scale purification of two major plasma proteinase inhibitors, α_1-acid glycoprotein and α_2-macroglobulin, from pooled plasma under non-denaturing conditions.

A solid-phase enzyme-linked immunosorbent assay (ELISA) for the assay of α_1-acid glycoprotein has been developed, using alkaline phosphatase covalently attached to the glycoprotein and α_1-acid glycoprotein-specific antibody coated to polystyrene tubes. No differences in the carbohydrate composition of normal and variant human α_1-proteinase inhibitors have been observed. The glycoprotein (mol. wt. 5.4×10^4) contains three carbohydrate chains, two of which contain the biantennary structure (30) and one the triantennary structure (31). The major proelastase 2 binding factor in human plasma is α_1-protease inhibitor. Proelastase 2 reacts directly with the inhibitor *via* a partial active site in a manner similar to that of *endo*-peptidases. Guinea-pig α_1-microglobulin, isolated from plasma or urine, is homologous to human α_1-microglobulin, on the basis of molecular weight, carbohydrate composition, and cross-reaction with antisera.

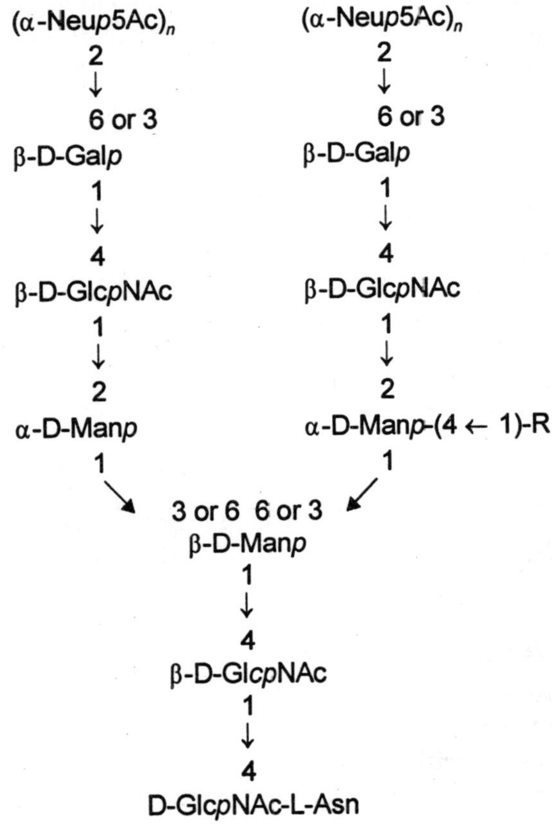

n = 0, 1
(30) R = H
(31) R = α-Neu*p*5Ac-(2 → 6)-β-D-Gal*p*-(1 → 4)-β-D-Glc*p*NAc

Traces of concanavalin A have been detected in samples of purified α_1-antitrypsin after affinity chromatography on immobilized concanavalin A. The interfering lectin may be removed by adsorption of the α_1-antitrypsin on thiolactivated macroporous agarose. Unequivocal evidence has been presented for the existence of a recognizable pattern in the complex multiple interactions between trypsin and human serum which allows the classification of human α_2-macroglobulin into seven distinct categories.

The molecular interpretation of the binding of trypsin with α_2-macroglobulin and the potential clinical value in recognizing the nature of such binding are discussed. The native tetramers of human plasma α_2-macroblogulin are reported to exist in two forms that differ in electrophoretic mobility as well as in other properties. The irreversible transition of the two forms is associated with the binding of proteases. Fragmentation of human α_2-macroglobulin (mol. wt. 1.85×10^5) has been achieved by heating in the presence of sodium dodecyl sulphate and dithiothreitol, two fragments (mol. wt. 1.2×10^5 and 6.2×10^4 being produced). On the basis of amino-acid sequencing studies, the native glycoprotein is composed of four identical polypeptide chains. The contents of Zn^{2+} in human α_2-macroglobulin in normal, hyperzincemic, and hypozincemic sera have been reported.

In cystic fibrosis patients the plasma α_2-macroglobulin contains normal amounts of total hexose, but as much as 40% decreased amounts of neuraminic acid. Compared with control preparations, the binding of these preparations to concanavalin A and wheat-germ agglutinin is markedly reduced. Alterations in the carbohydrate moiety of the glycoprotein in cystic fibrosis may be due to a defective post-translational process. An α_2-acute phase macroglobulin has been isolated from rat under conditions avoiding damage to biological activity.

The molecular weight and carbohydrate composition of the glycoprotein were similar to those of α_2-macroglobulins from other sources. A pregnancy-associated α_2-glycoprotein (mol. wt. 3.65×10^5) dissociates into two sub-units (mol. wt. 1.44×10^5). Physicochemical similarities, but no antigenic similarities, have been demonstrated between this glycoprotein and human α_2-macroglobulin. Pregnancy-specific β_1-glycoprotein has been measured using an enzyme-linked immunoassay, and a radioactive immunoassay system. An agglutinin, identified by physical and serological methods as β_2-glycoprotein, has been isolated from human serum by affinity chromatography on immobilized anti-(β_2-glycoprotein) antibody. A treatise on chemistry, biology, and clinical applications of the carcinoembryonic proteins, α-foetoprotein, isoferritin, pregnancy-specific β_1-glycoprotein, and carcinoembryonic antigen has been published. Studies on carcinoembryonic antigen and related glycoproteins as tumour markers have been reviewed. The chemical and immunological properties of carcinoembryonic antigen purified after sodium chloride extraction are almost identical to those of the glycoprotein purified after perchloric acid treatment.

Glycopeptides have been isolated from carcinoembryonic antigen after its treatment with neuraminidase and digestion with proteolytic enzymes. Large peptide fragments are required to demonstrate the antigenicity of the molecule. The glycoprotein has been deglycosylated by treatment with hydrogen fluoride revealing a single polypeptide chain. The polypeptide, which retains 15% of its original antigenic activity and is able to inhibit up to 85% of the binding of intact carcinoembryonic antigen to and-(carcinoembryonic antigen) antibody, has been sequenced through the first thirty amino-acid residues.

An improved indirect and direct radioimmunoassay for carcinoembryonic antigen has been developed. Blood group A activity has been observed in two preparations of carcinoembryonic antigen even after passage of these preparations through columns of immobilized anti -A, -B, and -H antibodies. Normal human bile contains a major carcinoembryonic antigen-cross-reactive glycoprotein, biliary glycoprotein I, which can be purified by a procedure similar to that used for carcinoembryonic antigen and non-specific cross-reacting antigen. The same sugars and amino-acids were found in similar molar proportions in the three glycoproteins. The structure of the component proteins of complement have been reviewed. All nine components of the human classical complement pathway bind to immobilized Cibacron Blue F3GA, and are eluted without contamination of substrates by albumin or immunoglobulin G.

The sub-component $C1_q$ of the first component of bovine complement shows many similarities to the corresponding human and rabbit component. Almost all of the L-hydroxylysine residues are substituted with D-galactosyl-D-glucosyl disaccharide units. These glycosyl L-hydroxylysine residues are only detected in the collagen-like regions of the molecu

Bovine sub-component $C1_q$ contains hexose, amino-sugar mainly as 2-amino-2-deoxy-D-glucose, and neuraminic acid. The carbohydrate is associated with the heavy chain in this glycoprotein. Like human component C4 of complement, the corresponding bovine glycoprotein (mol. wt. 2×10^5) contains three polypeptide chains, α, β, and γ (mol. wts. 9.8×10^4, 8.2×10^4, and 3.2×10^4, respectively). The carbohydrate is located in the α- and β-chains. Erythropoietin has been isolated by affinity chromatography on immobilized wheat-germ agglutinin, and on immobilized phytohaemagglutinin. Human plasma cold-insoluble globulin, which appears to exist in a slightly modified form (fibronectin) on the cell surface of fibroblasts, has been purified by affinity chromatography on immobilized gelatin, as well as by chromatography on either immobilized heparin or immobilized and-(cold-insoluble globulin) antibody.

The glycoprotein agglutinates trypsin-treated erythrocytes from certain species, and the agglutination is inhibited by polyamines, basic amino-acids, non-acetylated amino-sugars, and antibodies to the glycoprotein. The structure and stability of cold-insoluble globulin have been investigated by ultracentrifugation, c.d., and by fluorescence measurements. Bovine plasma cold-insoluble globulin contains six *N*-glycosidic chains per dimer molecule. These have been released from the polypeptide backbone by hydrazinolysis followed by reduction with sodium borohydride, and their structures [(32)-(35)] established using conventional methods. The preparation and properties of anti-freeze glycoproteins from Arctic and Antarctic fish blood have been reviewed. In a c.d. spectroscopic study of protein and glycoprotein anti-freezes, two glycoproteins appeared to have an unordered conformation in solution, whereas two proteins had a right-handed α-helical conformation. Thus the freezing-point depressing activity of the anti-freezes does not require them to have a particular secondary structure.

R¹-β-D-Gal*p*-(1 → 4)-β-D-Glc*p*NAc-(1 → 2)-α-D-Man*p*
1
↓
3
β-D-Man*p*-(1 → 4)-β-D-Glc*p*NAc-(1 → 4)-D-GlcNAc-ol
6
↑
1
R²-β-D-Gal*p*-(1 → 4)-β-D-Glc*p*NAc-(1 → 2)-α-D-Man*p*

(32) R¹ = H, R² = α-Neu*p*5Ac-(2 → 6)

(33) R¹ = α-Neu*p*5Ac-2(2 → 4 or 6), R² = α-Neu*p*5Ac-(2 → 6 or 4)

Factor VIII and fibrinogen can be removed from normal human plasma by affinity chromatography on immobilized rabbit anti-Factor VIII. In human Factor VIII neither the size distribution nor its ristocetin cofactor activity is related to the carbohydrate content. The binding affinity of bovine Factor VIII for concanavalin A is gradually decreased with increasing aggregate size, suggesting an impaired accessibility of reactive sugar residues in the large aggregates of the bovine glycoprotein. Factor VIII-related protein aggregates are multimers of pairs of basic sub-unit chains. When prepared from lipid-poor plasma, the size heterogeneity ranges from 8.0×10^5 to 2.0×10^7. Heparin and dextran sulphate can bind to Factor VIII, an essential factor for the maintenance of weak interplatelet bonds, and either inhibits or reverses platelet aggregation.

```
α-Neup5Ac-(2 → 6)-β-D-Galp-(1 → 4)-β-D-GlcpNAc-(1 → 2)-α-D-Manp
                                                          1
                                                          ↓
                                                          6
                                                      β-D-Manp-(1 → 4)-β-D-Glcp-NAc-(1 → 4)-D-GlcpNAc-ol
                                                          3
                                                          ↑
                                                          1
α-Neup5 Ac-(2 → 4)-β-D-Galp-(1 → 3)-β-D-GlcpNAc-(1 → 2)-α-D-Manp
                                     6
                                     ↑
                                     1
                                  α-NeupAc
                                    (34)

                                  α-NeupAc
                                     2
                                     ↓
                                     6
α-Neup5 Ac-(2 → 4)-β-D-Galp-(1 → 3)-β-D-GlcpNAc-(1 → 2)-α-D-Manp
                                                          2
                                                          ↓
                                                          6
                                                      β-D-Manp-(1 → 4)-β-D-Glcp-NAc-(1 → 4)-D-GlcpNAc-ol
                                                          3
                                                          ↑
                                                          1
α-Neup5Acm,-(2 → 4)-β-D-Galp-(1 → 3)-β-D-GlcpNAc-(1 → 2)-α-D-Manp
                                     6
                                     ↑
                                     2
                                  α-NeupAc
                                            (35)
```

The Ca^{2+}-binding characteristics of the products of Factor IX activation are similar to those of the native zymogen. A low α-helical content and a high level of β-structure and random coil conformations were the main conformational features of all the products as shown by c.d. spectroscopic studies. Factor V is cleaved by thrombin to form either an activation intermediate or a stable form of fully activated Factor V. Both the intermediate and the activated Factor V can be dissociated by chelating agents. Bovine anti-thrombin is cleaved by thrombin at only one site, which is at the peptide bond in the L-arginyl-L-serine sequence near the C-terminal end of the chain.

The modified anti-thrombin thus formed loses all or most of its ability to inhibit thrombin, and has a reduced affinity for heparin. Two polypeptide fragments (mol. wt. 5.0×10^4 and 5.0×10^3) are formed. Limited proteolysis of anti-thrombin by thrombin may be involved in formation of an inactive complex. The inhibitor may therefore exist in the complex as a proteolytically modified two-chain glycoprotein. Subsidiary forms of the anti-thrombin-thrombin complex may arise as a result of complex formation between anti-thrombin and autolysed thrombin. Bovine prothrombin contains three *N*-glycosidically linked chains per molecule and these have been isolated from the glycoprotein by hydrazinolysis. One of the oligosaccharide chains has the typical complex-type widely found in other glycoproteins, however, the other two are unique in possessing a 2-acetamido-2-deoxy-3-O-β-D-galactosyl-D-glucosyl disaccharide moiety in the outer chains. The covalent modification of human α-thrombin with pyridoxal 5'-phosphate involves phosphopyridoxylation at two sites.

Attachment of the cofactor to the first site was accompanied by a loss in the fibrinogen clotting activity of the thrombin, whereas attachment of the cofactor to the second site resulted in a decrease in the sensitivity to heparin in the anti-(thrombin III)-thrombin reaction. The oligosaccharide side chain of fragment E of rat plasma fibrinogen contains residues of neuraminic acid, D-galactose, D-mannose, and 2-amino-2-deoxy-D-glucose (in the molar ratio 1 : 1 : 2 : 2). Removal of the neuraminic acid residue has no effect on the clearance rate of fragment E from the circulation and so no role is played by the oligosaccharide chain in the clearance of fibrinogen degradation products. The carbohydrate composition of two human plasminogen variants has been reported.

Plasminogen 1 possesses two oligosaccharide units, one of which is similar to the single 2-acetamido-2-deoxy-D-galactosyl-containing oligo-saccharide in plasminogen 2, but the second chain of residues consists of neuraminic acid, D-galactose, D-mannose, and 2-acetamido-2-deoxy-D-glucose (in the molar ratio 1 : 2 : 3 : 4). A D-mannosyl-containing oligosaccharide attached to L-asparagine at residue 288 was shown to have a structure identical to that of oligosaccharide chains proposed for glycopeptides isolated from transferrin and thyroxine-binding globulin. Both variants contain tri- and tetra-saccharide units linked to L-threonine with structures identical to those of *O*-glycosidically linked oligosaccharides (27 and 28) of *K*-casein. Greater structural heterogeneity apparently exists in the carbohydrate moiety of human transferrin than previously envisaged. This heterogeneity can be reflected in several molecular forms, which after desialylization differ significantly in their affinities for the hepatic lectin.

Three types of asialotransferrin have been isolated after affinity chromatography of human transferrin on immobilized rabbit-liver lectin specific for the asialoglycopeptides. Type 1 contains two biantennary glycans, but Types 2 and 3 are assumed to possess one biantennary and one triantennary glycan. Bovine transferrin contains only one heterosaccharide chain per molecule.

The asialoglycopeptide from bovine transferrin has been compared with asialoglycopeptides from other serum proteins for affinity to rat-liver hepatic lectin. Rabbit transferrin contains two glycopeptide segments structurally identical to those from plasminogen and thyroxine-binding globulin. Mouse α-foetoprotein is heterogeneous with respect to binding to concanavalin A owing to the derivation of the glycoprotein from the yolk sac and the liver.

The glycoprotein derived from the yolk sac, and therefore the major α-foetoprotein before birth, does not bind to the lectin. The major α-foetoprotein, which is synthesized in the liver after birth binds to concanavalin A. Human α-foetoprotein from cord serum has been purified to homogeneity after affinity chromatography on immobilized concanavalin A followed by polyacrylamide gel electrophoresis. Two variants of bovine α-foetoprotein with different affinities for concanavalin A show antigenic identity suggesting that the polypeptide chain rather than the carbohydrate moiety is the antigenic determinant. The technique of supplementation counter-immunoelectrophoresis is more sensitive than immuno-diffusion, but less sensitive than radioimmunoassay, for the detection of α-foetoprotein. The *O*-glycosically linked carbohydrate chains of fetuin have been cleaved from the protein by alkaline borohydride treatment.

Structural analysis revealed the existence of tri- and tetra-saccharides linked *O*-glycosidically to L-serine or L-threonine and having structures identical to those previously identified for bovine K-casein and human chorionic gonadotrophin [(27) and (28)]. The *N*-glycosidically linked chains of fetuin have been assigned structures (36) or (37). Glycosylated haemoglobins have been isolated following ion-exchange chromatography of haemolysed erythrocytes. Bound hexoses, after release from the protein by hydrolysis with oxalic acid, were measured as 5-hydroxymethylfurfural derivatives.

Four glycosylated minor components, possessing unique functional properties, have been isolated from human adult haemoglobin. Two of these components are low affinity haemoglobins similar to certain rare human haemoglobin variants but present in every human haemolysate. Rhesus monkey haemoglobin Aqc contains a carbohydrate content identical to that of human haemoglobin A_{1c}. Heterogeneity of the major haemoglobin component Hb A_0 was demonstrated, reflecting a slow non-enzymic glycosylation at sites other than the *N*-terminus of the β-chain. The non-enzymic glycosylation of haemoglobin in normal and diabetic subjects has been reviewed. Non-enzymically D-glucosylated albumin represents approximately 6-15% of total serum albumin from normal adults.

The post-translational modification appears to occur by a process analogous to that responsible for D-glucosylation of haemoglobin A through Schiff base formation and Amadori rearrangement to a ketoamine derivative. Elevated levels of D-glucosylated albumin have been detected in the sera of diabetic patients. The level of D-glucosylated albumin may be a sensitive indicator of moderate hyperglycemia and of early D-glucose intolerance.

IMMUNOGLOBULINS

Reviews dealing with the structure of immunoglobulins, the organization of immunoglobulin genes, the antibody-combining site, and the structural basis of antibody complementarity have been published. Marine anti-[high α-(1 → 3)-content dextran] antibody production by hybrid cells has been reported. Binding studies of methyl 6-*O*-benzoyl-β-D-galacto-pyranoside to the Fab' fragment of immunoglobulin J539 have indicated that the phenyl group appears to participate

```
α-Neup5Ac-(2 → 3)-β-D-Galp-(1 → 4)-β-D-GlcpNAc-(1 → 2)-α-D-Manp-(1 → 6)-β-Manp-(1 → 4)-β-D-Glcp-NAc-(1 → 4_-b-D-GlcpNAc-L-Asn
                                                                           3
                                                                           |
                                                                           1
            α-Neup5Ac-(2 → 6)-β-D-Galp-(1 → 4)-β-D-GlcpNAc-(1 → 2)-α-D-Manp
                                                                           4
                                                                           ↑
                                                                           1
                         α-Neup5 Ac-(2 → 3)-β-D-Galp-(1 → 4)-β-D-GlcpNAc
```

(36)

```
                   α-Neup5 Ac-(2 → 6)-β-D-Galp-(1 → 4)-β-D-GlcpNAc
                                                              1
                                                              ↓
                                                              2
α-Neup5Ac-(2 → 3)-β-D-Galp-(1 → 4)-β-D-GlcpNAc-(1 → 4)-α-D-Manp-(1 → 6)-β-Manp-(1 → 4)-β-D-Glcp-NAc-(1 → 4)-β-D-GlcpNAc-L-Asn
                                                                              3
                                                                              ↑
                                                                              1
            α-Neup5 Ac-(2 → 3)-β-D-Galp-(1 → 4)-β-D-GlcpNAc-(1 → 2)-α-D-Manp
```

(37)

in interactions with the antibody. Immunogens have been prepared from various oligosaccharides covalently attached to edestin in order to study the specificities of the anti-carbohydrate antibodies that they elicit in rabbits.

The radioimmunoassay method used to define the specificity of the antibodies offers a sensitive means for detecting specific carbohydrate linkages in various complex carbohydrates. Several oligo-D-mannosides and glycoproteins have been tested for their ability to inhibit the binding of D-[^{3}H]mannotetraitol (38) to anti-D-mannan antibodies. The antibodies are highly specific for the 3-*O*-α-D-mannosyl-D-mannose units and react poorly with 2-*O*- and 6-*O*-α-D-mannosyl-D-groups. Antisera with specificities for chito-oligosaccharides have been obtained by immunization of animals with the oligosaccharides covalently attached to bovine serum albumin or to cytochrome *c* as carrier proteins. Differences in the electrophoretic mobility of serum lysozyme from humans may be due in some cases to its interaction with anti-lysozyme antibody.

$$\alpha\text{-D-Man}p\text{-}(1 \rightarrow 3)\text{-}\alpha\text{-D-Man}p\text{-}(1 \rightarrow 2)\text{-}\alpha\text{-D-Man}p\text{-}(1 \rightarrow 2)\text{-D-Mannitol}$$

(38)

By monitoring various fluorescent properties of both a constant and a variable region in γ- and L-chains, reassociation studies have identified a two-stage transition of immunoglobulin chains. There is a fast step involving the two polypeptides and a slow step confined to the variable region. In MPCH (immunoglobulin G_{2b}) and MOPC (immunoglobulin G_1) cells, the initial step in the formation of intermolecular disulphide bonds between heavy and light, or heavy and heavy, chains occurs to a substantial extent on nascent heavy chains as well as on the completed chains.

The nascent heavy chains must have molecular weights of approximately 3.8×10^4 before light chains become covalently bound. Some immunoglobulin molecules have neuraminic acid-containing oligosaccharides linked to variable segments of the light chains in addition to their Fc fragments. These Fab oligosaccharides are potential determinants of antibody specificity for the formation of immunoglobulin G-immunoglobulin G complexes. The lack of binding of the pFc′ fragment (corresponding to the CH_3 domain of immunoglobulin G) or of the Facb derivative (which is derived from CH_3) of rabbit immunoglobulin to staphylococcal protein A, suggests that the locus of protein A binding is at the interface between the CH_2 and CH_3 domains. The interpretation of ^{13}C n.m.r. spectroscopic results agrees with *X*-ray data in considering that the carbohydrate moiety of human immunoglobulin G is as immobile as the protein itself.

In contrast the results of spin-labelling of the carbohydrate moiety imply mobile carbohydrate in the Fc fragment. The carbohydrate content of immunoglobulin G, the apoprotein of low density lipoprotein, and normal sera from mucolipidosis II and III patients show no differences in the relative ratios and total amounts of D-mannose, D-galactose, neuraminic acid, and 2-acetamido-2-deoxy-*D*-glucose. If the postulated post-translational defect in these disorders involves changes in the carbohydrate composition, it is not a general defect with glycosylation of these glycoproteins, but may be specific for the lysosomal hydrolases. In addition to their Fc oligosaccharides, some immunoglobulin molecules have oligosaccharides linked to variable segments of heavy or light chains, making the Fab fragments potential determinants of antibody specificity.

Examination of the immunoglobulin anti-globulin antibody from a patient with immunoglobulin G-immunoglobulin G complexes showed that a neuraminic acid-containing oligosaccharide on the light chain of the anti-globulin antibody is required for the binding action. Horse-radish peroxidase has been coupled to immunoglobulin G with retention of high enzymatic and immunological activity. Immunoglobulins G from antisera specific for human chorionic gonadotrophin, carcinoembryonic antigen, α-foetoprotein, and casein have been isolated by affinity chromatography on columns of immobilized staphylococcal protein A, before being themselves immobilized on supports. The resulting coupled antibodies provide a greater sensitivity and convenience in radioimmunoassay studies than the conventional double antibody-precipitation methods. A two-point kinetic turbidimetric technique using commercially available monospecific antisera has been used for the determination of immunoglobulins G, A, and M in sera

The accuracy of the method when measuring idiotypic monoclonal proteins is greater than that of the radial diffusion method. Polystyrene particles coated with immunoglobulin G are agglutinated by guinea-pig and human skin collagen. The interaction appears to be specific for the F_c region of the immunoglobulin and also for native collagen, which is unexpected since other workers have suggested that the *C*-terminal regions of the $C1_q$ sub-units of complement and not the collagen-like segments react with immunoglobulins. The murine anti-(insulin myeloma immunoglobulins A47N and E109) anti-bodies have an affinity for the D-fructofuranosyl residues of inulin, with the D-glucosyl residues contributing little if at all to the binding of the whole poly-saccharide to the immunoglobulins. The specific binding of [^{125}I]-protein A to the Fc region of rabbit immunoglobulin G anti-immunoglobulin E, has been used in the determination of specific anti-immunoglobulin E antibody bound to immobilized immunoglobulin E. A synthetic immunogen, poly-(L-tyrosine-L-glutamic acid)-poly-DL-alanine-poly-L-lysine has been used to raise immunoglobulin M in rabbits and cattle.

The immobilized antigen has then been used in the purification of the immunoglobulin. Immobilized concanavalin A and immobilized *Ricinus communis* lectin have been used in the separation of Fc from Fab fragments of a myeloma immunoglobulin M. The F_c fragment binds to both lectins, but the Fab fragment only interacts with concanavalin A. *N*-Glycosidically-linked oligosaccharides having a high proportion of D-mannosyl residues have been identified in two glycopeptides (I and II) isolated from a human immunoglobulin M myeloma protein. The basic structure (39) of the oligosaccharide has been established. The larger oligosaccharides present have additional α-D-mannosyl groups linked (1 → 2) to terminal D-mannosyl residues. Glycopeptide I contains D-*manno*-pentaosyl and *D-manno*-hexaosyl species, whereas glycopeptide II contains D-*manno*-hexaosyl and D-*manno*-octaosyl species. Glycopeptide I also contains four minor oligosaccharides which are all related to the basic structure (39). In contrast to human membrane-derived immunoglobulins M and D, those from murine lymphocyte membrane display different affinities for lentil lectin. Some properties of a cell-bound homogeneous murine immunoglobulin M anti-dextran have been reported. Immunoglobulin antibodies invoked by *Mycoplasma pneumoniae* are cross-reactive with native erythrocyte antigens, which share determinants with *M. pneumoniae*.

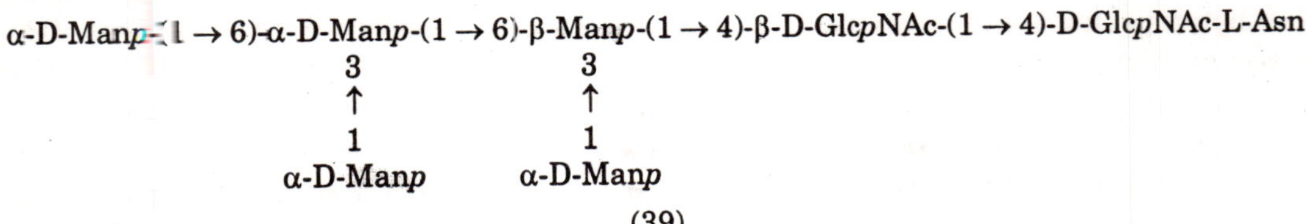

(39)

ERYTHROCYTE GLYCOPRATEINS

Residues of neuraminic acid bearing *O*-acetyl groups have been detected in glycoproteins from murine and rat erythrocyte ghosts, but not in similar preparations from human, rabbit, or guinea-pig erythrocytes. A modified periodate-Schiff reagent has been developed for the removal of these *O*-acetyl groups. Two sialoglycoproteins, glycophorin B and C, have been isolated from erythrocyte membranes by extraction with lithium di-iodosalicylate. Glycophorin B is found in two forms, corresponding to monomer and dimer, on electrophoresis of normal erythrocyte membranes.

The amino-acid sequence of thirty-six residues of the *C*-terminal segment and nineteen residues of an internal glycosylated segment of porcine glycophorin A have been shown to differ from the corresponding human glycoprotein. The behaviour of glycophorin in aqueous solution and some of its structural characteristics have been studied by ^{1}H n.m.r. spectroscopy. Carbohydrate chains were found to be conformationally mobile and no evidence was found for tight carbohydrate-protein interactions. Despite its extensive glycosylation (60%), the carbohydrate moieties of glycophorin A contribute little to the c.d. spectrum.

The native glycoprotein contains 27% helix, 10% β-form, and 63% unordered form. This transmembrane glycoprotein has three distinct chemical domains-, the heavily glycosylated *N*-terminal portion located externally, a central portion thought to span the lipid bilayer, and a *C*-terminal portion consisting of a heavily charged, L-proline-rich segment that extends into the cytoplasm. The human leukaemic cell line K562 synthesizes a surface membrane which is identical to or closely similar to glycophorin A. Anti-glycophorin A antiserum precipitated the glycoprotein from both surface-labelled and metabolically-labelled K562 cells. The biosynthesis of glycophorin A by this cell line resembles that described for viral membrane glycoproteins. *X*-Ray diffraction studies on sheep erythrocyte ghost membranes, which are prepared by agglutination of the ghosts with phytohaemagglutinin M, have revealed side-to-side packed transmembrane α-helices. The content of 2-amino-2-deoxy-D-galactose in human erythrocyte membranes of the A, B, and AB blood groups is higher than that of the O group.

Similarly significant differences in the amounts of D-galactose in the membranes derived from different pairs (A/O, B/O, AB/O, A/B, A/AB) of blood groups were observed. Basic differences in the core oligosaccharide chains may exist between the different erythrocyte membranes. An abnormal blood group S-s active sialoglycoprotein has been detected in the membrane of Miltenberger class III, IV, and V human erythrocytes.

Interactions among the major sialoglycoproteins and peripheral proteins of human erythrocytes have been studied by fluorescence energy transfer. Divalent cations are necessary for the direct association between membrane sialoglycoproteins and spectrin. Vesicles, enriched in glycophorin, are obtained by selective extraction of human erythrocyte membranes with a non-ionic detergent. Glycoproteins of the human erythrocyte membrane, including glycophorin A and 'band 4.5' have been identified as two major classes of methyl-acceptor polypeptides for the enzyme, protein methylase II (S-adenosylmethionine : protein carboxy *O*-methyl transferase, EC 2. 1.1. 24). The structure of band 3 glycopeptide, prepared from erythrocytes of normal adult (blood group OI) contains the branched repeating unit (40), whereas the corresponding glycopeptide from an *i* adult variant who fails to develop I antigen contains a linear repeating unit (41).

The lack of a branching enzyme in the *i* individual may give rise to the differences in the structures. Under appropriate conditions, band-3 glycoprotein can be cross-linked with Cu^{2+}-2-phenanthroline primarily to give a dimer with the resultant loss of only one reactive thiol group per band-3 monomer. The formation of one disulphide bond precludes the formation of others, suggesting that a conformational change occurs in the glycoprotein on cross-linking.

$$\rightarrow\text{-}\beta\text{-D-Gal}p\text{-}(1 \rightarrow 4)\text{-}\beta\text{-D-Glc}p\text{NAc-}(1 \rightarrow 3)\text{-D-Gal}p$$

$$\begin{array}{c} 6 \\ \uparrow \\ 1 \\ \text{R} \end{array}$$

(40) R = glycosyl residue

$$\rightarrow\text{-}\beta\text{-D-Gal}p\text{-}(1 \rightarrow 4)\text{-}\beta\text{-D-Glc}p\text{NAc-}(1 \rightarrow 3)\text{-D-Gal}p$$

(41)

Changes in the cell-surface carbohydrates of erythrocytes during their life span include changes in the neuraminic acid, 2-acetamido-2-deoxy-D-glucose, 2-acetamido-2-deoxy-D-galactose, and D-galactose contents of the membranes. Extensive heterogeneity exists in the cell-surface carbohydrates of the circulating population of erythrocytes. Several surface components of human erythrocytes are sensitive to hydrolysis by *endo*-β-D-galactanase. The removal of residues of neuraminic acid from the glycoproteins of human erythrocytes decreases the degree of solubilization of membrane glycoproteins by non-ionic detergents.

Solubility of these glycoproteins with these detergents may be achieved by addition of borate, implying that a charge on the membrane protein is an important consideration. Human erythrocytes having blood group I antigen have been surface-labelled with ^{125}I, and with ^{3}H by treatment with D-galactose oxidase-sodium borotritide. Using human serum containing monoclonal anti-blood group I antibody, the immunoprecipitates having predominant ^{125}I-label were in band-3 protein (mol. wt. $0.9 \times 10^5 - 1.0 \times 10^5$), whereas the tritium label was in a diffusely migrating component (mol. wt. $4.0 \times 10^4 - 7.0 \times 10^4$) on electrophoresis.

The tritium-labelled component has a similar mobility to a sub-population of erythrocyte poly(glycosyl) ceramides with blood group I activity. Human erythrocyte membranes have been labelled with a photosensitive hydrophobic probe, l-azido-4-iodo-[^{3}H]benzene. Characterization of purified tryptic and chymotryptic peptides show that the probe is covalently attached only to the transmembrane region of the glycoprotein. The aggregation of bovine erythrocytes by concanavalin A is highly sensitive to the level of functionally identical concanavalin A receptors.

Threshold densities of these receptors are required for aggregation to occur. Glycoproteins isolated from the erythrocytes of rhesus monkeys inhibit haemagglutination by measles virus.

SALIVARY AND MUCOUS GLYCOPROTEINS

The structure and biosynthesis of mucins have been reviewed. Mucin glycoproteins and protein components of human-lung musosal gel have been resolved. In contrast to meconium, the purified lung mucin glycoproteins are capable of aggregation and gel formation without either addition of albumin or of the lower molecular weight proteins of lung mucosal gel. Purified human asthmatic bronchial mucin has a minimum molecular weight of 1.8×10^6 with aggregates of 1×10^7 or greater. Oligosaccharide chains are linked N-glycosidically *via*

2-acetamido-2-deoxy-D-galactosyl residues to L-serine or L-threonine residues, which account for approximately one of every five amino-acids in the protein. Human bronchial mucous glycoproteins have been fractionated by ion-exchange and gel permeation chromatography.

The glycoproteins (mol. wt. 4.0×10^5) have an extended rod-like polypeptide conformation with about one amino-acid residue in three bearing a glycosidic side-chain attached through an L-serine- or L-threonine linkage. The major component of goblet cell mucin in human intestinal scrapings contains a polydisperse glycoprotein component containing residues of D-galactose, L-fucose, 2-acetamido-2-deoxy-D-galactose, 2-acetamido-2-deoxy-D-glucose, and neuraminic acid. A minor monodisperse glycoprotein component contains the same sugar residues with the exception of neuraminic acid. The extraparotid saliva of Macaque monkeys contains a glycoprotein (mol. wt. 9.6×10^3) capable of inhibition of the precipitation of calcium phosphate.

The major glycoproteins obtained from Bonnet monkey cervical mucous of the peri-ovulatory and of the pre-menstrual phases differ mainly in the point of linkage of the terminal neuraminic acid residues and of the substitution of the 2-acetamido-2-deoxy-D-galactosyl residues linked to the protein chain. Differences in the levels of heterogeneity and in N-terminal amino-acid residues were also detected in the two groups of glycoproteins. On the basis of the structural identity of the newly formed oligosaccharide unit arising from the *in vitro* D-galactosylation of desialylized ovine sub-maxillary mucin, the porcine submaxillary gland galactosyltransferase has been defined as a UDP-D-galactose : α-D-2-acetamido-2-deoxygalactosyl-protein 3-β-D-galactosyl transferase. Sulphated glycoproteins from porcine fundic mucosa inhibit pepsin activity by becoming bound to proteins.

The physiological role of these sulphated glycoproteins is speculated to be in the protection of the components of the gastric gland from peptic digestion. Canine tracheal pouch secretions contain glycoproteins (mol. wt. $\geq 3.0 \times 10^6$), which dissociate into three sub-units on treatment with reducing agents.

Sub-fractions containing 87% carbohydrate, high levels of L-serine and L-threonine, but variable amounts of 2-acetamido-2-deoxy-D-glucose may then be isolated by ion-exchange chromatography. Similarities were observed between the properties of the canine tracheal pouch preparations and tracheobronchial secretions from human patients. The presence of hydrophobic sites in fetuin, ovine submaxillary mucin, and two canine trachael mucins has been established by a fluorescence probe technique which involves the interaction between the glycoproteins and sodium mansate (*N*-methyl-2-anilinonaphthalene-7-sulphonate). In each case the interaction is accompanied by an enhancement in fluorescence and a shift in the fluorescence maximum to a shorter wavelength.

A tracheal mucin (mol. wt. 5.8×10^5) containing 80% carbohydrate exhibited the highest fluorescence enhancement of all the glycoproteins and contained the largest number of binding sites for these fluorescent probes. A mucin isolated from rat tracheal transplants contained protein (16.5%) and had D-galactose, 2-acetamido-2-deoxy-D-glucose, 2-acetamido-2-deoxy-D-galactose, and neuraminic acid as the principal sugars. The presence of *O*-glycosidic linkages was established by β-elimination in alkaline borohydride solution. Cleavage of the O-glycosidic chains of armadillo submandibular mucin releases 30% disaccharide (42) and 75% monosaccharide (43) linked to L-serine or L-threonine.

Although similar oligosaccharide chains are found in the salivary glycoproteins of other species, those glycoproteins also contain a considerable proportion of oligosaccharides of larger size.

α-Neu*p*5Ac-(2 → 6)-α-D-Gal*p*NAc-(1 → L-Ser or L-Thr
(42)

α-D-Gal*p*NAc-(1 → L-Ser or L-Thr
(43)

URINARY GLYCOPROTEINS

Gel filtration has been used in conjunction with specific automated analyses for the separation of high and low molecular weight urinary material with a new potential for diagnosis based on the elution profiles obtained. An increased urinary excretion of the D-glucosyl tetrasaccharide (44) has been detected in patients with Duchenne muscular dystrophy. This oligosaccharide was originally reported in the urine of a patient with glycogen storage disease type II (Pompe's disease). A glycoprotein composed of sub-units (mol. wts. 1.5×10^4) and exhibiting marked gastric antisecretory activity has been isolated from human urine.

High yields of this inhibitor have been recovered by sub-unit exchange chromatography, in which protein sub-units are immobilized on a solid matrix and interact with sub-units in solution. An increased urinary excretion of free *N*-acetylneuraminic acid has been detected in patients with Salla disease, which is a newly reported type of lysosomal storage disorder. Patients appear to lack an *N*-acetylneuraminyl lyase or some lysosomal enzyme responsible for the degradation of *N*-acetylneuraminic acid. The urine of a fucosidosis patient contained large amounts of L-fucosyl glycopeptides. By use of a combination of gel chromatography, methylation analysis, and digestion with (*exo*)-glycoside hydrolases, the structures [(45)-(49)] have been established for these glycopeptides. The related L-fucosyl glycopeptide analogue (50) has also been characterized from the urine of a patient with fucosidosis. The structures of twenty-two urinary glycopeptides from a fucosidosis patient have been established. All contain one L-fucosyl residue at either O-3 or O-6 of the 2-acetamido-2-deoxy-D-glucosyl residue linked to L-asparagine.

The accumulation of glycopeptides in the urine may be caused by the inability of human *endo-β*-D-2-acetamido-2-deoxy-glucanase to cleave the L-asparagine-linked sugar chains that bear an L-fucosyl residue on the innermost 2-acetamido-2-deoxy-D-glucosyl residue. Although homogeneous on gel filtration, human urinary β_2-microglobulin can be separated into three zones on gel electrophoresis. The partial sequences of amino-acids in rabbit and rat urinary β_2-microglobulins have been reported. Tamm-Horsfall glycoprotein has been isolated from calf urine. Mild acid hydrolysis cleaved the glycoprotein into two lower molecular weight fragments (mol. wts. 6.6×10^4 and 5.1×10^4). By the theoretical addition of the structural sequences [(51) and (52)] to the oligosaccharides isolated from human urine from mannosidosis, fucosidosis, or sialidosis patients, numerous structures of L-asparaginyl glycans can be proposed.

It is postulated that these structures, which have not yet been characterized, pre-exist in glycans of human glycoproteins, probably in the cytoplasm and/or the cell membrane, and that they are the products of the action of endo-β-D-2-acetamido-2-deoxy-glucanases and accumulate in the cells and then in the urine.

α-D-Glc*p*-(1 → 6)-α-D-Glc*p*-(1 → 4)-α-D-Glc*p*-(1 → 4)-D-Glc
(44)

α-L-Fuc*p*
1
↓
3

β-D-Gal*p*-(1 → 4)-β-D-Glc*p*NAc-(1 → 2 or 4)-α-D-man*p*
1
↓
6
β-D-Man*p*-(1 → 4)-D-Glc*p*NAc
3
↑
1
β-D-Gal*p*-(1 → 4)-β-D-Glc*p*NAc-(1 → 2 or 4)-α-D-man*p*
3
↑
1
α-L-Fuc*p*

(45)

β-D-Gal*p*-(1 → 4)-β-D-Glc*p*NAc-(1 → 2)-α-D-man*p*-(1 → 3)-β-Man*p*-(1 → 4)-D-Glc*p*NAc
3
↑
1
α-L-Fuc*p*

(46)

β-D-Gal*p*-(1 → 4)-β-D-Glc*p*NAc-(1 → 4)-α-D-man*p*-(1 → 3)-β-Man*p*-(1 → 4)-D-Glc*p*NAc
3
↑
1
α-L-Fuc*p*

(47)

β-D-Gal*p*-(1 → 4)-β-D-Glc*p*NAc-(1 → 2)-α-D-man*p*-(1 → 6)-β-D-Man*p*-(1 → 4)-D-Glc*p*NAc
3
↑
1
α-L-Fuc*p*

(48)

β-D-Gal*p*-(1 → 4)-β-D-Glc*p*NAc-(1 → 4)-α-D-Man*p*-(1 → 6)-β-D-Man*p*-(1 → 4)-D-Glc*p*NAc
3
↑
1
α-L-Fuc*p*

(49)

β-D-Gal*p*-(1→4)-β-D-Glc*p*NAc-(1→2)-α-D-Man*p*-(1→3 or 6)-β-Man*p*-(1→4)-β-D-Glc*p*NAc(1→4)-D-Glc*p*NAc-(1→L-Asn
3 | 3 or 6
↑ | ↑
1 | 1
α-L-Fuc*p* | α-L-Fuc*p*

(50)

→ 4)-β-D-Glc*p*NAc-(1 → Asn

(51)

→ 4)-β-D-Glc*p*NAc-(1 → L-Asn
3 or 6
↑
1
α-L-Fuc*p*

(52)

AVIAN GLYCOPROTEINS

Interactions of an ovalbumin glycopeptide with concanavalin A have been measured using solvent proton relaxation rates over a wide range of magnetic fields. The binding of the glycopeptide to the lectin reduces the solvent proton relaxation rates in the same manner, but with a lower magnitude, as simple saccharides such as methyl α-D-glucopyranoside. Data from a Raman difference spectroscopic investigation have led to the suggestion that the conversion of ovalbumin into the heat-stable form S-ovalbumin involves a conformational change in a small part (3-4%) of the protein from α-helix to anti-parallel β-sheet geometry. This small difference in the three dimensional arrangement of the peptide chain may contribute to the large difference in the thermodynamic stabilities of the two forms.

```
                    β-D-GlcNAc
                        1
                        ↓
                        4
β-D-GlcpNAc-(1 → 2)-α-D-Manp
                        1
                        ↓
                        3
β-D-GlcpNAc-(1 → 4)-α-D-Manp-(1 → 4)-β-D-GlcpNAc-(1 → 4)-β-D-GlcpNAc-(1 → L-Asn
                        6
                        ↑
                        1
β-D-GlcpNAc-(1 → 2)-α-D-Manp
```

(53)

```
β-D-GlcpNAc-(1 → 2)-α-D-Manp
                        1
                        ↓
                        3
β-D-GlcpNAc-(1 → 4)-α-D-Manp-(1 → 4)-β-D-GlcpNAc-(1 → 4)-β-D-GlcpNAc-(1 → L-Asn
                        6
                        ↑
                        1
β-D-GlcpNAc-(1 → 2)-α-D-Manp
                        4
                        ↑
                        1
                    β-D-GlcpNAc
```

(53)

Two structures I(53) and (54)] are proposed for the unique glycan of hen egg-white ovotransferrin. A comparative study of this glycan with those of human serotransferrin and lactotransferrin reveals profound differences that could form the basis for the specificity of recognition of target cells by these glycoproteins. In addition, the structures of the four avian glycoproteins of egg white (ovo-transferrin, ovalbumin, ovomucoid, and ovomucin) are distinct even though they are all synthesized by the oviduct. It is assumed that each of them is formed in specialized cells containing different glycosyltransferases.

The protein moiety may also exert a control over the activity of the glycosyltransferases through the mechanism of polypeptide signals specifically recognized by each of these enzymes.

MISCELLANEOUS GLYCOPROTEINS AND CHITIN

A heterogeneous polysaccharide, containing residues of D-galactose, D-mannose, L-fucose, D-glucose, 2-acetamido-2-deoxy-D-glucose, and 2-acetamido-2-deoxy-D-galactose, which activates the classical complement pathway has been isolated from tropical ant *(Pseudomyrmex* species) venom. A low molecular weight fraction (3×10^5), which may be a cleavage product of a larger polysaccharide, was isolated by gel chromatography. A glycoprotein (mol. wt. 4.5×10^4) composed of carbohydrate (36%) and protein (5 3%) has been isolated from the articular lubricating fraction of bovine synovial fluid.

It is likely that this glycoprotein is synthesized by the joint tissues and is not derived from serum. Crab chitin, and chitin from the diatom *Thalassiorira fluviatilis* dissolves in hexafluoro-2-propanol; far-u.v. and c.d. spectra of solutions, gels, and films of the chitin have been recorded. A trans-amide conformation and intermolecular hydrogen bonding are proposed as the important determinants of the observed solid state c.d. spectra. Chitins and chitosans from a variety of sources have simple e.s.r. spectra, provided they have been isolated and purified under controlled conditions to avoid excessive oxidative oxidation.

The primary product of the thermal degradation of chitin, 2-acetamido-1,6-anhydro-2-deoxy-β-D-glucopyranose, has been detected and isolated on a preparative scale. Chitin has been treated with isocyanate or acid chloride derivatives of pesticides to produce controlled-release polymeric pesticide systems. The action of lysozyme on partially deacetylated chitin and chitin oligosaccharides has been studied. Chitin was deacetylated to an extent of about 70% of its 2-acetamido-2-deoxy-D-glucosyl residues, and lysozyme digests of the deacetylated chitin were fractionated by gel chromatography, paper chromatography, and paper electrophoresis.

ANALYSIS OF GLYCOPROTEINS

Wheat-germ agglutinin has been immobilized in the presence of chito-oligosaccharides to xive an affinity support for the purification of glycoproteins. Although the resulting immobilized lectin has a high capacity for glycoproteins, possible hydrophobic interactions as well as non-specific adsorption effects may limit the yield and degree of purification achieved with this support. Glycoproteins, after electrophoresis in polyacrylamide or agarose gels, have been visualized after incubation with fluorescein-labelled concanavalin A. The method has been used to detect less than 100 ng of hexose bound to fibrinogen.

Glycoproteins have been detected in polyacrylamide gels using sequential treatment which includes treatment with thymol and staining with alcoholic solutions of H_2SO_4. Colours are produced in proportion to the carbohydrate content of the glycoprotein. A laser light scattering technique has been used for the measurement of the concentration of glycoproteins when complexed with lectins. A high degree of sensitivity has allowed less than 25 pmol glycoprotein or 4.5 pmol lectin to be detected. Methods have been reported for the ultrastructural localization of complex carbohydrates, including glycoproteins, in rat macrophages and monocytes, by employing the reactivity of vicinal glycols to periodate ions and the presence of carboxylate and sulphate ester groups. The neutral sugar components in human serum glycoproteins have been identified and determined by hydrolysis of the glycoprotein with ion-exchange resin followed by g.l.c. as the alditol acetates.

Hydrolysis conditions were optimized for cleavage of glycosidic bonds with only minimal destruction of L-fucose, D-mannose, and D-galactose. Tritium-labelled alditol acetates, derived from monosaccharides released by acid hydrolysis of glycoproteins and lipopoly-saccharides, have been subjected to g.l.c. analysis. By trapping the gas effluent on silicone-coated glass-beads and determining the level of radioactivity, 0.2 mmol of monosaccharide have been detected with an accuracy of ±10 – 15%. After methanolysis of glycoproteins and *O*-trimethylsilylation of the mono-saccharides, re-*N*-acetylation of amino-sugars has been performed directly on gas liquid chromatographic columns, by injecting acetic anhydride together with the sample.

Quantitative *N*-acetylation occurs whilst the *O*-trimethylsilyl ethers remain intact. By using a two column system on a commercially available analyser, amino-acids and the α- and β-anomers of 2-amino-2-deoxy-D-glucose and 2-amino-2-deoxy-D-galactose have been separated in the 100-1000 pmol range.

The separation of the anomeric forms of the amino-sugars is critically dependent on pH. In a method used for glycoprotein analysis, amino-sugars and several L-cysteine derivatives have been resolved on a single column amino-acid analyser. Monosaccharides, derived from polysaccharides, glycolipids, and glycol-proteins, have been converted into the corresponding 1-amino-1-deoxy-glycitols by treatment with ammonia and sodium cyanoborohydride, prior to determination using an amino-acid analyser. The content of L-fucose and D-mannose in glycoproteins has been determined by h.p.l.c. of the perbenzoyl-L-[^{3}H]fucitol and D-[^{3}H]mannitol derivatives. Since D-galactitol and D-glucitol cannot be separated from D-mannitol using this system, samples containing these polyols must be subjected to electrophoresis prior to h.p.l.c. The direct quantitative comparison of sugars and amino-acids by h.p.l.c. has been reported. Stepwise elution systems have been developed for the ion-exchange liquid chromatographic separation of these components. A modified screening method using t.l.c. has been reported for the rapid diagnosis of glycoprotein storage disorders which are identifiable by their secretion of excess oligosaccharides.

The method has been used in the diagnosis of mannosidosis, G_{M1}-gangliosidosis, mucopolysaccharidoses type I and type III, aspartylglycosaminuria, and fucosidosis. Oligosaccharides derived from glycoproteins have been separated by t.l.c. Apart from its diagnostic value in detecting glycoprotein-derived storage oligo-saccharides in glycohydrolase deficiency diseases, the system may be used in the analysis of oligosaccharides released from glycoproteins or glycolipids by either chemical or enzymic methods. Oligosaccharides partly degraded at the reducing end can be separated from low molecular weight peptide degradation products of glycoproteins which have been treated with mixtures of trifluoroacetic anhydride and trifluoroacetic acid. Carbohydrate chains attached to the protein by *N*-glycosyl linkages are cleaved by transamidations of the N^4-(2-acetamido-2-deoxy-β-D-glucopyranosyl) hydrogen L-asparaginate-linkage and carbohydrate chains attached by *O*-glycosyl linkages are cleaved, probably by acid-catalysed elimination, from L-serine or L-threonine residues. Radiolabelled glycopeptides have been analysed by polyacrylamide gel electrophoresis as their borate complexes.

Heterogeneity, which is not apparent after gel filtration, is observed for glycopeptides derived from immunoglobulin G and Sindbis virus glycoprotein. An analytical method, applicable to the assay of glycoprotein glycosyltransferases, has been developed.

BIOSYNTHESIS OF GLYCOPROTEINS

The thyroid contains a microsomal enzyme that can selectively release D-glucose from dolichyl-linked D-gluco-D-manno-oligosaccharides. This D-glucosidase may carry out the first step in a series of processing reactions of the kind which have been proposed for vesicular stomatitis viral glycoprotein. Since the thyroid D-glucosidase can act on lipid-linked as well as protein-bound oligo-saccharides, it may have an additional physiological function involving the regulation of the amount of appropriate oligosaccharide-lipid donors available for the glycosylation of proteins. The transfer of the D-glucosyl oligosaccharide from its lipid carrier to endogeneous proteins *via* an L-asparagine linkage is carried out by a calf thyroid particulate enzyme classified as a dolichyl pyro-phosphoryl oligosaccharide : protein oligosaccharide transferase.

Although the entire D-glucosyl oligosaccharide moiety is initially transferred to the protein, a selective loss of D-glucosyl residues for the newly biosynthesized glycoprotein occurs shortly afterwards owing to the D-glucosidase, which may represent the first of a series of processing enzymes. In glycoprotein biosynthesis, two D-glucosidases from rat-liver microsomes are responsible for the cleavage of D-glucosyl residues from the pre-formed oligosaccharides (55). The transfer of D-glucose, from UDP-D-glucose to dolichyl D-glucopyranosyl phosphate, a glucosylated oligosaccharide lipid and at least one D-glucoprotein, catalysed by membranes of central nervous tissue has been reported.

$$\text{(D-Glc}p)_n\text{-(D-Man}p)_9\text{-D-Glc}p\text{NAc}$$

$$(55)\ n = 1, 2, \text{or } 3$$

Amphomycin specifically inhibits the formation of dolichyl D-mannosyl phosphate in pig aorta microsomal preparations, but does not inhibit the further transfer of D-mannose from dolichyl D-mannosyl phosphate or from lipid-linked oligosaccharides. However, some of the D-mannosyl residues in the lipid-linked oligosaccharides come directly from GDP-D-mannose without the involvement of dolichyl D-mannosyl phosphate. The enzyme responsible for the transfer of oligosaccharide from lipid-linked intermediates to endogenous proteins of rat liver is a structural component of the rough and smooth liver microsomes.

Rat-liver mitochondria contain the highest specific activity of glycosyltransferase activity for D-mannose, D-glucose, and 2-acetamido-2-deoxy-D-glucose 6-phosphate from nucleotide donors to dolichol monophosphate in vesicles derived from rough and smooth endoplasmic reticulum and mitochondria. In the smooth vesicles, the D-mannosyltransferase and D-glucosyltransferases are located at both the inner and outer faces of the membrane, whereas in the rough vesicles and mitochondria the two enzymes occur only at the outer face. The primary structural requirements for the enzymic formation of *N*-glycosidic bonds in glycoproteins have been investigated. Much of the primary structure of the acceptor protein is not required for an oligosaccharide transferase to interact with the acceptor site. The tripeptide (56) is the minimal peptide active as an acceptor of oligosaccharide from oligosaccharide-lipid and only if both its *N*- and *C*-termini are blocked.

The accessibility of this sequence to the oligosaccharide transferase is probably the predominant structural feature of a protein that determines whether or not it is glycosylated. Low molecular weight synthetic peptides, containing the sequence (57) and larger peptides, can be *O*-glycosylated with a crude preparation of UDP-2-acetamido-2-deoxy-D-galactose: mucin polypeptide 2-acetamido-2-deoxy-D-galactosyltransferase from porcine submaxillary glands and UDP-2-acetamido-2-deoxy-D-galactose. The peptides have sequences similar to the region of L-threonine-98 of bovine basic myelin protein.

L-Asn-X-L-Ser/L-Thr
(56)

L-Thr-L-Pro-L-Pro-L-Pro
(57)

Cancer-associated D-galactosyltransferase (GT-II) and the normal isoenzyme (GT-I) have been purified from pooled effusions from patients with various cancers. The purified cancer-associated enzyme appears to be structurally and kinetically distinct from the normal isoenzyme. Optimal conditions for the determination of UDP-D-galactosyltransferase in the Golgi system of rat liver have been reported. In contrast to other reports the enzyme is not stimulated by non-ionic detergents. The α-2- and α-3-L-fucosyltransferase activities of human cells and plasma are inhibited by N-ethylmaleimide, but activities are restored by treatment with dithiothreitol, indicating the necessity of a free thiol group for activity. CMP-*N*-Acetylneuraminate : D-galactosylglycoprotein *N*-acetylneuraminyl-transferase (sialyltransferase) activity is released in large amounts by two hepatoma cell lines.

The enzyme has been resolved into two forms (mol. wts. 6.5×10^4 and 8.0×10^4) by gel chromatography. Sialyltransferase activity has been studied in renal cortex tissue using asialofetuin and Tamm-Horsfall glycoproteins as exogenous acceptors and rat asialo-glomerular basement membrane as an endogenous acceptor. The activities of sialyltransferase and D-galactosyltransferase of rat-liver Golgi membranes are enhanced to different levels by binding to wheat-germ agglutinin. The stimulation of the transferase activities is similar to that reported for concanavalin A.

Differences in behaviour of the transferases support earlier conclusions that the two enzymes occupy distinct membrane environments.

The substrate requirements, linkage specificity, and kinetic mechanism of an α-2,3-sialyltransferase have been compared with an α-2,6-sialyltransferase from porcine submaxillary mucin. The former enzyme will specifically modify glycoproteins and glycolipids which contain 2-acetamido-2-deoxy-3-*O*-β-D-galactosyl-D-galactose sequences. The only acceptor substrates reported for the α-2,6-sialyltransferase are those glycoproteins containing the structure (58).

R-(1 → 3)-D-Gal*p*NAc-α-(1 → L-Thr/Ser
(58) R = H or β-D-Gal*p*

CHAPTER

4 Proteins Diversity

Now that some of the major types of protein structures have been described it is appropriate to turn to the question of how protein structure relates to protein function. To explore this question, two protein systems, hemoglobin and the actin-myosin complex, are examined in detail.

Hemoglobin

Hemoglobin is the principal soluble protein in the red blood cells (erythrocytes). It is an oxygen-binding protein, the main task of which is to pick up oxygen as the blood passes through the lungs and to release it in the capillaries of other tissues. A closely related protein, myoglobin, serves to store oxygen temporarily in the muscles and other peripheral tissues, where the oxygen is consumed by cellular metabolism. Myoglobin consists of a single polypeptide subunit with a molecular weight of 16,900 and a bound cofactor, heme. Hemoglobin has an $\alpha_2\beta_2$ subunit

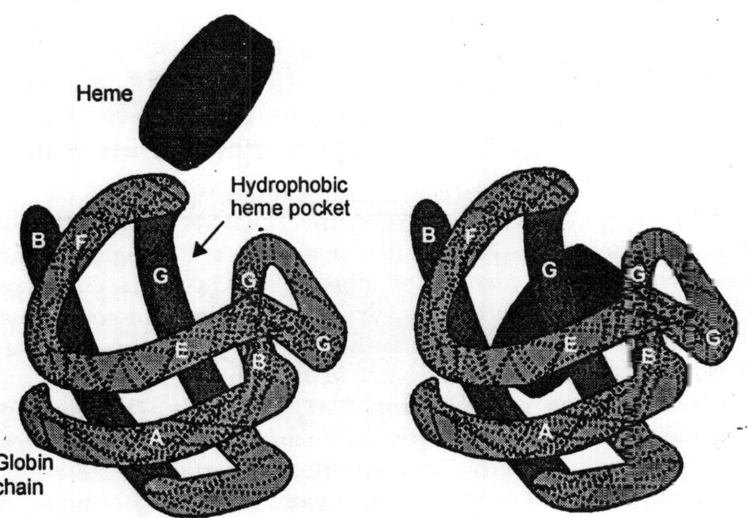

Fig. 4.1. The heme pocket. The helices of hemoglobin (and myoglobin) form a hydrophobic pocket for the heme and provide an environment where the iron atom reversibly binds oxygen.

structure, with a heme bound to each of the four subunits. The α and β subunits are structurally very similar to each other and to the single subunit of myoglobin. (The α subunits have 141 amino acids; the β subunits, 146.) The heme cofacter in each subunit consists of a porphyrin (protoporphyrin IX) with a central iron atom. (The structure of heme is shown in fig. 4.1. Fig. 4.1 shows the overall folding plan of the polypeptide chains of hemoglobin and myoglobin. Each chain has eight stretches of α helix, which are labeled A–H. These helical segments fold together to form a pocket that surrounds the heme.

The amino acid side chains lining the heme-binding pocket are predominantly apolar, and thus are well suited to binding the hydrophobic porphyrin ring. This apolar, water free environment allows the Fe^{2+} iron atom to bind O_2 reversibly while minimizing the opportunity for oxidation of the iron to the +3 state. Although the components of myoglobin and hemoglobin are remarkably similar, their physiological responses are very different.

On the basis of weight, each molecule binds about the same amount of oxygen at high oxygen tensions (pressures). At low oxygen tensions, however, hemoglobin releases bound oxygen much more readily. These differences are reflected in the oxygen-binding curves of the purified proteins in aqueous solution (Fig. 4.2). The oxygen-binding curve for myoglobin (Mb) is hyperbolic in shape, as is expected for a simple one-to-one association of myoglobin and oxygen :

$$Mb + O_2 \rightleftharpoons MbO_2$$

$$K_f = \frac{[MbO_2]}{[Mb][O_2]} = \text{equilibrium formation constant}$$

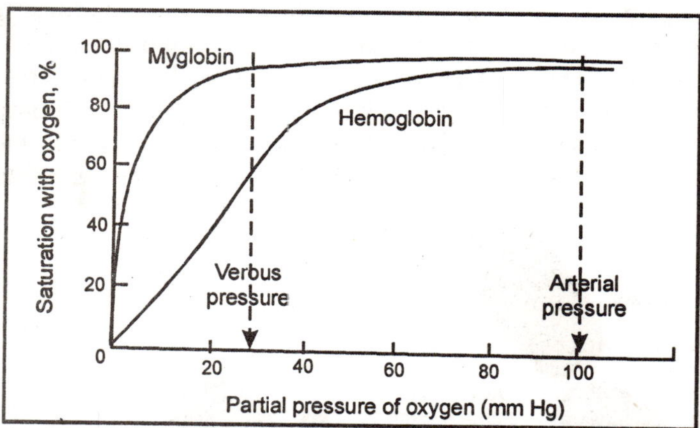

Fig. 4.2. Equilibrium curves measure the affinity for oxygen of hemoglobin and of the simpler myoglobin molecule. Myoglobin, a protein of muscle, has just one polypeptide chain and resembles a single subunit of hemoglobin. The vertical axis gives the amount of oxygen bound to one of these proteins, expressed as a percentage of the total amount that can be bound. The horizontal axis measures the partial pressure of oxygen in a mixture of gases with which the solution is allowed to reach equilibrium. For myoglobin, the equilibrium curve is hyperbolic. Myoglobin absorbs oxygen readily but becomes saturated at a low pressure. The hemoglobin curve is sigmoidal. Initially, hemoglobin is reluctant to take up oxygen, but its affinity increases with oxygen uptake. At arterial oxygen pressure, both molecules are nearly saturated, but at venous pressure, myoglobin gives up only about 10% of its oxygen, whereas hemoglobin releases roughly half. At any partial pressure, myoglobin has a higher affinity than hemoglobin, which allows oxygen to be transferred from blood to muscle.

If y is the fraction of myoglobin molecules saturated, and if the oxygen concentration is expressed in terms of the partial pressure of oxygen $[O_2]$,[a] then

$$K_f = \frac{y}{[1-y][O_2]} \quad \text{and} \quad y = \frac{K_f O_2^{\,n}}{1+K_f[O_2]} \qquad ...(4.1)$$

This is the equation of a hyperbola, as shown in figure 4.2. Hemoglobin (Hb) behaves differently. Its sigmoidal binding curve can be fitted by an association-constant expression with a greater-than-first-power dependence on the oxygen concentration:

$$K_f = \frac{[HbO_2]}{[Hb][O_2]^n} \quad \text{and} \quad y = \frac{K_f O_2^{\,n}}{1+K_f O_2^{\,n}} \qquad ...(4.2)$$

Under physiological conditions the value of n is around 2.8. This indicates that the binding of oxygen molecules to the four hemes in hemoglobin does not occur independently: /Binding to any one heme is affected by the state of the other three. The first oxygen attaches itself with the lowest affinity, and successive oxygens are bound with a somewhat higher affinity, and so on. The exact value of n in equation 2 is a function of other factors discussed later on.

In general, a value of $n > 1$ indicates *cooperative binding* of O_2 (or *positive cooperativity*), a value of $n < 1$ indicates anticooperative binding (or *negative cooperativity*), and a value of $n = 1$ indicates no cooperativity. The cooperative binding of oxygen by hemoglobin is ideally suited to the conditions involved in oxygen transport. In the lungs, where the oxygen tension is relatively high, hemoglobin can become nearly saturated with oxygen.

In the tissues, however, where the oxygen tension is relatively low, hemoglobin can release about half its oxygen (fig. 4.2). If myoglobin were used as the oxygen transporter, less than 10% of the oxygen would be released under similar conditions. The positive cooperativity associated with oxygen binding to hemoglobin is a special case of *allostery* in which the binding of "substrate" to one site stimulates or inhibits the binding of "substrate" to another site on the same multisubunit protein. Allostery is also commonly observed in regulatory proteins.

The Binding of Certain Factors to Hemoglobin

The combination of Hb with O_2 is not only a function of the oxygen tension but also depends on the pH, CO_2 concentration, and *glycerate-2,3-bisphosphate* (GBP or BPG). GBP (fig. 4.3) binds preferentially to the deoxygenated form of hemoglobin with a dissociation constant of about $10^{-3}M^{-1}$. Its dissociation constant with HbO_2 is about $10^{-3}M^{-1}$. Since the concentration of GBP and hemoglobin are both about 5 mM in the erythrocyte, we expect most of the deoxy form to be complexed with GBP and much of the oxyhemoglobin to be free of GBP.

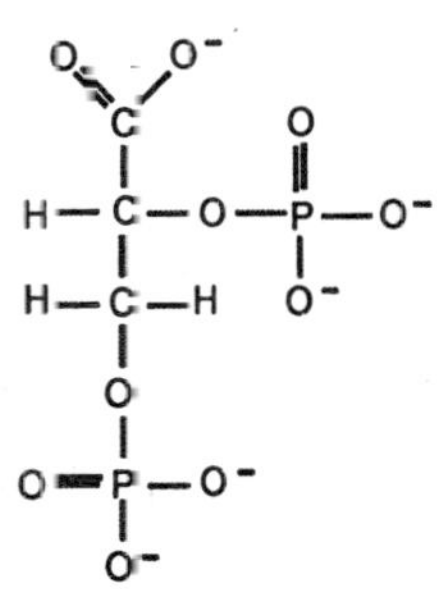

Fig. 4.3. The structure of glycerate-2,3-bisphosphate, an allosteric effector for hemoglobin oxygen release.

The net effect of the GBP is to shift the oxygen-binding curve to higher oxygen tensions (fig. 4.4). This shift is not sufficient to lower the binding of oxygen at the high oxygen tensions in the capillary tissues of the lungs, but it is sufficient to cause a substantially greater release of oxygen at the lower oxygen tensions that exist in the peripheral tissues.

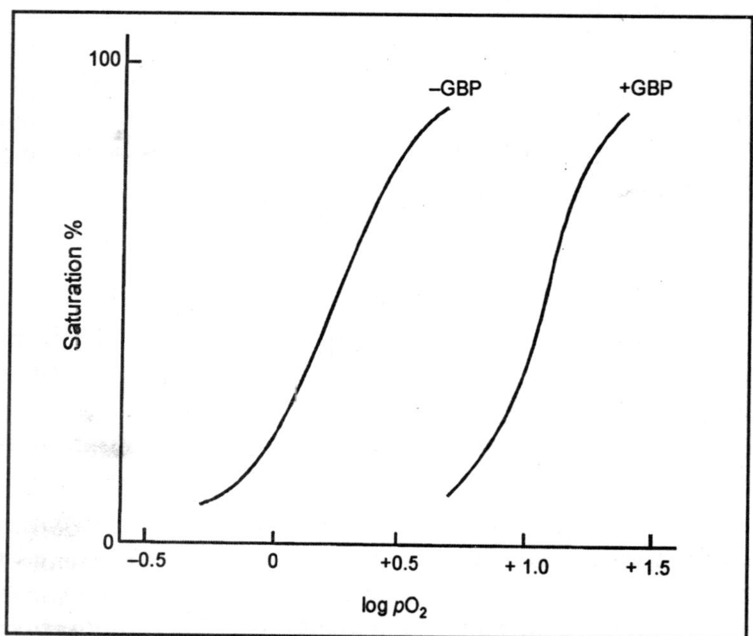

Fig. 4.4. Oxygen-binding curve for hemoglobin as a function of the partial pressure of oxygen. Two curves are shown: one in the absence and one in the presence of glycerate-2,3-bisphosphate (GBP). GBP decreases the affinity between oxygen and hemoglobin, as shown by the displacement of the binding curve to high oxygen concentrations in its presence.

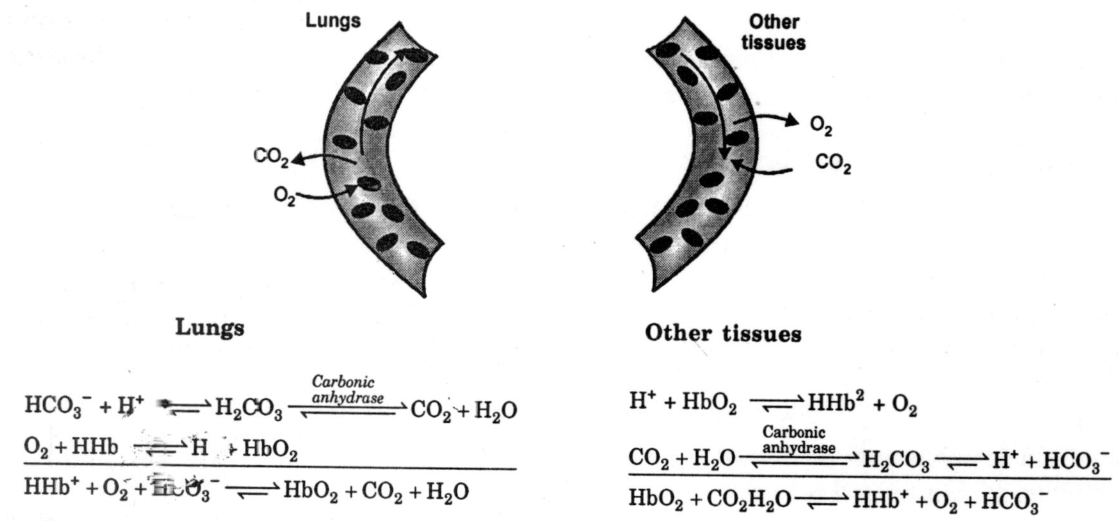

Fig. 4.5 Transport of oxygen (O_2) and carbon dioxide (CO_2) in the circulatory system. In most tissues O_3 is released and CO_2 is with drawn by the red blood cells; in the lungs these processes are reversed.

Carbon dioxide (CO_2) and hydrogen ions (H^+) both have negative effects on oxygen binding also. These effects are closely related and together are known as the Bohr effect. A substantial amount of CO_2 generated by decarboxylation reactions of intermediary metabolism diffuses from cells through interstitial fluid into the blood plasma, with only a small fraction becoming hydrated as carbonic acid. This is because the nonenzymatic hydration of CO_2 to form H_2CO_3 is a slow reaction, with a half-time of about 10 s. On entry into erythrocytes, hydration of CO_2 occurs in a reaction that is catalyzed rapidly by carbonic anhydrase, an enzyme that is localized in the erythrocytes. At the pH of blood the carbonic acid largely dissociates into H^+ and HCO_3^-:

$$CO_2 + H_2O \xrightleftharpoons[]{\text{Carbonic anhydrase}} H_2CO_3 \rightleftharpoons H^+ + HCO_3^-$$

The release of protons through these two consecutive reactions explains how increasing concentrations of CO_2 cause a lowering of pH. Lowering the pH decreases the affinity of hemoglobin for oxygen, because protons, like GBP, bind preferentially to deoxyhemoglobin. For example, at pH 7.6 and an oxygen tension of 40 mm Hg, hemoglobin retains more than 80% of its oxygen; if the pH is lowered to 6.8, only 45% of the oxygen is retained.

The negative effect of CO_2 on oxygen binding is mainly due to the tendency of CO_2 to lower the pH (i.e., raise the H^+ concentration); however, HCO_3^- also binds preferentially to deoxyhemoglobin, and this binding also promotes the release of O_2. The Bohr effect is closely related to the major roles that hemoglobin plays in disposing of the CO_2 produced in tissues, and in controlling the blood pH. While oxygen is being delivered to the tissues in the venous blood, the CO_2 is being removed from the tissues (Fig. 4.5). The CO_2 that diffuses into the erythrocytes is rapidly converted into carbonic acid, which in turn dissociates into H^+ and HCO_3.

The protons produced by this dissociation would lower the pH and reverse the dissociation if it were not for the buffering action of the hemoglobin. Loss of oxygen decreases the acid dissociation constant of the hemoglobin so that it picks up the excess protons. This serves two purposes: it helps to keep the pH constant and it enables the blood to absorb more CO_2. This CO_2 is ultimately released when the blood reaches the lungs and the hemoglobin again becomes oxygenated. Another important point is that the majority of the HCO_3^- produced in the erythrocyte diffuses into the venous blood. Indeed, the pH of the blood system is controlled within narrow limits by the buffering action of the bicarbonate and the hemoglobin with minor assistance from other proteins in the blood. One must marvel at the way various factors work in concert so that hemoglobin can be useful in multiple roles: oxygen deliverer, carbon dioxide remover, and pH stabilizer.

From the explanation we have given it should be clear why the hemoglobin is confined to cells in the blood rather than being present as a free plasma protein. The intracellular carbonic anhydrase and GBP of the erythrocytes are essential for efficient hemoglobin function.

X-Ray Diffraction Studies

X-ray diffraction studies on fully oxygenated hemoglobin and deoxygenated hemoglobin have shown that the molecule is capable of existing in at least two states, with significant differences in tertiary and quaternary structures (Fig. 4.6). Further studies on partially oxygenated hemoglobin may indicate additional intermediate structures between these two extremes. Presently the two-state model serves as a useful conceptual framework for explaining the cooperative nature of oxygen binding.

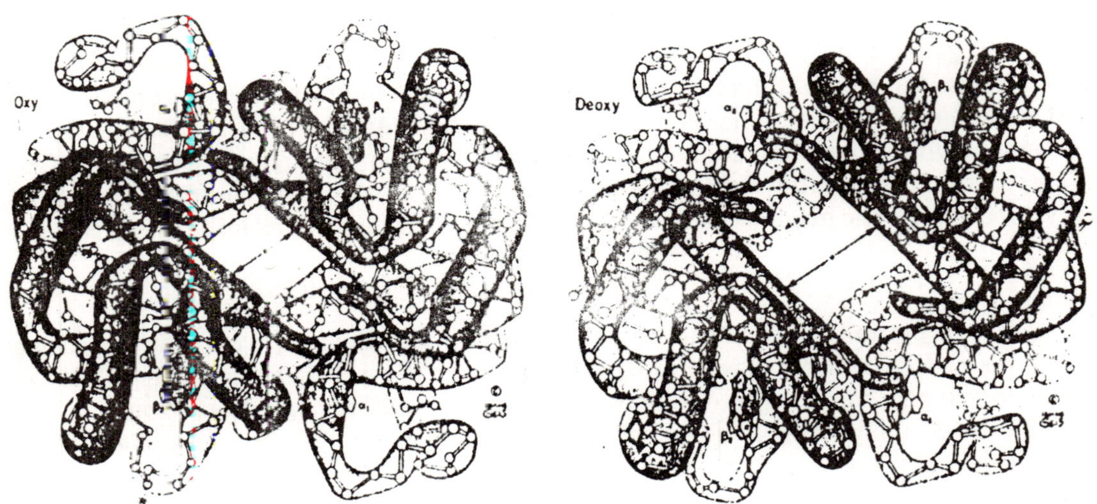

Fig. 4.6. Three-dimensional structure of oxy- and deoxyhemoglobin as determined by x-ray crystallography. This is a view down the twofold symmetry axis, with the β chains on top. In the oxy-deoxy transformation (quaternary motion) $\alpha_1\beta_1$ and $\alpha_2\beta_2$ dimers move as units relative to each other. This allows glycerate-2,3-bisphosphate to bind to the larger central cavity in the deoxy conformation.

The hemoglobin tetramer is composed of two identical halves (dimers), with the $\alpha_2\beta_2$ subunits in one dimer and the $\alpha_2\beta_2$ subunits in the other. The subunits within the dimers are tightly held together while the dimers themselves are capable of motion with respect to each other (Fig. 4.7). The interface between the movable dimers contains a network of salt bridges and hydrogen bonds when hemoglobin is in the deoxy conformation (Fig. 4.8). The quaternary transformation that takes place on binding of oxygen causes the breakage of these bonds. The effects of H^+, HCO_3^-, and *glycerate-2,3-bisphosphate on oxygen binding can be understood in terms of their stabilizing effect on the deoxy conformation.*

The decreased oxygen binding as the pH is lowered from 7.6 to 6.8 suggests the involvement of histidine side chains, because these are the only side-chain groups in proteins that commonly have a pK_a in this range. Certain histidines in the charged form make salt linkages that contribute to the stability of the deoxy form (See Fig. 4.8). As the pH is lowered, these histidines tend to become charged, which increases the stability of the deoxy form. Such a change should inhibit a structural transition to the oxy form and thereby lower the affinity of the protein for oxygen. Similarly, glycerate-2,3-bisphosphate binds most strongly to the deoxy form (Fig. 4.9) and thereby discourages the transition to the oxy form, which lowers the affinity for oxygen.

Bicarbonate ions bind to the amino terminal α-amino groups in hemoglobin; this binding also favours the deoxy conformation. Binding of HCO_3^- is favoured by the high HCO_3^- concentration created by the carbonic anhydrase reaction in the red cells. Hemoglobin thus serves as a carrier of HCO_3^- from tissues to the lungs, where CO_2 is discharged.

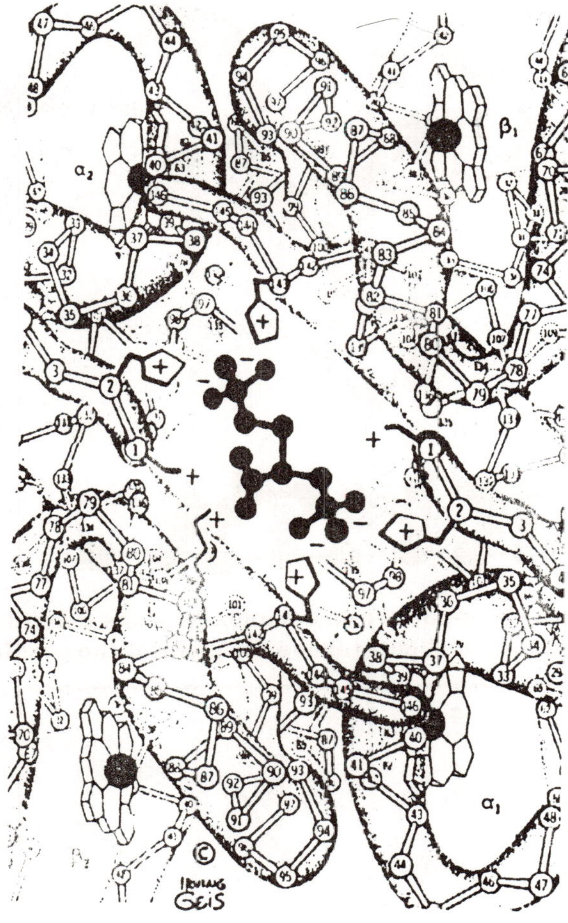

Fig. 4.7. The binding of glycerate-2,3-bisphosphate in the central cavity of deoxyfiemoglobin between β chains. The surrounding positively *charged* residues are the amino terminal, His 2, Lys 82, and His 143.

Oxygen Binding

The oxygen binding at the heme group itself initiates the tertiary- and quaternary-structure changes that are responsible for the cooperative effects seen in hemoglobin. The heme group contains an Fe^{2+} ion located near the center of a porphyrin. This Fe^{2+} makes four single bonds to the nitrogens in the porphyrin ring, and a fifth bond to a histidine side chain of the F helix, histidine F8 (fig. 4.10). When oxygen is present it binds at the sixth coordination position of the iron, on the other side of the heme. Movement of the iron on oxygen binding (or release) pulls the F8 histidine and the F helix to which it is covalently attached (Fig. 4.11). The tertiary-structure change in the F helix of one subunit causes strains elsewhere in this subunit and in the other three subunits as well. The resulting structural reorganization favors the binding of additional oxygen at other unoccupied sites in the tetramer.

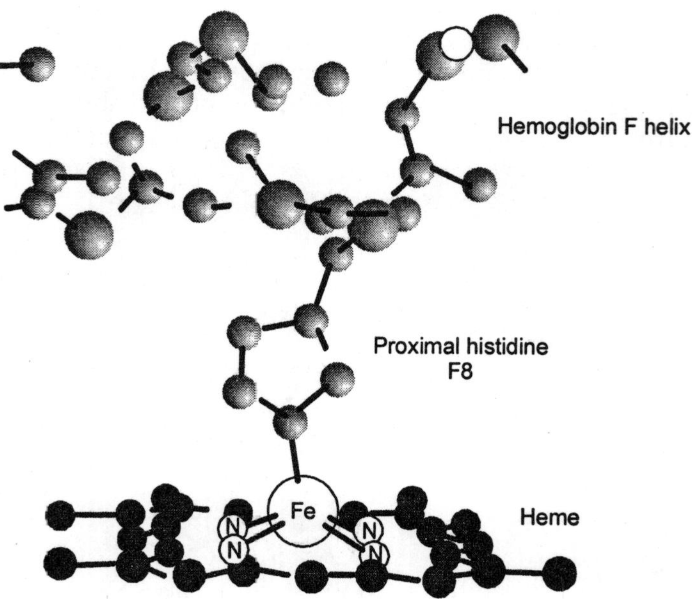

Fig. 4.8. A close-up view of the iron-porphyrin complex with the F helix in deoxyhemoglobin. Note that the iron atom is displaced slightly above the plane of the porphyrin.

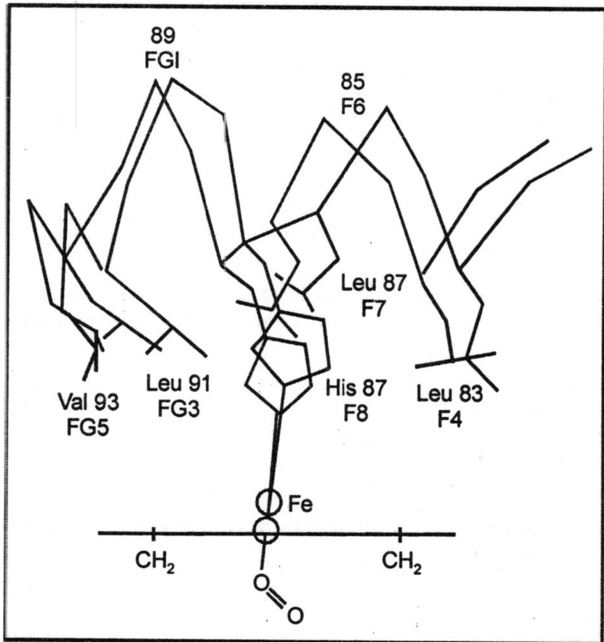

Fig. 4.9. Downward movement of the iron atom and the complexed polypeptide chain on binding oxygen. The structure is shown before and after binding oxygen. Movement of His F8 is transmitted to valine FG5, straining and breaking the hydrogen bond to the penultimate tyrosine. Only the α chain is shown here.

The Fe^{2+} ion is well suited to its job in hemoglobin, not only because it has a natural affinity for O_2, but also because it changes its electron structure in a significant way when it binds O_2. Fe^{2+} is normally paramagnetic as a result of the four unpaired electrons in its outer *(3d)* electron orbitals. As such, it is too large to sit precisely in the plane of a porphyrin, as studies with model compounds have shown. Fe^{2+} is also paramagnetic when it is pentacoordinated in deoxyhemoglobin, and as expected, it is displaced from the plane of the porphyrin by a few tenths of an angstrom.

When O_2 binds, the Fe^{2+} becomes hexacoordinated and dia-magnetic (no unpaired electrons). This change reflects a major reorganization of its *3d* orbitals, which decreases the radius of the Fe^{2+} so that it can move to an energetically more favourable position in the center of the porphyrin.

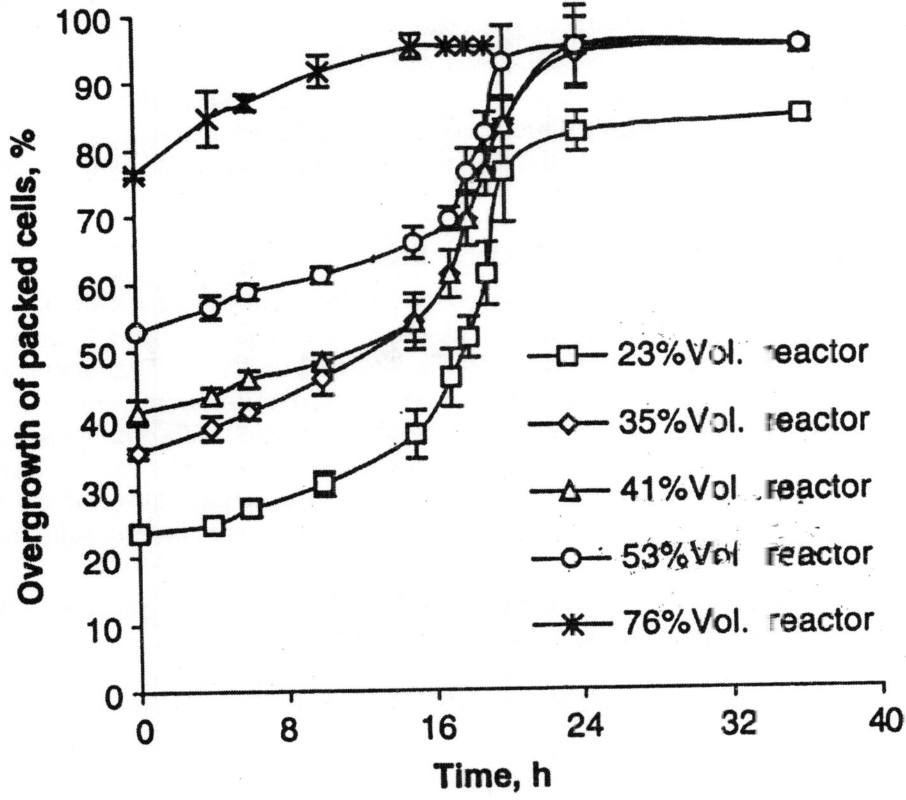

Fig. 4.10. Positions of mutations in hemoglobin that produce a pathological condition. Comparison shows that these mutations, in general, show the same pattern as the distribution of the invariant positions. Dark circles around a numbered position indicate positions of abnormal residues, solid black dot indicates the valine β_6 mutation in sickel cell anemia. Heavy black circles indicate hemoglobin that (is easily) oxidized to the Fe^{3+} from (methemoglobin), and a jagged black perimeter indicates unstable hemoglobin.

The structural arguments advanced here to explain oxygen binding by hemoglobin are supported by comparisons of the amino acid sequences of α and β chains for a large number of hemoglobins from different species. Data from 60 species of a chains and 66 species of β chains reveal 43 invariant positions in the hemoglobin molecule. These invariants are plotted on a map of the hemoglobin structure in figure 4.11. Here the invariant positions are shown by blue dots. In a sense, the positions of these dots provide a diagram of the working machinery of hemoglobin, because they line the heme pockets, where oxygen is bound, and the crucial $\alpha_1\beta_2$ interface, which changes its orientation when oxygen binds.

Electrostatic forces and hydrogen bonds stitch this inferface together in the deoxy conformation when the molecule gives up its oxygen to the tissues. If changes occur, resulting from mutation at any of these positions, then trouble is likely to develop. Fig. 4.10 shows the positions of pathological mutations in hemoglobin that result from changes in the heme pockets or in the $\alpha_1\beta_2$ interface. As a rule, the invariant amino acids are the only critical loci for hemoglobin function. One striking exception occurs at position 6 of the β chain. If a hydrophobic valine residue is substituted for glutamic acid (a charged side chain), disaster results.

The specific consequence of this alteration is to cause hemoglobin tetramers to aggregate when they are in the deoxy state. The aggregates form long fibers that stiffen the normally flexible red blood cell. The resulting distortion of the red cells leads to capillary occlusion, which prevents proper deliveryof oxygen to the tissues. This pathological condition is known as *sickle cell anemia*.

How Hemoglobins and Other Allosteric Proteins Work?

We have spent a great deal of time describing the function of hemoglobin because it is the best understood regulatory protein and provides us with a model system for understanding in general terms how other allosteric proteins work. The first indication that hemoglobin was an allosteric protein came from the sigmoidal shape of its oxygen-binding curve (See Fig. 4.2). Allosteric proteins are usually composed of two or more subunits. Different ligands may bind to quite different sites or to quite similar sites as in the case of hemoglobin. Two quite different models were proposed about 25 years ago to explain the unusual nature of the hemoglobin oxygen-binding curve (Fig. 4.14). These models can also be used as a starting point for discussing other allosteric proteins.

The first model, introduced by Jacque Monod, Jeffery Wyman, and Pierre Changeux in 1965, is called the *symmetry model*. In this model hemoglobin can exist in only two conformations, one with all four of the subunits within a given tetramer in the low affinity form and one with all four subunits in the high-affinity form. In this model, the hemoglobin molecule is always symmetrical, meaning that all the subunits are either in one state or the other and that all of the binding sites within a given tetramer have identical affinities.

The binding of oxygen to one of the subunits favours the transition to the high-affinity form. The greater the number of oxygens binding to the tetramer. the more likely is the transition from the low-affinity form to the high-affinity form. The second model, proposed by Koshl and, Nemethy, and Filmer in 1966, is referred to as the sequential model. In this model the binding of an oxygen molecule to a given subunit causes that subunit to change its conformation to the high-affinity form. Because of its molecular contacts with its neighbours, the change increases the probability that another subunit in the same molecule will switch to the high-affinity form and bind a second oxygen more readily. The binding of the second oxygen has the same type of enhancing effect on the remaining unoccupied oxygen binding sites.

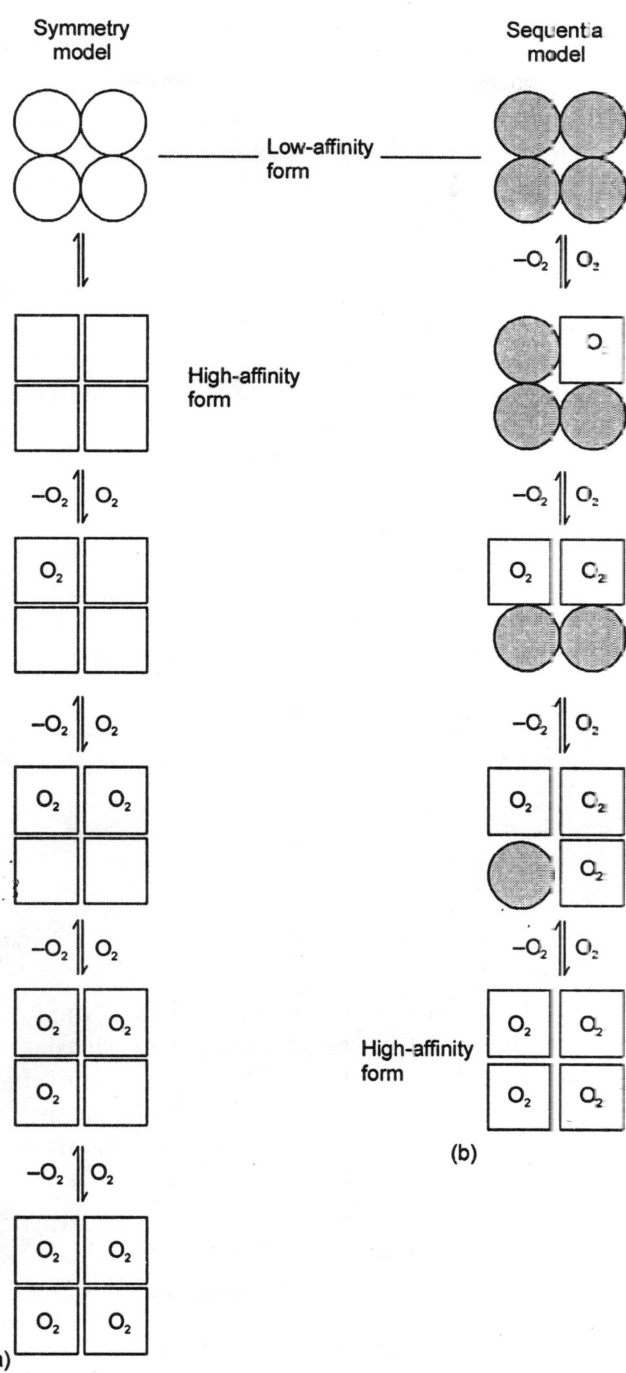

Fig. 4.11. Alternative models for hemoglobin allostery. ***(a)*** **In the symmetry model hemoglobin can exist in only two states,** ***(b)*** **In the sequential model hemoglobin can exist in a number of different states. Only the subunit binding oxygen must be in the high-affinity form.**

Either of these models (or something in between) could account for the sigmoidal oxygen-binding curve of hemoglobin, and either is consistent with the fact that deoxygenated and fully oxygenated hemoglobin have different conformations. The only way to discriminate rigorously between the two models is to obtain structural information on partially oxygenated hemoglobin. This information is still lacking, so no final judgment can yet be made. Even when the situation is fully resolved for hemoglobin, there is no assurance that other aljosteric proteins work in the same way.

Muscle Is an Aggregate of Proteins Involved in Contraction

Vertebrate skeletal muscle represents a remarkable example of a supermolecular aggregate capable of undergoing a reversible reorganization. Voluntary muscle tissue is arranged into *fibers* that are surrounded by an electrically excitable membrane called the *sarcolemma* (Fig. 4.15). Each fiber is composed of many *myofibrils*, which when viewed in the light microscope present a striated and banded appearance. As shown in figure 4.16a myofibril exhibits a longitudinally repeating structure called the *sarcomere*. This 23,000-Å long repeating unit is characterized by the appearance of several distinct bands: the less optically dense band being referred to as the *I band*, and the denser one as the A band. Furthermore, a dense line appears in the center of the *I band*, called the *Z line*; and a dense narrow band somewhat similar in appearance also occurs in the center of the A band, called the *M line*. Adjacent to the M line are regions of the A band that appear less dense than the remainder and are termed the *H zone*.

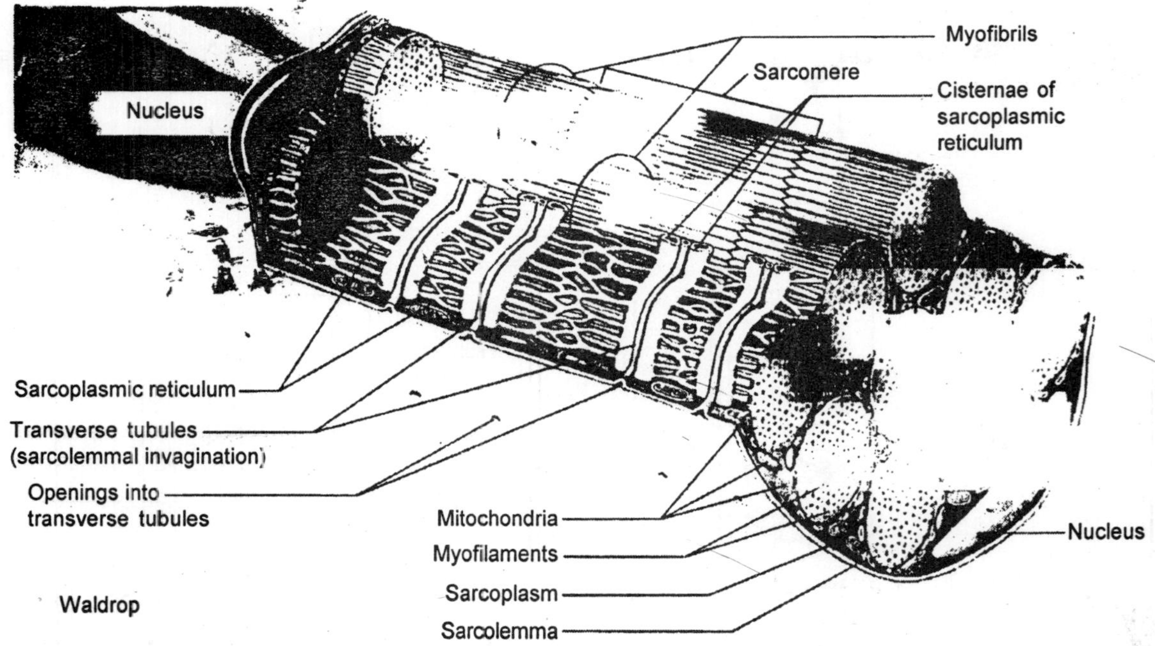

Fig. 4.12. The hierarchy of muscle organization. A voluntary muscle such as the bicep is a composite of many fibres connected to tendons at both ends. Each muscle fiber is composed of several myofibrils that are surrounded by an electrically excitable membrane (sarcolemma). Myofibrils exhibit longitudinally repeating structures called sarcomeres.

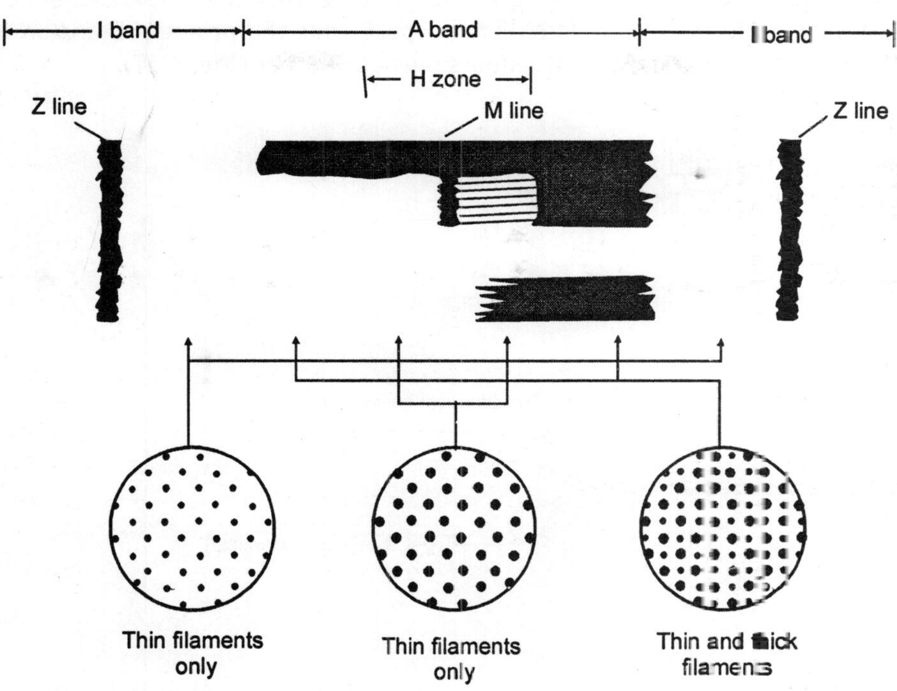

Fig. 4.13. Electron micrograph of a striated muscle sacromere showing the appearance of filamentous structures when cross-sectioned at the locations illustrated below.

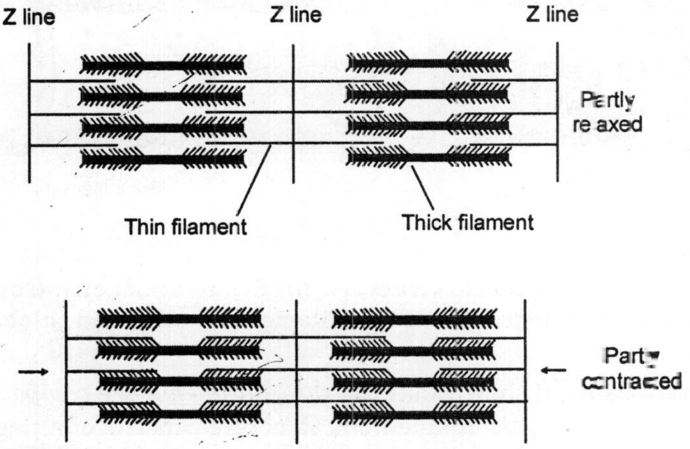

Fig. 4.14. The sliding-filament model of muscle contraction. During contraction, the thick and thin filaments slide past each other so that the overall length of the sarcomere becomes shorter.

Transverse sections of the sarcomere reveal that these patterns result from the interdigitation of two sets of filaments. For example, when a sarcomere is sectioned in the I band, a some what disordered arrangement of thin filaments (each about 70 A in diameter) is seen. In contrast, when sectioned in the H zone, a hexagonal array of thick filaments (each about 150 A in diameter) is apparent. The substantive observation is that a transverse section in the dense region of the

A band shows a regularly packed array of interdigitating thick and thin filaments. It was this observation that led Hugh Huxley (1990) to propose that the process of muscle contraction involved sliding of the thick and thin filaments past each other (Fig. 4.17).

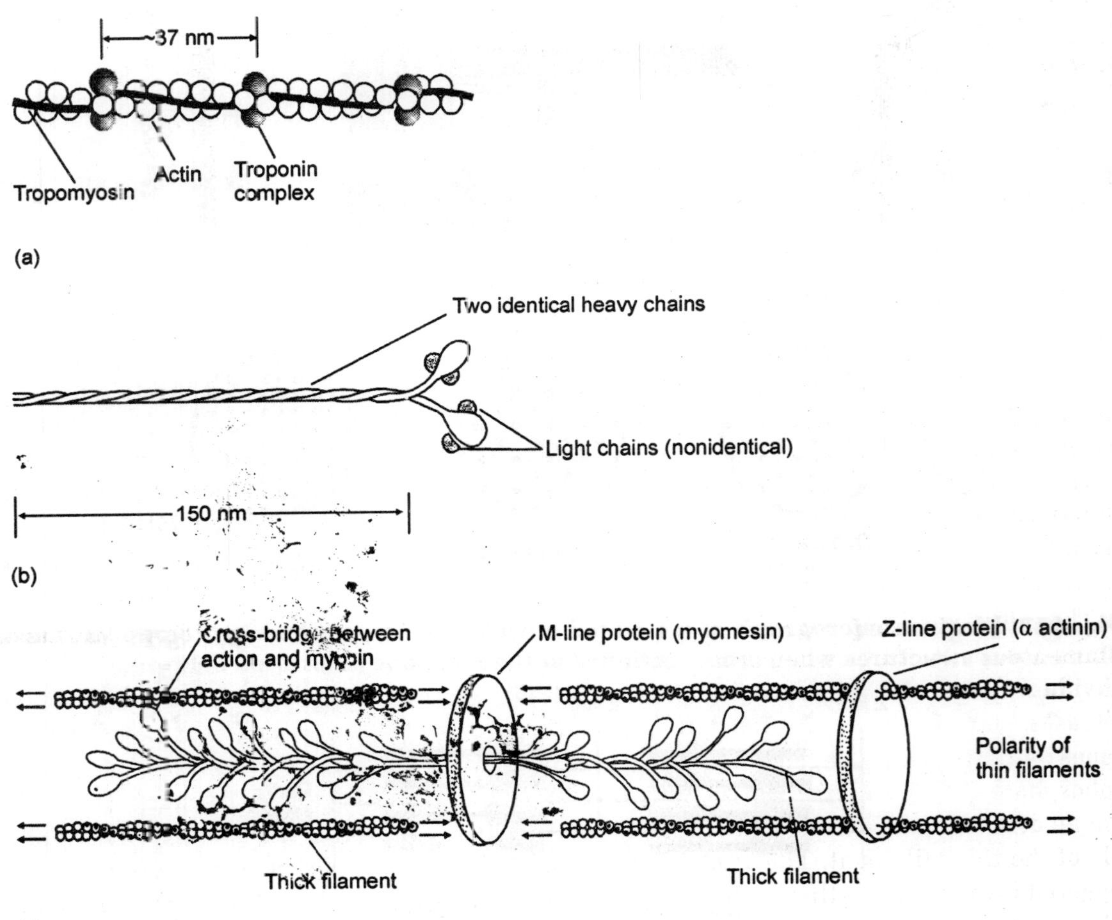

Fig. 4.15. A molecular view of muscle structure. (*a*) Segment of actin-tropomyosin-troponin. (*b*) Segment of myosin. (*c*) Integration of thin filaments (actin) and thick filaments (myosin) in a muscle fiber.

Subsequent analyses have shown that the *thin Filaments are composed of three proteins* (Fig. 4.16 and table 4.1). The main filamentous structure consists of an aggregate of globular actin molecules that takes on the form of a right-handed double helix. The individual actin molecules have a molecular weight of 42,000. Every turn of the actin helix incorporates 14 actin molecules and two molecules of a 360-A long filamentous molecule of tropomyosin (TM) that fits into the two grooves of the actin double helix.

The TM molecule is a dimer of two identical α-helical chains, which wind around each other in a coiled coil. Each TM dimer spans seven actin monomers, and a succession of TM dimers extends the full length of the thin filament. Two molecules of troponin (TN) bind to the actin filament at each helical repeat. Troponin is a complex of three nonidentical subunits: TN-C (a calcium-binding subunit) TN-T (a TM-binding subunit), and TN-A (an "inhibitory" subunit). The TN-TM proteins form a regulatory complex the properties of which are discussed below.

Table 4.1. Principal Proteins of Vertebrate Skeletal Muscle

Protein	*Mr*	*Subunits*	*Function*
Myosin	5,10,000	2 × 223,000 (heavy chains) 22,000 – 18,000 (light chains)	Major component of thick filaments
Actin	42,000	One type	Major component of thin filaments
Tropomyosin	64,000	2 × 32,000	Rodlike protein that binds along the length of actin filaments
Troponin	69,100	30,500 (TN-T) 20,800 (TN-I) 17,800 (TN-C)	Complex of three protein subunits involved in the regulation of muscel contraction

Thick filaments are composed of myosin, a large molecule containing two identical heavy chains (223 kD) and two different light chains (22 and 18kD). The structural organization of myosin is illustrated in Figure 4.16. The molecule has two identical globular head regions that incorporate the light chains and a significant fraction of the heavy chains. The tails of the heavy chains form very long α helices that wrap around each other to form left-handed coiled coils. The long a-helical structure is favoured by the absence of proline over a region of more than a thousand residues and by the abundance of helix formers, such as leucine, alanine, and glutamate.

The coiled coil is favoured by repeating seven-residue units in which every first and fourth residue has hydrophobic side chains that interact best in the coiled-coil conformation. The individual myosin molecules can be cleaved into fragments by partial degradation with various proteases. Separation of such fragments has made it possible to demonstrate that globular head regions contain binding sites for actin and that the head pieces catalyze the hydrolysis of adenosine triphosphate (ATP). The a-helical coiled coils form the backbone of the thick filament. They also form an arm that can provide a flexible extension or hinge connecting the globular head to the body of the thick filament. The thick filament contains many myosin molecules oriented in a staggered fashion. Given this marvelous piece of molecular architecture, the question remains as to how it works.

The answer lies in the observation that actin cyclically binds the globular myosin head group to form cross-bridges in a reaction that depends on the myosin-catalyzed hydrolysis of ATP. The cyclic binding and release of actin from myosin is driven by the energy-releasing hydrolysis of ATP in a manner that causes rearrangement of the actin-myosin cross-bridges. When muscle is completely relaxed, there is a minimum number of cross-bridges and the muscle is fully stretched. When the muscle is activated and under tension, it contracts and more cross-bridges are formed as the region of overlap between actin and myosin increases. At each stage of the contraction process it is essential to break the existing bridges with the help of ATP hydrolysis before new ones can be formed.

It is important to realize that although ATP allows new bridges to be formed, the ATP is required to break the existing bridges, not to form the new ones. Cross-bridge formation is energetically favored in the absence of ATP. A likely scenario for the contraction process is shown in Fig. 4.17. The ATP that binds to myosin causes the bridges to break or weaken. This

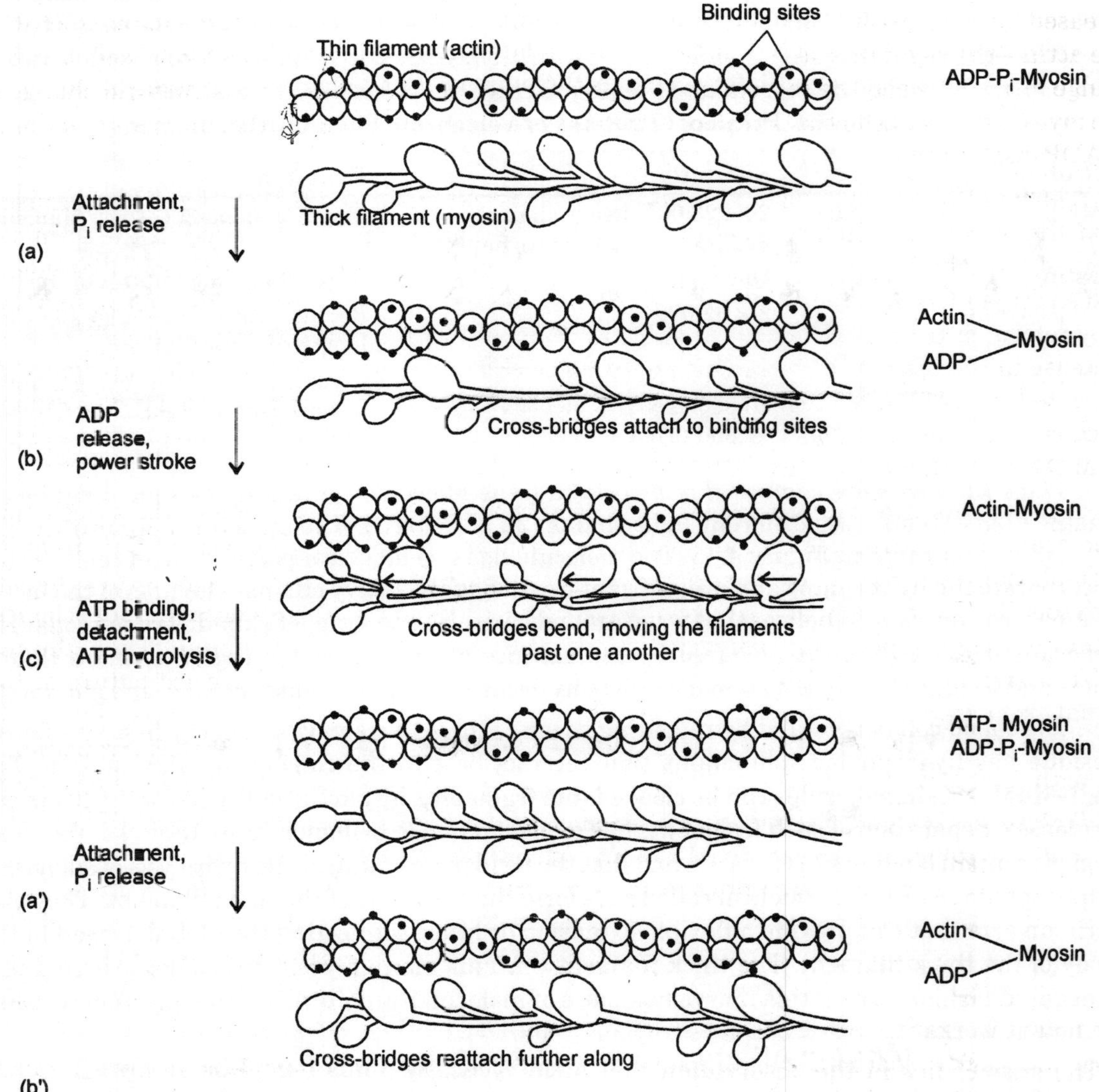

Fig. 4.16. Steps in the contraction process. Because contraction is a cyclical process, the choice of a starting point is somewhat arbitrary. Five s are shown; the first two and the last two frames are identical the cyclical nature of the process clear. In the first frame (a), die rayosin head groups contain the hydrolysis products of a single ATP molecule, ADP and P_i. A structural transition in the actin leads io contact between the actin and the myosin and the release of P_i. The release of the P_i is the rate-limiting step in muscular contraction. In the second frame *(b)*, strong bridges form between actin and myosin. This is followed by a structural alteration in the myosin molecules and an effective translocation of the thick filament relative to the thin filament in *(c)*. During this process the ADP is released. After the translocation step, the bridge structure is broken by the binding of ATP, which is rapidly hydrolyzed to ADP and P_i. Each thick filament has about 500 myosin heads, and each head cycles about five times per second in the course of a rapid contraction.

bound ATP is rapidly hydrolyzed to ADP and P_i, but the hydrolysis products are not immediately released by the myosin. The P_i is released first. but its release requires effective contact with the actin—the regulated step in muscular contraction. Once the P_i has been released, a strong bridge forms between the myosin and the actin. This step is followed by a structural change in the myosin that leads to the translocation of the myosin relative to the actin filament and finally to ADP dissociation.

The translocation step is referred to as the power stroke in muscular contraction because it is at this point that the energy ultimately donated by the ATP is expended in the form of a complex structural change. Further contraction requires fresh ATP followed by bridge dissociation and ATP hydrolysis. If conditions are right, the P_i dissociates and the bridges reform; this time they reform at points further along the actin. Cyclic repetition of this process results in a net increase in the number of actin—myosin bridges and further contraction of the sarcomere. As indicated above, the process of contraction is regulated or triggered by the TN-TM system. Because voluntary muscles are under the conscious control of the animal, one expects a signal from the central nervous system to initiate the process of contraction.

A nerve impulse communicated to the muscle causes a depolarization of the sarcolemma membrane that surrounds the muscle fibers, which in turn causes a release of Ca^{2+} from the endo-plasmic reticulum in cytoplasm of the muscle cell (Fig. 4.17). The Ca^{2+} ions form a complex with the TN-C component of the troponin molecule (see table 4.1). This process induces changes within the TM complex that overcome the inhibitory effect of the TN-I subunit. Then, through TN-T, a signal is sent to TM that triggers the contraction event. The precise nature of this signal from troponin to tropomyosin is unclear; it appears to involve a movement of the tropomyosin on the actin surface. This movement evidently leads to an allosteric transition of the actin that encourages more favourable contact between the complementary binding sites on actin and myosin. As a result, a strong bridge is formed and the P_i is released from the myosin.

TN — TM — Actin $\xrightarrow{Ca^{2+}}$

Ca^{2+}
|
TN — TM — Actin*
|
Myosin — ATP ⟶ Myosin — ADP — P_i ⟶ Myosin — ADP
↘ P_i

Fig. 4.17. The effect of calcium on muscle contraction. Binding of calcium to the TN-TM-actin complex produces a shift in the location of TM, which produces an allosteric transition in actin. The allosteric transition in actin facilitates the release of P_i from myosin, which strengthens the interaction between actin and myosin.

It is noteworthy that for some time after death, muscles enter a state of rigor in which they are fully contracted and the maximum number of bridges are formed. Rigor is probably due to the depletion of ATP and a considerable discharge of calcium from the sarcoplasmic reticulum. In living tissue, the cytosolic Ca^{2+} concentration is restored to resting levels within 30 ms of receiving a signal, and the myofibrils relax.

Concluding Remarks

In this chapter the relationship between structural and functional properties for two protein systems have been considered.

1. Hemoglobin is a tetramer made of two almost identical subunits. The function of hemoglobin is threefold: to transport O_2 from the lungs to the tissues where it is consumed, to transport CO_2 from the tissues where it is produced to the lungs where it is expelled, and to maintain the blood pH over a narrow range. Cooperative interactions among the subunits allow hemoglobin to pick up the maximum amount of oxygen at high oxygen tensions in the lung tissue and to deliver the maximum amount of oxygen to the oxygen-consuming tissues.
2. Muscle is an aggregate of several different proteins. Its main protein components are organized as overlapping filaments of two types: thin filaments, composed mainly of actin molecules, and thick filaments, composed of myosin molecules. The process of muscular contraction entails a sliding of the two types of filaments past each other. In a fully contracted myofibril the actin and myosin filaments show a maximum overlap with each other. The contraction process involves the breakage and reformation of bridges between the actin and myosin molecules in a reaction that requires the expenditure of ATP.

CHAPTER

5 Biochemical Adaption

The endocrine system includes special glands whose cells function to secrete chemical regulators, commonly referred to as *hormones,* into the internal media of the organism, *i.e.* blood and lymph. The notion "hormone" (from the Greek *hormaion,* to incite, to set in motion) was coined by Beyliss and Starling in 1905. Currently, hormones are called compounds that are produced in the gland cells and secreted into the blood or lymph to exercise control over metabolism and development of the organism.

The following general biological signs are characteristic of hormones:

1. *remote action,* i.e. the hormones control metabolism and the functions of effector cells at a distance;
2. *strict specificity of biological action,* i.e. functionally, no hormone can be entirely replaced by another' one;
3. *high biological activity,* i.e. quite small amounts of hormone, in some instances of the order of 10^{-5} g, are sufficient to provide for the vital activity of the organism.

The hormone-secreting glands are occasionally subdivided into the *central glands* (anatomically related to specialized portions of the central nervous system) and the *peripheral glands.*

Table 5.1 Hormones of Endocrine Glands

Endocrine Gland	*Hormones*	*Hormonal funtions*
	Hormones of central glands	
Hypothalamus (neuroenocrine cell)	1. Neuropepetides : (a) releasing hormones (liberines) (b) inhibitory hormones (statins)	Control of secretion of tropic hypophyseal haromens
	2. Vasopressin and oxytocin	Control of metabolism and functions of peripheral tissues and organs
Pituitary gland	1. Gonadotropins (a) Follitropin (b) lutropin (c) prolactin (Lactotropin)	Control of formation and secretion of hormones in peripheral endocrine glands; partial involvement in direct metabolism in peripheral tissues and

	2. Somatotropin 3. Corticotropin 4. Thyrotropin 5. α- and β-Lipotropins 6. Melanotropin 7. Vasopressin and oxytocin supplied from hypothalamus	organs
Epiphysis	1. Melatonin	Control of production of hypophyseal gonadotropins
	2. Adrenoglomerulotropin	Presumbly, control of aldosterone secretion in adrenal cortex
	Hormones of peripheral glands	
Thyroid gland	1. Iodothyronines (a) thyroxine (b) triiodothyronine 2. Calcitonin	Action of all peripheral gland hormones on metabolism and functions of peripheral tissues and organs
Parathyroid gland	1. Parathyrine 2. Calcitonin	
Pancreas (insular cells)	1. Insulin 2. Glucagon	
Adrenal glands	1. Corticosteroids : (a) corticosterone (b) cortisol (c) aldosterone (d) estrogens (e) androgens 2. Adrenalin	
Six glands :		
(a) testes	1. Androgens (a) testosterone (b) 5–α-dihydrotestoterone	
(b) overies	1. Estrogens (a) estradiol (b) estrone (c) estriol 2. Gestagens (progesterone) 3. Relaxin	
Placenta (temporal endocrine gland in pregnancy period) gland	1. Estrogens 2. Gastagens 3. Testosterone 4. Chorionic gonadotropin 5. Placental lactogen 6. Thyrotropin 7. Relaxin	
Thymus	Thymosin	

In addition to the glands summarized in Table 5.1, endocrine functions are also exercised by other cells capable of secreting biologically active compounds whose properties resemble those of hormones. For this reason, these agents are conventionally referred to as *hormone-like compounds,* or *hormanoids* (local hormones, parahormones). Their action is, as a rule, confined to the site where they are produced; they are secreted by cells dispersed about different tissues.

Hormonoids are secreted by the cells of gastrointestinal tract (they control digestion), by the intestinal enterochromaffin cells (which produce serotonin—a regulator of intestinal function), by the mast cells of connective tissue (heparin and histamine), and by the cells of kidneys, seminal vesicles, and other organs (prostaglandins). Presumably, these endocrine cells are specific peripheral representatives of the endocrine system. This is borne out by the occurrence of certain true hormones in the peripheral organs, for example, somatostatin in the pancreas and liver. By their *chemical structure*, hormones are subdivided into:

1. protein-peptide hormones (those of hypothalamus, pancreas, pituitary and parathyroid glands; thyroidal calcitonin);
2. amino acid derivatives (adrenalin, derived from amino acids phenylalanine-.and tyrosine; iodothyronines, derived from tyrpsine; melatonin, derived from tryptophan);
3. steroids (sex hormones: androgens, estrogens, and gestagens; corticosteroids).

NEUROENDOCRINE RELATIONSHIP SCHEME

Information on the slate of internal and external media of the organism is transmitted to the nervous system to be processed therein; in response, regulatory signals are transmitted to the peripheral tissues and organs. The nervous regulation of metabolism and effector functions is effected not only via the nervous impulses transmitted by the centripetal nerves, but also indirectly, via the endocrine system. Both information channels—nervous and hormonal—meet at the level of hypothalamus.

The nervous impulses transmitted from various parts of the brain exert influence on the secretion of neuropeptides by the hypothalamic cells which control the secretion of pituitary tropic hormones, the latter affecting the hormonal secretion in the peripheral glands. The production and secretion of hormones by the peripheral glands are accomplished incessantly. This is essential for maintaining the hormones at appropriate level in the blood, since they are liable to fast inactivation and elimination from the organism.

The required level of a hormone in the blood is maintained owing to the self-regulation mechanism. This mechanism is based on interhormonal relationships which are referred to as *"plus-minus" interactions,* or *mutually exclusive relationships.* Schematically, these interrelationships are shown in Fig. 5.1. Tropic hormones stimulate the production and secretion of hormones by the peripheral glands (the sign "plus"), and the latter, by the mechanism of negative feedback inhibit (the sign "minus") the production of tropic hormones by acting through the pituitary cells (short feedback, Fig. 5.1, 1) or through the neurosecretory cells of the hypothalamus (long feedback, Fig. 5.1, *2*). In the latter case, the secretion of releasing hormones (liberins) is inhibited in the hypothalamus.

In addition, the metabolite-hormonal feedback is also operative (Fig. 5.1.3): a hormone, by acting on tissue metabolism, evokes a change in the concentration of a metabolite in the blood, and the metabolite, by feedback mechanism, affects the hormonal secretion in peripheral glands (either directly, or via pituitary gland and hypothalamus). Candidates for the metabolites capable of acting as indicators of hormonal action on the cell may be glucose (carbohydrate metabolism indicator); amino acids (protein metabolism indicators); nucleotides and nucleosides (indicators for nucleic and, presumably, protein metabolism); fatty acids and cholesterol (lipid metabolism indicators); Ca^{2+}, Na^+, K^+, Cl^-, and water (salt-water balance indicator). Some of the metabolites (glucose, certain amino acids, Ca^{2+}, Na^+, K^+, Cl^-, and H_2O) have been shown to be able to exercise hormonal secretion control by the feedback mechanism.

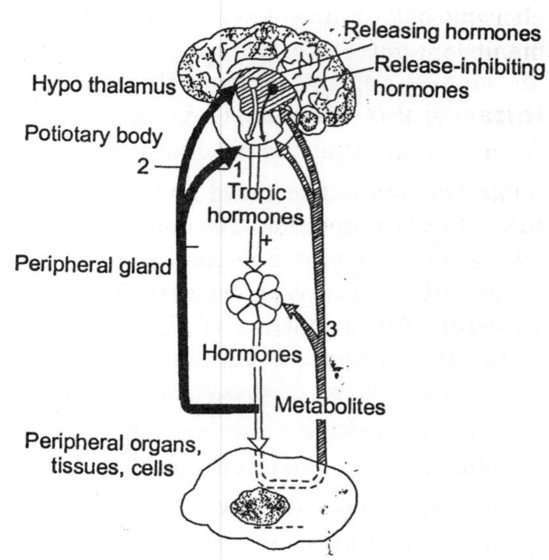

Fig. 5.1. Scheme for the re-lationships in the endocrine system

HORMONAL CONTROL: GENERAL OUTLINE

Hormones, which are secreted by the glands into the blood, usually become bound to specific blood plasma transport proteins (or, in certain cases, adsorbed on blood cells) to be carried to the peripheral tissues, where they exert influence on metabolism and tissue function. Various tissues respond to hormonal action differently. Tissues (cells, organs) exhibiting a high sensitivity to hormones, i.e. the tissues in which the hormones elicit pronounced changes in metabolism and function, are called "targets" for a given hormone, while other, less responsive, tissues are assigned to "nontarget" tissues; it should be admitted that this classification is rather arbitrary.

The hormonal influence extends, in a wider sense, to all metabolic manifestations. However, in order to get a better understanding of the mechanism of action of any hormone or, in fact, any extracellular regulator, it is worthwhile to consider the general principles of hormonal metabolic control.

The following possible types of action for extracellular regulators (often referred to as first messengers), including hormones, are distinguished:

1. membrane, or local, action;
2. membrane intracellular, or indirect, action;
3. cytosolic, or direct, action.

Hormonal Action

The membrane type of hormonal action consists in that the hormone, at the site of its binding with the cell membrane, renders the membrane permeable to glucose, amino acids, and certain ions. In this case, the hormone acts as an allosteric effector for membrane transport systems. The glucose and amino acids supplied exert, in turn, influence on biochemical cellular processes, while a change in ion partition on both sides of the membrane affects the electric potential and function of the cell. The membrane type of hormonal action in pure form is rarely observed. As a typical example, insulin may be cited which exhibits both a membrane type (by eliciting local changes in the transport of certain ions, glucose, and, possibly, amino acids) and a membrane-intracellular type of action.

Intracellular Mechanism

The membrane-intracellular type of action is characteristic of hormones or functionally related to them extracellular regulators called first messengers. These are not capable of entering the cell and cannot therefore influence the intracellular processes directly. Their action is effected indirectly, through the agency of an intracellular mediator (or second messenger) which triggers a chain of successive biochemical reactions leading to a modification of cellular functions. The sequence of information transfer from diverse external stimuli into the cell may be outlined within the following general scheme: external signal (first messenger)—membrane receptor—transducer—chemical amplifier—second messenger—internal effector—cell response. The first messenger (for example, a hormone) becomes bound to the receptor on the outer side of plasma membrane.

The membrane receptors are a kind of "start buttons" on a control desk of cellular metabolism: the hormone-receptor complex thus formed acts on a protein, the membrane transducer, which, in turn, transmits the signal to an enzyme (chemical amplifier) acting as a catalyst for the production of a second messenger inside the cell. The second messenger binds to a special protein (internal effector) which exerts influence on the activity of definite enzymes or on the properties of nonenzymic proteins conducive to alterations of cellular processes, viz. chemical conversion rates, permeability, secretion, structural contractility, activity of genes, etc. The merit of second messenger discovery is linked to the name of E. Sutherland.

In 1957, he found a thermostable factor acting as a mediator in the stimulation of glycogenolysis by adrenalin. In 1958, E. Sutherland and T. Roll provided evidence that this factor might be 3′,5′-AMP or cAMP. Currently, there are known several groups of compounds that are candidates for the function of a second messenger. These are: (1) nucleotides: cyclic nucleotides (cAMP and cGMP), 2,5-oligo(A)$_n$ (adenyl oligonucleotide); (2) lipids: diacylglyceride, inositol triphosphate; (3) Ca^{2+} ions; (4) peptides. Let us consider the general principles on which the action of these messengers is based.

Metabolic Control Effected via Cyclic Nucleotides. The intracellular regulators affect the production of cyclic nucleotides via either of the two signal systems: adenylate cyclase or guanylate cyclase. The former controls the production of cAMP, and the latter, of cGMP :

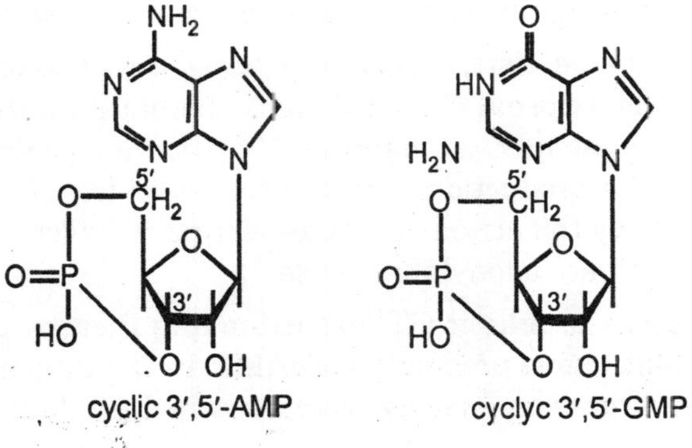

NH2, N, N, N, N, 5′, O—CH2, O, O=P, 3′, HO, O, OH

cyclic 3′,5′-AMP

O, HN, N, H2N, N, N, 5′, O—CH2, O, O=P, 3′, HO, O, OH

cyclyc 3′,5′-GMP

Adenylate cyclase is built into the membrane and is composed of : (1) discriminator (R) which is represented by a set of membrane receptors of stimulating (R_s) and inhibitory (R_i) types; (2) transducer (N), represented by a special protein which takes an intermediate position in the cell membrane between the receptor and chemical amplifier (catalytic moiety of adenylate cyclase). This transducer protein is called N-protein or G-protein (since it binds GMP also). N- (or G-) proteins are of two types: stimulating (N_S) and inhibitory (N_i); (3) chemical amplifier, or catalytic moiety (C), which is an enzyme whose active portion is facing into the cell's interior. When activated, this enzyme catalyzes the production of cAMP according to the scheme: ATP → 3′,5′-AMP + $H_4P_2O_7$.

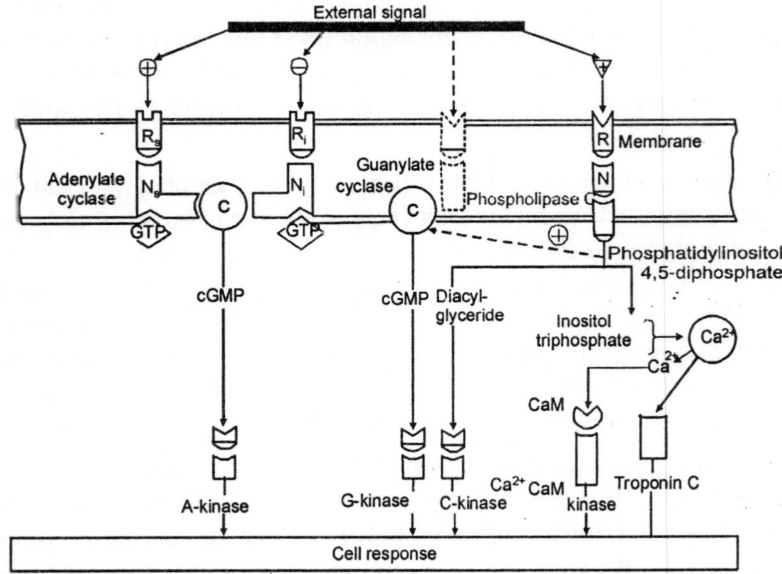

Fig. 5.2. Scheme for transfer of hormonla signal into cell

Now, how does the adenylate cyclase work? The interaction of the external regulator with R_S-type receptors brings about a conformational change in the latter, and this effect is transmitted to the N_S-protein which thus acquires an ability to bind GTP. The N_S-GTP complex acts as an allosteric activator towards adenylate cyclase. Its activation initiates the production of cAMP from the intracellular ATP. Thus, at the moment as the N_S-GTP complex is formed, information is transferred from the external signal molecule to an internal molecule, the second messenger.

Until the N_S-GTP complex persists, adenylate cyclase continuously catalyzes the cAMP synthesis, owing to which the external chemical signal becomes multiply amplified: per one molecule of R_s-bound first messenger, as many as 100-1000 molecules of cAMP, the second messenger, are generated. The production of cAMP ceases as the N_S-GTP complex becomes inactivated, which is effected by the action of GTPase acting to hydrolyze GTP to GDP. In this case, N_S loses its ability to activate adenylate cyclase.

The'cholera toxin is known to inhibit GTPase extending thereby the lifespan of N_S-GTP complex which elicits the continuous production of cAMP. In the intestinal cells, cAMP makes the secretion of fluid in the organism sharply increase, which explains the acute diarrhea in patients with cholera. If the external regulator becomes bound to a receptor of the other type, Rh, then its complex with R, exerts action on the $N_r$protein which, on binding with GTP, acquires the properties of an allosteric adenylate cyclase inhibitor. The production of cAMP thus ceases. It is commonly recognized that the pertussis toxin blocks the N_i-protein and prevents the adenylate cyclase inhibition. From the roots of *Coleus forskohlii,* a plant encountered in India, a substance forskohlin was isolated, which is capable to directly activate adenylate cyclase and stimulate the production of cAMP. The other cyclic nucleotide, a potential candidate for the second messenger, is cGMP. It is produced from GTP with the participation of guanylate cyclase.

In distinction from adenylate cyclase, this enzyme is loosely bound to the plasma membrane. There are good reasons to believe that the transmission of external signal to guanylate cyclase is effectuated not via receptor. Moreover, a form of guanylate cyclase, soluble in the cell sap, has been identified and found to be responsive only to intracellular factors. Interestingly, guanylate cyclase is activated by products of peroxide lipid oxidation generated by the action of certain compounds in the plasma membrane. Further transmission of the signal through the aid of cAMP and cGMP is effected with the participation of protein kinases as internal effectors.

There are known several types of protein kinases (their rational classification remains as yet a matter of dispute) remarkable for their ability to specifically bind to various second messengers and to be activated by them. The cAMP is activated by protein kinase of A type. This protein is a tetramer composed of two regulatory (R) and two catalytic (C) subunits. The addition of a cAMP molecule to each of the two R-sub-units produces dissociation of the tetramer; in this process, the C-subunits combine to form a dimer which is the active form of A-protein kinase. The G-type protein kinase is activated by cGMP. This enzyme is a dimer.

Either of its subunits possesses a catalytic site (C) and a regulatory site (R) for binding cGMP. As cGMP becomes bound to the two R-sites, the G-protein kinase undergoes a conformational change to be activated. The protein kinase function is phosphorylation of definite proteins according to the scheme: protein + ATP $\rightarrow$ phosphoprotein + ADP. Since the cyclic nucleotides transmit information to protein kinase which is a catalyst, an added, 10-to-100-fold increase in chemical signal occurs; otherwise stated, each of the protein kinase molecules, while persisting in an activated state, is able to phosphorylate and, therefore, to functionally modify, to about a hundred diverse proteins.

Thus a single hormone molecule is potentially capable of modifying the structure and affecting the function of nearly a million molecules in the cell. The A-type protein kinases act as phosphorylation agents on a wide variety of proteins including: proteins that affect the template activity of chromatin (histones and nonhistone proteins, RNA-polymerase sigma factor); enzymes that control the synthesis and breakdown of energetic materials (triacylglyceride lipase, glycogen-phosphorylase kinase, glycogen synthase, pyruvate dehydrogenase, and others); proteins that control myofibrillary contractions (troponin); membrane proteins involved in ion permeability control, etc.

The phosphorylation of some proteins elicits their activity, the phosphorylation of others produces an inhibitory effect on activity. Therefore, as might have appeared; a protein kinase reaction of the same type can produce a whole spectrum of biological effects determinative of the specificity of cell response. This specificity is defined, on the one hand, by the set of membrane receptors responsive to an external signal, and, on the other hand, by the manifold of proteins in a given cell subject to phosphorylati'r on through the agency of protein kinases. The functions of proteins liable to phosphorylation by G-kinase are less clear.

To date, the concepts of the biological role of cGMP as a second messenger are in many details rather hypothetic than substantiated experimentally. Thus, it is commonly recognized that the effects produced by cAMP and cGMP in the cell are opposite. However, this is not always the case. Presumably, other cyclic nucleotides found inthecell (for example, cIMP and cCMP) may play a certain role, but this presumption awaits further verification. In the cell, there are systems that offset the influence of cyclic nucleotides on biochemical processes. Be such systems nonexistent, the regulatory action of a hormone on its contact with the cell could extend over a longer period of time.

The suppression of chemical signal is effected by special enzymes which destroy both the cyclic nucleotides and the products to which these are conducive, i.e. phosphoproteins. The hydrolysis of cyclic nucleotides is carried out by phosphodi-esterases according to the scheme

$$3',5'\text{-AMP (or } 3',5'\text{-GMP)} \xrightarrow[\text{Phosphodiesterase}]{+H_2O} 5'\text{-AMP (or } 5'\text{-GMP)}$$

The AMP and GMP thus formed are incapable of activating the protein kinases. Alongside phosphodiesterases, there is available in the cells an enzyme, phospho-protein phosphatase, which dephosphorylates the phosphoproteins by the scheme: phosphoprotein $+H_2O \rightarrow$ protein $+ H_3PO_4$. Numerous regulators act to increase cAMP concentration. They encompass all the hypothalamic and hypophyseal hormones, calcitonin, parathyrine, glucogen, catecholamines (acting via β-adrenergic receptors), vasopressin, histamine, certain prostaglandins.

The action exerted by hormones is reversible, it is operable within a few seconds or minutes after the contact of hormone with a membrane receptor. The hormones whose action is mediated by cAMP produce not only short-term metabolic changes (i.e. stimulate lipolysis, glycogenolysis, ion transport, etc.), but long-term ones also. These latter bear relation to alterations in protein biosynthesis. There has been reported evidence that the complex "cAMP—protein kinase regulatory subunit" is able to penetrate the cell nucleus, to bind to chromatin, and to induce the synthesis of definite proteins.

Quite probable, that the catalytic protein kinase subunit, upon entering the nucleus, becomes either involved in the chromatin protein modification, or affects the activity of methylases responsible for DNA methylation. In this manner, this subunit exerts influence on protein synthesis in the cell. The binding of external regulators with other membrane receptors inhibits the production of cAMP (with an eventual increase in cGMP concentration). Such receptors include α_2-adrenergic and M_2-cholinergic receptors, ADP receptors, activation factors for thrombocytes, thromboxane A_2 and certain prostaglandins, D_2-dopamine and A_1-adenosine receptors as well as angiotensin II, delta opiate and somatostatin receptors. Insulin produces a decrease in cAMP concentration in the cell by activating phosphodiesterase.

2,5-Oligo (A) as Inlracellular Messenger: The involvement of this nucleotide in intracellular regulatory metabolism has been little studied. As has been ascertained, the antiviral protein interferon facilitates the formation of 2,5-oligo(A) in lymphoid cells. A similar effect has been observed due to a number of external factors which stimulate the activity of the enzyme *oligo (A) synthetase,* a catalyst for the synthesis of 2,5-oligo(A). This oligoadenyl nucleotide inhibits cAMP-phos-phodiesterase and produces an increased concentration of cAMP. Quite probable that the influence of 2,5-oligo(A) on cellular functions is mediated by cAMP.

Lipid Molecules as Intracellular Messengers: The participation of lipids in the transmission of external signals into the cell has come as a surprise. The first mention of this effect dates back to 1953 when M.N. Hockin and L.E. Hockin from the Montreal Hospital showed that acetylcholine enhanced the metabolism of a membrane phospholipid, phosphatidylinositol.

The studies initiated by P. Mitchell at the Birmingham University in 1975 have revealed that the substrate for production of intracellular messengers is phosphatidylinositol 4,5-diphosphate, a basic constituent of biomembranes. The information transfer through the intermediacy of lipid messengers is effectuated in the following manner. An external signal molecule becomes bound to a membrane receptor; the complex thus formed acts through the agency of N-protein on an amplifying enzyme, phospholipase C. This enzyme hydrolyzes the membrane phosphatidylinositol 4,5-diphosphate into two second messengers: inositol triphosphate and diacylglyceride.

The former diffuses into the cytoplasm and elicits a release of Ca^{2+} ions from the intracellular calcium depots. Ca^{2+} ions bind to calcium-binding proteins, for example, troponin C (implicated in contractile processes) or to cadmodulin (CaM for short). The Ca^{2+}CaM complex activates a protein kinase responsive to it, which is a catalyst for phosphorylation of a definite family of proteins. Diacylglyceride, a weakly hydrophilic molecule, remains in the juxtamembranous layer and, through the aid of another membrane lipid, phosphatidylserine, activates a special protein kinase, C-kinase, responsible for phosphorylation of a definite class of proteins.

Besides, the lipid messenger's activate guanylate cyclase and increase the cGMT concentration in the cell. Apparently, Ca^{2+} and cGMP determine the specificity of cell response to external signals, including the specific production of lipid messengers.

Mechanism of Metabolic Control via Ca^{2+} Ions: The Ca^{2+} ion was known to be an intracellular regulator for muscular contraction as early as the 19th century. However, its involvement in the control of a vast variety of processes in the capacity of a major intracellular messenger has been acknowledged quite recently. In contrast to other mediators, Ca^{2+} ions are not subject to a chemical conversion; for this reason, the calcium signal membrane systems, responsive to external stimuli, can only control the Ca^{2+} ion supply to cytoplasm.

The intracellular concentration of Ca^{2+} ions is negligeably small, 10^{-7} mol/litre, while outside the cell, their concentration is as high as 10^{-3} mol/litre. The Ca^{2+} ions are delivered into the cell from the external medium via several "calcium channels" in the cell membrane. Potential-operated and receptor-operated Ca^{2+} channels are distinguished. The former operate in the manner as to allow opening the channels for Ca^{2+} supply down the concentration gradient into the cell as the membrane potential is changed. The latter channels respond to the complex "external regulator—receptor", allowing Ca^{2+} ions to penetrate into the cell. The Ca^{2+} flow is controlled by the Ca^{2+}-ATPase of cell membrane which pumps Ca^{2+} ions out at the expense of ATP energy.

Inside the cell, there are calcium depots whose function is to entrap the Ca^{2+} ions and maintain the Ca^{2+} concentration in the cell sap lowered to an appropriate level. These depots are mitochondria and vesicles of endoplasmic reticulum. The elimination of Ca^{2+} ions from the intracellular depots is accomplished in a manner similar to their supply from the exterior via calcium channels. The calcium supplied from the external medium or from-intracellular depots by the action of various external stimuli interacts with cytoplasmic calcium-binding proteins. As has been mentioned earlier, the most important of them is cadmodulin. The Ca^{2+}-CaM complex, by activating the protein kinase sensitive to it or other enzymes, exerts control over biochemical processes and other functional activities of the cell related to them.

Peptides as Intracellular Messengers: The potential participation of peptides as mediators in the transmission of external signals may be envisioned as sequent upon a study of the mechanism of insulin action. J. Larner and coworkers at the Medical College of—Virginia are of opinion that the formation of membrane insulin-receptor complex activates a membrane trypsin-like proteinase which effects the hydrolytic cleavage of a peptide from the membrane protein. This peptide is believed to be involved in a variety of mechanisms of insulin action on metabolism. At the same time, another route of signal transmission is known for insulin and certain growth factors (for example, those of epidermis).

In certain cells, the receptor for these regulators spans the plasma membrane. The inner portion of this receptor exhibits an activity similar to that of protein kinase (tyrosine kinase) participating in the addition of phosphate to proteinic tyrosine (commonly, phosphate adds to serine or threonine). Thus, the specific action of hormones is effected through the agency of a small set of intracellular mediators. At first glance, this fact appears to be rather puzzling. Still, on a close look, it is amenable to explanation. Firstly, the functional specificity of a cell is defined by a set of genes and proteins.

The cell is responsive to hormones only in the presence of the corresponding receptors. The same hormone acts, through the mediacy of a single messenger (for example, cAMP) on different cells in an unlike manner, depending on the specific set of proteins and the function associated therewith. Secondly, a hormone acts only on the cell that contains receptors specific for this hormone. For this reason, the biological effect due to any hormone is highly specific. At the same time, the hormones exhibiting a similar type of signal transmission can exert a similar effect on tissues with simplified functions.

For example, in the fat tissue containing as many as eight receptors for activation of adenylate cyclase, different hormones can produce a similar fatmobilizing effect. In addition to the natural substances that affect the formation of cAMP and cGMP via adenylate and guanylate cyclases, a target for potential attack by drugs is phosphodiesterase. When cell-invading drugs inhibit either of the phosphodiesterases cAMP or cGMP, they may simulate a metabolic effect due to natural hormones and mediators operative via cyclic nucleotides.

The phosphodiesterase activators inhibit the action of these hormones and mediators. This possibility is made use of in practical medicine by applying drugs that can inhibit phosphodiesterase. These include xanthine derivatives, such as *caffeine, theopkylline, theobromine,* and *euphylline.* These drugs mainly increase the concentration of cAMP, while such a drug as *trental* increases chiefly the cGMP concentration.

Cytosolic Mechanism of Action

Cytosolic mechanism is typical of the hormones that can penetrate through the lipid layer of plasmic membrane, i.e. are related, by their physico-chemical properties, to lipophilic compounds, for example, steroid hormones (vitamin D, by its metabolic effect, is also close to steroid hormones, and for this reason it is often assigned to hormonal agents). Hormones with cytosolic type of action penetrate the cell to be complexed with cytosolic or nuclear receptors. The receptor-complexed hormones control the enzyme concentration in the cell by selectively affecting the gene activity of nuclear chromosomes and exerting thereby influence on metabolism and functions of the cell.

Since the cell-invading hormone is involved in the mechanism of enzyme concentration control, this type of action is called *direct action* in contrast to the intracellular membrane mechanism, when a hormone controls metabolism *indirectly,* through the agency of second messengers. In terms of lipophilicity, iodothyronines are found midway between steroids and other water-soluble hormones. Probably, for this reason they exhibit, *n combined type of action* on cell metabolism, i.e. both intracellular membrane and cytosolic.

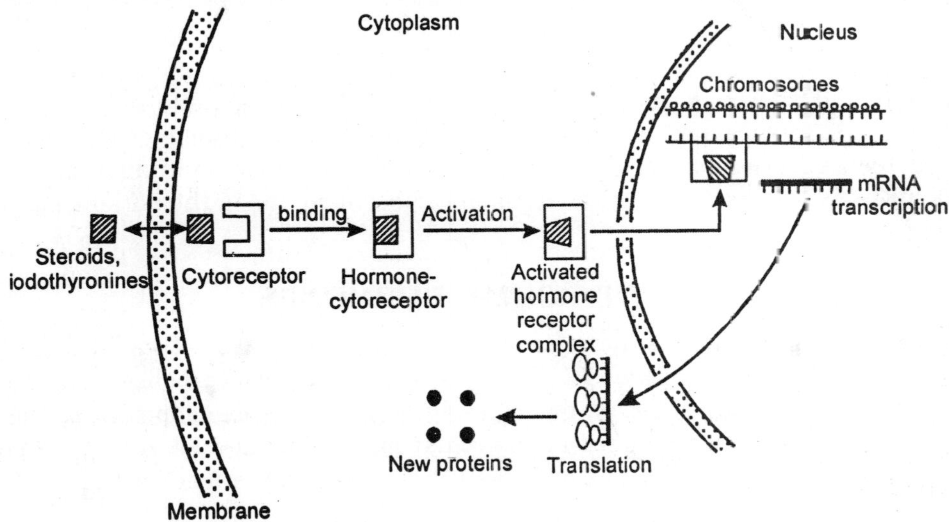

Fig. 5.3. Scheme for cytosolic mechanism of hormonal action

Schematically, the cytosolic mechanism is presented in Fig. 5.3. Steroid hormones and iodothyronines, as they penetrate through the cell membrane, become bound with cytosolic receptors through whose agency the hormones exert a regulatory action on the cell metabolism. Cytoreceptors are proteins with a molecular mass of 60000-120000. They are characterized by binding stereospccificity and thence by a high affinity to the intrinsic hormone. In the cytoplasm, the primary complex hormone-cytoreceptor is subject to activation manifested by a rearrangement of cytoreceptor molecule. In its "activated form", the hormone-receptor complex is capable of penetrating through the nucleus membrane to come into interaction with the nuclear chromosomes.

A given hormone-receptor complex, on binding to the chromatin regulatory proteins (histones and nonhistone proteins) or to DNA, controls either the cell division, or the transcription of its "intrinsic" genes in nonproliferating cells and exerts influence on the synthesis of specific proteins. This actually determines the specific effect produced by any hormone. Since steroid hormones exert influence only on the chromosomal gene activity, they are more operative in growth control and cell differentiation, *i.e.* in the organism's development, as compared to the hormones with intracellular membrane type of action. Such hormones, when administered in large doses, are capable of a direct action on enzymes and intracellular membranes, since the cytosolic receptors may happen to be in short supply for binding a large quantity of the invaded hormone.

PRACTICAL APPLICATION OF HORMONES

For their practical applications, hormones are produced by extraction from biological materials, by chemical synthesis, or by genetic engineering methods. The first of the above techniques is employed in the production of insulin and glucagon . from bovine pancreas, of corticotropin and melanotropin from bovine pituitary, of follitropin (serumal gouadotropin) from blood serum, and of lutropin (chorionic gonadotropin) from the urine of pregnant mares. Currently, chemical synthesis is widely used for production of all steroid hormones, their analogues and derivatives, iodothyronines and other hormones—amino acid derivatives and peptide hormones of oxytocin type.

Practically all the protein hormones can be prepared by laboratory synthesis. Insulin, somatostatin and related hormones have been prepared by the genetic engineering method in the laboratory. In medical practice, hormones are used for *substitution therapy* and *pathogenetic therapy.* In the latter instance, certain specific properties of hormones (antiinflam-matory, anabolic, etc.) are made use of, even though the hormonal concentrations in the patient's organism may be at normal level.

THYROIDAL HORMONES

The thyroid gland secretes hormones of two groups, which affect metabolism differently. Iodothyronines—*thyroxine* and *triiodolhyronine*—control energy metabolism and exert influence on cell division and differentiation, determining there by the development of the organism. *Calcitonin* (polypeptide hormone with a molecular mass of about 30000) controls phosphorus-calcium metabolism; it appears more expedient to discuss calcitonin action in parallel to trial of parathyroid glands.

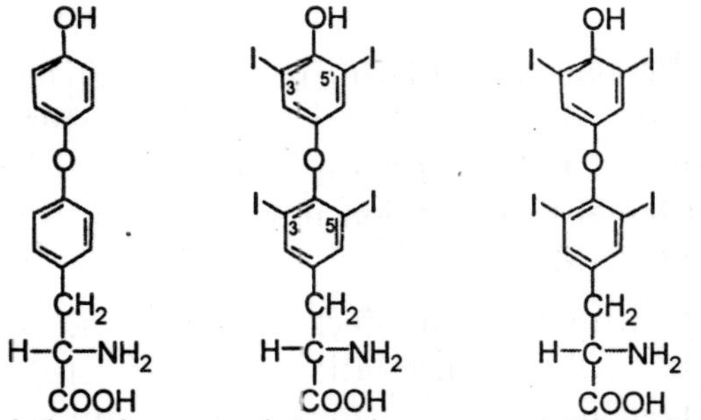

L-thyronine | L-thyronine (3,5,3'5'-tetraiodothyronine) | L-3,5,3'-triiodothyronine

Synthesis of Iodothyronines

Iodothyronines make part of the protein thyroglobulin contained in the colloid of the thyroid gland follicles. For the synthesis of these hormones, essential are iodide as supplied by active transport from blood to thyroidal epithelium, and thyrogiobuiin which is produced in the epithelium and fills the follicular cavity to form colloid.

Iodothyronine synthesis proceeds by several steps which are:

1. Formation of an "active" iodine from iodide through the aid of *iodide per-oxidase* by reaction

$$I^- — e^-(2e^-)—I^0(I^+)$$

H_2O_2 serves as an electron acceptor. The active iodine is capable of iodinating tyrosine.

2. Iodination of tyrosine constitutive of thyroglobulin with the participation of *tyrosine iodinase*. The products formed are monoiodotyrosine or diiodotyrosine.

3. Oxidative condensation of mono- and diiodotyrosines with the formation of triiodothyronine and thyroxine within the thyroglobulin molecule. The process is carried out on the surface of tyrosine iodinase.

4. Uptake of thyroglobulin from the colloid by epithelium cells and its trans-location to the outer membrane surface bathed by the extracellular fluid. This process is reminescent of endocytosis.

5. Secretion of iodothyronines, which is accomplished by hydrolysis of thyroglobulin by proteases enabling the release of thyroxine (T_4) and triiodothyronine (T_3) into blood.

The secretion and synthesis of T_4 and T_3 are controlled by thyrotropin. In humans, the daily secretion of T_3 is about 55 μg and that of T_4, twice as large. In blood, iodothyronines form complexes with a thyroxine-binding globulin (albumin) and prealbumin of the plasm, and are transported to peripheral tissues. The binding of T_3 with plasmic albumins is about 3-5 times less strong than that of T_4. Because of this, the biological effect due to T_3 is accordingly stronger by a factor of 3-5 as compared to that of T_4.

Mechanism of Iodothyronine Action

Iodothyronines exert action on numerous tissues of the organism, the highest sensitivity being exhibited by the tissues of liver, heart, kidney, skeletal muscles, and, to a lesser extent, adipose and nervous tissues. In the organism, the thyroid hormones affect to a higher degree cell division, cell differentiation, and energy metabolism. Alterations in energy metabolism due to the so-called calorigenic property of thyroid hormones, outwardly show up as an increased oxygen consumption and production of heat. What is the relationship between the ability of iodothyronines to control cell growth and cell differentiation and their calorigenic properties? Iodothyronines exert influence on metabolism in a dual manner: via cytosolic receptors by affecting the nuclear chromosomes, and via cAMP.

As has been ascertained, the chromosomal effect manifests itself in an accelerated DNA replicatioa during mitosis and in a selective activation of transcription of certain genes. This stimulates the synthesis of corresponding proteins and favours enhancement in efficiency of specialized enzymic apparatus in the cells susceptible to iodothyronines. Hormones act as inducers for the synthesis of over 100 enzymes, most of these being energy metabolism enzymes. Iodothyronines activate oxidative mitochondrial enzymes and enzymes assisting in the "shuttle" transport of hydrogen from cytoplasm to mitochondrial matrix.

In the cells, the iodothyronine effect shows up in an increased number of mitochondria, and in many mitochondria the cristae carrying a large number of respiratory ensembles tend to increase in size. To briefly summarize, one may say that the involvement of iodothyronines in the genetic apparatus of the cell creates a basis for intense aerobic production of energy. This process is intimately related to wide-scale mobilization of energy resources in the organism. This mobilization is effected by iodothyrouine-induced activation of adenylate cyclase in the tissues and increased cAMP tissue concentration which, in turn, activates lipolysis in fat tissue and glycogenolysis in liver and muscles.

The intense combustion of lipid and carbohydrate metabolism sub-strates requires a large consumption of oxygen in the organism, which is actually observed when iodothyronines are administered. Iodothyronines increase not only the overall energy production in the mitochondria, but also the heat release. Formerly, this effect was believed to be due to the uncoupling action produced by iodothyronines on mitochondria. However, in the organism, the tissue iodothyronine concentrations never reach the uncoupling level and, nonetheless, thyroid hormones always increase heat production.

Currently, this energetic effect is attributed to the increased consumption of ATP in various synthetic processes in tissues and, especially, to active transport of materials involving Na^+, K^+-ATPase. Outwardly, the above-mentioned molecular processes that are triggered by iodothyronines are manifested by proliferation, growth, and differentiation of the cells, and at the level of the whole organism, by normal growth and regular development of tissues and organs. The specific action of these hormones on energy metabolism shows up in increased oxygen consumption and production of heat in the organism.

Functional Disturbances of the Thyroid Gland

The thyroid gland malfunction may entail an excess or deficiency of iodothyronines in the organism. In the hyperfunction of the thyroid gland, or hyperthyroidism, excessive production of iodothyronines is observed. Acute hyperfunctional forms have been named *thyrotoxicosis,* or *Rasedow's disease,* since the symptoms of disturbed metabolism and malfunction may be likened to iodothyronine intoxication, in the blood of patients with this disease, tri iodothyronine is the predominant thyroxine form. A clinical sign for thyrotoxicosis is an accelerated catabolism of carbohydrates and triacylglycerides, the latter being mobilized from fat depots.

Fast combustion of fatty acids, glycerols, and glycolyzates requires an increased consumption of oxygen. The mitochondria are swollen and increased in size. In acute thyrotoxicoses. the mitochondria become distorted in shape; hence a figurative name "mitochondrial disease" for thyrotoxicosis. Although an excess in iodothyronines accentuates their characteristic effects on the genetic apparatus and specific protein synthesis, with time the processes of protein breakdown and nitrogen loss become predominant in the organism.

In aggravated forms of thyrotoxicosis, a negative nitrogen balance sets in. Apparently excessive regulatory influence due to iodothyronines is manifested by increased basal metabolism, elevated body temperature (because of enhanced production of heat), loss in body weight, distinct tachycardia, hyperexcitability, and exophlhalmos (abnormal protrusion of the eyeball). These disturbances are relieved either through surgical intervention partial excision of the thyroid gland), or through the use of pharmaceuticals capable of .suppressing hormonogenesis in the thyroid follicles. In hypofunction of the thyroid gland, or hypothyrosis, the organism is in short supply of iodothyronines. Hypothyrosis in neonates or infants is called *cretinism,* or *infantile myxedema;* in adults, it is called simply *myxedema*. The symptoms of this disease are opposite to those of thyrotoxicosis.

Cretinism is characterized by pronounced physical and mental retardation. The afflicted persons are stunted in growth, of disproportionate constitution, and their extreme mental retardation renders them inept to learning and productive labour. Their basal metabolism is reduced, and the body temperature is below normal. The symptoms of this disease are chiefly explained by an ineffective action of iodothyronines on cell division and cell differentiation, with ensuing retarded and improper growth of bone tissue, and impaired differentiation of the neurons which are rendered incapable of performing their specific functions inherent in them. In the adult organism, the major processes of growth and tissue differentiation are essentially completed.

Therefore, myxedema is manifested in a reduced basal metabolism, lowered body temperature, less retentive memory, impaired renewal of dermal epithelium (dry, desquamative skin), and deposition of mucoid materials in subcutaneous fat. In the organism tissues, partially inhibited are aerobic oxidation of carbohydrates and fatty acids and overall energy metabolism. The removal of all the disease symptoms, including metabolic disturbances, is achieved by substitution therapy with iodothyronine preparations (triiodothyronine is most commonly applied).

HORMONES OF THE PARATHYROID GLANDS

The parathyroid glands secrete two protein hormones *calcitonin* (also secreted by the thyroid gland) and *parathyrin (parathormone)*. The latter is a protein with a molecular mass of 9500; it is composed of 84 amino acid residues. Both hormones are synthetized in the gland cells as preprohormones of large molecular mass. The preprohormones are subject to an attack by protein kinases which split off a polypeptide chain fragment from the preprohormones to convert them to prohormones; the latter are hydrolytically converted to active hormones to be stored in the secretory granules of Golgi's apparatus.

Calcitonin and parathyrin control the balance of Ca^{2+} ions and inorganic phosphate in the organism. In turn, the secretion of calcitonin and parathyrin, which lack appropriate tropic hormones, is feedback-controlled by Ca^{2+} ions. An increased concentration of Ca^{2+} ions in blood plasma elicits the secretion of calcitonin; on the contrary, a decreased percentage of Ca^{2+} in blood plasma is a stimulus for liberation of parathyrin from the glands.

Mechanism of Action for Parathyrin and Calcitonin

These two hormones control the phosphorus-calcium metabolism: parathyrin increases the calcium level and decreases the inorganic phosphate level in blood. Calcitonin lowers the concentration of both calcium and phosphates in blood. Let us now discuss the mechanism of these alterations and its effect on the phosphorus-calcium metabolism.

Parathyrin to a significant extent exerts influence on phosphorus-calcium metabolism through vitamin D. In the kidneys, parathyrin activates adenylate cyclase. The cAMP thus produced stimulates activity of 25-hy-droxycalciferol hydroxylase and, consequently, formation of 1,25-dihydroxycal-ciferol. The latter enhances the intestinal uptake of Ca^{2+} ions and P_i mobilizes Ca^{2+} and P_i from osseous tissue and increases the renal reabsorption of Ca^{2+} ions." All these processes lead to an increased Ca^{2+} level in the blood and, apparently, should have stimulated an increase in P_i level. However, this takes no place in reality, since parathyrin sharply inhibits the reabsorption of phosphates in tubules of the kidney and leads to urinary loss of phosphates (to phosphaturia). Because of massive phosphaturia. the phosphate level in blood is lowered by parathyrin. There is a reported evidence that parathyrin, through the agency of Ca^{2+} ions, affects DNA synthesis and proliferation of lymphocytes which are a source of antibodies. After surgical removal of gastrointestinal tract and kidneys, calcitonin produces a hypocalcemic effect, *i.e.* a major target for calcitonin attack is the osseous tissue, which serves as a calcium depot in the organism.

The calcitonin effect on osseous tissue metabolism is reverse to that of parathyrin, since the former hormone elicits, in distinction from the latter, deposition of calcium phosphate salts on the collagen template of the bones. This results in a decreased level of calcium and phosphates in blood. Hypocalcemia is accompanied by a reduced urinary excretion of calcium; however, calcitonin produces, similar to parathyrin, an enhanced phosphaturia whose mechanism is independent of the calcium metabolism control exercised by calcitonin.

Abnormalities of the Parathyroid Glands

Hypofunction of the parathyroid glands, or hypoparathyrosis, is of rare occurrence and is seen in hyperexcitability of the neuromuscular system (convulsive contraction of muscles). The cause of this malfunction is low Ca^{2+} concentration in blood and extracellular fluid. Low Ca^{2+} concentration in the extracellular medium facilitates the membrane depolarization produced by the Na^+ ion flow within the cell and increases the excitability of nerve andjmuscle cells. This deleterious effect can be removed by administering calcium or parathyrin preparations or vitamin D. *Hyper function, or hyperparathyrosis,* occurs due to either an excessive production of parathyrin in the glands, or a prolonged administration of parathyrin preparations.

In hyperparathyrosis, massive mobilization of endogenic calcium from the bone depots occurs; in certain cases, a complete local deossification of the bones may take place. The hazard of spontaneous bone fracture is thus dangerously increased. The Ca^{2+} ion concentration in blood is very high, and the phosphorus level is lowered. Calcium, because of its poor solubility, becomes deposited in the internal organs and tissues, which leads to detrimental calcification of blood vessels, kidneys, gastrointestinal tract, and liver.

PANCREATIC HORMONES

Hystologically, in the pancreas distinguished are insular (Sobolev-Langerhans islands) and acinous tissues which contain cells responsible for synthesis and secretion of a number of hormones. The cells of A type (α-cells), B type (β-cells), and D type (all of them contained in insular tissue) secrete, respectively, *glucagon, insulin,* and *somatostatin.* The cells of PP type (or F cells), contained both in insular and in acinous tissues, secrete *pancreatic polypeptide.* Somatostatin, initially isolated from the hypothalamus, inhibits the. pituitary somatotropin secretion and also suppresses the pancreatic secretion of glucagon, gastrin, and, presumably, insulin.

Somatostatin inhibits the secretion of glucagon and somatostatin, exerting thereby a favourable effect on metabolism under diabetes mellitus. The pancreatic polypeptide, which is composed of 36 amino trcids, affects the gastrointestinal tract through stimulating the secretion of enzymes by gastric mucosa and secretion of pancreatic enzymes. In addition, this polypeptide inhibits the intestinal peristalsis and relaxes the urinary bladder.

Glucagon and Insulin

Glucagon is a protein with a molecular mass of 3485; it is composed of 29 amino acids. The a-cells produce proglucagon composed of 37 amino acids; proglucagon is then hydrolyzed by proteases to convert to the active glucagon. The glucagon secretion is enhanced as the concentrations of Ca^{2+} and arginine in blood increase, and is inhibited by glucose and somatostatin. Insulin is formed in the β-cells as preproinsulin which, when subjected to hydrolysis, yields proinsulin.

Proinsulin, containing 84 amino acid residues, suffers the cleavage of a polypeptide fragment (called C-peptide) composed of 33 amino acids. This results in the formation of insulin (of a molecular mass of about 6,000) composed of 51 amino acid residues. Insulin has two chains: a short one (A-chain) made up of 21 amino acids, and a long one (B-chain), of 30 amino acids; the two chains are cross-linked through disulphide bridges:

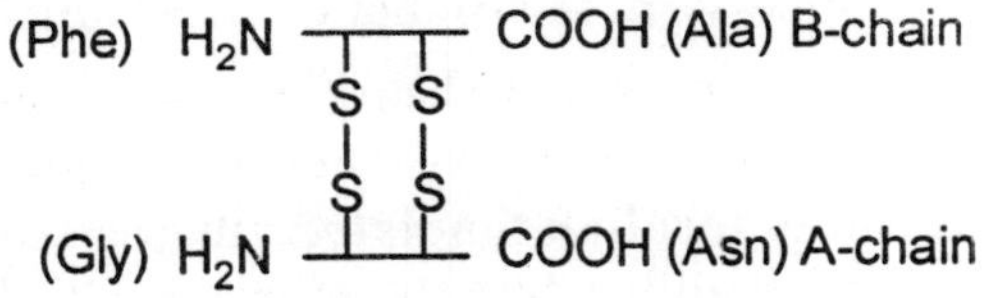

(Phe) H_2N — COOH (Ala) B-chain
S S
| |
S S
(Gly) H_2N — COOH (Asn) A-chain

In order to make insulin biologically active, its molecule must contain disulphide linkages and C-terminal asparagine. Insulin is inactivated by reducing S—S linkages and by proteolysis. The secretion of insulin is increased by glucose and Ca^{2+} ions as well as by amino acids arginine and leucine. Insulin secretion is stimulated by somatotropin and inhibited (although to a lesser extent, as compared with glucagon secretion) by somatostatin.

Mechanism of Glucagon Action

Glucagon binds to the membrane receptors of target tissues. The targets for glucagon are liver, fat tissue, and, to a lesser extent, muscles. By activating adenylate cyclase and by making increase the cAMP concentration, glucagon elicits mobilization of glucogen in iiver and, partly, in skeletal muscles, and of triacylglycerides in fat tissue.

Mobilization of these energy reserves leads to increased levels of glucose, fatty acids, and glycerol in blood. The hepatic combustion of fatty acids produces a large amount of acetyl-CoA and, ultimately, ketone bodies. For this reason, glucagon evokes moderate ketonemia and ketonuria. In the liver, this hormone inhibits ribosomal protein synthesis and facilitates protein catabolism. Amino acids thus formed are uSBd in urea production and in gluconeogenesis. Therefore, an increased glucose concentration in blood is to be attributed to two factors: activation of glycogenolysis (fast process) and activation of gluconeogenesis (slow process). On the whole, the alterations in metabolism produced by glucagon resemble those observed in diabetes mellitus (see below).

Mechanism of Insulin Action

Insulin supplied to the blood occurs in a free state or as bound to plasma proteins. Free insulin exerts influence on the metabolism of all insulin-sensitive tissues, while the bound insulin, on fat tissue only. The insulin-sensitive tissues include muscular and connective tissues (fat tissue is a variety of connective tissue). The liver is less sensitive to insulin; the nervous tissue is, in all likelihood, insensitive to it at all.

Membrane insulin receptors of glycoprotein nature have been found in the tissues. These receptors are numerous in the cells with a more pronounced susceptibility to metabolic insulin influence. The insulin-receptor complex is capable of drastically changing the cell membrane permeability for glucose, amino acids, Ca^{2+}, K^+, and Na^+ ions, or, to be more precise, of accelerating the transport of glucose, amino acids, and Ca^{2+} and K^+ ions into the cell.

A major cause of this effect is the local, *i.e.* membrane-oriented, action of insulin on active transport systems and the influence of this hormone on the generation of second messengers. At present, it is believed that the effects due to insulin are mediated by one or more peptidic second messengers which activate cAMP phosphodiesterase and decrease cAMP concentration. These peptides stimulate the transport of Ca^{2+} and glucose and the function of Na^+, K^+ pump, activate the protein synthesis and perform sr number of other functions.

Currently, the peptidic insulin messengers are being extensively studied. The most distinctive feature of insulin is its ability to intensify the active transport of glucose to the cells of hormone-sensitive tissues. The mechanism of insulin-stimulated glucose transport into the cells is far from being clear. Presumably, insulin either directly interacts with the proteins constitutive of the glucose channels in the membrane and opens "pores" for glucose passage, or indirectly, through cyclic nucleotides, affects phosphorylation of membrane proteins and thereby the membrane permeability for glucose. The activation of the Na^+/K^+-pump leads to an increased Na^+/K^+ membrane gradient and to membrane hyperpolarization.

The Na^+/K^+ gradient facilitates the secondary active transport of amino acids into the cell and, in the fat tissue presumably favours glucose transport too. The action of insulin on intracellular metabolism is effectuated via second messengers. Insulin facilitates the passage of Ca^{2+} ions into the cell, which, apparently, increases the activity of soluble guanylate cyclase and intensifies cGMP synthesis. On the contrary, Ca^{2+} ions reduce cAMP concentration by activating phosphodiesterase which splits cAMP. The low cAMP concentration is conducive to inhibition of glycogenolysis, gluconeogenesis (through consumption of amino acids), and lipolysis and to suppression of ketone body formation.

Simultaneously, a lower cAMP/cGMP ratio, which can be observed for the action of insulin, facilitates glycogen synthesis, lipogenesis (synthesis of triacylglycerides), and protein synthesis (the induction of ribosomal protein synthesis is a cGMP-dependent process which is amplified by insulin). Moreover, insulin, presumably through the agency of cGMP and Ca^{2+} ions, accelerates the syntheses of DNA (replication) and RNA (transcription) and favours thereby proliferation, growth, and differentiation of the cells. Alterations in the quotients for carbohydrate, lipid, protein, and mineral metabolisms in the blood testify to the regulatory action of insulin on target tissues. They allow one to estimate the insulin effect on the organism.

Insulin evokes a drop in blood concentration of glucose, amino acids, fatty acids, glycerol, and K^+ ions, and lowers urinary losses of amino acids and K^+ ions. On the whole, the effect of insulin on metabolism may be characterized as an anabolic action with a positive nitrogen balance.

Disturbances in Hormonal Function of Pancreas

Metabolic disturbances due to an excess or deficit of insulin in the organism are of common occurrence in medical practice. The excessive insulin may be observed in tumoral islands (insulomes) or in overdosed insulin therapy. All the metabolic changes that have been described above are manifested on a larger scale. Pronounced hypoglycemia conducive to syncopal states frequently occurs.

In extreme hypoglycemia, convulsions are observed, with an eventual fatal outcome. Hyperinsulinism can be alleviated by administering glucose and hormones that elicit hyperglycemia (for example, glucagon and adrenalin). With *deficient insulin,* a widely spread disease, diabetes mellitus, develops (according to the world statistics, about 100 million people suffer from diabetes mellitus). Diabetes mellitus develops under true and false insulin deficiency. The true insulin deficiency is due to a disorder in production and secretion of insulin in the insular β-cells. Under false insulin deficiency, the overall insulin activity of blood plasma remains normal, with apparently manifest signs of metabolic disturbances typical of diabetes mellitus.

Similar forms of diabetes mellitus develop because of an imbalance between the free and the bound form of blood plasma protein, with a prevalence of the latter form; because of this, glucose is absorbed by fat tissue only to be converted-to fat. Diabetes meltitus may also develop due to the absence or a lowered level of membrane insulin receptors in the target tissues. In this case, the cells are little sensitive to insulin, although the insulin level in the organism is normal, or even above normal. In diabetes mellitus, the changes in carbohydrate, lipid, protein, and water-mineral metabolisms are exactly opposite to those described above for the excessive insulin case.

The major metabolic changes in diabetes mellitus are summarized below. On the whole, in diabetes mellitus catabolic pathways are distinctly predominant over anabolic ones. In place of glucose, which is poorly assimilated in insulin-sensitive tissues, lipids are mobilized in the organism, and the combustion of fatty acids is increased. Sequent to the metabolic disturbances described above, observed are: hyperglycemia and glucosuria; hyperaminoacidemia and hyperaminoaciduria;, increased concentrations of fatty acids, glycerol, and cholesterol in blood; and ketonemia and ketonuria. A high concentration of ketone bodies with distinctly acidic properties results in the low pH of blood, which may lead to fatal outcome.

Process in tissues	***Observable effect***
Transport into cells :	
glucose	inhibited
aminot acids	inhibited
K^+ ions	inhibited
Ca^{2+} ions	inhibited
Cyclic nucleotide concentration :	
cAMP	increased
cGMP	reduced
Carbohydrate metabolism :	
glycogenolysis	increased
glycogen synthesis	inhibited
gluconeogenesis (from amino acids)	increased

Process in tissues	*Observable effect*
Lipid metabolism :	
triacylglyceride synthesis (lipogenesis)	inhibited
triacylglyceride breakdown (lipolysis)	increased
ketone body synthesis (on fatty acid combustion)	
cholesterol synthesis (from acetyl-CoA	increased
produced on fatty acid combustion)	increased
Protein metabolism :	
protein synthesis	inhibited
protein breakdown	increased
amino acid metabolism and	
production of urea	increased

Practical Applications of Insulin

Insulin preparations are used in the therapy of diabetes mellitus, as well as anabolic stimulators in dystrophy of organs, in malnutrition and inanition, and for restoration of metabolism after heavy muscular work.

HORMONES OF THE ADRENAL GLANDS

Hormones of Adrenal Medulla

In the medulla of the human adrenal glands, *adrenalin and,* to a lesser extent, *norodrenalin* are formed which are stored as secretory granules in the chromaffin cells. An increased adrenalin secretion occurs as the glucose concentration in blood becomes lowered, as well as in a state of the organism called *stress* (stress occurs when the physiologic activity of the organism increases faster than the adaptive responses). Adrenalin exerts a dual effect on the metabolism of target tissues depending on the predominant occurrence in them of either α-, or β-adrenoreceptors to which the hormone becomes bound.

The binding of adrenalin to β-adrenoreceptors stimulates adenylate cyclase and produces metabolic alterations characteristic of cAMP. The adrenalin binding with a-adrenoreceptors stimulates guanylate cyclase and produces metabolic alterations typical of cGMP. On the whole, adrenalin, similar to glucagon, exerts a cAMP-dependent action on the fat tissue metabolism as well as on skeletal muscles and liver, which are targets for the hormone The alterations that occur in carbohydrate and lipid metabolisms are very much the same as those produced by glucagon.

Moreover, adrenalin affects the function of cardiovascular system. It enhances the amplitude and frequency of systole, elevates blood pressure, and expands capillary arterioles. Adrenalin also relaxes the smooth muscles of intestine, bronchi, and uterus. In practice, adrenalin is' rarely used as a metabolic regulator (occasionally, to compensate for an insulin overdosage, adrenalin is applied simultaneously with glucose to increase the glucose level in blood). Most frequently, it is employed for stimulating systole and elevating blood pressure.

ADRENAL CORTEX HORMONES

In the adrenal cortex, steroid harmones, or corticosteroids, are formed from cholesterol. Corticosteroids, by the physiological effect they produce, are subdivided into three groups: glucocorticoids chiefly affecting carbohydrate metabolism, mineralocorticoids chiefly affecting mineral metabolism, and sex hormones (male hormones androgens, and female hormones estrogens).

The sex hormones are secreted in small amounts; their action will be discussed separately in greater detail. Normally, the human adrenal glands secrete glucocorticoids (**hydrocortisone** and **corticosterone**) and a mineralocorticoid (**aldosterone**).

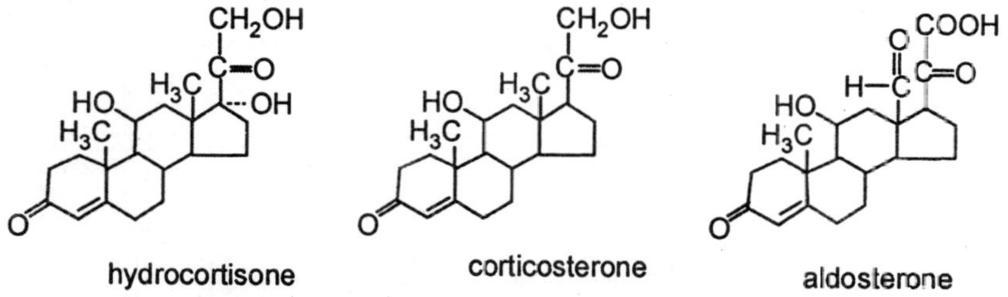

hydrocortisone corticosterone aldosterone

The secretion of glucocorticoids is controlled by corticotropin bound to the adrenocortical cell membrane; corticotropin stimulates production of cAMP and, through the intermediacy of the latter, triggers the delivery of cholesterol esters for the synthesis of glucocorticoids. The release of corticotropin from the pituitary gland is a typical response to stress. This entails the secretion of glucocorticoids into the blood to facilitate the release of adrenalin.

Glucocorticoids inhibit corticotropin secretion by a negative feedback mechanism. Aldosterone secretion is controlled by Na^+ and K^+ cations. At low Na^+ and high K^+ concentrations in blood, the synthesis and secretion of aldosterone become increased. As is commonly believed, the epiphysis produces a tropic hormone called adrenoglomerulotropin acting as a stimulator for aldosterone secretion. However, a reliable evidence for the existence of this hormone is lacking.

Mechanism of Glucocorticoid Action: Glucocorticoids become bound to α_1-globulin of blood plasma, called *transcortin,* to be transported in the complexed state to peripheral tissues. The targets for glucocorticoids are liver, kidney, lymphoid tissue (spleen, lymph nodes, lymphoid plaques of the intestine, lymphocytes, and thymus), connective tissues (bones, subcutaneous connective tissue, and adipose tissue), and skeletal muscles. These tissues contain cytosolic receptors for binding glucocorticoids. To be noted, a hormone-cytoreceptor complex may exert an entirely opposite effect on protein synthesis in different tissues.

In the liver and kidneys, this complex enhances the transcription of specific genes and the synthesis of corresponding proteins; in other tissues, on the contrary, it inhibits protein synthesis, while in the lymphoid tissue, this complex elicits lymphocytolysis, (or degradation of lymphoid tissue). The blocking of protein synthesis in the lymphoid tissue and the active proteolysis in it increase the fund of free amino acids that arc supplied in large amounts to the blood. The amino acids are used in liver and kidneys for protein synthesis; they serve also as substrates for gluconeogenesis. In the liver and kidneys, glucocorticoids favour the utilization of amino acids in gluconeogenesis, since they act as specific inducers of the synthesis of gluconeogenesis enzymes (pyruvate carboxylase, phosphopyruvate carboxylase, .glucose 6-phosphatase, and -fructose bisphosphatase).

The glucose produced by gluconeogenesis is consumed in the synthesis of hepatic glycogen (since glucocorticoids stimulate synthesis of the enzyme glycogen synthetase) as well as in the production of glycogen in muscles. Since glucocorticoids enhance the secretion of adrenalin from the adrenal medulla, the action of glucocorticoids becomes "augmented" by the metabolic effect of adrenalin. Thus, glucocorticoids mobilize triacylglycerides from the adipose tissue at the expense of adenylate cyclase activation, although the membrane in-tracellular activity is not typical of them.

Apparently, the mobilization of fat from the fat depots is associated with adrenalin. As a result, glycerol and fatty acids are supplied to the blood; glycerol is used in gluconeogenesis, while fatty acids are consumed in the liver to produce ketone bodies which are excreted in the blood.

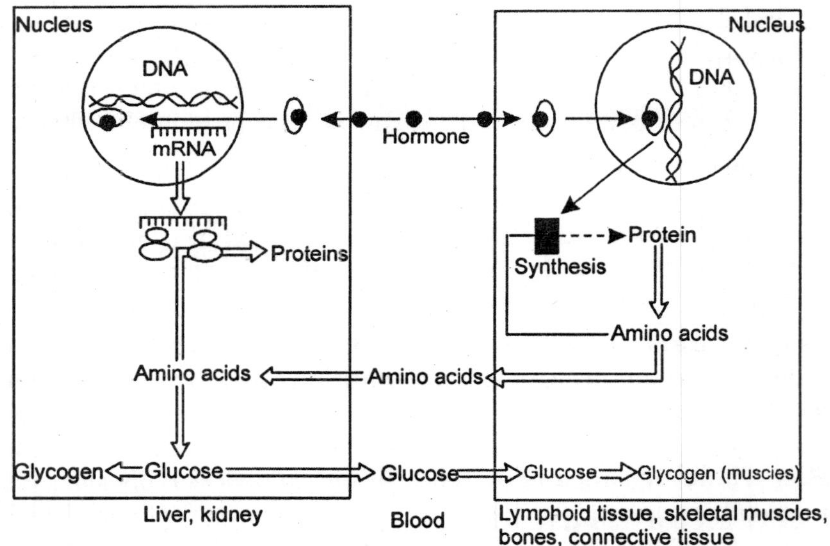

Fig. 5.4. Scheme for glucorticoid control of metabolism in organism.

Glucosuria, aminoaciduria, and ketonuria set in, as the concentrations of glucose, amino acids, fatty acids, glycerol, and ketone bodies in the blood increase. On the whole, these metabolic changes resemble a picture of diabetes mellitus. To be noted, this diabetic state is of different nature and for this reason is referred to as "steroid" diabetes. Glucocorticoids produce changes in the water-salt metabolism: they increase the Na^+ ion reabsorption and renal excretion of K^+; they retain sodium and water in the extracellular space of the organism tissues (which may lead to oedemas). This action is similar to the effect due to mineralocorticoids, only less pronounced.

Inhibition of the bone tissue protein synthesis leads to local deassification of the bones. Calcium and phosphorus are eliminated from the affected bone tissue into the blood and thernare excreted in the urine. Glucocorticoids and their analogues are widely applied in clinics. It stands to reason that this is not solely because of their ability to produce a diabetes-like state (metabolic diabetes-like disturbances are in fact side effects in therapy with" giucocorticoid preparations).

The medicinal effect of glucocorticoids is based on their ability to affect, lymphoid and connective tissues. The lymphoid tissue is involved in the generation of antibodies and in the defence of the organism from extraneous agents. In response to an infection or invasion of foreign substances, antibodies are produced in the organism that specify a state of hypersensitivity, or *sensitization,* to a given extraneous agent.

On repeated contact of the organism with the same invader, the antibodies interact with the invading agent, which is manifested by a vigorous reaction called *allergic response,* or simply *allergy.* Allergy leads to an inflammation accompanied by local disorders of vascular permeability and by tissue damage. The destroyed portions of tissue are replaced by connective tissue, and a connective cicatrbc is formed which deforms the affected organ. Glucocorticoids inhibit the formation of antibodies in the lymphoid tissue to reduce the state of sensitization towards invaders, and thus prevent further development of allergic response and inflammation.

The glucocorticoid inhibition of collagen formation by connective tissue fibroplasts prevents an excessive growth of connective fibres at the sites of tissue damaged by inflammation. Thus the hormones retard the development of vicious cicatrices or cicatricial adhesions that lead to a deformity of organs and impair their normal function. Mechanism of Mineralocorticoid Action. Aldosterone controls the balance of Na^+, K^+, Cl^- ions and water in the organism; for this reason, the normal function of this hormone is of utmost importance for the vital activity of the organism. Aldosterone is transported in the blood to tissues by using plasma albumins as carrier adsorbents.

The targets for aldosterpne are epithelial cells of distal tubules of the kidney, which contain a large number of cytoreceptors for binding this hormone. The aldosterone-cytoreceptor complex penetrates the nuclei of the renal tubule ceils and activates the transcription of chromosomal genes that carry information on the proteins involved in the transport of Na^+ ions across the membranes of tubular epithelium. Owing to this, the reabsorption of Na^+ and its counterion Cl^- from the urine into intercellular fluid and further into blood becomes increased. Simultaneously, K^+ ions are excreted (in exchange for Na^+) in the urine from the epithelium of renal tubules. On the whole, the aldosterone effect is manifested by a retention of Na^+, Cl^-, and water in the tissues and by urinary loss of K^+ ions.

Disturbances of Hormonal Function of the Adrenal Glands

Hyperfunction of the adrenal cortex, or hypercorticoidism, can manifest itself as an enhanced secretion of all the corticosteroids, or as a prevalent secretion of a group of hormones. For example, in such forms of hypercorticoidism as *Gushing's disease* (which occurs due to the impaired hypothalamohypophyseal.

System conducive to corticotropin hypersecretion) and *corticosteroma* (a tumor active chiefly in the synthesis of hydrocortisone), a hyperproduction of glucocorticolds is observed, which explains the symptoms of these disturbances in the organism: atrophy of subcutaneous connective tissue, development of steroid diabetes, osteoporosis (abnormal rarefication of bone), and hypertension (because of the secondary enhancement of adrenalin and noradrenalin secretion by substantia medullaris). There occurs hypercorticoidism attended by excessive secretion of aldosterone *(hyperaldo-steronism,* or *Konn's disease).*

In this disease, the symptoms of influence of aldosterone excess on the water-salt balance are observed, viz. oedemas, high blood pressure, and myocardial hyperexcitability. An excessive dietary intake of salt may lead to the so-called "salt" hypertension. *Hypocorticoidism,* also called *Addisons disease,* or *bronze disease,* is manifested by a deficiency in all the corticosteroids and attended by manifold alterations in metabolism and functions of the organism. Glucocorticoid deficiency causes a reduced resistance of the organism to emotional stress and damage factors (infectious, chemical, and.mechanical) and leads to the development of pronounced hypoglycemia. This symptomatology is aggravated by water-salt metabolism disturbances produced by aldosterone deficiency.

The organism loses sodium and water and accumulates potassium, with the ensuing development of hypotension (relaxation of the smooth muscles of vascular wall), acute myasthenia, progressive fatiguability leading to a total importance. These symptoms are associated with a disturbed myoblast membrane potassium-sodium gradient (hyperpolarization), with the ensuing low muscular excitability. In hypocorticoidism, the fatal outcome is due to the disturbed water-salt balance.

Practical Applications of Corticosteroids

Glucocorticoids and their numerous analogues are widely used in treatment of allergic and autoimmune diseases (rheumatism, collagenoses, nonspecific artritides, bronchial asthma, dermatoses, etc.) as desensitizing, antiinflammatory, and immunodepressive agents. Their immunodepressive action (inhibition of antibody synthesis by lymphoid cells) is used in the prophylaxy of transplanted organ rejection. In clinical practice, a synthetic analogue of natural mineralocorticoids, deoxycorticosterone, is applied in the substitution therapy of hypocorticoidism and, occasionally, in treatment of hypotension.

SEX HORMONES

The sex glands (gonads) are paired organs represented by the testes in males and by the ovaries in females. The testes and the ovaries are glands of combined secretion that produce sex cells (spermatozoa and ova) needed for reproduction (perpetuation) of the species, and sex (gonadal) hormones The male sex hormones, *androgens,* are produced by Leidig's cells (also called interstitial cells), and the spermatozoa, by the seminiferous tubules of the testes. The female sex hormones (*estrogens* and *gestagens*) and the ova arc produced in the ovarian follicles.

The male and female hormones are synthelized from their common precursor, cholesterol; since their synthetic routes in part overlap, the male and female sex hormones are found, in small amounts, in females and males, respectively. Sexual development is predetermined not only by sex chromosomes, but also by the specific features of sex hormone secretion by the gonads in the embryonal period. The gonadal differentiation occurs at the early stage of embryogeny.

The crucial moment is the action of androgens, which are produced by the embryonal testes, on the sex differentiation of the hypothalamus at the so-called "critical" developmental period. If, at this period, the hypothalamus is subject to the action by androgens, later, on reaching the sexual maturity, it ensures, through releasing hormones (liberins) and inhibitory hormones (stating), the male-type (acyclic) gonadotropin secretion; if otherwise, the gonadotropin secretion follows the female (cyclic) type. In sexually mature human males, *follitropin* controlling the spermatogenesis, and *lutropin* controlling the production of androgens (chiefly testosterone) are secreted.

Testosterone inhibits the hypophyseal secretion of lutropin by the negative feedback mechanism. *Prolactin* in males remains bound in the pituitary gland and normally is not secreted. In human females, all of the three gonadotropins are secreted, in a cyclic mode. In the female organism, gonadotropin secretion is the pacemaker in cyclic processes that are encompassed by the notion "sex cycle". The sex cycle includes two closely interrelated processes: the ovarian cycle, *i.e.* cyclic processes occurring in the ovaries, and the menstrual cycle, *i.e.* cyclic alterations occurring in the uterus. Both cycles are of equal duration. The cycle duration may vary from 21 to 35 days, but most commonly, it spans a period of 26-28 days.

Female Sex Hormones

Sex Cycle Scheme. The ovarian cycle has three phases: (1) follicular phase; (2) luteal phase; and (3) corpus luteum involution. In the 26-28-day sex cycle, the duration of the first phase is 13-14 days; of the second phase, 9-10 days; and of the third phase, 3-4 days. Schematically, the ovarian cycle presents itself in the following manner. Follitropin elicits the growth of the primordial ovarian follicles. The follicle cells start, at a definite developmental stage, to secrete estrogens, mainly, estradiol; other estrogens (estrone and estriol) are products of estradiol conversion.

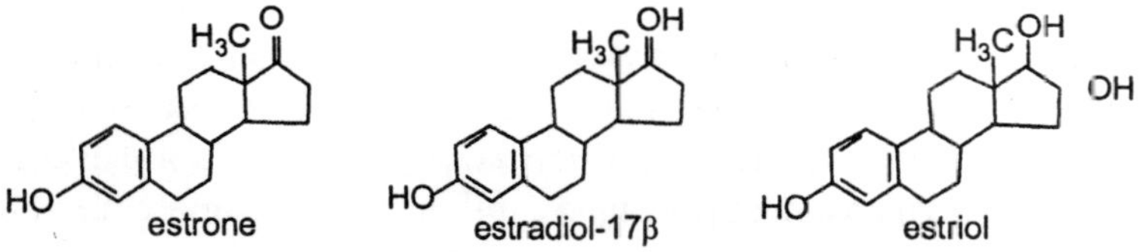

H_3C O HO estrone — H_3C OH HO estradiol-17β — H_3C OH OH HO estriol

Secretion of estrogens into the blood inhibits the secretion of follitropin (negative feedback mechanism) and stimulates the pituitary secretion of lutropin (positive, feedback mechanism). The released lutropin, in co-action with follitropin, elicits ovulation.

The ovum enters the abdominal cavity and becomes entrapped into the villi of the fallopian tube and moves towards the endometrium (inner mucous membrane of the uterus); the disrupted follicle from which the ovum has been released is transformed into the corpus luteum. Ovulation signals the termination of the follicular phase of the cycle; at this phase, the female organism develops a high level of estrogens and a low level of gestagens. Lutropin favours the development of the corpus luteum and secretion of progesterone by it. Progesterone, supplied to the blood, inhibits lutropin secretion and stimulates the release-of prolactin from the anterior pituitary.

Prolactin maintains progesterone secretion and simultaneously facilitates progesterone the development of milk ducts in the mammary glands.

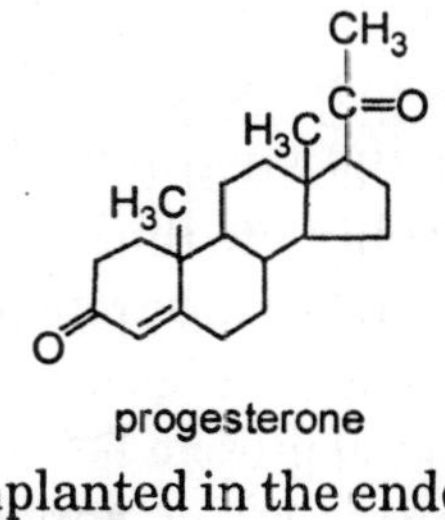

CH_3 C=O H_3C H_3C O

progesterone

If the ovum fails to be fertilized and implanted in the endometrium, the corpus luteum ceases its function.

The luteal phase which is supported by the high concentration of progesterone and by the low concentrations of estrogens in the female organism is thus terminated. In the third phase, the corpus luteum is liable to a retrograde development (involution) and secretes no progesterone.

The third phase is manifested by a decline in hormonal support of the endometrium by progesterone and estrogens. Subsequently, the ovarian cycle is repeated, since the low level of estrogens in the blood becomes a stimulus to increased secretion of pituitary follitropin. The cyclic secretion of hormones in the ovary is concomitant with alterations in the uterus and other organs and tissues that are targets for these hormones.

In the uterus, hormone-responsive are the myometrium (smooth muscle coat of the uterus) and, especially manifestly, the endometrium (the mucosa of the uterus). In the follicular ovarian phase, the released estrogens elicit an extensive proliferation of the endometrial epithelium (*proliferative phase* of the uterine cycle). The excitability and contractile activity of myometrium are enhanced. In the luteal phase, the increased secretion of progesterone leads to a loosening of the endometrium whose glandular epithelium discharges a mucous secretum.

For this reason, this phase of uterine cycle is referred to as *secretory,* and the attendant changes of the endometrium, as *pregravidic, i.e.* bearing resemblance to those typical of the early stages of pregnancy. In addition, progesterone inhibits the excitability and contractility of myometrium. All progesterone-induced uterine alterations facilitate the fixation of the fertilized ovum and its implantation in the endometrium. The rapid decline in progesterone secretion by the corpus luteum (at its ovarian involution stage) leads to a shrinkage of the proliferated endometrium, to the angiospasm of the blood capillaries of endometrium (endometrial ischemia), followed by the desquamation of endometrium and physiologic bleeding of the uterus (menstruation). This phase of uterine cycle is called *menstrual.* The correspondence between the phases of ovarian and uterine cycles is shown below:

Ovarian cycle (phases)	**Uterine cycle (phases)**
Follicular (secretion of estrogens)	Proliferative
Luteal (Secretion of progesterone	Secretory
Involution of corpus luteum (no secretion of estrogens and progesterone	Menstrual

If the ovum becomes fertilized and implanted in the endometrium, the trophoblast, within 24 hours after the fertilized ovum implantation, starts secreting chorionic gonadotropin which, acting like the hypophyseal lutropin, stimulates growth, development, and secretory function of the ovarian corpus luteum. No corpus luteum involution occurs, and pregnancy sets in.

Endocrine Function of Placenta. During pregnancy, a specific endocrine organ, placenta, is formed, which secretes protein and steroid hormones in the organism of expectant mother. The placenta is a producer of a number of protein hormones: chorionic gonadotropin (exhibiting lutropin activity); placental lactogen, or so-matomammotropin (exhibiting somatotropic activity, alongside lactotropic and luteotropic action, similar to that of hypophyseal prolactin); and thyrotropin (whose secretion is believed to stimulate an increased function of the thyroid glands in pregnant females). Biosynthesis of steroid hormones exhibits a number of specific features.

The major one of these is that in this process, both placenta and fetal tissue are involved. This cognizance has allowed Disfalusi to suggest the notion of a *fetoplacental unit* (from fetus + placenta) involved in the synthesis and secretion of steroid hormones. The fetoplacental unit produces progesterone, estradiol, estrone, estriol, and testosterone. Other steroid compounds are synthetized in smaller amounts. In the placenta, cholesterol is converted via pregnenolone to progesterone which is secreted into the maternal blood.

Precursors for the placental biosynthesis of other steroid hormones can be formed in the fetal tissues only. These precursors include dehydroepiandrosterone sulphate (DHEAS) produced from pregnenolone in the adrenal glands of the fetus, and 16-α-hydroxydehydroepiandrosterone sulphate, which is a product of DHEAS hydroxylation in the fetal liver. The former hormone is a precursor for the placental synthesis of estrone, estradiol, and testosterone, and the latter, of estriol.

Mechanism of Action and Biological-Functions of Female Sex Hormones: Estrogens and progesterone may be said to subserve one another in their regulatory effects on metabolism, growth, and development of tissues and organs. Most commonly, the progesterone effect is operative against the background of prior action produced by estrogens on tissues.

Estrogens ensure the normal effectuation of the following physiological processes:

1. Development of the genital Organs (ovaries, uterus, and vagina) involved in the puerperal function of the female.
2. Formation of the secondary sex characteristics in the puberty period (female-type hair growth; development of laryngeal cartilage and vocal apparatus characteristic of female type; developments of the mammary glands; ossification of tubular bone epiphyses and formation of a typically "female" skeleton).
3. Regulation of proliferative encfometrial changes and coordinated contractions of fallopian tubes and uterus in the follicular phase of the ovarian cycle.
4. Formation of sexual instinct and psychic status of human female.
5. Gestation, lactational development of the mammary glands in pregnancy, and parturition.

Most physiological effects due to estrogens are basically determined by the hormonal influence of estrogens on the activity of specific chromosomal genes. These effects ultimately induce the synthesis of specific proteins which are decisive for characteristic alterations in metabolism, cell growth, and cell differentiation.

The pronounced anabolic action of estrogens, *i.e.* their ability to stimulate protein synthesis in target organs, provides for a positive nitrogen balance in the organism. In the epiphyses of the bones, estrogens ensure collagen synthesis and osseous deposition of calcium and phosphorus. Moreover, estrogens in their capacity of enzymic inducers of glycolysis and pentose phosphate cycle, facilitate the aerobic generation of energy from carbohydrates and the reductive syntheses involving NADP-H_2 and ribose 5-phosphate. The estrogen effect on lipid metabolism has also been reported.

The estrogen hormones accelerate the renewal of lipids, inhibit accumulation of lipids in liver and fat tissues, favour the elimination of cholesterol from the organism and lower the cholesterol level in blood. Presumably, this is the reason for a lesser incidence rate for the atherosclerosis of coronary and other blood vessels in females than in males.

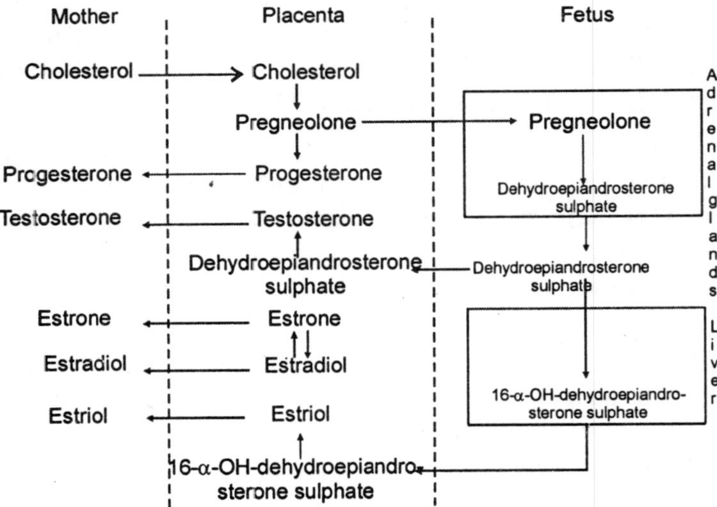

Fig. 5.5. Scheme for the biosynthesis of steroid hormones by phetoplacental system

Estrogens have been observed, on binding to the myometrium, to inhibit Na^+, K^+-ATPases of muscle cell membranes. This leads to the retention of Na^+ and, consequently, water in the myometrium and to the loss of K^+ ions, that is, the membrane depolarization sets in, conducive to enhanced excitability and contractility of the myometrium.

Progesterone exerts physiological and biochemical effects only during the luteal phase of ovarian cycle. This hormone provides for:

1. contractile inhibition of the uterus and fallopian tubes;
2. pregravidic alterations of the endometrium during the sex cycles and implantation of the fertilized ovum;
3. milk duct growth (sequent to prior estrogen action on the mammary glands) and lactation;
4. reduction of the excitability of hippocampus and heat centre as well as of sexual reactivity.

The mechanism of progesterone action on the growth and development of endometrium and mammary glands is the same as that of estrogens. As is commonly believed, the inhibitory progesterone action on the myometrial contractile function is associated with persistent depolarization of the muscle cell membranes conducive, after the initial short-term excitation, to paralysis of the smooth muscles. The smooth muscles become nonresponsive to mediators. Progesterone, when administered in large doses to the organism, produces a glucocorticoid-like effect.

Hormonal Functional Disturbances of the Ovary: Estrogen deficiency in the prepubertal period leads to developmental retardation of the genital organs (genital infantilism), delayed formation of secondary sex characteristics, and late ossification of eptphyseal cartilage as well as to sex cycle disturbances. These effects arc commonly accompanied by the negative nitrogen balance, loss of calcium and phosphate, and hyperlipemia.

In the adult female, progesterone deficiency affects (lie normal course of sex cycles and is conducive to habitual abortions.

Practical Applications of Female Sex Hormones: Estrogens and their synthetic analogues are used, simultaneously with progesterone, for restoration of disturbed sex cycles, and in treatment of ovarian insufficiency; progesterone is used for pregnancy preservation.

Male Sex Hormones

Androgens: Normally, prior to be supplied to the tissues, testosterone becomes bound to glycoprolein of blood plasma called testosterone-estradiol-bindiug globulin (which is a specific hinder for testosterone, dihydrotestosterone, and estradiol). In the cells, *testosterone* is reduced to *5α-diliydrotestosterone* with NADP-11-specilic 5α-steroid reductase:

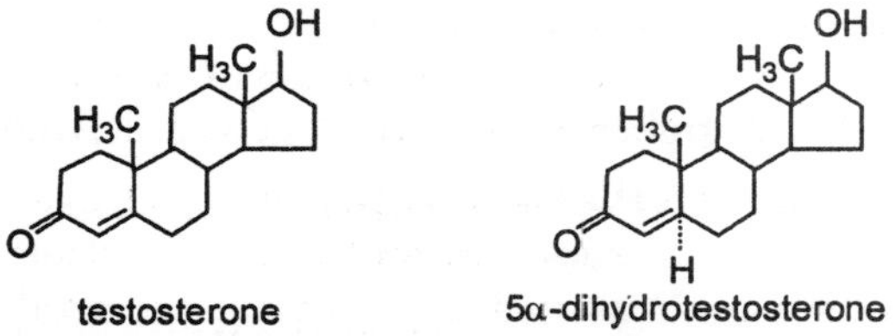

testosterone   5α-dihydrotestosterone

Both androgens are capable of specific activity; however, in certain tissues (prostate and seminal vesicles), 5α-dihydrotestosterone is more active, while in others (muscles), testosterone. The androgens become bound to androgenic receptors and act on the nuclear chromatin of target cells to facilitate DNA synthesis activation during replication and to accelerate the specific gene transcription.

The response to the regulatory androgen effect on the genetic apparatus is a sharp enhancement of tissue protein biosynthesis and; consequently, the positive nitrogen balance in the organism. The anabolic action of testosterone is much superior to that of estrogens. In males, this results in {lie development of powerful skeletal musculature, in the development and mineralization of the epiphyseal growth zones of the bones which become massive like the whole male skeleton in the period of pubertal development.

Androgens sharply increase protein synthesis in kidney and liver. In addition, they stimulate the development of male genitals and accessory sex glands (prostate and seminal vesicles), and at puberty, provide for the development of secondary sex characteristics—facial hair, pubic and axillary hair, size of larynx and the buildup of typically masculine vocal apparatus. Cooperatively with follitropin, androgens activate spermatogenesis. Androgens exercise a marked effect on the brain, in whose different portions the androgenic receptors are found.

Androgens affect the cerebral development, the sexual differentiation of hypothalamus in embryogenesis, behavioural reactions and libido; they are also involved in the formation of psychophysiological features of the male character. The induction of protein synthesis, including enzymes, is the major mechanism by which androgens provide an additional stimulus to the aerobic combustion of carbohydrates, fatty acids, and energy generation, has been ascertained that androgens activate phosphoglyceride synthesis and lower the total concentrations of lipids and cholesterol, although to a lesser extent as compared with estrogehs. From the practical standpoint, the anabolic androgen effect merits great attention. In all evidence, it is coupled to androgenic, or masculinizing, action.

Attempts to suppress the androgenic and to retain the anabolic properties have led to the manufacture of androgen derivatives named *anabolic steroids.* Viewed chemically, these are norsteroids with the methyl group missing from the C_{19} carbon of stefafie ring. In high-quality androgen preparations, the anabolic-to-androgenic activity ratio (which- has arbitrarily been taken for a unit in testosterone) is 5-12 times superior to that of testosterone. In addition to exhibiting a pronounced effect on protein synthesis in muscles, bones, kidneys, and liver, norsteroids facilitate, similar to testosterone, deposition of calcium phosphate salts in bones and mobilization of fat from the fat depots.

Disturbance of Androgenic Function in Testes. Androgen deficiency in the male organism (the state referred to as *eunuchoidism)* is commonly manifested by a genital underdevelopment (hypogenitalism), by failure to develop male secondary sex characteristics, absence of libido, retardation in epiphyseal ossification of the bones (conducive toMengthening of the limbs), atrophy of skeletal musculature, excessive deposition of subcutaneous fat, and by disturbed cortical inhibitory processes. Practical Applications of Androgens and Anabolic Steroids.

Testosterone preparations and their synthetic analogues are used clinically in treating the seminal hypofunction, disturbed sexual differentiation, functional sexual disorders in males, etc. Anabolic steroids (methyl-androstenediol, nerobolil, and retabolil) are used to treat various dystrophies, diabetes mellitus, thyrotoxicosis, and steroid diabetes (which are attended by the negative nitrogen balance); they are also used to stimulate growth and physical development in children as well as to stimulate fractured bone consolidation.

THYMUS HORMONES

The thymus may be denned as a gland of mixed secretion, since its function is the production of lymphoid cells and their distribution over lymph nodes and the spleen; in addition, the thymus is also a hormone producer. Five polypeptides— *thymosin, homeostatic thymic hormone, thymopoietins I and II, humoral thymus factor,* and a steroid-like compound, *thymosterine*— have been isolated from the thymus. Functionally, they are recognized as hormones affecting the rate of development and maturation of lymphoid cell precursors. Thus, the thymus is engaged in the formation and activity of the immune system of the organism.

The thymic function is closely related to other, nonthymic, hormones that exert influence on lymphocyte production and hormonal secretion of the gland. For example, iodothyronines, estrogens, somatotropin, and, possibly, insulin as well as other compounds that enhance their secretion stimulate lymphopoiesis and production of thymic hormones. In contrast, glucocorticoids, androgens, and progesterone exert an opposite effect and inhibit humorai immunity.

The detailed involvement of thymic hormones in regulatory metabolism, in addition to their immunologic functions, has not as yet been elucidated. In the congenital absence of thymus, a combined immune insufficiency is observed owing to the missing producer of lymphoid cells and thymic hormones controlling the development and maturation of lymphocytes.

The thymus underdevelopment in children leads to a disordered synthesis of humoral antibodies (agammaglobulinemia), or to a failure of cellular immunity (under normal production of antibodies).

EPIPHYSEAL HORMONES

In the epiphysis, *melatonin* is synthetized from tryptophan. A synthetic intermediate in this process is serotonin subject to methyltransferase-assisted methylation and to further acetylation with the involvement of serotonin N-acetyltrans-ferase. Serotonin synthesis is liable to diurnal variations and is dependent on sunlight irradiation.

In the dark, the synthesis of methyltransferase and production of melatonin are increased. In the light, the nervous signals transmitted from the visual analyzer via sympathic fibres to the epiphysis inhibit acetyltransferase activity and melatonin synthesis. Melatonin inhibits the hypophyseal secretion of gonadotropins either through the agency of hypothalamic releasing hormones, or directly, inhibiting there by sexual maturation. The extended daylight inhibits melatonin synthesis, which activates the secretion of gonadotropins stimulating the gonadal growth, gonadal secretion, and sexual activity. The shortening of daylight produces reverse alterations.

HORMONES OF HYPOTHALAMUS-HYPOPHYSEAL SYSTEM

Production and secretion of hormones by hypothalamus and hypophysis are closely related and for this reason it appears expedient to discuss the relevant processes in parallel. In the anterior lobe of the pituitary gland (adenohypophysis), tropic hormones are produced, while the posterior pituitary lobe (neurohypophysis) releases only neurohormones (vasopressin and oxytocin), which are produced in the hypothalamus nuclei. By their chemical structure, thyrotropin, follitropin, and lutropin are glycoproteins. They are composed of two subunits, α and β. In all these glycoproteins, the α-subunits are identical, while the β-subunits are structurally different and determinative of the specificity of hormonal action.

Other relevant hormones are simple proteins possessing a single polypeptide chain; vasopressin and oxytocin are cyclic octapeptides. The secretion of tropic hormones is controlled by hypothalamic peptides. To date, the following hypothalamic neuropeptides, which are regulators of hypophyseal hormonal secretion, have been isolated:

Tropic (pituitary) hormone	*Hypothalamic hormone*
Somatotropin	Somatoliberin, somatostatin
Corticotropin	Corticoliberin
Thyrotropin	Thyroliberin
Follitropin	Folliliberin
Lutropin	Luliberin
Protaction	Prolactoliberin, prolactostatin
Melanotropin	Melanoliberin, melanostatin

Mechanism of Action and Functions of Hypophyseal Hormones

All the tropic hormones exert their action either on the function of peripheral glands, or directly on the peripheral tissues by binding to the membrane receptors iind liy activating adenylate cyclase. cAMP effects the hormone production or metabolism in the target cells.

The effects that the hypophyseal hormones can produce may be divided into four groups:

1. control of biosynthesis and hormonal secretion by peripheral glands (thy-rolropin, follitropin, lutropin, prolactin, corticotropin, and somatotropin);
2. control of sex cell production (follitropin);
3. functional and metabolic control of effector tissues and organs (somatotropin, α- and β-lipotropins, corticotropin, lutropin, follitropin, melanotropin, prolactin, oxytocin, and vasopressin);
4. functional control of the nervous system (corticolropin, β-lipotropin, and others).

The two former functions of hypophyseal hormones have been dealt with previously, in the discussion of peripheral glands.

Direct Effect of Hypophyseal Hormones on Peripheral Tissues. *Corticotropin* exerts a direct action on the fat tissue by stimulating the tissue's glucose absorption and release of fatty acids and glycerol. The hormonal fat-mobilizing effect is related to adenylate cyclase activation; the cAMP formed stimulates triacylglycerde lipase, which splits triacylglycerides into glycerol and fatty acids.

In addition, the hormonal action of corticotropin on melanin production and skin pigmentation is similar to that of melanotropin. α- *and* β-*Lipotropini* exert a specific fat-mobilizing action By the mechanism typical of all cAMP-stimulating hormones. *Gonadotropins* produce a fat-mobilizing effect similar to that of lipotropins. Moreover, prolactin stimulates protein biosynthesis and lactose production by the mammary gland epithelium. *Vasopressin,* or *antidiuretic hormone,* in addition to its fat-mobilizing action, exerts a selective control of water reabsorption in the distal tubes and collecting ducts of the kidneys and activates adenylate cyclase. cAMP activates protein kinases, which phosphorylate the cell membrane proteins to increase their permeability for water.

The water reabsorption reduces diuresis, makes increase the urine density and urinary concentrations of sodium and chlorides. Vasopressin stimulates the contraction of the muscular tissue of the capillaries and arterioles and produces a moderate rise in blood pressure. Deliciency in vasopressin develops a disease called *diabetes insipidus.* It is manifested by a large discharge of urine (4 to 10 litres per day) of low density (1.002–1.006); excessive thirst (polydipsia) develops.

Administration of vasopressin preparations removes the symptoms of this disease. *Oxytocin* stimulates contraction of uterine muscles; this action is associated with an increased intracellular Ca^{2+} concentration and with cGMP production. Oxytocin accelerates protein synthesis in the mammary glands during the lactation period (in part, this effect isalso exhibited by vasopressin) and stimulates the release of breast milk owing to an enhanced contractile activity of the milk duct myoepithelium. This hormone produces an insulin-like effect on the fat tissue, *i.e.* increases glucose consumption and triglyceride synthesis in it. *Melanotropin.* α- and β-Melanotropins are distinguished.

The two hormones are secreted by the anterior pituitary and affect the production of melanin in skin, iris of the eye, and the epithelial pigment of the retina. They produce a fat-mobilizing action on the fat tissue by stimulating cAMP generation in it.

Somatotropin, or growth hormone, is the only hormone that exhibits a biological species-specific effect. The animal somatotropin produces no effect on the humans. Somatotropin stimulates the cartilage cell division and the growth of bones in length. It also controls the growth of internal organs and soft tissues of the face arid oral cavity. Somatotropin exerts both direct and indirect effects on the peripheral tissues. Its direct action is associated with the activation of adenylate cyclase and formation of cAMP, for example, in muscles and insular pancreatic tissue. In the pancreatic islands, the hormone stimulates the secretion ..oi.gluc agon and insulin, the former being produced in larger amounts than the latter.

Presumably, this may be the cause of the diabetogenic effect seen in somatotropin. The indirect effect of somatotropin consists in the generation of growth-regulating polypeptides called *somatomedins.* These are secreted into the blood and xhibit properties characteristic of somatotropin. Seven somatomedins with molecular masses about 7,000 have been isolated from the human blood plasma: somatomedins A (A, and A_2), somatomedins B, represented by a group of four polypeptides, and somatomedin C.

Somatomedins A and C intensify the cartilage cell division, the syntheses of DNA, RNA, and proteins, and facilitate the insertion of sulphate into proteoglycans. In addition, somatomedin C acts, similar to insulin, on the adipose and muscle tissues, *i.e.* stimulates glucose uptake and inhibits lipolysis in adipose tissue. Somatomedins B stimulate DNA and protein syntheses in the nervous system cells. On the whole, somatotropin produces a pronounced anabolic effect attended by the positive nitrogen balance. Somatotropin deficiency in the young organism is conducive to a premature growth cessation, with an eventual development of dwarfism. The stature of an adult dwarf is 100-120 cm.

Hypophyseal dwarfs, in distinction to those afflicted with hypothyroid dwarfism, have a proportionate constitution, with no signs of mental retardation. Somatotropin hypersecretion in a juvenile period shows up as gigantism; in a maturity period, it is conducive to a state called acromegalia. Acromegalia is manifested by an enlargement of the prominent parts of the face (nose, chin, superciliary eyebrows) and the soft tissues of the face and oral cavity (for example, the tongue); occasionally, separate fingers, toes, or the hand or foot are increased in size. Physiologically, a weakly pronounced acromegalia develops in pregnant females in the gestation period, when the placenta produces somato-mammotropin; it disappears after parturition.

Involvement of Hypothalamohypophyseal Hormones in the Functional Regulation of the Nervous System

In 1975, two pentapeptides—leucine-enkephalin and methionine-enkephalin— have been isolated from the nervous tissue and found to exhibit the capacity to bind to opioid receptors and to act like morphine. Later, other endogenic opiates: α-, β-, and γ-endorphins, which are also polypeptides, have been found in the pituitary gland. All these compounds with opiate-like activity, including the earlier discovered enkephalins, have been grouped under a common name endorphins, or endogenous morphines.

Endorphins are produced by limited proteolysis from the liypophyseal hormones, *i.e.* they are of hormonal origin, in the pituitary gland, a large-size prehormonal protein is presumed to form which is the parent compound for β-lipotropin and corticotropin. The limited proteolysis of β-lipotropin and corticotropin leads to endorphins (peptides that are operative in the processes of learning and memorization) and to a- and p-melanotropins. The two latter also contain an amino acid peptide sequence assistant in learning and memorizing. All the neuropeptides appear to act as mediators or modulators in the synapses through affecting the neuron function. Endorphins are believed to produce an analgesic effect and are conducive to states of euphoria and psychic aberrations as sequent to endorphin metabolic disturbances in'schizophrenia. Endorphins are more potent analgetics than morphine. The analgesic effect in acupunctural therapy is also believed to be due to endorphins.

Practical Applications of Hypophyseal and Hypothalamic Hormones

Of the hypothalamic peptide regulators, *somatostatin* has found medicinal application. It is used in therapy of diabetes mellitus owing to its property to deter endogenous somatotropin from intervention in glucagon secretion. Vasopressin-containing preparations (e.g. *adiurecrine)* are used in treatment of diabetes insipidus; *ozytocin* is used in stimulating the contraction of the uterus during labour. All the tropic hypophyseal hormones, except somatotropin and thyrotropin, find application in practical medicine.

Corticotropin is used to stimulate the adre-nocortical function as well as in treating the states responsive to glucocorticoids whose secretion it stimulates. Serumal gonadotropin and chorionic gonadotropin, which are analogues of follitropin and lutropin, are used for normalization of cyclic ovarian activity in females and in treatment of hypospermia in males. Prolactin is used for stimulating the lactation in females. Attempts have been reported to use melanotropin for stimulating an increased production of melanin in affected retina of the eye, and β-lipotropin, in treatment of adiposis.

ANTIHORMONES

Antihormones are compounds that exhibit antihormonal activity through binding to cytosolic receptors. While being active antagonists towards other hormones, they may either exhibit or not the intrinsic hormonal activity. The molecular mechanism of antihormonal action is based on the competition for binding sites with the corresponding cytosolic receptors. Owing to this, the complex "antihormone—cytosolic receptor" is incapable of acting as a protein synthesis inducer in the cells.

Antihormones exhibit a lesser affinity to receptors than true hormones. An antihormone displaces a true hormone only when its concentration in the cell is very high. Natural antihormones include estrogens and androgens which compete with each other for binding with the receptors of opponen hormone: estrogens block the androgenic receptors, and androgens, the estrogenic ones. Antihormones may be exemplified by *17-α-methytestosterone* for glucocorticoid receptors, *spirolactones* for mineralocorticoid receptors, *flutamlde* for androgen receptors, and *nafoxidine* for estrogen receptors. Specific behaviour of certain antihormones should be emphasized.

If, after their binding to a receptor, the complex formed is incapable of being transported to the nucleus and to interact with chromatin, the true antihormonal effect is observed, *i.e.* the given hormone remains silent. If the antihormone-receptor complex has to a certain extent retained its activity, a simulated hormonal action by the antihormone involved is observed. By way of example, the behaviour of nafoxidine extends over both antiestrogenic and estrogenic activities.

In the uterus, on binding to the receptors, it acts as estradiol; simultaneously, in tumors of the mammary gland and liver, nafoxidine, on interacting with the estrogen receptors, exhibits antiestrogenic activity. Antihormones are employed both in experimental laboratory studies of the mechanism of hormonal action and in medical practice. Testosterone and estradiol are used in medication of genital tumors in the patients of opposite sex: testosterone, in females, and estradiol, in males. Antihormones are used in treating hormone-dependent tumors when a need arises to prevent the action of hormone on-proliferating tumoral cells, in therapy of abnormal sexual behaviour (hyper-sexuality), etc.

PROSTAGLANDINS

Prostaglandins are hormone-like compounds {hormonoids) derived from C_{20}-polyene fatty acids containing a cyclopentane ring. They were first isolated by Euler from extracts of the prostate (Latin *prostata*); hence the name coined for them. Currently, prostaglandins are known to occur in all cells and organs of the human organism, except erythrocytes. Prostaglandins are short-lived species that are synthetized at need in small amounts to exert a local biological effect at the site of their formation.

Structure and Nomenclature. Prostaglandins (PG for short) may be regarded as derivatives of prostanic acid; this compound does not occur in nature, but has been obtained synthetically. Depending on the type and the number of oxygen substituents, and the location of double bonds in the cyclopentane ring, the prostaglandins fall into several types conventionally denoted by Latin characters: A, B, C, D, E, F, G, and H. Within each prostaglandin type, there are distinguished several series differing in the number of double bonds in the side chains of the molecule and denoted by numeral subscripts (for example, PGE_1, PGE_2, and PGE_3). In addition, the OH group and the cyclopentane ring may be differently oriented in space, which is designated by additional subscripts α and β (for example, $PGF_{2\alpha}$ and $PGF_{1\beta}$). All naturally occurring prostaglandins have an α-configuration.

Biosynthesis and Metabolism. Starting substrates for the synthesis of prostaglandins are polyene fatty acids such as eucosa-8,ll,14-trienic (homo-γ-linolic), eucosa-5,8,ll,14-tetraenic (arachidonic), and eucosa-5,8,ll,14,17-pentaenic acids. In tissues, these acids are produced from linolic acid which is not synthetized in the animal organism and must therefore be supplied in food, *i.e.* linolic acid is an essential food factor. In the organism, linolic acid is chiefly converted into arachidonic acid, which is incapable of free existence and immediately enters compositionally into phosphoglycerides of practically all tissues.

Arachidonic acid makes up a major fraction of all C_{20}-polyene fatty acids of intracellular phosphoglycerides and is the starting substrate for production of prostaglandins. Prostaglandin biosynthesis is started by releasing arachidonic acid from phosphoglycerides under the action of tissue phospholipase A_2 (occasionally C). Subsequently, arachidonic acid is subject to the action of fatty acid cyclooxygenase making part of a multienzyme complex called *prostaglandin synthetase.* This results in the production of biologically active intermediates—prostaglandin endoperoxides—also referred to as PGG, and PGH_2.

The end products of prostaglandin biosynthesis are not the same in different cells and tissues. In most of them, prostaglandin endoperoxides are converted to prostaglandins of E or F type (of which, PGE_2 and PGF_{2d} constitute a major portion) and also to PGD. The A-type prostaglandins are formed by dehydration of PGE, and B- and C-type prostaglandins, by double bond isomerization in the PGA cyclopentane ring. However, in the vascular wall, a special type of prostaglandin, denoted 1_2 and most commonly known as *prostacyclin,* is formed from the PGG_2-type endoperoxide. In thrombocytes and mast cells, compounds with a six-membered oxane ring, which have been named *thromboxanes* (TX for short), are synthetized from PGH_2. There are known thromboxanes of A and B types that fall into a number of series by the same principle as prostaglandins.

In leucocytes, araciridonic acid metabolism follows another route. Assisted by the enzyme lipooxygenase, this acid is converted to noncyclic and unsaturated derivatives, which have been named *leucotrienes* (LT for short).

Depending on their structural specificity, the LTs are subdivided into A, B, C, D, and E types, while by the number of double bonds, they fall into series 3, 4, and 5. A scheme for generation of prostaglandins, prostacyclins, thromboxanes, and leucotrienes is presented below :

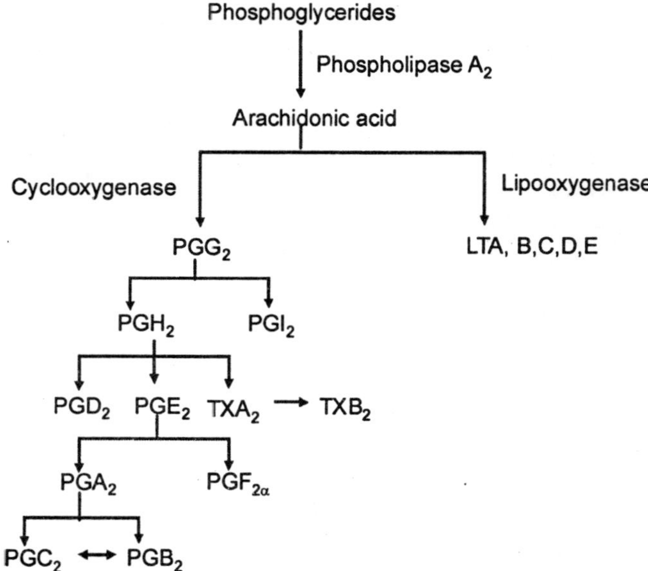

All these compounds have a short half-life (a few seconds for thromboxanes to 20 minutes for prostaglandins), are liable to fast inactivation, and are eliminated from the organism in the urine.

Biological Action : Different prostaglandin types, for example, E and F, exercise different, and even opposite, effects; moreover, the prostaglandins of one type but of different series, for example, E_1 and E_2, in certain instances are opposite by the action they produce. The uncommon diversify of effects and high activity manifested by prostaglandins are attributable to their influence on the metabolism exercised through the agency of intracellular mediators cAMP, cGMP, and Ca^{2+} ions. As has been observed, prostaglandins most commonly increase the cAMP con centration, which results in an enhanced cAMP-mediated effect on the metabolism and function of a given tissue. Thus, by increasing the cAMP level in the endocrine glands, prostaglandins stimulate production and secretion of hormones.

For example, in the adrenal glands, prostaglandins stimulate the formation of steroid hormones and secretion of catecholamines; in the thyroid gland, the synthesis of iodothyronines; in the pancreas, the release of insulin. On the other hand, prostaglandins in fat tissue make decrease the production of cAMP and, similar to insulin, inhibit lipolysis, *i.e.* they act as antagonists towards catecholamines and polypeptide hormones, which accelerate lipolysis by increasing the concentration of cAMP in fat tissue. $PGF_{2\alpha}$ controls the contraction of smooth muscles of the uterus, bronchi, and intestine acting chiefly through the agency of cGMP or Ca^{2+} ions, Certain cells have membrane receptors for binding prostaglandins.

For this reason, in some tissues, the prostaglandins formed can affect the second messenger concentration from within (by directly varying the activity of the enzymes involved in the conversion of these mediators), while in other instances they act, like hormones, through specific receptors. Prostaglandins exhibit a strikingly broad spectrum of biological activity and extremely high efficiency (a millionth of a gram is sufficient to produce a marked effect). Prostaglandins, especially $F_{2\alpha}$, stimulate the contraction of uterus and fallopian tubes, and elicit the involution of corpus luteum to produce parturifacient effect facilitating thereby childbirth. Various types of prostaglandins differ in their action on the myogenic tonus of other organs.

For example, PGD_2, PGG_2, PGH_2, TXA_2, and LT elicit contraction of the bronchi, while PGE makes them relax; $PGF_{2\alpha}$ and thromboxane A_2 contract the blood vessels and build up the arterial pressure, while prostacyclin arid PGE_2 produce a vasodilative effect and make the arterial pressure drop. Prostacyclin and PGE_2 increase the volume of urinary discharge and the urinary concentration of sodium; this facilitates the decrease in the blood vessel tension and prevents the development of hypertension. Prostaglandins intensify the intestinal motility, but affect in a different manner the gastric juice secretion. In particular, PGE inhibits and $PGF_{2\alpha}$, on the contrary, stimulates the gastric juice secretion.

Role of Prostaglandins in Pathology: An excessive production of prostaglandins or their deficiency may lead to pathologic processes such as inflammation, thrombosis, gastric ulcer, and others.

Typical inflammation symptoms (reddening, oedema, elevated temperature, and pain) can be attributed to the effect by prostaglandins and leucotrienes produced from arachidonic acid in tissues. Prostaglandins produce dilation of blood vessels and increase their permeability, which results in the development of reddening and edematous processes at the inflammation focus. Leucotrienes favour the leucocyte chemotaxis, *i.e.* migration of leucocytes towards the site of formation of leucotrienes, adhesion of leucocytes to the vascular wall, and their accumulation at the site of inflammation.

Leucocytal enzymes take part in the destruction of the bacteria that have evoked inflammation. If leucotrienes fail to be formed, the localization of inflammation becomes inhibited, which may lead to the development of bacteriemia. Elevated temperature and fever observed in inflammation are caused by prostaglandins affecting the hypothalamic heat centres, while the algesic symptom is due to that the prostaglandins enhance the sensibility of nerve endings to the excitatory histamine action.

Moreover, it has been ascertained that the slow reacting substance of anaphylaxis (SRS-A) is a mixture of leucotrienes, while the substance RCS formed in pulmonary anaphylaxis (substance that produces aortic contraction in rabbit) has been shown to be merely a mixture of PGG,, PGH_2, and TXA,. Consequently, prostaglandins, leucotrienes, and thromboxanes act as mediators in inflammatory and allergic reactions. Natural and synthetic glucocorticoid preparations, by blocking the phospholipase A_2, reduce the formation of prostaglandins from arachidonic acid and produce an antiinflammatory effect.

Nonhormonal antiinflammatory agents (acetylsalicylic acid, indometacin, diclofenac, and others) prevent prostaglandin synthesis by inhibiting cyclooxygenase, while rutin (an inhibitor to lipooxygenase) decelerates the formation of leucotrienes and alleviates the inflammation symptoms. Thromboxane and prostacyclin play an important role in the thrombogenesis of blood vessels. Thromboxane A_2 assists in the development of thrombosis, since it induces the agglutination of thrombocytes and facilitates thereby the participation of thrombocytic factors in blood clotting.

Prostacyclin, on the contrary, is a powerful natural inhibitor of thrombocytic aggregation and acts as an antithrombic agent. The prostacyclin-to-thromboxane ratio in the vascular wall is of utmost importance for the development of thrombosis. The antithrombic effects observed in acetylsalicylic acid and indometacin are explained by their ability to inhibit the formation of thromboxane and the aggregation of thrombocytes. It should also be noted that prostaglandin E_2 prevents ulceration of intestinal and gastric mucosae; for this reason, preparations inhibiting prostaglandin biosynthesis (especially glucocorticoids) may be conducive to ulcers with concomitant gastrointestinal hemorrhage.

Applications in Practice : Prostaglandin $F_{2\alpha}$ (dinoprost and ensoprost F) is used in obstetric practice for pregnancy termination and as a parturifacient agent, while prostaglandin E_2 (dinoprostone and prostin E_2) is applied in arresting bronchial spasmodic attacks, hypertension, and peptic ulcer.

BIOCHEMICAL ADAPTATION

Adaptation is the sum total of all the processes occurring in the organism that provide for the organism's stability under variable conditions of its existence. During adaptation, a purposeful rearrangement of physiological and biochemical functions takes place to offset the effects caused by detrimental factors, enabling the organism not only to survive, but also to continue to exist under extreme environmental conditions. Depending on the level of adaptive reactions, it is expedient to single out *physiological* (systemic) and *biochemical* (cellular) *adaptation.*

The physiological adaptation is concerned with an operative rearrangement of the systemic functions of the organism (for example, blood circulation and nervous system), which permits maintaining the constancy of the internal milieu of the organism (or homeostasis) and facilitates the activity of tissues and cells by, for example, improving their supply with nutrients and oxygen and thus accelerating the eliminative processes of vital activity of the organism. However, while making part of the organism, the cells possess intrinsic mechanisms for metabolic rearrangement which counteract the variable conditions of vital activity and assist the cells to adapt. In essence, the cellular adaptation is based on the biochemical mechanism of metabolic rearrangement—hence its name *biochemical adaptation.*

The interactive union of physiological (systemic) and cellular adaptive mechanisms provides the organism with the possibility to adapt to unfavourable environment. Adaptation and regulation are closely interconnected, since metabolic processes can be directed along an appropriate route only with the aid of a system of extracellular regulators that trigger the intrinsic regulatory mechanisms of the cell. Biochemical adaptation, like regulation, can be *short-term* and *long-term*.

The short-term adaptation is concerned with a fast rearrangement of metabolism occurring at the onset of a critical situation. The concomitant metabolic changes are conditioned by the intervention of short-term regulatory mechanisms of cellular metabolism or, to be more exact, by the action of neurohormonal impulses on the cell membrane permeability and enzymic activity. The long-term adaptation is manifested by a steady metabolic reorganization that develops in response to a prolonged action of disfavourable factors. The short-term adaptation is oriented to the survival of the cell, while the long-term adaptation is oriented to the maintenance of the cell's vital activity under unfavourable conditions.

During the long-term adaptation, the metabolic reorganization is accomplished through the intervention of long-term regulatory mechanisms, *i.e.* by the action of neurohormonal stimuli on the synthesis- of enzymes and other functional proteins. The induction of "needed" proteins generates another, distinct from the usual, metabolic regime, which in the best manner corresponds to the cell function in a given unfavourable situation. If, owing to certain reasons, the needed composition of functional proteins fails to be produced, the cells and the organism as a whole cannot adapt to the new environmental conditions, and the organism must pass through a period of adaptation disease, or acclimatization.

CHAPTER

6

Biosynthesis of Organic Acid & Ethanol

Microorganisms are biocatalysts actively used in the fermentation process where product and biomass are obtained from fermentable sugars. The cells are harvested and reused. The immobilisation of cells has been practised in the production of vinegar. Enzymes and whole cells can be bound to solid support, fixed in the form of an active layer. When substrate passes over the surface, enzymatic reactions change the substrate to the desired product. Enzymes and suitable microorganisms are used in manufacturing food flavours, additives, medicines, and other goods are produced by variety of microbial metabolites. Immobilised cells may perform differently to an equivalent mass of freely suspended cells. They may produce extracellular polymers, which are usually lumped together with cells in dry biomass measurements, resulting in overestimation of the activity of immobilised biomass.

In addition, cell location in the form of biofilm can affect its quality and activity, because profiles of environmental conditions such as effect of substrate concentration often exit. However, in many studies results for freely suspended cultures have been used to model immobilised cell bioreactors, usually without a lot of attention in assessing the quality of the biomass in the system. Typically these models allow for the effect of difficult limitations in and around the biofilm, and assume that the immobilised cells have the same quality and activity is the suspended cell culture, regardless of the position in the biofilm. In fact significant substrate concentration gradients may exist for cells immobilised in biofilm.

Cells located close to the nutrient supply are likely to maintain higher quality and activity compared with cells located relatively further away, leading to differentiation in the quality or activity of the immobilised cell population. This differentiation is more pronounced if there are starvation regions. In practice, zero substrate concentration may exist inside the biofilm, because in these regions the cell physiology may be markedly different from that of the freely suspended cells. The cells activities have been described based on a multi-species biofilm model, and the microbial kinetics by a mathematical model. Using this model predicts that the biomass on the external surface of the biofilm has higher activity than the biomass near the solid support surface, and that condition may occur, after the biofilm has reached a critical dept or formed a thick multi-layers of the biofilm. The model was based on microbial growth kinetics determined by Wang *et al.*

However, no experimental results were presented in either of the immobilization studies to verify the model predictions, nor was the predicted immobilised biomass activity compared with that of freely suspended cells in a comprehensive way. Wang *et al.*, *and* Najafpour *et al.*, worked with immobilised microbial cells of *Nitrobacer agilis, Saccharomyces cerevisiae* and *Pseudomonas aeruginosa* in gel beads, respectively. They found separately that the cells retained more than 90% of their activity after immobilisation by using specific oxygen uptake rate (SOUR) [mg $O_2 \cdot g^{-1}$ (dry biomass) h^{-1}] as the biomass activity indicator. Such differences in immobilised biomass and activity between free and immobilised biomass activities depend strongly on the particular characteristics of the microbial systems and their interaction with the support matrix.

IMMOBILISED MICROBIAL CELLS

Since 1978, several papers have examined the potential of using immobilised cells in fuel production. Microbial cells are used advantageously for industrial purposes, such as *Escherichia coli* for the continuous production of L-aspartic acid from ammonium furmarate. Enzymes from microorganisms are classified as extracellular and intracellular. If whole microbial cells can be immobilised directly, procedures for extraction and purification can be omitted and the loss of intracellular enzyme activity can be kept to a minimum.

Whole cells are used as a solid catalyst when they are immobilised onto a solid support. There are three general methods available for the immobilisation of whole microbial cells: carrier binding, entrapment and cross-linking.

Carrier Binding

As shown in the figure 6.1a, the carrier binding method is based on binding microbial cells directly to water-insoluble carriers. The binding is due to ionic forces between the microbial cells and the water-insoluble carriers. This technique has rarely been used, however, because of lyses during the enzyme reactions. Microbial cells may leak from the carrier, thereby disrupting the immobilisation. Therefore, this method has not been applied successfully.

Entrapping

In applying entrapment, microbial cells are directly entrapped into polymer matrices. This method has been applied to several microorganism by using gelatin, agar, polyacrylamide gel, calcium alginate, etc., as the entrapping agents. The following matrices are used extensively to immobilise microbial cells:

- Collagen
- Gelatin
- Agar
- Alginate
- Carrageenan
- Cellulose triacetate
- Polycarylamide
- Polystyrene

In this method, microbial cells are apparently lysed within the entrapping agent, but they retain the desired enzymatic activity. To prepare an efficient immobilization, the type and concentration of a bifunctional reagent for entrapment should be optimised. The advantage of this method is that cell leakage may act as a diffusion barrier, thus causing difficulty in transferring substrate and product through the matrix.

Cross-Linking

Microbial cells immobilised by cross-linking with bi- or multifunctional reagents such as glutaraldehyde have been reported to be successful, but active immobilisations have not been obtained with other reagents such as toluenediisocyanate. The microbial cell wall is composed of lipoproteins, with lipopolysaccharide extending from the cell membrane. Glutaraldehyde reacts with lysine within the protein in the lipid bilayer of the cell membrane. Furthermore, gelatin is physically absorbed on a solid support, which provides a covalent link between the microbial cells which is tightly bound to a solid support by absorption. Glutaraldehyde has been used as cross-linking reagent for ethanol production.

Maximum achieving rates was reported as high as 7.4 $g \cdot l^{-1} \cdot h^{-1}$ ethanol produced from glucose. The major advantage of the cross-linking method is that the immobilisation reagent does not act as diffusion barrier. Actually, a thin film of cells is provided in this system, making it ideal for many applications.

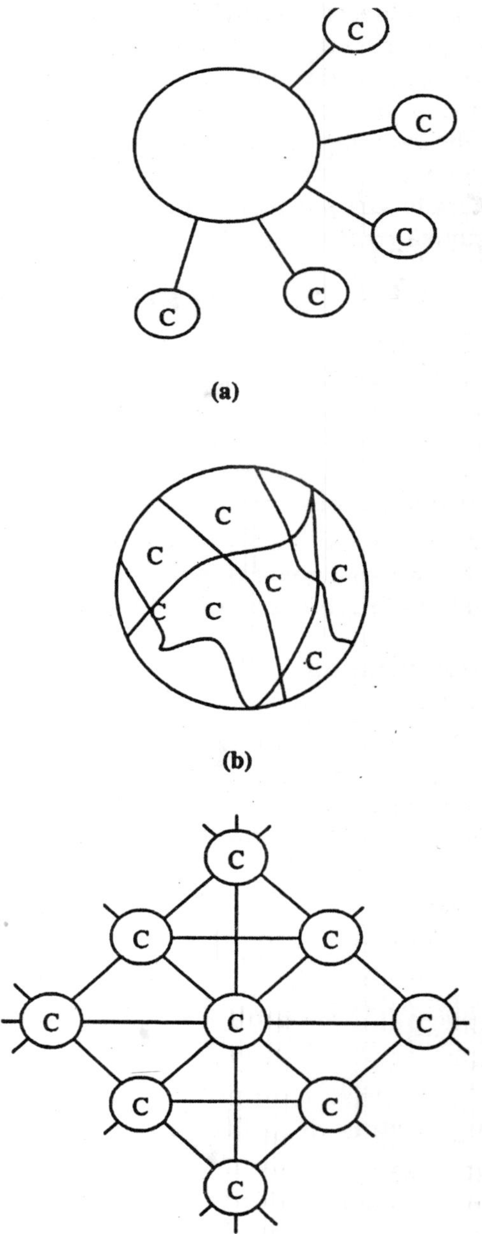

Fig. 6.1. Alternatives methods used to immobilize cells. (*a*) Carrir binding; (*b*) entrapment; (*c*) cross-linking.

Advantages and Disadvantages of Immobilised Cells

There are several advantages of an immobilised cell system over a batch, or CSTR system. The first and most obvious benefit is the capability of recycling or reusing the microorganisms since they are retained inside the reactor as the product leaves the reactor. Secondly, immobilisation can be easily used for a continuous process maintaining high cell density, thus providing a high productivity. Thirdly, nutrient depletion and any inhibitory compounds do not, in general, have a large effect on the immobilised cells because the cells are fixed by immobilisation.

In batch and CSTR fermentation, nutrient depletion, inhibition and accumulation of toxic by-products are major problems, but immobilised cells are usually unaffected by toxic by-products. However, there are disadvantages to using immobilised cells. The cell may contain numerous catalytically active enzymes, which may catalyse unwanted side reactions. Also, the cell membrane itself may serve as a diffusion barrier, and may reduce productivity. The matrix may sharply reduce productivity if the microorganism is sensitive to product inhibition. One of the disadvantages of immobilised cell reactors is that the physiological state of the microorganism cannot be controlled.

IMMOBILISED CELL REACTOR EXPERIMENTS

The immobilised cell reactor (ICR) experiments were undertaken to determine the performance of immobilised *Propionibacterium acidipropionici* in a plug-flow tubular reactor. The productivity of the ICR was evaluated by measuring glucose and xylose consumption, and propionic and acetic acid production, along the length of the reactor. The Rasching rings were sterilised and dip-coated in sterile 20% gelatin and 1.5% agar n solution. The coated rings were randomly packed in the column. After the coated rings had dried, the packing was sprayed with a 2.5% glutaraldehyde solution.

An alternative procedure to prepare the packing was also tested and proved to be satisfactory. In using the latter method, the reactor column was filled with clean Rasching rings. A hot gelatin agar solution was passed through the column, and allowed to wet the entire packing surface. When the coating was dried, the glutaraldehyde solution was passed through the column in a similar fashion. The column was allowed to stand for 24 hours and then washed with sterile, distilled water. The reactor was sterilised by passing ethylene oxide through the column. The ethylene oxide was allowed to stand in the column for eight hours before the system was purged with sterile nitrogen. The feed and product tubing were autoclaved by using steam at 15 psig. After sterilisation a 24-hour-old seed culture was pumped through the column.

An adaptation time of about 48 hours was allowed for the establishment of a film of microorganisms cross-linking to the Rashing rings. Feed media was then pumped into the reactor. An accurate calculation of the retention times in the reactor was difficult, owing to the fact that growth and carbon dioxide evolution may take a major fraction of void volume of the ICR, resulting in a false retention time. The microbial film thickness could be controlled by periodically passing sterile nitrogen or carbon dioxide through the column, to stop cell over-growth. A feed concentration of 15 g glucose and 15 g xylose per litre was used over a feed rate of 20-200 ml/hr.

Samples were taken at successive points along the reactor length, and the usual analysis for glucose and xylose consumption, organic acid production and cell density were done. A kinetic model for the growth and fermentation of *P. acldipropionici* was obtained from these data. Analytical procedures were set up to measure cell density, sugar concentration and organic acid concentration. The turbidity of the culture (optical density) was determined by reading the absorbance with spectrophotometer and cell counts. Sugar concentration as a single substrate was measured by an industrial digital analyser. Total sugar was evaluated by a reducing agent such as dinitrosalicylic acid in alkaline solution. An orange colour was produced which was read on a colourimeter at 540 nm. Organic acids were determined by a gas chromatograph, with a flame ionisation detector.

ICR RATE MODEL

Table 6.1 presents the results of the ICR retention time studies, sugar concentration (dual substrate) studies and cell density. The kinetic model for ICR was derived on the basis of a first order reaction, plug flow and steady-state behaviour.

$$r_A = kC_A \qquad ...(6.1)$$

Table 6.1. Continuous Fermentation of Dual Substrate (Glucose, Xylose) in ICR at 36° *C*.

Retention Time (h)	*Length of reactor (inches)*	*Substrate concentration (g l⁻¹)*		*Organic acid concentration (g l⁻¹)*		*Reaction rate of glucose (g l⁻¹h⁻¹)*	*Cell density (number of cells / ml)*
		Glucose	*Xylose*	*Propionic acid*	*Acetic acid*		
28	0	15	15	—	—		
	6	6.28	10	1.3	7.78	0.31	9×10^{11}
	14	4.52	8.6	1.98	8.45	0.37	9×10^{11}
	24	2.13	7.2	2.7	12.4	0.46	9×10^{11}
	34	1.36	5.4	3.25	17.2	0.49	9×10^{11}
	44	1.14	3	3.95	19.1	0.495	9×10^{11}
20	0	15	15	—	—		
	6	7 1.	11	1.2	6.45	0.395	9×10^{11}
	14	4.72	9.8	1.46	8.1	0.514	9.5×10^{11}
	24	2.21	7.9	1.87	12	0.64	9.5×10^{11}
	34	1.5	7.15	2.12	16.4	0.675	9.5×10^{11}
	44	1.2	5.8	2.4	18	0.69	9.5×10^{11}
12	0	15	1.5	—	—		
	6	9.3	14.5	2.59	1.33	0.475	1×10^{10}
	14	6.02	14	3.26	2.37	0.75	2×10^{10}
	24	3.62	12	4.5	2.45	0.95	3×10^{11}
	34	2.65	13	5.07	2.5	1.03	3×10^{11}
	44	2.3	10	5.36	2.54	1.06	5×10^{11}
5	0	15	15	—	—		
	6	9.8	15	0.5	0.58	1.04	6×10^{10}
	14	7.27	14.5	0.6	0.76	1.55	1.8×10^{10}
	24	6.05	14	0.65	1.2	1.79	5×10^{10}
	34	5.84	14	0.7	1.36	1.83	5×10^{10}
	44	5.2	13.5	0.8	1.65	1.96	6×10^{10}

$$u\frac{dC_A}{dz} = r_A \qquad ...(6.2)$$

where r_A is the reaction rate, k is the rate constant, C_A is the sugar concentration, z is the axial reactor length, and u is the bulk fluid velocity. Substituting (6.1) into (6.2) yields.

$$u\frac{dC_A}{dz} = kC_A \qquad ...(6.3)$$

Equation (6.3) is a linear first-order differential equation of concentration and reactor length. Using the separation of variables technique to integrate (6.3) yields.

$$\ln\left(\frac{C_A}{C_{A_o}}\right) = \frac{kz}{u} \qquad ...(6.4)$$

Fig. 6.2 shows a plot of ln C_A/C_{A_o} as a function of dimensionless reactor length for the fermentation data obtained at a feed concentration of 15 g·l^{-1} glucose and 15 g·l^{-1} xylose at different retention times. A straight line is obtained at each retention time. The value of the slope increased with increasing retention time owing to decreasing velocity through the column Table 6.2 also indicates that the rate constant in the first-order reaction is fixed with different retention times.

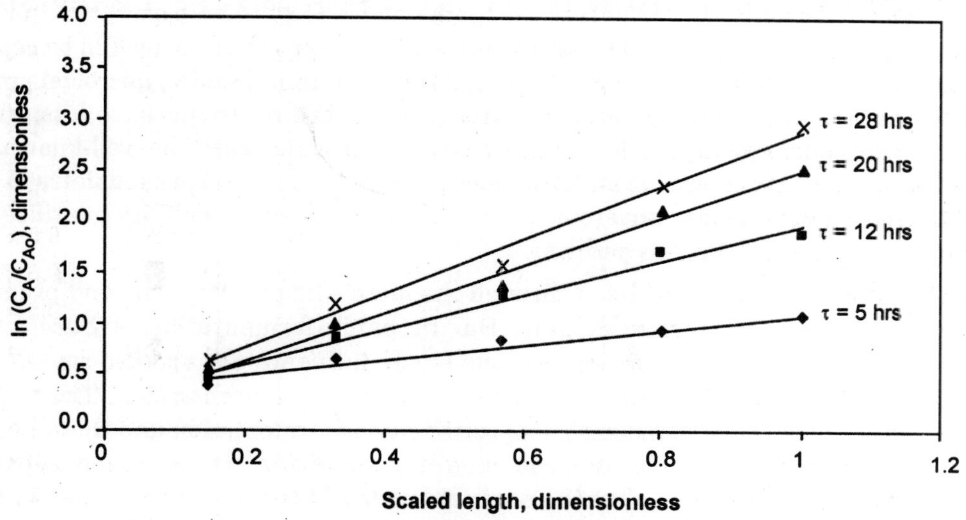

Fig. 6.2. Model test for ICR using *Propionibacterium acidipropionici.*

Table 6.2. ICR Kinetic Model for Immobilized Propionibacterium Acidipropionici.

Retention time τ, hour	*Flow rate ml / h*	*Rate constant h^{-1}*
5	214	0.70
8	135	0.69
12	90	0.68
20	55	0.68
28	38	0.51

The ICR flow rate was five to eight times faster than the CSTR. The overall conversion of sugars in the ICR at a 12 hour retention time was 60%, At this retention time, the ICR was eight times faster than CSTR, but in the CSTR an overall conversion rate of 89% was obtained. At the washout rate for the chemostat, the ICR resulted in a 38% conversion of total sugars. Also, the organic acid production rate in the ICR was about four times that of the CSTR. At a higher retention time of 28 hours, the conversion of glucose in the ICR and CSTR are about the same, but the conversion of xylose reached 75% in the ICR and 86% in the CSTR.

Nomenclature

- r_A Reaction rate, $g{\cdot}l^{-1} \cdot h^{-1}$
- k Rate constant, $l{\cdot}h^{-1}$
- C_A Sugar concentration, $g{\cdot}l^{-1}$
- z Axial reactor length, cm
- u Bulk fluid velocity, $cm \cdot h^{-1}$.

ETHANOL FERMENTATION USING *SACCHAROMYCES CEREVISIAE*

Owing to diminishing fossil fuel reserves, alternative energy sources need to be renewable, sustainable, efficient, cost-effective, convenient and safe. In recent decades, microbial production of ethanol has been considered as an alternative fuel for the future because fossil fuels are depleting. Several microorganisms, including *Clostridium* sp. and yeast, the well-known ethanol producers *Saccharomyces cerevisiae* and *Zymomonas mobilis,* are suitable candidates to produce ethanol. Microorganisms under anaerobic growth conditions have the ability to utilise glucose by the Embden–Mereyhof–Parnas pathway.

Carbohydrates are phosphorylated through the metabolic pathway; the end products are two moles of ethanol and carbon dioxide. During batch fermentation of *Saccharomyces cerevisiae,* other influential parameters can adversely influence the specific rate of growth, and inhibition can be caused either by product or substrate concentration. The viability of the *Saccharomyces cerevisiae* population, its specific rate of fermentation and the sugar uptake rate are directly related to the desired medium condition. It has been reported by Nagodawithana and Steinkraus, that the addition of ethanol to the cultured media was less toxic for *Saccharomyces cerevisiae* than ethanol produced by cell bodies. This indicates that there are other metabolic by-products that can cause inhibition, and may show impurity of ethanol produced in the fermentation system. Also, the death rates were lower with addition of pure ethanol than the similar condition with the endogenously produced ethanol concentration. Use of biofilm reactors for ethanol production has been investigated to improve the economics and performance of fermentation processes.

Immobilisation of microbial cells for fermentation has been developed to eliminate inhibition caused by high concentrations of substrate and product, also to enhance productivity and yield of ethanol. Recent work on ethanol production in an immobilised cell reactor (ICR) showed that production of ethanol using *Zymomonas mobilis* was doubled. The immobilised recombinant Z. *mobilis* was also successfully used with high concentrations of sugar (12% –15%). The potential use of immobilised cells in fermentation processes for fuel production has been described previously. If intact microbial cells are directly immobilised, the removal of microorganisms from downstream product can be omitted and the loss of intracellular enzyme activity can be kept to a minimum level. Recently, immobilised biomass activity has been given more attention, because it has been acknowledged to play a significant role in bioreactor performance.

Frequently, immobilised cells are subjected to limitations in the supply of nutrients. Thus, because of the presence of heterogeneous materials such as immobilised cells, there is no convective flow inside the beads and the cells can receive nutrients only by diffusion. Immobilisation of cells to a solid matrix is an alternative means of high biomass retention. The cells divide within and on the core of the matrix. Alginate is widely used in food, pharmaceutical, textile and paper products. Alginate is used in these products for thickening, stabilizing, gel and film forming.

Sodium alginate is a linear polysaccharide, normally isolated from many strains of marine brown seaweed and algae, hence the name 'alginate'. The copolymer consists of two uronic acids or polyuronic acid. It comprises primarily D-mannuronic acid (M) and L-glucuronic acid (G). Alginic acid can be either water-soluble or non-water soluble depending on the type of the associated salt. Interchange of sodium ions with calcium ions in the solution may follow solidification of sodium alginate in calcium chloride solution. The sodium salt, other alkaline metals and ammonia are soluble in water, whereas the polyvalent cations salts, *e.g.* calcium, are not water-soluble, except for magnesium ions. The purpose of this research was to obtain high ethanol production with high yields of productivity and to lower the high operating costs.

The ICR column fermenter was used-with *Saccharomyces cerevisiae* by an entrapment technique utilising alginate as a porous wall to retain the yeast cells. The effect of initial glucose concentration on the production of ethanol by *S. cerevisiae* was evaluated. The yield for large-scale ethanol production was compared with the ethanol produced in batch fermentation using the same microorganism.

MATERIALS AND METHODS

Experimental Reactor System

The immobilised cell reactor was a plug flow tubular column, constructed with a nominal diameter of 5 cm, internal diameter of 4.6 cm, Plexiglas of 3 mm wall thickness and 85 cm length. The medium was fed to the ICR column from a feed tank located above the column. A variable-speed Masterflex pump, model L/S Easyload (Cole-Parmer, Vernon Hills, IL, USA) was used to transfer feed medium from a 20 litre polypropylene autoclaveable Nalgene carboy, the carboy serving as reservoir. The effluent from the column was collected in a 20 litre polypropylene autoclaveable carboy serving as product reservoir. A flow breaker was installed between the column and feed pump, which prevented growth of microorganisms and contamination of feed line and feed tank. Samples from the ICR column were taken from the inlet and outlet compartments of the column. A 16 hour culture was harvested at exponential growth phase and mixed with 2% sodium alginate.

The slurry of yeast culture was converted to droplet form while it was dripped into a 6% bath of calcium chloride using a 50 ml syringe. Once the slurry was added to the bath, beads of calcium alginate with entrapped cells were formed. The bed consisted of uniformly packed 5mm beads. The solidified beads were transferred to the column. About 70% of the column was packed. The extra space was counted for bed expansion by the fresh media. The void volume was measured by volume of distilled water pumped through the bed. The packed ICR column was used in continuous mode for the duration of fermentation. The fresh feed was pumped an in upflow manner, while sugar and ethanol concentration was monitored during the course of continuous fermentation. The working volume of ICR after random packing was 740 ml. The bed volume was about 660 ml. The experimental set up of the ICR is shown in figure 6.3. There was no evidence of cell leakage from the beads to the surrounding media, the matrix was permeable to substrate and product, cell growth and glucose conversion were monitored in the ICR. Over-growth of beads after a few days of operation was controlled. Carbon dioxide was purged to eliminate over-growth of beads.

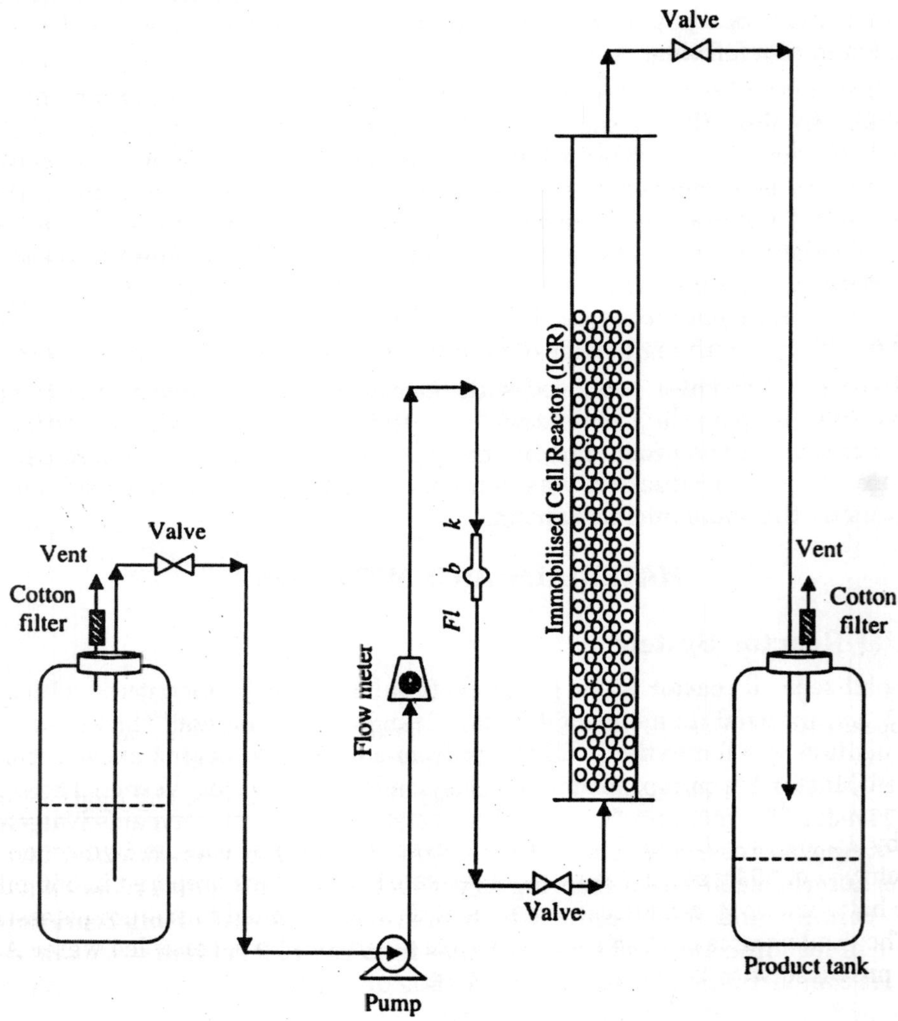

Fig. 6.3. Schematic diagram of ICR experimental setup.

The major over-growth occurred at the entrance region, in about the first 30 cm length of the column where the sugar concentration was very high. High sugar concentrations of 25, 35, 50 and 150 g·l^{-1} were used. Microbial over-growth was controlled with carbon dioxide passed through the bed. There was a maximum 30% increase in the beads' diameter at the lower part of column, where the glucose concentration was maximum. The void volume was measured by passing sterilized water. In addition to the carbon source, the feeding media consisted of 1 $g \cdot l^{-1}$ yeast extract pumped from the bottom of the reactor, while the flow rate was constant for a minimum duration of 24 hours. A seed culture of *S. cerevisiae* ATCC 24860 was grown in a media of 5 g glucose, and 0.5 g yeast extract, respectively, 1.5 g KH_2PO_4 and 2.25 g Na_2PO_4 phosphate buffer up to a total volume of distilled water, 500 ml.

The media was sterilized at 121 °C for 15 min. The stock culture of the microorganisms was transferred to the broth media for preparation of seed culture. Sodium alginate was prepared by dissolving 10 g of powder form in 500 ml of distilled water. A separate solution of 120 g of calcium chloride was dissolved in 21 of distilled water. Sodium alginate and calcium chloride solution were autochaved at 121 °C for 15 min. the sterilized sodium alginate solution and the high cell density of the grown seed culture were thoroughly mixed. Beads were prepared by droplet from pipette with about 5 mm diameter in a sterilized calcium chloride solution. The cell density of the seed culture for bead preparation was 3.1 $g \cdot l^{-1}$. The wet and dry weight of beads incubated for 16 hours were measured : 3.28 and 0.5 g, respectively. The moisture content of the beads was 85%. The physical properties and appearance of the *S. cerevisiae* beads are summarized in Table 6.3.

Determinaiton of Glucose Concentration

To determine the concentration of inlet and outlet glucose in the ICB, a reducing chemical reagent, 3, 5-dinitrosalicylic (DNS) acid 98% solution, was used. The DNS solution was

Table 6.3. Physical Properties and Appearance of S. Cerevisiae Beads

Alginate, wt %	*1.5*	*2*	*3*	*6*
Beads diameter, mm	5	4.9–5	4.8–4.9	4.5
Growth, diameter expansion after 72 h	50–60%	25–30%	20–25%	No expansion
Diffusion problem	Nil	Nil	Nil	May exists
Cell activity	Active	Fully active	Fully active	Partly active
Physical appearance	Easy to break	Flexible, hard enough to stand	Flexible and hard	Very hard
Stability	Not stable	Good stability	Stable and rigid	Very rigid

prepared by dissolving 10 g of 3, 5-dinitrosalicylic acid in 2 M sodium hydroxide solution. A separate solution of 300 g sodium potassium tartrate solution was prepared in 300 ml of distilled water. The hot alkaline 3, 5-dinitrosalisylate solutions were added to sodium potassium tartrate solution. The final volume of DNS solution was made up to 11 with distilled water. A calibration curve was prepared with 2 $g \cdot l^{-1}$ glucose solution.

Detection of Ethanol

Ethanol production in the fermentation process was detected with gas chromatography, HP 5890 series II (Hewlett-Packard, Avondale, PA, USA) equipped with a flame ionisation detector (FID) and GC column Porapak QS (Alltech Associates Inc., Deerfield, IL, USA) 100/120 mesh. The oven and detector temperature were 175 and 185 °C, respectively. Nitrogen gas was used as a carrier. Isopropanol was used as an internal standard

Yeast Cell Dry Weight and Optical Density

In batch fermentation, approximately 2ml sample was harvested every 2 hours. The absorbance of each sample during batch fermentation was measured at 620 nm using spectrophotometer, Cecil 1000 series (Cecil Instruments, Cambridge, England). The cells dry weight was obtained using a calibration curve. The cell dry weight was proportional to cell turbidity and absorbance at 620 nm. The cell concentration (dry weight) of 2.1 g·l^{-1} was obtained from the 24 hours culture broth, the free cell samples with absorbance of 1.6 at 620 nm.

Electronic Microscopic Scanning of Immobilised Cells

For electronic micrographs, samples were taken from fresh beads and 72 hour beads from the ICR column. The samples were dipped into liquid nitrogen for 10 minutes, then freeze dried for 7 hours (EMITECH, model IK750, Cambridge, UK). The sample was fixed on an aluminium stub and coated with gold-palladium by a Polaron machine model SD515 (EMITECH, Cambridge, UK) at 20 nm coating thickness. Finally the sample was examined under scanning electron microscope (SEM) by using a Stereoscan model S360 (SEM-Leica Cambridge, Cambridge, UK).

Statistical Analysis

The size of beads was uniform and consistent, the mean size of beads with 3% alginate and based on measurement of 20 samples; the mean value for the beads' diameter was 4.85 mm, with a standard deviation of 0.3 mm and calculated variance of 0.1 mm. The standard deviation was less than 5%. The data for the batch fermentation experiment with 50 g·l^{-1} glucose are presented and were replicated three times. The data for ICR with 25, 35, 50 and 150 g·l^{-1} glucose that were conducted at a wide range of flow rates are shown in figure 6.5 to 6.9, which were repeated in additional runs. The standard deviations of collected data for the batch experiments were approximately 5%, and the standard deviation for the ICR experimental data with a sugar concentration of 50 g·l^{-1} was approximately 10%. Data were analysed in a spreadsheet, Microsoft Excel 2000. The error analysis for the ICR data with a substrate concentration of 150 g·l^{-1} was slightly higher but it was in the range of 10-12%.

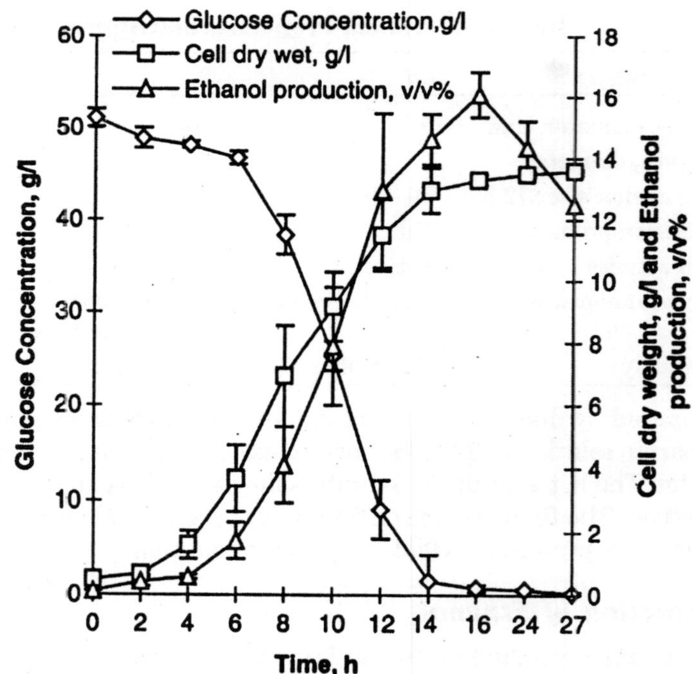

Fig. 6.4. Glucose concentration, cell density and production of ethanol in batch fermentation with intial concentration of 50 g·l^{-1} glucose versus time.

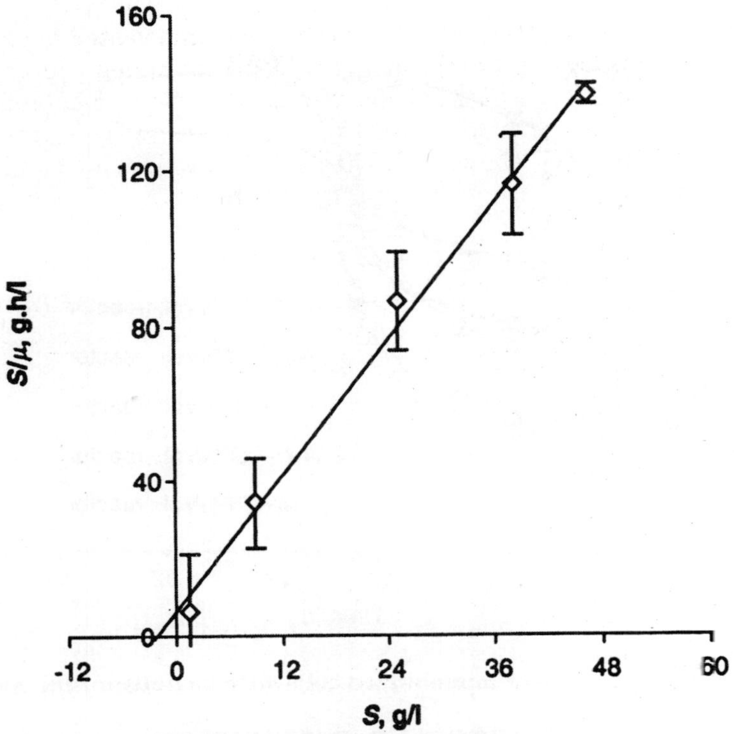

Fig. 6.5. Kinetic model for batch fermentation, Langmuir–Hanes plot.

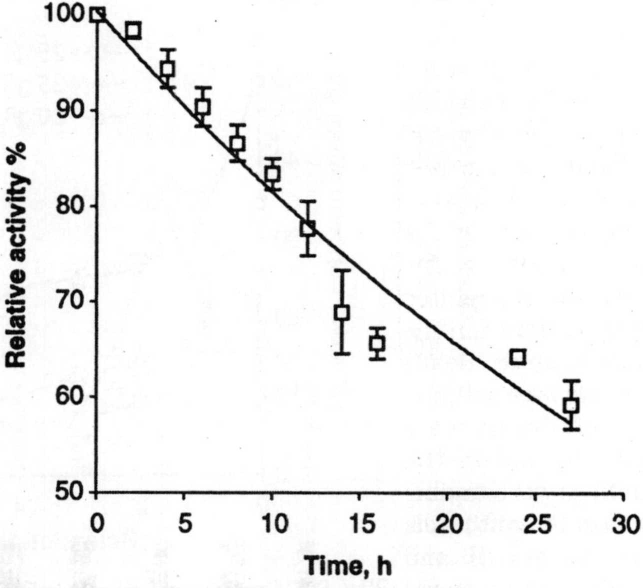

Fig. 6.6. Relative activities of *S. cerevisiae* in batch fermentation.

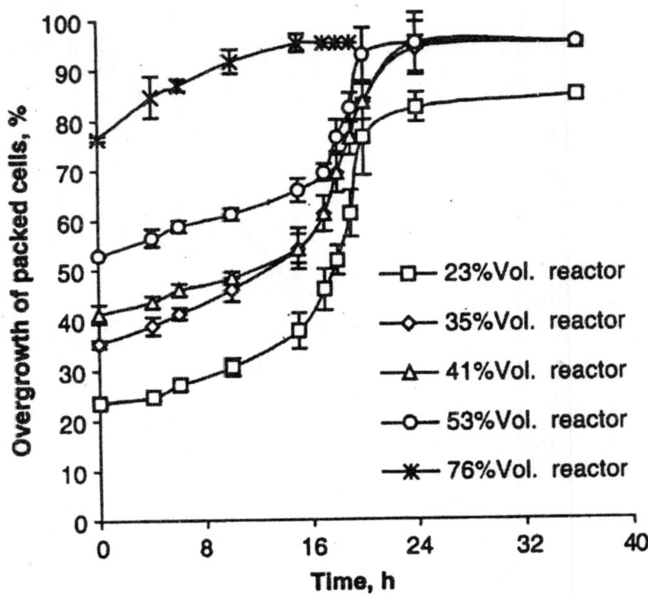

Fig. 6.7. Percentage, growth of immobilized cells with various initial beads loading.

RESULTS AND DISCUSSION

Evaluation of Immobilised Cells

In preparing immobilized cells, 1.5, 2, 3 and 6% alginate was used. By pressing them manually, the hardness and rigidity of the beads were tested. The physical criteria of the prepared beads summarized in Table 6.1. The suitable alginate concentration based on activity of the beads for ethanol production was 2%. The weight percentage of alginate was related to substrate and product penetration the beads and return to bulk of fluid. Beads with low alginate (1.5%) were too soft and easily breakable. The soft beads were pressed once they were loaded in the ICR column. Therefore they were unable to be used successfully; also the soft beads faced problems such as overgrowth and expansion of their diameter when grown in sugar solution. The high alginate

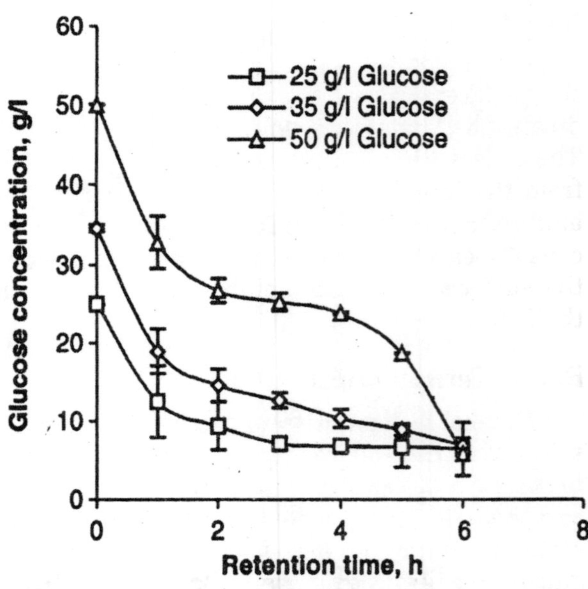

Fig. 6.8. Consumption of glucose in the immoblised cell column.

beads (6%) were very hard and almost unbreakable by pressing manually. They were very rigid, therefore diffusion was the most probable cause since ethanol production declined. The 2% alginate was used because the beads were strong enough to hold the weight of packing in the ICR column. The packed beads were used in the ICR column for 10 days. The prepared beads were refrigerated for 2 weeks and activated in sugar solution.

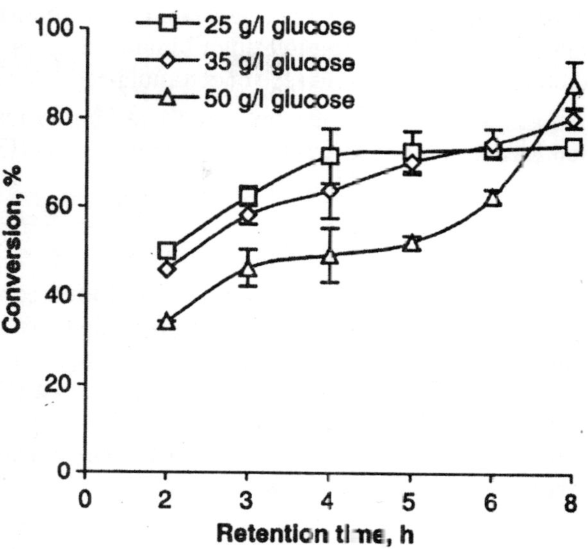

Fig. 6.9. Conversion versus retention time in the immobilized cells column.

The storage condition and ICR column were non-sterile; therefore, there was the possibility of contamination in the packing of the ICR and in the activation process. A series of electron micrographs were taken from the fresh and 72 hour immobilised beads. The outer and inner surfaces of the beads are respectively. These micrographs were used as a comparative indicator for yeast growth on the surface of the solid support (2% calcium alginate). There was no apparent leakage of cells from the beads into the bulk of the fluid. It was also apparent that by 72 hours in the ICR, the yeast was growing on the outer surface. Therefore the active sites were potentially available for ethanol production without diffusion problems. The outer surface of fresh and 72 hour immobilised cell beads at magnifications of 300 and 2000 can easily be seen.

A comparison between fresh and 72 hour beads indicates that the cells were present on the surface, with a few contaminants from a bacillus-type organism, This may have occurred during the transfer stage. The inner surface of the beads before and after use was compared. The cells were initially trapped inside the beads; after 72 hours the cells apparently migrated from the inner side to the surface. The micrographs of the inner sides of the beads before and after use, at magnifications of 300 and 2000 are shown in Fig. 6 5. After 72 hours the cells appeared to have formed new colonies on the surface of the alginate layer. By contrast, the surfaces were completely covered with colonies after 72 hours of ethanol production in the ICR.

Batch Fermentation

Ethanol fermentation in batch experiments was carried out in triplicate with 50 $g{\cdot}l^{-1}$ glucose solution as the sole carbon source for *S. cerevisiae.* The purpose of the batch experiment was to compare the amount of glucose concentration and ethanol production in batch fermentation and the ICR. The concentration of glucose was gradually decreased while the cell density and ethanol production were increased for 27 hours. There was a lag phase of 4 –6 hours; the glucose consumption was low at this stage. Subsequently the concentration profile drastically decreased during batch fermentation after 8 – 16 hours. Since the cell density was initially low, sugar consumption was also low.

The resulting cell growth curve from the batch experiment was a typical sigmoidal (S-)shape. The maximum cell density of S. *cerevisiae* in batch fermentation was 13.7 $g{\cdot}l^{-1}$ with 50 $g{\cdot}l^{-1}$ glucose concentration. The average yield of biomass and product on substrates ($Y_{X/S}$ and $Y_{P/S}$) were 33 and 32%, respectively. The rate of ethanol productivity for 24 hours was 1.4 $g{\cdot}l^{-1}\cdot h^{-1}$.

A Langmuir-Hanes plot based on the Monod rate equation is presented in figure 6.5. The Monod kinetic model can be used for microbial cell biocatalyst and is described as follows :

$$\frac{S}{\mu} = \frac{K_s}{\mu_m} + \frac{S}{\mu_m} \quad ...(6.5)$$

The terms K_s and μ_m are defined as the Monod constant and maximum specific growth rate, respectively. The data generated in this study were linearly fitted with the model, as a function of concentration produced during the exponential phase versus time. From the plot, the maximum specific growth rate and Monod constant were determined to be 0.35 h^{-1} and 2.23 $g{\cdot}l^{-1}$, respectively. The large value of the Monod constant may suggest that at high concentrations of substrate, more influence on the cell-substrate complex [C.S] dissociation occurred than [C.S] formation.

Relative Activity

The economics of an immobilised cell process depend on the lifetime of the microorganism and a continued level of clean product delivered by the fixed cells. It is important to eliminate the free cells from the downstream product without the use of any units such as centrifuge or filtration processes. Since the cells are retained in the ICR, the activity of intracellular enzymes may play a major role. It is assumed that the deactivation of the enzyme at constant temperature follows a first-order equation as shown below.

$$A_t = A_0 \exp(-k_d t) \quad ...(6.6)$$

where A_t and A_0 are activities of enzymes at time, t and initial time zero, respectively. Also, k_d is the dissociation constant. The plot of relative activity versus time in batch fermentation with free cells is shown in figure 6.6. The value of k_d was 0.36 h^{-1}. According to the first-order dissociation rate constant, the half-life of the *S. cerevisiae,* (τ_d = ln $2/k_d$), in batch fermentation suspended cells with 50 $g{\cdot}l^{-1}$ of glucose was 1.95 h. The free cells were apparently completely deactivated after 60 hours in batch operation. The data indicate that the free cells were deactivated rapidly compared with immobilised cells when used in a continuous mode for more than 7 days.

Reactor Set-up

The volume of reactor without beads was 1.41. The column was loaded with the solidified uniform beads of *S. cerevisiae*. The void volume of the reactor was 660ml when it was packed with immobilised beads. The growth of beads with different proportions of column packing is shown in figure 6.7. A fresh feed of 10 $g{\cdot}l^{-1}$ glucose solution was pumped from the bottom of the reactor. The optimum amount of packing obtained was 65-70% of the reactor volume. The trend of the collected data resembles the growth curve of yeast in suspended cell culture. The diameters of the beads increased with the increase in *S. cerevisiae* cell density, which indicates that S. *cerevisiae* was growing within the solid support. Under these circumstances substrate would be easily consumed at the solid surface coated with immobilised yeast cells.

Thus, it was expected that substrate concentration at the surface would be less than the concentration of substrate in the bulk fluid. The main objective was to determine whether substrate penetrated into the beads.

Since the matrix of beads was quite porous, it was assumed that the concentration gradient was the major force that influenced the mass transfer process in immobilised cells of *S. cerevisiae*. Therefore, the immobilised yeast system was preferred compared with free cells in the solution. Moreover, economic aspects of immobilised yeast must be considered as it eliminates the need for the extra unit for free-cell removal from the product stream. The other advantages of the immobilisation system are that the substrate may not accumulate on the surface of the beads and that there was no evidence of cell leakage from the beads.

Effect of High Concentration of Substrate on Immobilised Cells

The fermentation was performed with various sugar concentrations to increase product concentrations. The initial sugar concentrations were 25, 35 and 50 g·l^{-1}. The sugar consumption profile in the ICR is presented. The sugar consumption trends of various glucose concentrations were similar, with a sharp reduction of substrate occurring within the first 3 hours. A range of 55 – 75% of the sugars was reduced within a 3 hour retention time. A 6 hour retention time indicated that this was the most suitable time to achieve high sugar consumption. A longer retention time was required for higher sugar concentrations of up to 150 g·l^{-1} in the ICR column (7 hours).

The amount of cell immobilized with calcium alginate was determined by the cell dry weight of immobilised cells and the assumption that 98% of the beads were considered to be loaded with active cells. Beads were measured before loading in the ICR column. Conversion of glucose versus dilution rate, used in the continuous fermentation process with immobilised *S. cerevisiae* is shown in figure 6.9.

As the sugar concentration increased the conversion may have decreased. At very high sugar concentrations, the conversion of sugar decreased. The maximum sugar conversion at 6 hours' retention time was obtained from 74.3, 80.5 and 88.2% for 25, 35 and 50 g·l^{-1} glucose concentrations, respectively. At high concentrations of sugar, conversion was highly influenced by dilution rate. The conversion decreased to less than 10% once the dilution rate approached wash out. The sugar conversion at low sugar concentrations appeared to be much

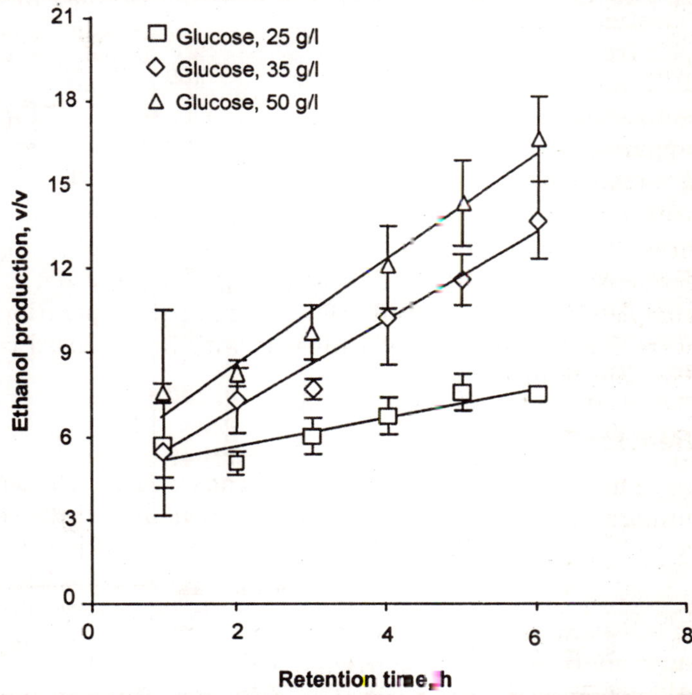

Fig. 6.10. Ethanol production versus retention time in the immobilised cell column.

less sensitive to feed dilution rate. The conversion for 25 g·l^{-1} glucose was constant at 74% for a retention time of greater than 3 hours. The trend of the data with 50 g·l^{-1} sugar showed that conversion increased with retention time.

At a higher sugar concentration, it required a longer retention time to achieve higher conversion. Reactor productivity was obtained by dividing final ethanol concentration with respect to sugar concentration at a fixed retention time. It was found that the rates of 1.3, 2.3 and 2.8 g·l^{-1} · h^{-1} for 25, 35 and 50 g·l^{-1} glucose concentrations were optimal. Ethanol productivities with various substrate concentrations were linearly dependent on retention time. The proportionality factor may have increased while the substrate concentration increased. As the sugar concentration doubled, the slope of the line for ethanol productivity with 50 g·l^{-1} sugar increased fivefold. These results indicate that the ICR column has a high capacity to produce very high concentrations of ethanol.

The final ethanol concentrations with 25 and 50 g·l^{-1} of glucose were 7.6 and 16.73 v/v, respectively. A high glucose concentration of 150 g·l^{-1} was used in continuous fermentation with immobilised *S. cerevisiae;* the obtained data for sugar consumption and ethanol production with retention time are shown in figure 6.11. As the retention time gradually increased the glucose concentration dropped, while the ethanol concentration profile showed an increase. The maximum ethanol concentration of 47 g·l^{-1} was obtained with a retention time of 7 hours. The yield of ethanol production was approximately 38% compared with batch data, where only an 8% improvement was achieved.

Continuous ethanol production in an ICR was successfully done with high sugar concentrations. In batch fermentation, when the concentration of glucose was 50 g·l^{-1} substantial substrate inhibition occurred. The advantage of the ICR was that the inhibition of substrate and product were not apparent even with a 150 g·l^{-1} glucose solution in the fresh feed (data not shown). The ICR system exhibited a higher yield of ethanol production (38%) than the batch system. The ICR column gave a high performance to processing feed with concentrated sugar. Most of ICR experimental runs resulted in a glucose consumption of 82-85%.

The results indicate that the immobilisation of *S. cerevisiae* possesses the capacity not only to utilise high concentrations of sugar but also to yield higher ethanol productivities during the course of continuous fermentation.

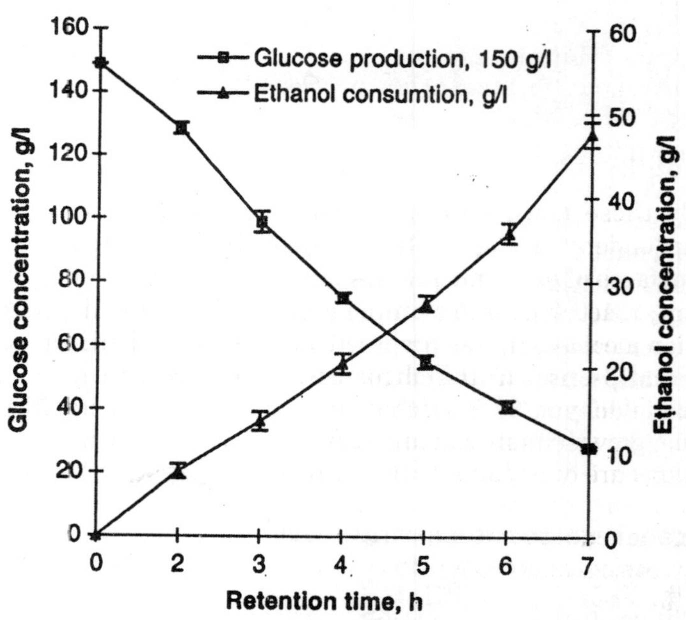

Fig. 6.11. Glucose concentration and ethanol production versus retention time in immobilised cell reactor with initial substrate concentration of 150 g·l^{-1} glucose.

Ethanol production in the ICR column increased by fivefold, as the glucose concentration was doubled from 25 to 50 $g{\cdot}l^{-1}$. It is clear: the new findings in the present investigation would be the application of concentrated feed with a higher rate of ethanol production, as the cell loaded into the gel matrices of sodium alginate shown in the SEM micrographs had eliminated the free cells from the ethanol product stream.

Nomenclature

S Substrate concentration, $g{\cdot}l^{-1}$
C Microorganisms cell concentration, $g{\cdot}l^{-1}$
K_S Monod constant, $g{\cdot}l^{-1}$
μ Specific growth rate, $g{\cdot}l^{-1} \cdot h^{-1}$
μ_m Maximum specific growth rate, h^{-1}
T Time, h
A_t Activity of bacterium at time *t*, U/g cell
A_o Initial activity of bacterium, U/g cell
k_d Dissociation constant, h^{-1}.

FUNDAMENTALS OF IMMOBILISATION TECHNOLOGY

It is well known that pure enzymes change their behaviour and stability when they are immobilised. In the past two decades the immobilisation of microorganisms, cells and parts of cells has gradually been introduced into microbiology and biotechnology. The cell immobilisation techniques are modifications of the techniques developed for enzymes. However, the larger size of microbes has influenced the techniques. As for immobilised enzymes, two broad types of method have been used to immobilise microorganisms: attachment to a support and entrapment.

Immobilisation of Microorganisms by Covalent Bonds

By these methods microorganisms are cross linked by chemical substances, *e.g.* by glutardialdehyde. The surfaces (especially the proteins) of microorganisms are linked with the surfaces of other microorganisms by aldehyde groups of glutardialdehyde. Yeast cells, for instant, react with free ε-amino group or N-terminal amino groups to form imines. Another reaction mechanism was proposed for a conjugated addition of amino groups to double linkages of α- and β-unsaturated oligomers, which are present in commercial aqueous solutions of glutardialdehyde. This mechanism may explain the stability of the linkages. By this chemical linking, growth inhibition and toxic influences on the microorganisms are very intensive. These reactions are only partly understood and can lead to decay or death of the microorganisms.

Oxygen Transfer to Immobilised Microorganisms

Availability of oxygen is one of the most important parameters that are different for immobilised and free microorganisms. Free organisms can get oxygen directly from the surrounding air or, in most technical processes, from the liquid, especially water, which contains dissolved oxygen. The molar transfer of oxygen with respect to time, dO_2/dt, can be described by the following equation:

$$\frac{dO_2}{dt} = K_L a(C_g - C_L) \quad ...(6.7)$$

$K_L a$ is the volumetric oxygen mass transfer coefficient, owing to the oxygen transfer from the gas phase or air, c_g, the surface of the cells, c_x or to the transfer of oxygen dissolved in water to the surface of the cells. In principle, the same formula can be applied to immobilised microoraganisms, but here the conditions are quite different.

The adsorbed cells form microbial films or varying thickness during a short incubation time. Oxygen has to be transferred into these microfilms, and because of this, zones of different oxygen concentration in the films exist, in which the growth of the microorganisms varies in direct relation to the oxygen concentration. By using the unsteady-state model, the maximum oxygen penetration depth for highly packed immobilised cells has been reported to be in the range of 50-2001 μm.

Substrate Transfer to Immobilised Microorganisms

In an immobilised cell bioreactor, significant substrate concentration gradients may exist. Cells are located close to the nutrient supply, which are likely to maintain higher quality and activity than cells located relatively further away, leading to differentiation in the quality or activity of the immobilised cell population. This differentiation is more pronounced if there are starvation regions (practically zero substrate concentration) inside the reactor. Typical approaches for measuring diffusivities in immobilized cell systems include bead methods, diffusion chambers and holographic laser interferometry. These methods can be applied to various support materials, but they are time consuming, making it onerous to measure effective diffusivity (D_{eff}) over a wide range of cell fractions.

Owing to the mathematical models involved, the deconvolution of diffusivities can be very sensitive to errors in concentration measurements. There are mathematical correlations developed to predict D_{eff} as a function of diffusivities in broth and inside the cells (D_0, D_c). The relation can also be used to extrapolate D_{eff} measurements from one cell fraction to any other cell fraction. Various values of D_0/D_c have been compiled in Table 6.4 for a variety of immobilised cell systems. These diffusivity ratios for specific systems (*i.e.* diffusing species, cell type and gel material) facilitate the prediction of D_{eff} values for a wide range of operating conditions.

Table 6.4. Methods of immobilisation

Attachment	
Without support	Aggregation of floc for formation of cross-linking
With support	Covalent binding
	Adsorption to ion-exchangers or inorganic
	Biofilm formation
Entrapment	Organic polymer
	Inorganic polymer
	Semi-permeable membrane

For a few systems, the diffusivity ratios are very large (∞). The infinite diffusivity ratio corresponds to a diffusing solute that does not enter the cells (*i.e.* $D_c = 0$); such cases were observed with large molecules of disaccharide (galactose), polysaccharides and cells of Z. *mobilis*.

Growth and Colony Formation of Immobilised Microorganisms

It was pointed out by several scientists that immobilised microorganisms differ in their growth rates and show altered morphological forms of colonies. Adsorbed cells form micro- and macro-films in which the microorganisms in the outer region have a different morphology than in the inner region. These microfilms often show different colony forms in relation to their density. Thick films have a slimy character, thin film often show the presence of individual microorganisms. These characteristics can be observed with bacteria, yeast and with moulds. They are caused by differing oxygen concentrations and limited concentrations of nutrients.

By referring to our previous work on ICR, a mathematical model for ICR performance may be obtained by applying a mass balance over a differential of the column:

$$\varepsilon\frac{\partial C_A}{\partial t} + u\frac{\partial C_A}{\partial z} = r_A \qquad \text{...(6.8)}$$

where ε is the void volume of the packed column (ml), C_A the substrate concentration ($g{\cdot}l^{-1}$), u the bulk fluid velocity ($cm{\cdot}hr^{-1}$), z the axial reactor length (cm), and r_A the rate of substrate transfer to microbial film ($g{\cdot}l^{-1} \cdot h^{-1}$).

Assuming plug-flow and steady-state behaviour (6.8), reduces to $u\frac{\partial C_A}{\partial z} = r_A$ Plug-flow behaviour has been shown in this type of reactor by the use of tracer studies.

Table 6.5. Microbial Cells Covalently Linked to Various Supports.

Species	*Support*	*Product*
Actobacter	Metal hydroxides	Acetic acid
Aspergillus niger	Glycidyl methacrylate	Gluconic acid
Micrococcus luteus	CM-cellulose	Urocanic acid
Saccharomyces cerevisiae	Aminopropyl silica	Ethanol
Saccharomyces cerevisiae	Hydrocaylkyl methacrylate	Killer toxin
Saccharomyces cerevisiae	Cellulose	Ethanol
Zygosacharomyces lactis	Hydroxyalkyl methacrylate	β-galactosidase

The reaction rate for simple fermentation systems is normally given by the Monod equation. This model indicates that the specific conversion rate is constant when applied to an immobilised cell system. If a first-order rate equation for sugar consumption is used, yields:

$$u\frac{\partial C_A}{\partial z} = kC_A \qquad \text{...(6.9)}$$

Table 6.6. Experimental Studies of Diffusion in Immobilised Cell Systems and their Associated Values of D_0/D_c.

Cell type	*Immobilisation*	*Solute*	D_0/D_c
Saccharomyces cerevisiae	Ca-alginate	glucose	0.1
Baker's yeast	Ca-alginate	glucose	10.6
Ehrlich ascites tumor	agar, collagen	glucose	2.4
Zymomonas mobilis	K-carrageenan, Ca-alginate	glucose	∞
Pseudomonas aeruginosa	Ca-alginate	glucose	2.3
Saccharomycess cerevisiae	Ca-alginate	glucose	2.8
Plant	Ca-alginate	sucrose	∞
Baker's yeast	Ca-alginate	galactose	15.8
Zymomonas mobilis	Ca-alginate	galactose	∞
Baker's yeast	Ca-alginate	lactose	∞
Ehrlich ascites tumor	agar, collagen	lactic acid	6.7
Clostridium butyricum	polyarylamide, agar collagen	hydrogen	3.2
Escherichia coli	natural aggregates	nitrous oxide	3.9
Saccharomyces cerevisiae	fermentation media	oxygen	2.3
Saccharomyces cerevisiae	Ca-alginate, Ba-alginate	oxygen	0.1
Escherichia coli	fermentation media	oxygen	2.2
Penicillium chrysogenum	fermentation media	oxygen	5.6
Bacillus amilaliquefaciens	Ca-alginate, PVA-SbQ gel	oxygen	1.8
Saccharomyces cerevisiae	Ca-alginate	ethanol	0.1
Baker's yeast	Ca-alginate	ethanol	∞

Equation (6.9) is linear first-order differential equation for concentration and reactor length. After separation of variables, the equation can be integrated as

$$\int_{C_{A0}}^{C_A} \frac{dC_A}{C_A} = \int_0^z \frac{kdz}{u} \quad ...(6.10)$$

Integration of the above differential equation yields:

$$\ln(C_A/C_{A_0}) = \frac{kz}{u} \quad ...(6.11)$$

Thus a linear relation between ln (C_A/C_{A_0}) and the reactor length should exist if the model accurately describes the immobilised cell reactor. The experimental data fitting the model was discussed earlier.

Immobilised Systems for Ethanol Production

The most significant advantage of immobilised cell systems is the ability to operate with high productivity at dilution rates exceeding the maximum specific growth rate (μ_{max}) of the microbe. Several theories have been proposed to explain the enhanced fermentation capacity of microorganisms as a result of immobilisation. A reduction in the ethanol concentration in the immediate microenvironment of the organism owing to the formation of a protective layer or specific adsorption of ethanol by the support may act to minimise end product inhibition.

Table 6.7. Ethanol Productivity from Immobilised Systems.

System	*Feed sugar conc* *($g l^{-1}$)*	*% of Feed sugar utilization*	*Dilution rate* *(h^{-1})*	*Max. ethanol productivity* *($g l^{-1} h^{-1}$)*
S. Cerevisiae Carrageenan	Glucose 100	86	1.0	43
S. Cerevisiae Ca-alginate	Glucose 127	63	4.6	53.8
S. Cerevisiae Ca-alginate	Molasses 175	83	0.3	21.3
S. Cerevisiae Carrier A	Molasses 197	74	0.35	25
Z. mobilis Ca-alginate	Glucose 150	75	0.85	44
Z. mobilis Ca-alginate	Glucose 100	87	2.4	102
Z. mobilis Carrageenan	Glucose 150	85	0.8	53
Z. mobilis Flocculation	Glucose 100			120
Z. mobilis Borosilicate	Glucose 50			85
Z. mobilis Carrageenan-locust bean gum	Whey-lactose 50	98		178

Alternatively, substrate inhibition may be diminished in the case of a gel matrix, if the rate of fermentation meets or exceeds the rate of glucose diffusion to the cell. A third possibility is that alteration of the cell membrane during immobilisation provides improved transfer of substrate into and product out of the microbe. The effect of temperature on the rate of ethanol production is markedly different for free and immobilised systems. Thus while a constant increase in rate is observed with free *S. cerevisiae* as temperature is increased from 25 to 42 °C, a maximum occurs at 30 °C with cells immobilised in sodium alginate. The lower temperature optimum for immobilised systems may result from diffusional limitations of ethanol within the support matrix. At higher temperatures, ethanol production exceeds its rate of diffusion so that accumulation occurs within the beads).

The achievement of inhibitory levels then causes the declines observed in the ethanol production rate. Significant differences are also apparent for the effect of pH on the fermentation rate. The narrow pH optimum characteristic of a free cell system is replaced by an extremely broad range upon immobilisation. This effect stems from the gradient pH that exists within the bead.

CHAPTER

7

Bioenergetics and Biosynthesis

Metabolism involves a bewildering array of chemical reactions, many of them organized as complex cycles which may appear difficult to understand. Yet, there is logic and orderliness. With few exceptions, metabolic pathways can be regarded as sequences of the reactions which are organized to accomplish specific chemical goals. In this chapter we will examine the chemical logic of the major pathways of catabolism of foods and of cell constituents as well as some reactions of biosynthesis (anabolism).

THE OXIDATION OF FATTY ACIDS

Hydrocarbons yield more energy upon combustion than do most other organic compounds, and it is, therefore, not surprising that one important type of food reserve, the fats, is essentially hydrocarbon in nature. In terms of energy content the component fatty acids are the most important. Most aerobic cells can oxidize fatty acids completely to CO_2 and water, a process that takes place within many bacteria, in the matrix space of animal mitochondria, in the peroxisomes of most eukaryotic cells, and to a lesser extent in the endoplasmic reticulum. The carboxyl group of a fatty acid provides a point for chemical attack. The first step is a priming reaction in which the fatty acid is converted to a water-soluble acyl-CoA derivative in which the a hydrogens of the fatty acyl radicals are "activated" (step *a,* Fig. 7.1.). This synthetic reaction is catalyzed by *acyl-CoA synthetases* (fatty acid:CoA ligases). It is driven by the hydrolysis of ATP to AMP and two inorganic phosphate ions using the sequence.

There are isoenzymes that act on short-, medium-, and long-chain fatty acids. Yeast contains at least five of these. In every case the acyl group is activated through formation of an intermediate acyl adenylate; hydrolysis of the released pyrophosphate helps to carry the reaction to completion.

Beta Oxidation

The reaction steps in the oxidation of long-chain acyl-CoA molecules to acetyl-CoA were outlined in Fig. 7.4. Because of the great importance of this P oxidation sequence in metabolism the steps are shown again in Fig. 7.1 (steps *b-e*). The chemical logic becomes clear if we examine the structure of the acyl-CoA molecule and consider the types of biochemical reactions available. If the direct use of O_2 is to be avoided, the only reasonable mode of attack on an acyl-CoA molecule is dehydrogenation. Removal of the a hydrogen as a proton is made possible by the activating effect of the carbonyl group of the thioester. The P hydrogen can be transferred from the intermediate enolate, as a hydride ion, to the bound FAD present in the *acyl-CoA dehydrogenases* that catalyze this reaction (step *b*, Fig. 7.1). These enzymes contain FAD, and the reduced coenzyme $FADH_2$ that is formed is reoxidized by an *electron transferring flavoprotein*, which also contains FAD. This protein carries the electrons abstracted in the oxidation process to the inner membrane of the mitochondrion where they enter the mitochondrial electron transport system. The product of step *b is* always a *trans-Δ^2-enoyl-CoA.*

One of the few possible reactions of this unsaturated compound is nucleophitic additional the β position. The reacting nucleophile is an HO^- ion from water. This reaction step (step *c*, Fig. 7.1) is completed by addition of H^+at the α position. The resulting *β-hydroxyacyl-CoA* (3-hydroxyacyl-CoA) is dehydrogenated to a ketone by NAD^+ (step *d*). This series of three reactions is the β oxidation sequence. At the end of this sequence, the β-oxoacyl-CoA derivative is cleaved (Fig. 7.1, step *e*) by a *thiolase*. One of the products is acetyl-CoA, which can be catabolized to CO_2 through the citric acid cycle. The other product of the thiolytic cleavage is an acyl-CoA derivative that is *two carbon atoms shorter than the original acyl-CoA*. This molecule is recycled through the P oxidation process, a two-carbon acetyl unit being removed as acetyl-CoA during each turn of the cycle (Fig. 7.1).

The process continues until the fatty acid chain is completely degraded. If the original fatty acid contained an *even* number of carbon atoms in a straight chain, acetyl-CoA is the only product. However, if the original fatty acid contained an *even* number of carbon atoms, *propionyl-CoA* is formed at the end. For every step of the β oxidation sequence there is a small family of enzymes with differing chain length preferences. For example, in liver mitochondria one acyl-CoA dehydrogenase acts most rapidly on H-butyryl and other short-chain acyl-CoA; a second prefers a substrate of medium chain length such as n-octanoyl-CoA; a third prefers long-chain substrates such as *n*-mitoyl-CoA; and a fourth, substrates with 2-methyl branches.

A fifth enzyme acts specifically on isovaloryl-CoA. Similar preferences exist for the other enzymes of the β oxidation pathway. In *Esclierichia coll* most of these enzymes are present as a complex of multifunctional proteins while the mitochondrial enzymes may be organized as a multiprotein complex.

Peroxisomal Beta Oxidation : In animal cells p oxidation is primarily a mitochondrial process, but it also takes place to a limited extent within peroxisomes and within the endoplasmic reticulum. This "division of labour" is still not understood well. Straight-chain fatty acids up to 18 carbons in length appear to be metabolized primarily in mitochondria, but in the liver fatty acids with very long chains are processed largely in peroxisomes. There, a very long-chain acyl-CpA synthetase acts on fatty acids that contain 22 or more carbon atoms. In yeast all (3 oxidation takes place in peroxisomes, and in most organisms, including green plants, the peroxisomes are the most active sites of fatty acid oxidation.

However, animal peroxisomes cannot oxidize short-chain acyl-CoA molecules; they must be returned to the mitochondria. The activity of peroxisomes in P oxidation is greatly increased by the presence of a variety of compounds known as *peroxisome proliferators*. Among them are chugs such as aspirin and clofibrate and environmental xenobiotics such as the plasticizer bis-(2-ethyl-hexyl)-phthalate. They may induce as much as a tenfold increase in peroxisomal P oxidation.

Several other features also distinguish β oxidation in peroxisomes. The peroxisomal flavoproteins that catalyze the dehydrogenation of acyl-CoA molecules to unsaturated enoyl-CoAs (step *b* of Fig. 7.1) are *oxidases* in which the FADH-, that is formed is reoxidized by O_2 to form H_2O_2. In peroxisomes the enoylhydratase and the NAD^+-dependent dehydrogenase catalyzing steps *c* and *d* of Fig. 7.1 are present together with an enoyl-CoA iuomerase (next section) as a trifunctional enzyme consisting of a single polypeptide chain.

As in mitochondrial p oxidation the 3-hydroxyacyl-CoA intermediates formed in both animal peroxisomes and plant peroxisomes (glyoxysomes) have the L configuration. However, in fungal peroxisomes as well as in *E. coli* they have the D configuration. Further metabolism in these organisms requires an epimerase that converts the D-hydroxyacyl-CoA molecules to. In the past it has often been assumed that percxisomnl membranes are freely permeable to NAD^+, NADH, and acyl-CoA molecules. However, genetic experiments with yeast and other recent evidence indicate that they are impermeable *in vivo* and that carrier and shuttle mechanisms similar to those in mitochondria may be required.

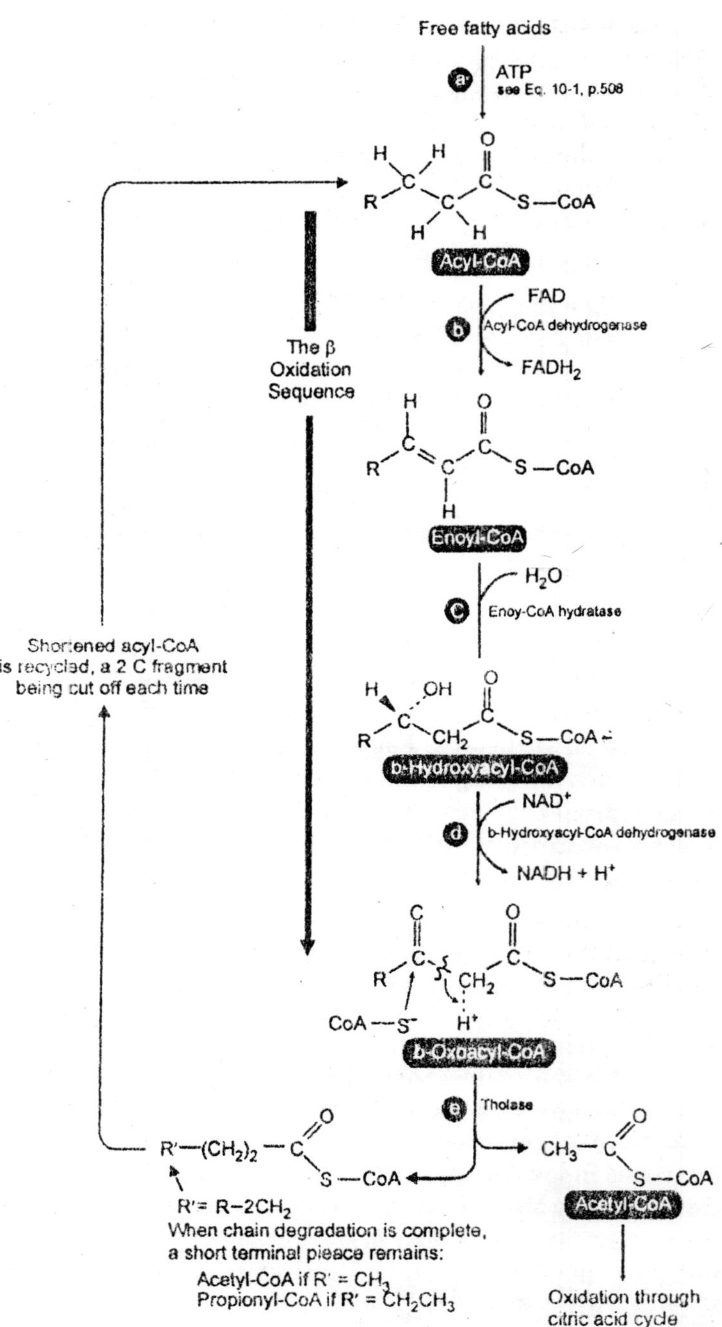

Fig. 7.1. The β oxidation cycle for fatty acids. Fatty acids are converted to acyl-CoA derivatives from which 2-carbon atoms are cut off as acetyl-CoA to give a shortened chain which is repeatedly sent back through the cycle until only a 2- or 3-carbon acyl-CoA remains. The sequence of steps *b*, *c*, and *d* also occurs in many other places in metabolism.

Unsaturated Fatty Acids : Mitochondrial P oxidation of such unsaturated acids as the Δ^9-oleic acid begins with removal of two molecules of acetyl-CoA to form a Δ^5-acyl-CoA. However, further metabolism is slow. Two pathways have been identified (Eq.7.1). The first step for both is a normal dehydrogenation to a 2-*trans*-5-*cis*-dienoyl-CoA. In pathway I this intermediate reacts slowly by the normal P oxidation sequence to form a 3-*cis*-enoyl-CoA intermediate which must then be acted upon by an auxiliary enzyme, a *cis*-Δ^3-trans-Δ^2-enoyl-CoA isomerase (Eq. 7.1, step c), before β oxidation can continue. The alternative reductase pathway (II in Eq. 7.1) is often faster.

It makes use of an additional isomerase which converts 3-*trans,* 5-*cis*-dienoyl-CoA into the 2-*trnns,* 4-*trnns* isomer in which the double bonds are conjugated with the carbonyl group. This permits removal of one double bond by reduction with NADPH as shown (Eq. 7.1, step *f*). The peroxisomal pathway is similar. However, the intermediate formed in step *e* of Eq. 7.1 may sometimes have the 2-*trans,* 4-*cis* configuration. The NADH for the reductive step *f* may be supplied by an NADP-dependent isocitrate dehydrogenase. Repetition of steps *a, d, e,* and *f* of Eq. 7.1 will lead to β oxidation of the entire chain of polyunsaturated fatty acids such as linoleoyl-CoA or arachidonoyl-CoA. Important additional metabolic routes for polyunsaturated fatty acid derivatives are described.

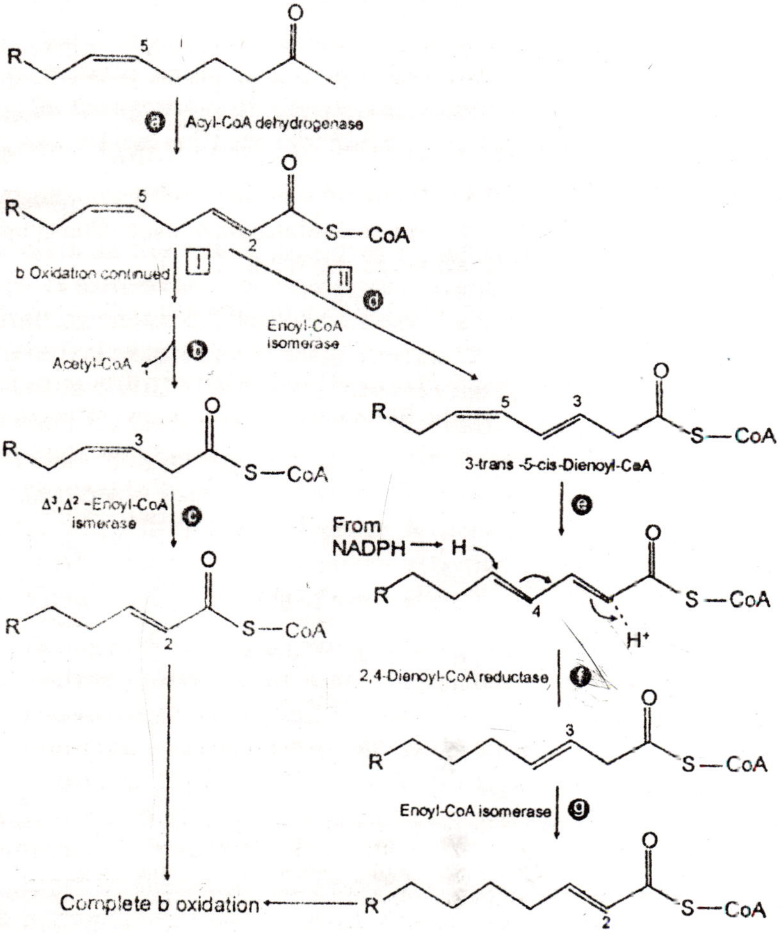

a Acyl-CoA dehydrogenase
b Oxidation continued I
II d Enoyl-CoA isomerase
Acetyl-CoA b
Δ^3,Δ^2 -Enoyl-CoA isomerase c
3-trans -5-cis-Dienoyl-CoA
e
From NADPH → H
2,4-Dienoyl-CoA reductase f
Enoyl-CoA isomerase g
Complete b oxidation

...(7.1)

Branched-chain Fatty Acids : Most of the fatty acids in animal and plant fats have straight unbranched chains. However, branches, usually consisting of methyl groups, are present in lipids of some microorganisms, in waxes of plant surfaces, and also in polyprenyl chains. As long as there are not too many branches and if they occur only in the even-numbered positions (*i.e.*, on carbons 2,4, etc.) β oxidation proceeds normally. Propionyl-CoA is formed in addition to acetyl-CoA as a product of the chain degradation. On the other hand, if methyl groups occur in positions 3, 5, etc., β oxidation is blocked at step *d* of Fig. 7.1.

A striking example of the effect of such blockage was provided by the synthetic detergents in common use until about 1966. These detergents contained a hydrocarbon chain with methyl

A CH_3 branch here blocks β oxidation (↓ at C-3)

$$\overset{4}{C}H_3 - \overset{3}{C}H_2 - \overset{2}{C}H_2 - \overset{1}{C}(=O) - S - CoA$$

A CH_3 branch here leads to formation of propionyl-CoA (↑ at C-2)

groups distributed more or less randomly along the chain. Beta oxidation was blocked at many points and the result was a foamy pollution crisis in sewage plants in the United States and in some other countries.

Since 1966, only biodegradable detergents having straight hydrocarbon chains have been sold. In fact, cells *are* able to deal with small amounts of these hard-to-oxidize substrates. The O_2-dependent reactions called a oxidation and 10 oxidation are used. These are related also to the oxidation of hydrocarbons which we will consider next.

Oxidation of Saturated Hydrocarbons : Although the initial oxidation step is chemically difficult, the tissues of our bodies are able to metabolize saturated hydrocarbons such as *n*-heptane slowly, and some microorganisms oxidize straight-chain hydrocarbons rapidly. Strains of *Pseudomonas* and of the yeast *Candida* have been used to convert petroleum into edible proteins. The first step in oxidation of alkanes is usually an O_2-requiring *hydroxylation* to a primary alcohol. Further oxidation of the alcohol to an acyl-CoA derivative, presumably via the aldehyde (Eq. 7.2), is a frequently encountered biochemical oxidation sequence.

$$\text{n-Octane} \xrightarrow[\text{Hydroxylation}]{O_2} \underset{\text{Octanol}}{C_7H_{15}-CH_2-OH} \rightarrow C_7H_{15}-CHO \rightarrow \underset{\text{Acyl-CoA}}{C_7H_{15}-C(=O)-S-CoA} \quad ...(7.2)$$

Alpha Oxidation and Omega Oxidation : Animal tissues degrade such straight-chain fatty acids as palmitic acid, stearic acid, and oleic acid almost entirely by β oxidation, but plant cells often oxidize fatty acids one carbon at a time. The initial attack may involve hydroxylation on the α-carbon atom (Eq. 7.3) to form either the D or the L-2-hydroxy acid. The L-hydroxy acids are oxidized rapidly, perhaps by dehydrogenation to the oxo acids (Eq. 7.3, step *b*) and oxidative decarboxylation, possibly utilizing H_2O_2. The D-hydroxy acids tend to accumulate and are normally present in green leaves.

However, they too are oxidized further, with retention of the a hydrogen as indicated by the shaded squares in Eq. 7.3, step *e*. This suggests a new type of dehydrogenation with concurrent decarboxylation. Alpha oxidation also occurs to some extent in animal tissues. For example, when β oxidation is blocked by the presence of a methyl side chain, the body may use a oxidation to get past the block. As in plants, this occurs principally in the peroxisomes and is important for degradation not only of polyprenyl chains but also bile acids. In the brain some of the fatty acyl groups of sphingolipids are hydroxylated to α-hydroxyacyl groups. Alpha oxidation in animal cells occurs after conversion of free fatty acids to their acyl-CoA derivatives (Eq. 7.3, step *a*). This is followed by a 2-oxoglutarate-dependent hydroxylation to form the 2-hydroxyacyl-CoA, which is cleaved in a standard thiamin diphosphate-requiring a cleavage (step *c*). The products are formyl-CoA, which is hydrolyzed and oxidized to CO_2, and a fatty aldehyde which is metabolized further by β oxidation. In plants α-dioxygenases convert free fatty acids into 2(*R*)-hydroperoxy

...(7.3)

derivatives (Eq. 7.3). These may be decarboxylated to fatty aldehydes but may also give rise to a variety of other products.

Compounds arising from linoleic and linolenic acids are numerous and include epoxides, epoxy alcohols, dihydroxy acids, short-chain aldehydes, divinyl ethers, and jasmomc acid. On other occasions, *omega (ω) oxidation* occurs at the opposite end of the chain to yield a dicarboxylic acid. Within the human body 3,6-dimethyloctanoic acid and other branched-chain acids are degraded largely via ω oxidation. The initial oxidative attack is by a hydroxylase of the cytochrome P450 group. These enzymes act not only on fatty acids but also on prostaglandins, sterols, and many other lipids. In the animal body fatty acids are sometimes hydroxylated both at the terminal (ω) position and at the next (ω -2 or ω 2) carbon. In plants hydroxylation may occur at the ω2, ω3, and ω4 positions as well.

Dicarboxylates resulting from co oxidation of straight-chain fatty acids can undergo p oxidation from both ends. The resulting short-chain dicarboxylates, which appear to be formed primarily in the peroxisomes, may be converted by further β oxidation into succinyl-CoA and free succinate. Incomplete β oxidation in mitochondria (Fig. 7.1) releases small amounts of 3(β)-hydroxy fatty acids, which also undergo ω oxidation and give rise to free 3-hydroxydicarboxylic acids which may be excreted in the urine.

Carnitine and Mitochondrial Permeability

A major factor controlling the oxidation of fatty acids is the rate of entry into the mitochondria. While some long-chain fatty acids (perhaps 30% of the total) enter mitochondria as such and are converted to CoA derivatives in the matrix, the majority are "activated" to acyl-CoA derivatives on the inner surface of the outer membranes of the mitochondria. Penetration of these acyl-CoA derivatives through the mitochondrial inner membrane is facilitated by *L-carnitine.*

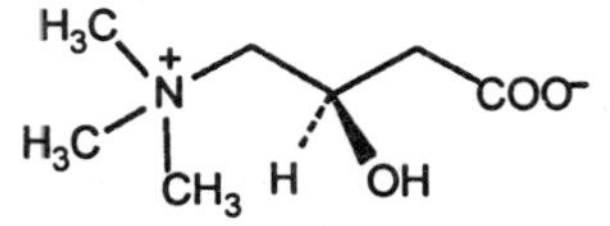

H_3C, H_3C, $\overset{+}{N}$, CH_3, H, OH, COO^-

L-Carnitine

Carnitine is present in nearly all organisms and in all animal tissues. The highest concentration is found in muscle where it accounts for almost 0.1% of the dry matter. Carnitine was first isolated from meat extracts in 1905 but the first clue to its biological action was obtained in 1948 when Fraenkel and associates described a new dietary factor required by the mealworm, *Tenebrio molitor.* At first designated *vitamin* B_t, it was identified in 1952 as carnitine.

Most organisms synthesize their own carnitine from lysine side chains. The inner membrane of mitochondria contains a long-chain acyltransferase (carnitine palmitoyltransferase I) that catalyzes transfer of the fatty acyl group from CoA to the hydroxyl group of carnitine (Eq. 7.4). Perhaps acyl carnitine derivatives pass through the membrane more easily than do acyl-CoA derivatives because the positive and negative charges can swing together and neutralize each other as shown in Eq. 7.4. Inside the mitochondrion the acyl group is transferred back from carnitine onto CoA (Eq. 7.4, reverse) by carnitine palmitoyltransferase II prior to initiation of β oxidation. Tissues contain not only long-chain acylcarnitines but also *acetylcarnitine* and other short-chain acylcarnitines, some with branched chains. By accepting acetyl groups from acetyl-CoA, carnitine causes the release of free coenzyme A which can then be reused.

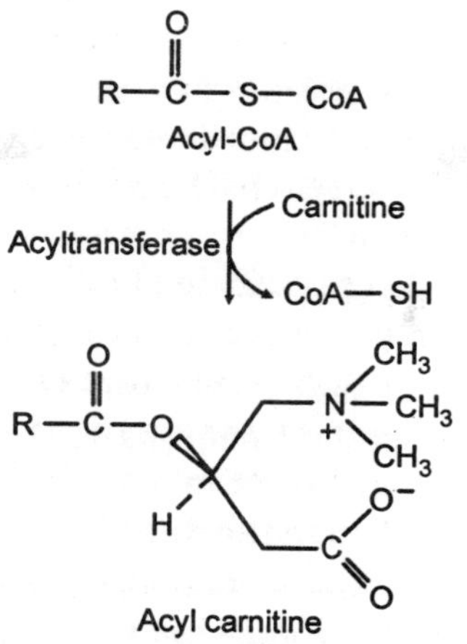

$$R-\overset{O}{\overset{\|}{C}}-S-CoA$$

Acyl-CoA

Acyltransferase (Carnitine → CoA—SH)

$$R-\overset{O}{\overset{\|}{C}}-O-CH(CH_2-\overset{+}{N}(CH_3)_3)(CH_2-COO^-)$$

Acyl carnitine ...(7.4)

Thus, carnitine may have a regulatory function. In flight muscles of insects acetylcarnitine serves as a reservoir for acetyl groups. Carnitine acyltransferases that act on short-chain acyl-CoA molecules are also present in peroxisomes and microsomes, suggesting that carnitine may assist in transferring acetyl groups and other short acyl groups between cell compartments. For example, acetyl groups from peroxisomal β oxidation can be transferred into mitochondria where they can be oxidized in the citric acid cycle.

Disorders of Fatty Acid Oxidation in Human Beings

Mitochondrial β oxidation of fatty acids is the principal source of energy for the heart. Consequently, inherited defects of fatty acid oxidation or of carnitineassisted transport often appear as serious heart disease (inherited cardiomyopathy). These may involve heart failure, pulmonary edema, or sudden infant death. As many as 1 in 10,000 persons may inherit such problems.

The proteins that may be defective include a plasma membrane carnitine transporter; carnitine palmitoyltransferases; carnitine/acylcarnitine translocase; long-chain, medium-chain, and short-chain acyl-CoA dehydrogenases; 2,4-dienoyl-CoA reductase (Eq. 7.1); and long-chain 3-hydroxyacyl-CoA dehydro-genase. Some of these are indicated in Fig. 7 2. Several cases of genetically transmitted carnitine deficiency in children have been recorded. These children have weak muscles and their mitochondria oxidize long-chain fatty acids slowly. If the inner mitochondrial membrane carnitine palmitoyltransferase II is lacking, long-chain acylcarnitines accumulate in the mitochondria and appear to have damaging effects on membranes.

In the unrelated condition of *acute myocardial ischemia* (lack of oxygen, *e.g.*, during a heart attack) there is also a large accumulation of long-chain acylcarnitines. These compounds may induce cardiac arrhythmia and may also account for sudden death from deficiency of carnitine palmitoyl-transferase II. Treatment of disorders of carnitine metabolism with daily oral ingestion of several grams of carnitine is helpful, especially for deficiency of the plasma membrane transporter. Metabolic abnormalities may be corrected completely. One of the most frequent defects of fatty acid oxidation is deficiency of a mitochondrial acyl-CoA dehydrogenase. If the long-chain-specific enzyme is lacking, the rate of β oxidation of such substrates as octanoate is much less than normal and afflicted individuals excrete in their urine hexanedioic (adipic), octanedioic, and decanedioic acids, all products of ω oxidation.

Much more common is the lack of the mitochondrial *medium-chain* acyl-CoA dehydrogenase. Again, dicarboxylic acids, which are presumably generated by ω oxidation in the peroxisomes, are present in blood and urine. Patients must avoid fasting and may benefit from extra carnitine. A deficiency of very long-chain fatty acid oxidation in peroxisomes is apparently caused by a defective transporter of the ABC type.

The disease, *X-linked adrenoleukodystrophy (ALD),* has received considerable publicity because of attempts to treat it with "Loreiizo's oil," a mixture of triglycerides of oleic and the C_{22} monoenoic *erucic acid.* The hope has been that these acids would flush out the very long-chain fatty acids that accumulate in the myelin sheath of neurons in the central nervous system and may be responsible for the worst consequences of the disease. However, there has been only limited success. Several genetic diseases involve the development of peroxisomes.

Most serious is the *Zellweger syndrome* in which there are no functional peroxisomes. Only "ghosts" of peroxisomes are present and they fail to take up proteins containing the C-terminal peroxisome-targeting sequence SKL. There are many symptoms and death occurs within the first year. Less serious disorders include the presence of catalaseless peroxisomes.

Ketone Bodies

When a fatty acid with an even number of carbon atoms is broken down through P oxidation the last intermediate before complete conversion to acetyl-CoA is the four-carbon *acetcacetyl-CoA:*

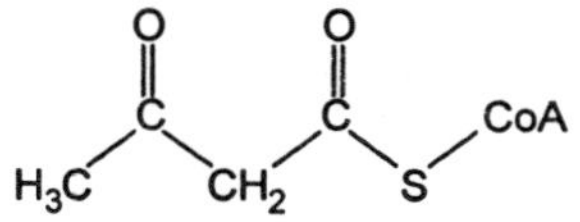

Acetoacetyl-CoA appears to be in equilibrium with acetyl-CoA within the body and is an important metabolic intermediate. It can be cleaved to two molecules of acetyl-CoA which can enter the citric acid cycle. It is also a precursor for synthesis of polyprenyl (isoprenoid) compounds, and it can give rise to free *acetoacetate,* an important constituent of blood. Acetoacetate is a β-oxoacid that can undergo decarboxylation to acetone or can be reduced by an NADH-dependent dehydrogenase to D-3-hydroxybutyrate. Notice that the configuration of this compound is opposite

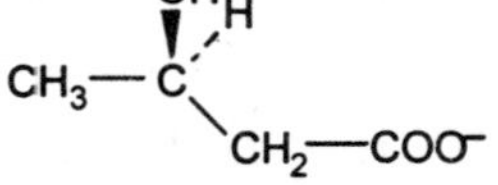

D-3-Hydroxybutyrate

to that of L-3-hydroxybutyryl-CoA which is formed during β oxidation of fatty acids (Fig. 7.1). D-3-Hydroxybutyrate is sometimes stored as a polymer in bacteria. The three compounds, acetoacetate, acetone, and 3-hydroxybutyrate, are known as *ketone bodies.*

The inability of the animal body to form the glucose precursors, pyruvate or oxaloacetate, from acetyl units sometimes causes severe metabolic problems. The condition known as *ketosis,* in which excessive amounts of ketone bodies are present in the blood, develops when too much acetyl-CoA is produced and its combustion in the critic acid cycle is slow. Ketosis often develops in patients with Type I *diabetes mellitus,* in anyone with high fevers, and during starvation. Ketosis is dangerous, if severe, because formation of ketone bodies produces hydrogen ions

(Eq. 7.5) and acidifies the blood. Thousands of young persons with insulin-dependent diabetes die annually from ketoacidosis.

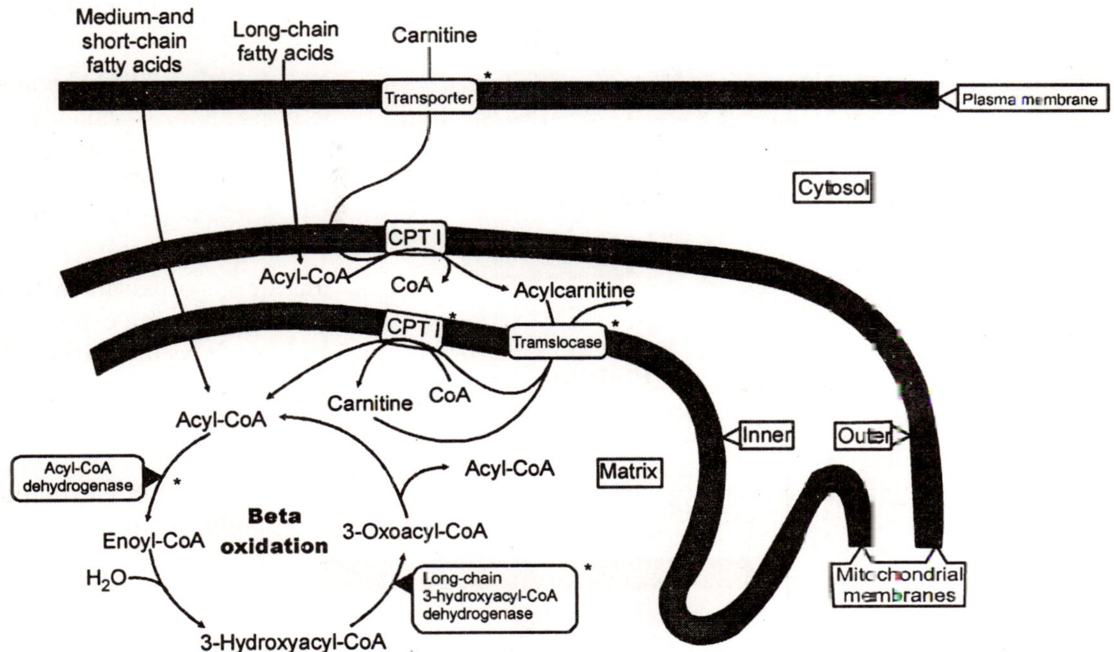

Fig. 7.2 Some specific defects in proteins of β oxidation and acyl-carnitine transport causing cardiomyopathy are indicated by the green asterisks. CPT I and CPT II are carnitine palmitoyltransferases I and II.

Rat blood normally contains about 0.07 mM acetoacetate, 0.18 mM hydroxybutyrate, and a variable amount of acetone. These amounts increase to 0.5 mM acetoacetate and 1.6 mM hydroxybutyrate after 48 h of starvation. On the other hand, the blood glucose concentration falls from 6 to 4 mM after 48 h starvation. Under these conditions acetoacetate and hydroxybutyrate are an important alternative energy source for muscle and other tissues.

Acetoacetate can be thought of as a transport form of acetyl units, which can be reconverted to acetyl-CoA and oxidized in the citric acid cycle. Some free acetoacetate is formed by direct hydrolysis of acetoacetyl-CoA. In rats, ~ 11% of the hydroxybutyrate that is excreted in the urine comes from acetoacetate generated in this way. However, most acetoacetate arises in the liver indirectly in a two-step process (Eq. 7.5) that is closely related to the synthesis of cholesterol and other polyprenyl compounds. Step *a* of this sequence is a Claisen condensation, catalyzed by 3-hydroxy-3-methyl-glutaryl CoA synthase (HMG-CoA synthase) and followed by hydrolysis of one thioester linkage. It is therefore similar to the citrate synthase reaction. Step *c* is a simple aldol cleavage.

The overall reaction has the stoichiometry of a direct hydrolysis of acetoacetyl-CoA. Liver mitochondria contain most of the body's HMG-CoA synthase and are the major site of ketone body formation (*ketogenesis*). Cholesterol is synthesized from HMG-CoA that is formed in the

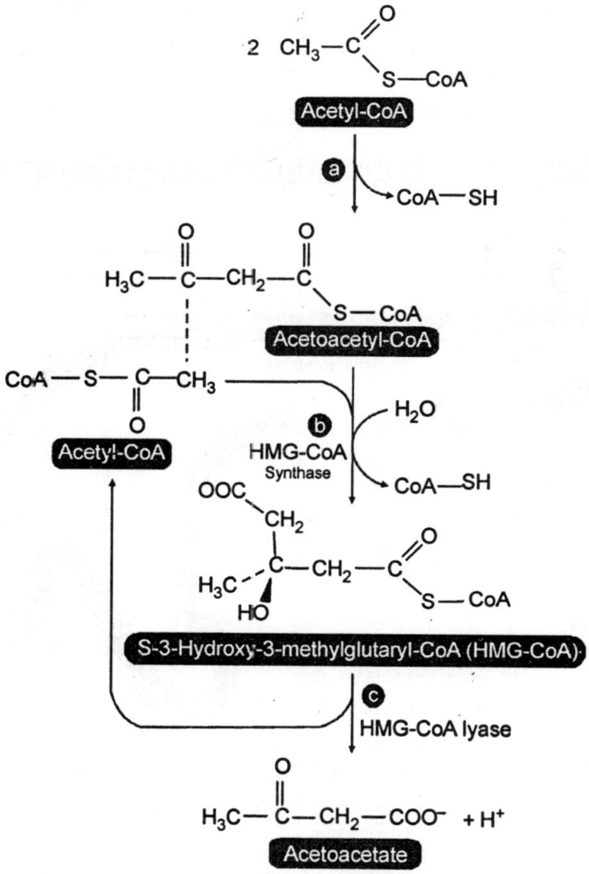

$2\ CH_3-C(=O)-S-CoA$ (Acetyl-CoA)

(a) → CoA—SH

$H_3C-C(=O)-CH_2-C(=O)-S-CoA$ (Acetoacetyl-CoA)

+ $CoA-S-C(=O)-CH_3$ (Acetyl-CoA)

(b) HMG-CoA Synthase; H_2O → CoA—SH

$^-OOC-CH_2-C(CH_3)(OH)-CH_2-C(=O)-S-CoA$

S-3-Hydroxy-3-methylglutaryl-CoA (HMG-CoA)

(c) HMG-CoA lyase (→ Acetyl-CoA)

$$H_3C-C(=O)-CH_2-COO^- + H^+ \quad \text{...(7.5)}$$

Acetoacetate

cytoplasm. Utilization of 3-hydroxybutyrate or acetoacetate for energy requires their reconversion to acetyl-CoA as indicated in Eq. 7.6.

All of the reactions of this sequence may be nearly at equilibrium in tissues that use ketone bodies for energy. Acetone, in the small amounts normally present in the body, is metabolized by hydroxylation to acetol (Eq. 7.7, step *a*), hydroxylation and dehydration to methylglyoxal

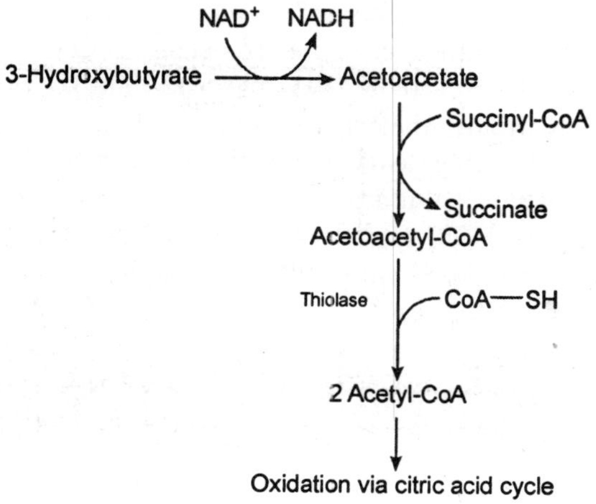

3-Hydroxybutyrate → Acetoacetate (NAD^+ → NADH)

Acetoacetate → Acetoacetyl-CoA (Succinyl-CoA → Succinate)

Acetoacetyl-CoA → 2 Acetyl-CoA (Thiolase; CoA—SH)

2 Acetyl-CoA → Oxidation via citric acid cycle ...(7.6)

(step *b*), and conversion to D-lactate and pyruvate. A second pathway via 1,2-propanediol and L-lactate is also shown in Eq. 7.7. During fasting the acetone content of human blood may rise to as much as 1.6 mM. As much as two-thirds of this may be converted to glucose.

Accumulation of acetone appears to induce the synthesis of the hydroxylases needed for methylglyoxal formation, and the pyruvate formed by Eq. 7.7 may give rise to glucose by the gluconeogenic pathway. However, at high acetone concentrations most metabolism may take place through a poorly understood conversion of 1,2-propanediol to acetate and formate or CO_2. No net conversion of acetate into glucose can occur in animals, but isotopic labels from acetate can enter glucose via acetyl-CoA and the citric acid cycle.

$H_3C-CO-CH_3$
Acetone
O_2 (a) Hydroxylation
$H_3C-CO-CH_2OH$
Acetol
ATP + 2[H] →
HO, H
$H_3C-CH(OH)-CH_2OH$
L-1,2-Propanediol
→ $HCOO^-$
CH_3-COO^-
O_2 (b) Hydroxylation
-4[H]
L-Lactate
$H_3C-CO-CHO$
Methylglyoxal
Glyoxalase
HO, OH
$H_3C-C(OH)-COO^-$
D-Lactate
-2[H] → Pyruvate → Further metabolism ...(7.7)

CATABOLISM OF PROPIONYL COENZYME A AND PROPIONATE

Beta oxidation of fatty acids with an odd number of carbon atoms leads to the formation of propionyl-CoA as well as acetyl-CoA. The three-carbon propionyl unit is also produced by degradation of cholesterol and other isoprenoid compounds and of isoleucine, valine, threonine, and methionine. Human beings ingest small amounts of free propionic acid, *e.g.*, from Swiss cheese (which is cultivated with propionk acid-producing bacteria) and from propionate added to bread as a fungicide.

In *ruminant* animals, such as cattle and sheep, the ingested food undergoes extensive fermentation in the *rumen,* a large digestive organ containing cellulose-digesting bacteria and protozoa. Major products of the rumen fermentations include acetate, propionate, and butyrate. Propionate is an important source of energy for these animals.

The Malonic Semialdehyde Pathways

The most obvious route of metabolism of propionyl-CoA is further p oxidation which leads to 3-hydroxypropionyl-CoA (Fig. 7.3, step *a*). This appears to be the major pathway in green plants. Continuation of the P oxidation via steps *a–c* of Fig. 7.3 produces the Co A derivative of malonic semialdehyde. The latter can, in turn, be oxidized to raalonyl-CoA, a β-oxoacid which can be decarboxylated to acetyl-CoA. The necessary enzymes have been found in *Clostridiuni kluyveri*, but the pathway appears to be little used. Nevertheless, malonyl-CoA is a major metabolite. It is an intermediate in fatty acid synthesis and is formed in the peroxisomal p oxidation of odd chain-length dicarboxylic acids.

Excess malonyl-CoA is decarboxylated in peroxisomes, and lack of the decarboxylase enzyme in mammals causes the lethal *malonic aciduria.* Some propionyl-CoA may also be metabolized by this pathway. The modified β oxidation sequence indicated on the left side of Fig. 7.3 is used in green plants and in many microorganisms. 3-Hydroxypropionyl-CoA is hydrolyzed to *free* β-hydroxypropionate, which is then oxidized to malonic semialdehyde and converted to acetyl-CoA by reactions that have not been completely described. Another possible pathway of propionate metabolism is the direct conversion to pyruvate via α oxidation into lactate, a mechanism that may be employed by some bacteria.

Another route to lactate is through addition of water to acrylyl-CoA, the product of step *a* of Fig. 7.3. The water molecule adds in the "wrong way," the OH~ ion going to the a carbon instead of the β (Eq. 7.8). An enzyme with an active site similar to that of histidine ammonia-lyase could presumably catalyze such a reaction. Lactyl-CoA could be converted to pyruvate readily. *Clostridium propionicum* does interconvert propionate, lactate, and pyruvate via acrylyl-CoA and lactyl-CoA as part of a fermentation of alanine. The enzyme that catalyzes hydration of acrylyl-CoA in this case is a complex flavoprotein that may function via a free radical mechanism.

$$H_2C{=}CH{-}C({=}O){-}S{-}CoA \xrightarrow{H_2O} H_3C{-}C(OH)(H){-}C({=}O){-}S{-}CoA \quad \text{...(7.8)}$$

D-Lactyl-CoA

The Methylmalonyl-CoA Pathway of Propionate Utilization

Despite the simplicity and logic of the β oxidation pathway of propionate metabolism, higher animals use primarily the more complex methylmalonyl-CoA pathway (Fig. 7.3, step *b*). This is one of the two processes in higher animals presently known to depend upon vitamin B_{12}. This vitamin has never been found in higher plants, nor does the methyl-malonyl pathway occur in plants. The pathway (Fig. 7.3) begins with the biotin- and ATP-dependent carboxylation of propionyl-CoA. The S-methylmalonyl-CoA so formed is isomerized to R-methylmalonyl-CoA, after which the methylmalonyl-CoA is converted to succinyl-CoA in a vitamin B_{12} coenzyme-requiring reaction step *d* (Table 7.1).

The succinyl-CoA is converted to free succinate (with the formation of GTP compensating for the ATP used initially). The succinate, by β oxidation, is converted to oxaloacetate which is decarboxylated to pyruvate. This, in effect, removes the carboxyl group that was put on at the beginning of the sequence in the ATP-dependent step. Pyruvate is converted by oxidative

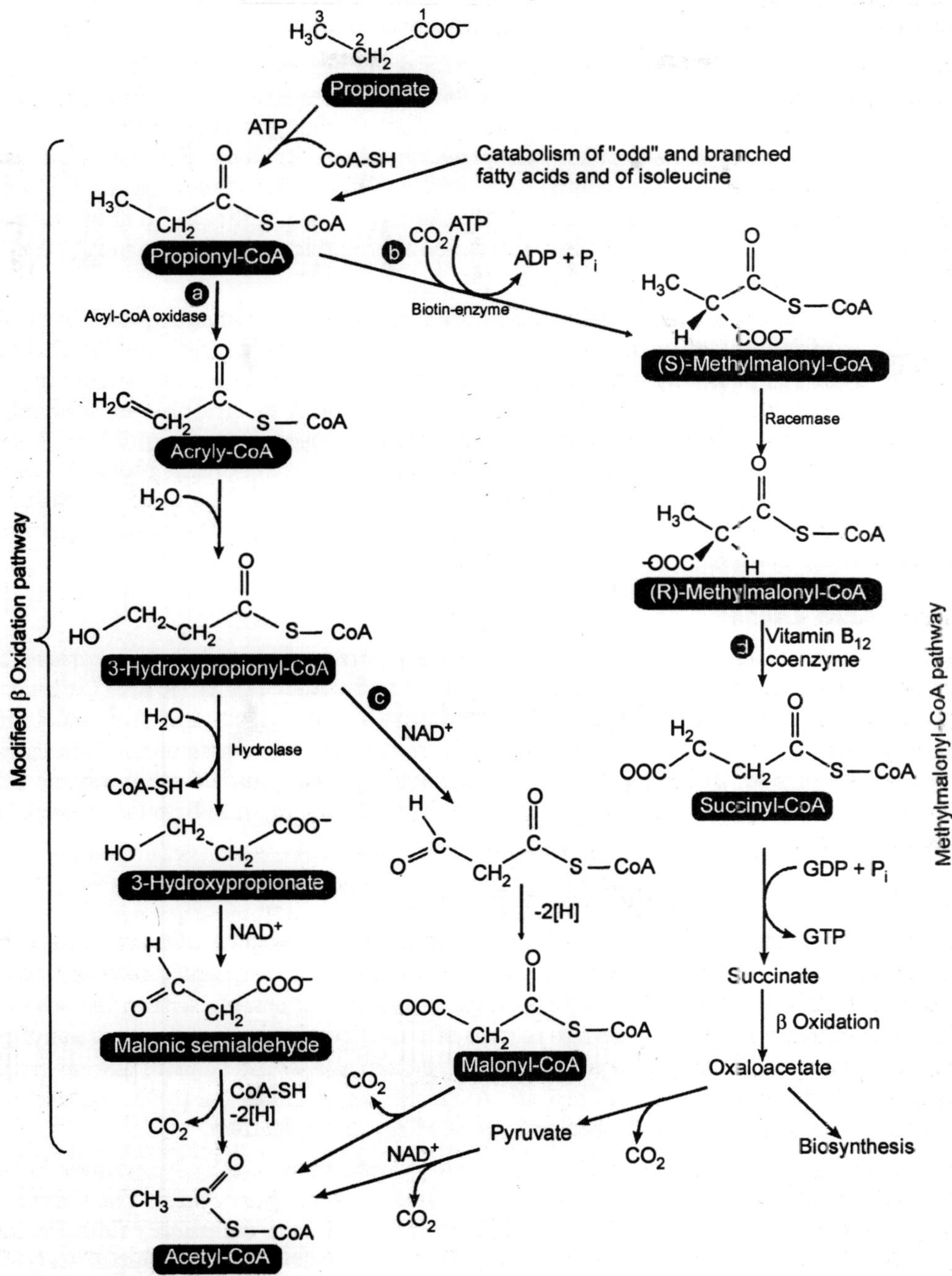

Fig. 7.3. Catabolism of propionate and propionyl-CoA. In the names for methylmalonyl-CoA the *R* and S refer to the methylmalonyl part of the structure. Coenzyme A is also chiral.

decarboxylation to acetyl-CoA. A natural question is "Why has this complex pathway evolved to do something that could have been done much more directly?" One possibility is that the presence of too much malonyl-CoA, the product of the P oxidation pathway of propionate metabolism (Fig. 7.3, pathways *a* and *c*), would interfere with lipid metabolism.

Malonyl-CoA is formed in the cytosol during fatty acid biosynthesis and retards mitochondrial β oxidation by inhibiting carnitine palmitoyltransferase I. However, a relationship to mitochondrial propionate catabolism is not clear. On the other hand, the tacking on of an extra CO_2 and the use of ATP at the beginning suggests that the *methylmalonyl-CoA pathway* (Fig. 7.3) *is a biosynthetic rather than a catabolic route*. The methyl-malonyl pathway provides a means for converting propionate to oxaloacetate, a transformation that is chemically difficult. In this context it is of interest that cows, whose metabolism is based much more on acetate than is ours, often develop a severe ketosis spontaneously.

A standard treatment is the administration of a large dose of propionate which is presumably effective because of the ease of its conversion to oxaloacetate via the methylmalonyl-CoA pathway. It is possible that this pathway was developed by animals as a means of capturing propionyl units, scanty though they may be, for conversion to oxaloacetate and use in biosynthesis. In ruminant animals, the pathway is especially important. Whereas we have 5.5 mM glucose in our blood, the cow has only half as much, and a substantial fraction of this glucose is derived, in the liver, from the propionate provided by rumen microorganisms. The need for vitamin B_{12} in the formation of propionate by these organisms also accounts for the high requirement for cobalt in the ruminant diet.

The Citric Acid Cycle

To complete the oxidation of fatty acids the acetyl units of acetyl-CoA generated in the p oxidation sequence must be oxidized to carbon dioxide and water. The citric acid (or tricarboxylic acid) cycle by which this oxidation is accomplished is a vital part of the metabolism of almost all aerobic creatures. It occupies a central position in metabolism because of the fact that acetyl-CoA is also an intermediate in the catabolism of carbohydrates and of many amino acids and other compounds. The cycle is depicted in detail in Fig. 7.6 and in an abbreviated form, but with more context, in Fig. 7.4.

A Clever Way to Cleave a Reluctant Bond

Oxidation of the chemically resistant two-carbon acetyl group to CO, presents a chemical problem. As we have seen, cleavage of a C-C bond occurs most frequently between atoms that are α and β to a carbonyl group. Such β cleavage is clearly impossible within the acetyl group. The only other common type of cleavage is that of a C-C bond adjacent to a carbonyl group (a cleavage), a thiamin-dependent process. However, a cleavage would require the prior oxidation (hydroxylation) of the methyl group of acetate. Although many biological hydroxylation reactions occur, they are rarely used in the major pathways of rapid catabolism.

Perhaps this is because the overall yield of energy obtainable via hydroxylation is less than that gained from dehydrogenation and use of an electron transport chain. The solution to the chemical problem of oxidizing acetyl groups efficiently is one very commonly found in nature; a catalytic cycle. Although direct cleavage is impossible, the two-carbon acetyl group of acetyl-CoA *can* undergo a Claisen condensation with a second compound that contains a carbonyl group. The condensation product has more than two carbon atoms, and a β cleavage to yield CO_2 is now possible.

Since the cycle is designed to oxidize acetyl units we can regard acetyl-CoA as the *primary substrate* for the cycle. The carbonyl compound with which it condenses can be

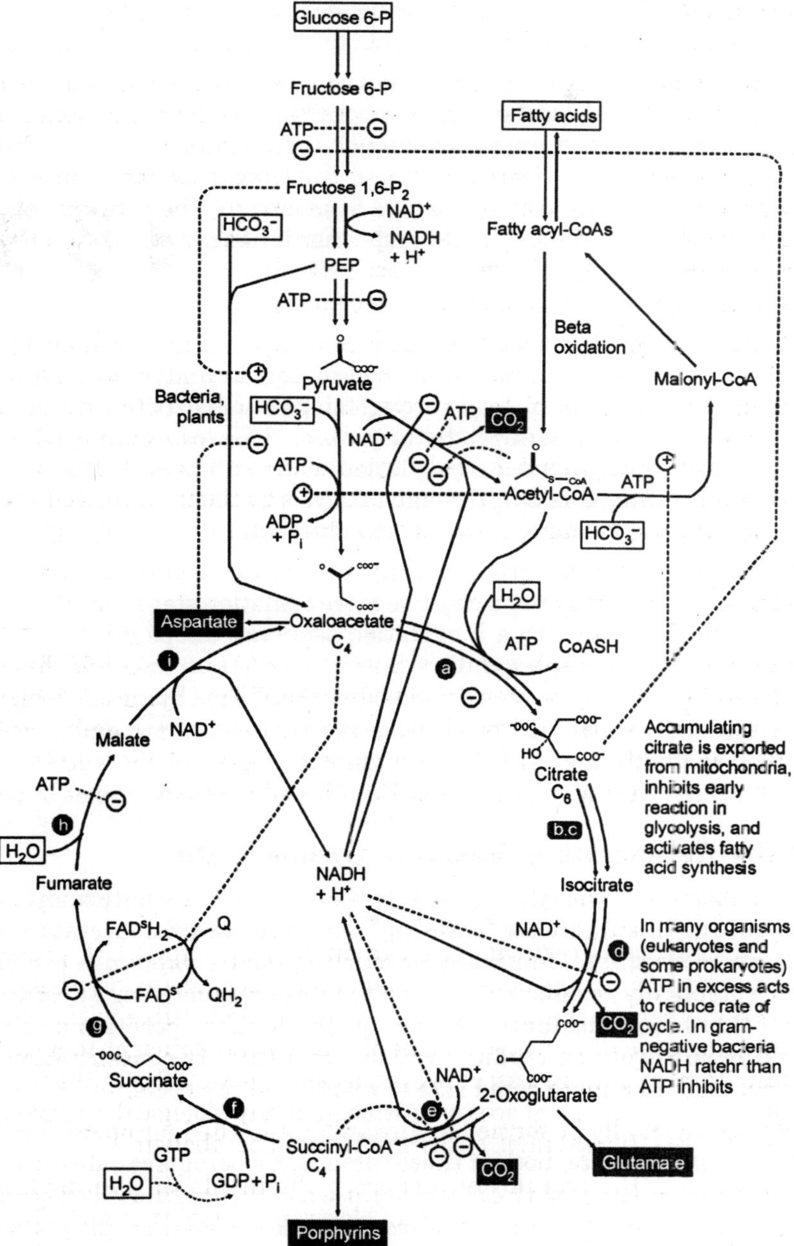

Fig. 7.4. The Krebs citric acid cycle. Some of its controlling interactions and its relationship to glycolysis. Positive and negative regulatory influences, whether arising by allosteric effects or via covalent modification, are indicated by ⊕ or ⊖. Some biosynthetic reaction pathways related to the cycle are shown in green. Steps are lettered to correspond to the numbering in Fig. 7.6, which shows more complete structural formulas. Three molecules of H_2O (boxed) enter the cycle at each turn, providing hydrogen atoms for generation of NADH + H^+ and reduced ubiquinone (QH_2). The covalently attached FAD is designated FAD^S.

described as the *regenerating substrate*. To complete the catalytic cycle it is necessary that two carbon atoms be removed as $CO_2$from the compound formed by condensation of the two substrates and that the remaining molecule be reconvertible to the original regenerating substrate.

The reader may wish to play a game by devising suitable sequences of reactions for an acetyl-oxidizing cycle and finding the simplest possible regenerating substrate. Ask yourself whether nature could have used anything simpler than *oxalo-acetate,* the molecule actually employed in the citric acid cycle. The first step in the citric acid cycle (step *a,* Fig. 7.4) is the condensation of acetyl-CoA with oxalo-acetate to form citrate. The synthase that catalyzes this condensation also removes the CoA by hydrolysis after it has served its function of activating a methyl hydrogen. This hydrolysis also helps to drive the cycle by virtue of the high group transfer potential of the thioester linkage that is cleaved.

Before the citrate can be degraded through p cleavage, the hydroxyl group must be moved from its tertiary position to an adjacent carbon where, as a secondary alcohol, it can be oxidized to a carbonyl group. This is accomplished through steps *b* and *c,* both catalyzed by the enzyme aconitase. Isocitrate can then be oxidized to the β-oxoacid *oxalosuccinate,* which does not leave the enzyme surface but undergoes decarboxylation while still bound. The second carbon to be removed from citrate is released as CO_2 through catalysis by the thiamin diphosphate dependent *oxidative decarboxylation of 2-oxoglutarate* (α-ketoglutarate).

To complete the cycle the four-carbon succinyl unit of succinyl-CoA must be converted back to oxaloacetate through a pathway requiring two more oxidation steps: Succinyl-CoA is converted to free succinate (step *f*) followed by a P oxidation sequence (steps *g-i;* Fig. 7.4). Steps *e* and *f* accomplish a substrate-level phosphorylation. Succinyl-CoA is an unstable thioester with a high group transfer potential. Therefore, step *f* could be accomplished by simple hydrolysis. However, this would be energetically wasteful. The cleavage of succinyl-CoA is coupled to synthesis of ATP in *E. coli* and higher plants and to GTP in mammals. Some of the succinyl-CoA formed in mitochondria is used in other ways, *e.g.,* as in Eq. 7.6 and for biosynthesis of porphyrins.

Synthesis of the Regenerating Substrate Oxaloacetate

The primary substrate of the citric acid cycle is acetyl-CoA. Despite many references in the biochemical literature to substrates "entering" the cycle as oxaloacetate (or as one of the immediate precursors succinate, fumarate, or malate), *these compounds are not consumed* by the cycle but are completely regenerated; hence the term *regenerating substrate,* which can be applied to any of these four substances. A prerequisite for the operation of a catalytic cycle is that a regenerating substrate be readily available and that its concentration be increased if necessary to accommodate a more rapid rate of reaction of the cycle.

Oxaloacetate can normally be formed in any amount needed for operation of the citric acid cycle from PEP or from *pyruvate,* both of these compounds being available from metabolism of sugars. In bacteria and green plants *PEP carboxylase* a highly regulated enzyme, is responsible for synthesizing oxaloacetate. In animal tissues *pyruvate carboxylase* plays the same role. The latter enzyme is almost inactive in the absence of the allosteric effector acetyl-CoA. For this reason, it went undetected for many years. In the presence of high concentrations of acetyl-CoA the enzyme is fully activated and provides for synthesis of a high enough concentration of oxaloacetate to permit the cycle to function. Even so, the oxaloacetate concentration in mitochondria is low, only 0.1 to 0.4×10^{-6} M (10-40 molecules per mitochondrion), and is relatively constant.

Common Features of Catalytic Cycles

The citric acid cycle is not only one of the most important metabolic cycles in aerobic organisms, including bacteria, protozoa, fungi, higher plants, and animals, but also *a typical catalytic cycle.* Other cycles also have one or more primary substrates and at least one regenerating substrate. Associated with every catalytic cycle there must be a metabolic pathway that provides for synthesis of the regenerating substrate. Although it usually needs to operate only slowly to replenish regenerating substrate lost in side reactions, the pathway also provides *a mechanism for the net biosynthesis of any desired quantity of any intermediate in the cycle.* Cells draw off from the citric acid cycle considerable amounts of oxaloacetate, 2-oxoglutarate, and succinyl-CoA for synthesis of other compounds.

For example, aspartate and glutamate are formed directly from oxaloacetate and 2-oxoglutarate, respectively, by transamination. Citrate itself is exported from mitochondria and used for synthesis of fatty acids. It is often stated that the citric acid cycle functions in biosynthesis, but when intermediates in the cycle are drawn off for synthesis the complete cycle does not operate. Rather, *the pathway for synthesis of the regenerating substrate, together with some of the enzymes of the cycle, is used to construct a biosynthetic pathway.* The word *amphibolic* is often applied to those metabolic sequences that are part of a catabolic cycle and at the same time are involved in a biosynthetic (anabolic) pathway. Another term, *anaplerotic,* is sometimes used to describe pathways for the synthesis of regenerating substrates. This word, which was suggested by H. L. Kornberg, comes from a Greek root meaning "filling up."

Control of the Cycle

What factors determine the rate of oxidation by the citric acid cycle? As with most other important pathways of metabolism, several control mechanisms operate and different steps may become rate limiting under different conditions. Factors influencing the flux through the cycle include (1) the rate of generation of acetyl groups, (2) the availability of oxaloacetate, and (3) the rate of reoxidation of NADH to NAD^+ in the electron transport chain. As indicated in Fig. 7.4, acetyl-CoA is a positive effector for conversion of pyruvate to oxaloacetate. Thus, acetyl-CoA "turns on" the formation of a substance required for its own further metabolism. However, when no pyruvate is available operation of the cycle may be impaired by lack of oxaloacetate. This may be the case when liver metabolizes high concentrations of ethanol. The latter is oxidized to acetate but it cannot provide oxaloacetate. Accumulating NADH reduces pyruvate to lactate, further interfering with formation of oxaloacetate.

In some individuals the accumulating acetyl units cannot all be oxidized in the cycle and instead are converted to the ketone bodies. A similar problem arises during metabolism of fatty acids by diabetic individuals with inadequate insulin. The accelerated breakdown of fatty acids in the liver overwhelms the system and results in ketosis, even though the oxaloacetate concentration remains normal. The rates of the oxidative steps in the citric acid cycle are limited by the rate of reoxidation of NADH and reduced ubiquinone in the electron transport chain which may sometimes be restricted by the availability of O_2. However, in aerobic organisms this rate is usually determined by the concentration of ADP and/or P_i available for conversion to ATP in the oxidative phosphorylation process. If catabo-lism supplies an excess of ATP over that needed to meet the cell's energy needs, the concentration of ADP falls to a low level, cutting off phosphorylation. At the same time, ATP is present in high concentration and acts as a feedback inhibitor for the catabolism of carbohydrates and fats. This inhibition is exerted at many points, a few of which are indicated in Fig. 7.4.

Important sites of inhibition are the *pyruvate dehydrogenase complex*, which converts pyruvate into acetyl-CoA; *isodtrate dehydrogenase*, which converts isocitrate into 2-oxoglutarate; and *2-oxoglutarate dehydrogenase*. The enzyme *citrate synthase*, which catalyzes the first reaction of the cycle, is also inhibited by ATP. Mitochondrial pyruvate dehydrogenase, which contains a 60-subunit icosohedral core of dihydrolipoyl-transacylase is associated with three molecules of a two-subunit kinase as well as six molecules of a structural *binding protein* which contains a lipoyl group that can be reduced and acetylated by other subunits of the core protein.

The binding protein is apparently essential to the functioning of the dehydrogenase complex but not through its lipoyl group. The specific pyruvate dehydrogenase kinase is thought to be one of the most important regulatory proteins involved in controlling energy metabolism in most organisms. Phosphorylation of up to three specific serine hydroxyl groups in the thiamin-containing decarboxylase subunit (designated El) converts the enzyme into an inactive form (Eq. 7.9). A specific phosphatase reverses the inhibition. The kinase is most active on enzymes whose core lipoyl (E2) sub-units are reduced and acetylated, a condition favoured by high ratios of [acetyl-CoA] to free [CoASH] and of [NADH] to [NAD^+].

Since the kinase inactivates the enzyme the effect is to decrease the pyruvate dehydrogenase action when the system becomes saturated and NADH and acetyl-CoA accumulate. Conversely, a high [pyruvate] inhibits the kinase and increases the action of the dehydrogenase complex. This system also permits various external signals to be felt. For example, insulin has a pronounced stimulatory effect on mitochondrial energy. One way in which this may be accomplished is through stimulation of the pyruvate dehydrogenase phosphatase, as indicated in Eq. 7.9. A *kinase activator protein* (Eq. 7.9) may also respond to various external stimuli and may be inhibited by insulin.

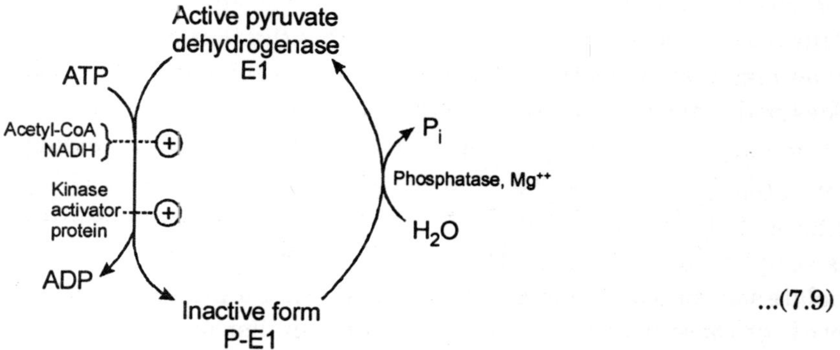

...(7.9)

The activities of 2-oxoglutarate dehydrogenase, and to a lesser extent of pyruvate and isocitrate dehydrogenases, are increased by increases in the free Ca^{2+} concentration. Calcium ions stimulate the phosphatase that dephosphorylates the deactivated phosphorylated pyruvate dehydrogenase and activate the other two dehydrogenases allosterically, increasing the affinities for the substrates. Phosphorylation of the NAD^+-dependent isocitrate dehydrogenase also decreases its activity. In *E. coli* the isocitrate dehydrogenase kinase and a protein phosphatase exist as a bifunctional protein able to both deactivate the dehydrogenase and restore its activity. For this organism, the decrease in activity forces substrate into the glyoxylate pathway (Section J,4) instead of the citric acid cycle. Acting to counteract any drop in ATP level, accumulating ADP acts as a positive effector for isocitrate dehydrogenases. Another way in which the phosphorylation state of the adenylate system can regulate the cycle depends upon the need for GDP in step *f* of the cycle (Fig. 7.4).

Within mitochondria, GTP is used largely to reconvert AMP to ADP. Consequently, formation of GDP is promoted by AMP, a compound that arises in mitochondria from the utilization of ATP for activation of fatty acids and activation of amino acids for protein synthesis. In *E. coli* and some other bacteria ATP does not inhibit citrate synthase but NADH does; the control is via the redox potential of the NAD^+ system rather than by the level of phosphorylation of the adenine nucleotide system.

Succinic dehydrogenase may be regulated by the redox state of ubiquinone. Another mechanism of regulation may be the formation of specific protein-protein complexes between enzymes catalyzing reactions of the cycle. This may permit one enzyme to efficiently have a product of its action transferred to the enzyme catalyzing the next step in the cycle.

Catabolism of Intermediates of the Citric Acid Cycle

Acetyl-CoA is the only substrate that can be completely oxidized to CO_2 by the reaction of the citric acid cycle alone. Nevertheless, cells must sometimes oxidize large amounts of one of the compounds found in the citric acid cycle to CO_2. For example, bacteria subsisting on succinate as a carbon source must oxidize it for energy as well as convert some of it to carbohydrates, lipids, and other materials. Complete combustion of *any citric acid cycle intermediate* can be accomplished by conversion to malate followed by oxidation of malate to oxaloacetate (Eq. 7.10, step *a*) and decarboxylation (β cleavage) to pyruvate, or (Eq. 7.10, step *b*) oxidation and decarboxylation of malate by the *malic enzyme* without free oxaloacetate as an intermediate. Pathway *b* is probably the most important. It is catalyzed by two different malic enzymes present in animal mitochondria. One is specific for $NADP^+$ while the other reacts with NAD^+ as well. They both have complex regulatory properties.

For example, the less specific NAD^+-utilizing enzyme is allosterically inhibited by ATP but is activated by fumarate, succinate, or isocitrate. Thus, accumulation of citric acid cycle intermediates "turns on" the malic enzyme, allowing the excess to leave the cycle and reenter as acetyl groups. Since the Michaelis constant for malate is high, this will not happen unless malate accumulates, signaling a need for acetyl-CoA. The $NADP^+$-dependent enzyme is activated by a high concentration of free CoA and is inhibited by NADH. Perhaps when glycolysis becomes slow the free CoA level rises and turns on malate oxidation. On the other hand, rapid glycolysis increases the NADH concentration which inhibits the malic enzyme. The result is a buildup of the oxaloacetate concentration and an increase in activity of the citric acid cycle. The malic enzymes are also present in the cytoplasm, where one of them functions as part of an NADPH-generating cycle.

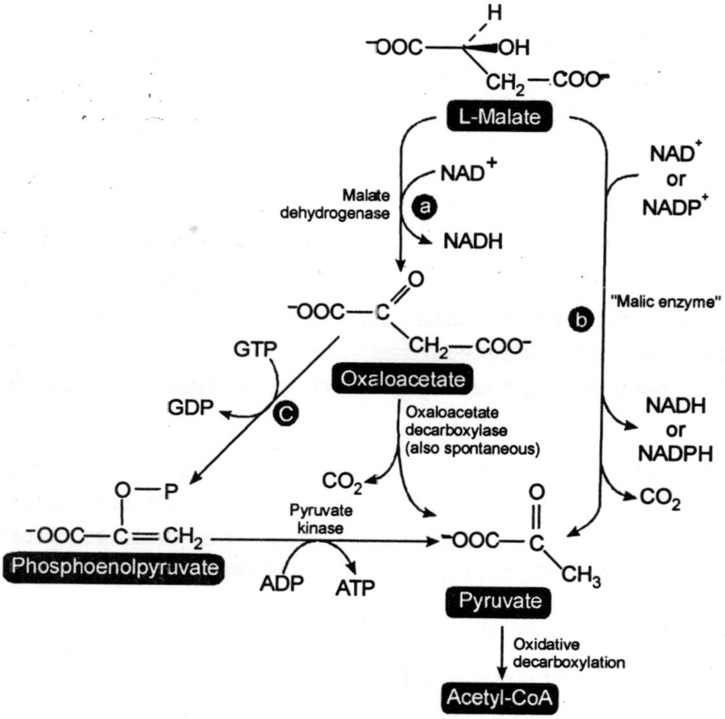

...(7.10)

OXIDATIVE PATHWAYS RELATED TO THE CITRIC ACID CYCLE

In this section we will consider some other catalytic cycles as well as some noncyclic pathways of oxidation of one- and two-carbon substrates that are utilized by microorganisms.

The γ-Aminobutyrate Cycle

A modification of the citric acid cycle which involves glutamate and gamma (γ) aminobutyrate (GABA) has an important function in the brain (Fig. 7.5). Both glutamate and γ-aminobutyrate occur in high concentrations in brain (10 and 0.8 mM, respectively). Both are important neurotransmitters, γ-aminobutyrate being a principal neuronal inhibitory substance. In the γ-aminobutyrate cycle acetyl-CoA and oxaloacetate are converted into citrate (step *a*) in the usual way and the citrate is then converted into 2-oxoglutarate. The latter is transformed to L-glutamate either by direct amination (*b*) or by transamination (*c*), the amino donor being 7-aminobutyrate. γ-Aminobutyrate is formed by decarboxylation of glutamate (Fig. 7.5, step *d*) and is catabolized via transamination (step *e*) to succinic semialdehyde, which is oxidized to succinate and oxaloacetate.

The two transamination steps in the pathways may be linked, as indicated in Fig. 7.5, to form a complete cycle that parallels the citric acid cycle but in which 2-oxoglutarate is oxidized to succinate via glutamate and γ-aminobutyrate. No thiamin diphosphate is required, but 2-oxoglutarate is reductively aminated to glutamate. The cycle is sometimes called the *γ-aminobutyrate shunt*, and it plays a significant role in the overall oxidative processes of brain tissue. This pathway is also prominent in green plants. For example, under anaerobic conditions the radish *Raphanus sativus* accumulates large amounts of γ-aminobutyrate. Most animal tissues contain very little γ-aminobutyrate, although it has been found in the oviducts of rats at concentrations that exceed those in the brain.

The Decarboxylic Acid Cycle

Some bacteria can subsist solely on glycolate, glycine, or oxalate, all of which are converted to glyoxylate (Eq. 7.11). Glyoxylate is oxidized to CO_2 and water to provide energy to the bacteria and is also utilized for biosynthetic purposes. The energy-yielding process is found in the *decarboxylic acid cycle* (Fig. 7.6), which catalyzes the complete oxidation of glyoxylate. Four hydrogen atoms are removed with generation of two molecules of NADH which can be oxidized by the respiratory chain to provide energy. In the dicarboxylic acid cycle glyoxylate is the principal substrate and acetyl-CoA is the regenerating substrate rather than the principal substrate as it is for the citric acid cycle.

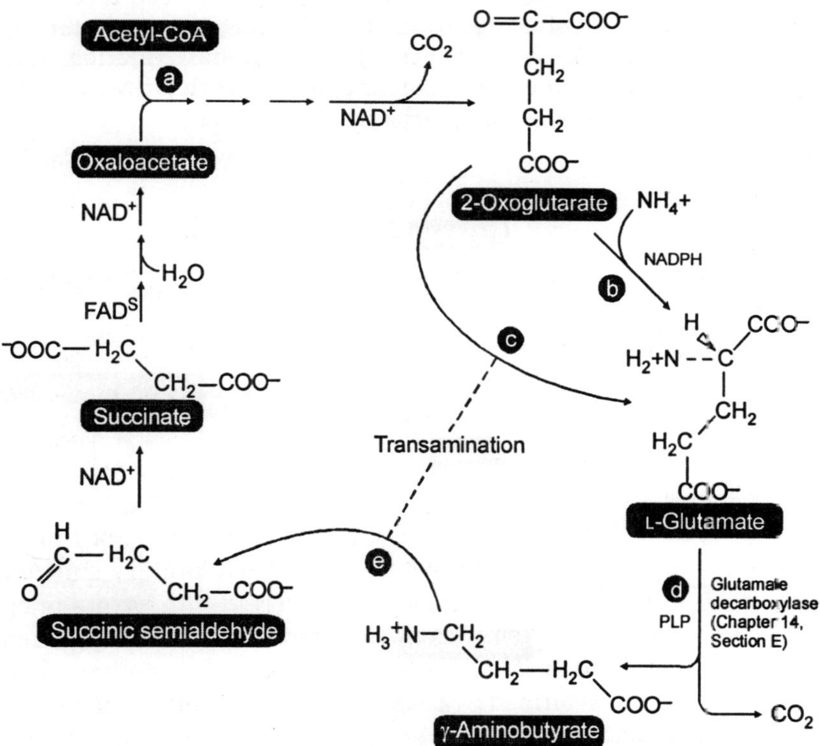

Fig. 7.5. Reactions of the γ-aminobutyrate (GABA) cycle.

The logic of the decarboxylic acid cycle is simple. Acetyl-CoA contains a potentially free carboxyl group. After the acetyl group of acetyl-CoA has been condensed with glyoxylate and the resulting hydroxyl group has been oxidized, the free carboxyl group appears in oxaloacetate in a position (3 to the carbonyl group. The carboxyl donated by the glyoxylate is still in the α position.

$$HO{-}CH_2{-}COO^- \text{ (Glycolate)} \xrightarrow{-2[H]} \text{Glyoxylate}$$
$$H_3^+N{-}CH_2{-}COO^- \text{ (Glycine)} \xrightarrow{\text{Transamination}} \text{Glyoxylate}$$
$$HO{-}CH_2{-}COO^- \text{ (Oxalate)} \longrightarrow \text{Oxalyl-CoA} \xrightarrow{2[H]} \text{Glyoxylate } (HC(=O){-}COO^-) \qquad ...(7.11)$$

A consecutive β cleavage and an oxidative a cleavage release the two carboxyl groups as carbon dioxide to reform the regenerating substrate. The cycle is simple and efficient. Like the citric acid cycle, it depends upon thiamin diphosphate, without which the α cleavage would be impossible. Comparing the citric acid cycle (Fig. 7.2) with the simpler dicarboxylic acid cycle, we see that in the former the initial condensation product citrate contains a hydroxyl group attached to a tertiary carbon atom.

With no adjacent hydrogen it is impossible to oxidize it directly to the carbonyl group which is essential for subsequent chain cleavage; hence the dependence on aconitase to shift the OH to an adjacent carbon. Both cycles involve oxidation of a hydroxy acid to a ketone followed by β cleavage and oxidative α cleavage. In the citric acid cycle additional oxidation steps are needed to convert succinate back to oxaloacetate, corresponding to the fact that the citric acid cycle deals with a more reduced substrate than does the dicarboxylic acid cycle.

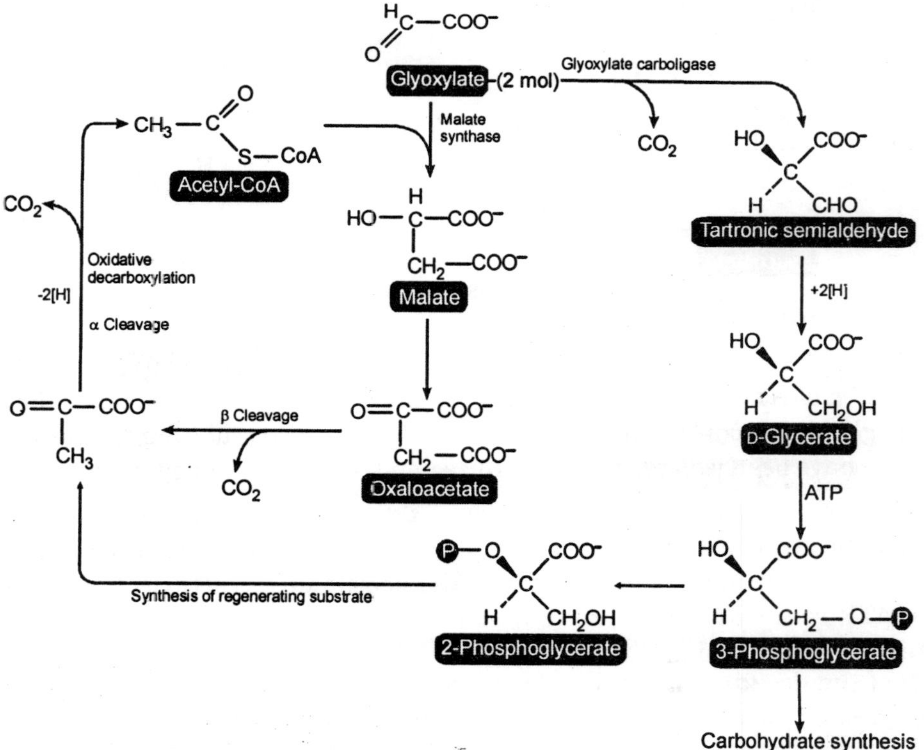

Fig. 7.6. The decarboxylic acid cycle for oxidation of glyoxylate to carbon dioxide. The pathway for synthesis of the regenerating substrate is indicated by green lines. This pathway is also needed for synthesis of carbohydrates and all other cell constituents.

The synthetic pathway for the regenerating substrate of the decarboxylic acid cycle is quite complex. Two molecules of glyoxylate undergo a condensation with decarboxylation by glyoxylate carboligase to form *tartronic semialdehyde*. The latter is reduced to D-glycerate, which is phosphorylated to 3-phosphoglycerate and 2-phosphoglycerate. Since the phosphoglycerates are carbohydrate precursors, this *glyceratc pathway* provides the organisms with a means for synthesis of carbohydrates and other complex materials from glyoxylate alone. At the same time, 2-phosphoglycerate can be converted to pyruvate and the pyruvate, by oxidative decarboxylation, to the regenerating substrate acetyl-CoA.

CATABOLISM OF SUGARS

In most sugars each carbon atom bears an oxygen atom which facilitates chemical attack by oxidation at any point in the carbon chain. Every sugar contains a potentially free aldehyde or ketone group, and the carbonyl function can be moved readily to adjacent positions by isomerases. Consequently, aldol cleavage is also possible at many points. For these reasons, the metabolism of carbohydrates is complex and varied. A sugar chain can be cut in several places giving rise to a variety of metabolic pathways. However, in the energy economy of most organisms, including human beings, the *Embden-Meyerhof-Parnas or glycolysis pathway* by which hexoses are converted to pyruvate (Fig. 7.7) stands out above all others. We have already considered this pathway, which is also outlined. Some history and additional important details follow.

The Glycolysis Pathway

The discovery of glycolysis followed directly the early observations of Buchner and of Harden and Young on fermentation of sugar by yeast juice. Another line of research, the study of muscle, soon converged with the investigations of alcoholic fermentation. Physiologists were interested in the process by which an isolated muscle could obtain energy for contraction in the absence of oxygen. It was shown by A. V. Hill that glycogen was converted to lactate to supply the energy, and Otto Meyerhof later demonstrated that the chemical reactions were related to those of alcoholic fermentation.

The establishment of the structures and functions of the pyridine nucleotides in 1934 coincided with important studies by G. Embden in Frankfurt and of J. K. Parnas in Poland. The sequence of reactions in glycolysis soon became clear. All of the 15 enzymes catalyzing the individual steps in the sequence have been isolated and crystallized and are being studied in detail.

Formation of Pyruvate : The conversion of glucose to pyruvate requires ten enzymes (Fig. 7.7), and the sequence can be divided into four stages: preparation for chain cleavage (reactions 1-3), cleavage and equilibration of triose phosphates (reactions 4 and 5), oxidative generation of ATP (reactions 6 and 7), and conversion of 3-phosphoglycerate to pyruvate (reactions 8-10). In preparation for chain cleavage, free glucose is phosphorylated to glucose 6-phosphate by ATP under the action of hexokinase (reaction 1).

Glucose 6-phosphate can also arise by cleavage of a glucosyl unit from glycogen by the consecutive action of glycogen phosphorylase (reaction 1*a*) and phosphoglucomutase, which transfers a phospho group from the oxygen at C-l to that at C-6 (reaction 1*b*). Why do cells attach phospho groups to sugars to initiate metabolism of the sugars? Four reasons can be given:

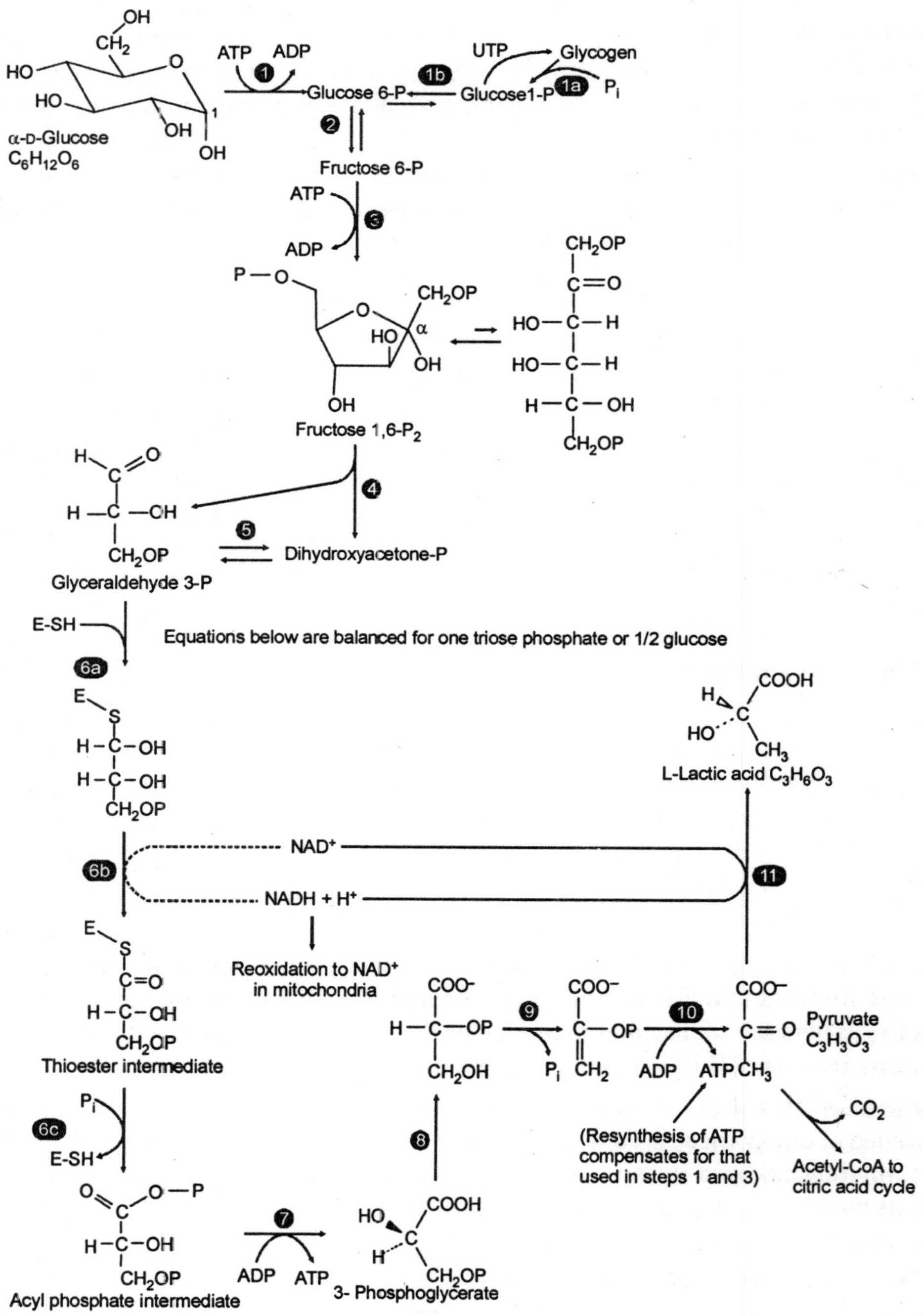

Equations below are balanced for one triose phosphate or 1/2 glucose

Fig. 7.7. Outline of the glycolysis pathway by which hexoses are broken down to pyruvate. The ten enzymes needed to convert D-glucose to pyruvate are numbered. The pathway from glycogen using glycogen phosphorylase is also included, as is the reduction of pyruvate to lactate (step *11*). Steps *6a-7*, which are involved in ATP synthesis via thioester and acyl phosphate intermediates, are emphasized.

(*a*) The phospho group constitutes an electrically charged handle for binding the sugar phosphate to enzymes.

(*b*) There is a kinetic advantage in initiating a reaction sequence with a highly irreversible reaction such as the phosphorylation of glucose.

(*c*) Phosphate esters are unable to diffuse out of cells easily and be lost.

(*d*) There is at least a possibility that the phosphor groups may function in catalysis.

Reaction 2 of Fig. 7.7 is a simple isomerization that moves the carbonyl group to C-2 so that β-cleavage to two three-carbon fragments can occur. Before cleavage a second phosphorylation (reaction 3) takes place to form fructose 1,6-bisphosphate. This ensures that when fructose bisphosphate is cleaved by aldolase each of the two halves will have a phosphate handle. This second priming reaction (reaction 3) is the first step in the series that is unique to glycolysis. The catalyst for the reaction, *phosphofructokinase*, is carefully controlled. Fructose bisphosphate is cleaved by action of an aldolase (reaction 4) to give glyceraldehyde 3-phosphate and dihydroxyacetone phosphate. These two triose phosphates are then equilibrated by triose phosphate isomerase.

As a result, both halves of the hexose can be metabolized further via glyceraldehyde 3-*P* to pyruvate. The oxidation of glyceraldehyde 3-*P* to the corresponding carboxylic acid, 3-phosphoglyceric acid (Fig. 7.7, reactions 6 and 7), is coupled to synthesis of a molecule of ATP from ADP and P_i. This means that two molecules of ATP are formed per hexose cleaved, and that two molecules of NAD^+ are converted to NADH in the process. The conversion of 3-phosphoglycerate to pyruvate begins with transfer of a phospho group from the C-3 to the C-2 oxygen (reaction 8) and is followed by dehydration through an α, β elimination catalyzed by *enolase* (reaction 9).

The product, phosphoenolpyruvate (PEP), has a high group transfer potential. Its phospho group can be transferred easily to ADP via the action of the enzyme *pyruvate kinase*, to leave the enol of pyruvic acid which is spontaneously converted to the much more stable pyruvate ion. Because two molecules of PEP are formed from each glucose molecule, the process provides for the recovery of the two molecules of ATP that were expended in the initial formation of fructose 1,6-bisphosphate from glucose. Several isoenzyme forms exist in mammals. Most of these are allosterically activated by fructose 1,6-bisphosphate. However, the enzyme from trypanosomes is activated by fructose 2,6-P_2.

The Further Metabolism of Pyruvate : In the aerobic metabolism that is characteristic of most tissues of our bodies, pyruvate is oxidatively decarboxylated to acetyl-CoA, which can then be completely oxidized in the citric acid cycle (Fig. 7.4). The NADH produced in reaction 6 of Fig. 7.7, as well as in the oxidative decarboxylation of pyruvate and in subsequent reactions of the citric acid cycle, is reoxidized in the electron transport chain of the mitochondria as described. An important alternative fate of pyruvate is to enter into fermentation reactions.

For example, the enzyme lactate dehydro-genase (Fig. 7.7, reaction 11) catalyzes reduction by NADH of pyruvate to L-lactate, or, for some bacteria, to D-lactate. This reaction can be coupled to the NADH-producing reaction 6 to give a balanced process by which glucose is fermented to lactic acid in the absence of oxygen. In a similar process, yeast cells decarboxylate pyruvate (a cleavage) to acetaldehyde which is reduced to ethanol using the NADH produced in reaction 6.

Generation of ATP by Substrate Oxidation

The formation of ATP from ADP and P_i is a vital process for all cells. It is usually referred to as "phosphorylation" and includes *oxidative phosphorylation* associated with the passage of electrons through an electron transport chain—usually in mitochondria; *photosynthetic phosphorylation*, a similar process occurring in chloroplasts under the influence of light; and *substrate-level phosphorylation*. Only the last is fully understood chemically. The dehydrogenation of glyceraldehyde 3-P and the accompanying ATP formation (reactions 6 and 7, Fig. 7.7; Fig. 7.6) is the best known example of substrate-level phosphorylation and is tremendously important for yeasts and other microorganisms that live anaerobically. They depend upon this one reaction for their entire supply of energy.

The conversion of glucose either to lactate or to ethanol and CO_2 is accompanied by a net synthesis of only two molecules of ATP and it is most logical tc view these as arising from oxidation of glyceraldehyde 3-P. The formation of ATP from PEP and ADP in reaction 10 of Fig. 7.7 can be regarded as the recapturing of ATP "spent" in the priming reactions of steps 1 and 3. With a gain of only two molecules of ATP for each molecule of hexose fermented, it is not surprising that yeast must ferment enormous amounts of sugar to meet its energy needs. Each glucose unit of glycogen stored in our bodies can be converted to pyruvate with an apparent net gain of *three* molecules of ATP.

However, two molecules of ATP were needed for the initial synthesis of each hexose unit of glycogen. Therefore, the overall net yield for fermentation of stored polysaccharide is still only two ATP per hexose. The fermentation of glycogen accounts for the very rapid generation of lactic acid during intense muscular activity. However, in most circumstances within aerobic tissues reoxidation of NADH occurs via the electron transport chain of mitochondria with a much higher yield of ATP. Substrate-level phosphorylation can also follow oxidative decarboxylation of an α-oxoacid. For example, in the citric add cycle GTP is formed following oxidative decarboxylation of 2-oxoglutarate (Fig. 7.4 step *e* and *f*).

The Pentose Phosphate Pathways

A second way of cleaving glucose 6-phosphate utilizes sequences involving the five-carbon pentose sugars. They are referred to as *pentose phosphate pathways*, the phosphogluconate pathway, or the hexose monophosphate shunt. Historically, the evidence for such routes dates from the experiments of Warburg on the oxidation of glucose 6-*P* to 6-phosphogluconate. For many years the oxidation remained an enzymatic reaction without a defined pathway. However, it was assumed to be part of an alternative method of degradation of glucose. Supporting evidence was found in the observation that tissues continue to respire in the presence of a high concentration of fluoride ion, a known inhibitor of the enolase reaction and capable of almost completely blocking glycolysis.

Some tissues, *e.g.*, liver, are especially active in respiration through this alternative pathway, whose details were elucidated by Horecker and associates. We now know that the pentose phosphate pathways are multiple as well as multipurpose. They function in catabolism and also, when operating in the reverse direction, as a reductive pentose phosphate pathway that lies at the heart of the sugar-forming reactions of photosynthesis. The oxidative pentose pathway provides *a means for cutting the chain of a sugar molecule one carbon at a time,* with the carbon removed appearing as CO_2.

The enzymes required can be grouped into three distinct systems, all of which are found in the cytosol of animal cells: (*i*) a dehydrogenation-decarboxylation system, (*ii*) an isomerizing system, and (*iii*) a sugar rearrangement system. The dehydrogenation-decarboxylation system cleaves glucose *6-P* to CO_2 and the pentose phosphate, ribulose 5-P (Eq. 7.12). Three enzymes are required, the first being glucose 6-P dehydrogenase (Eq. 7.12, step *a*). The immediate product, a lactone, undergoes spontaneous hydrolysis. However, the action of *gluconolactonase* (Eq. 7.12, step *b*) causes a more rapid ring opening.

A second dehydrogenation is catalyzed by *6-phosphogluconate dehydrogenase* (Eq. 7.12, step *c*), and this reaction is immediately followed by a β decarboxylation catalyzed by the same enzyme. The value of ΔG° for oxidation of glucose 6-P to ribulose 5-P by $NADP^+$ according to Eq. 7.12 is –30.8 kJ mol^{-1}, a negative enough value to drive the [NADPH]/[$NADP^+$] ratio to an equilibrium value of over 2000 at a CO_2 tension of 0.05 atm. The isomerizing system, consisting of two enzymes interconverts three pentose phosphates (Eq. 7.13). As a consequence the three compounds exist as an equilibrium mixture. Both xylulose 5-P and ribose 5-P are needed for further reactions in the pathways.

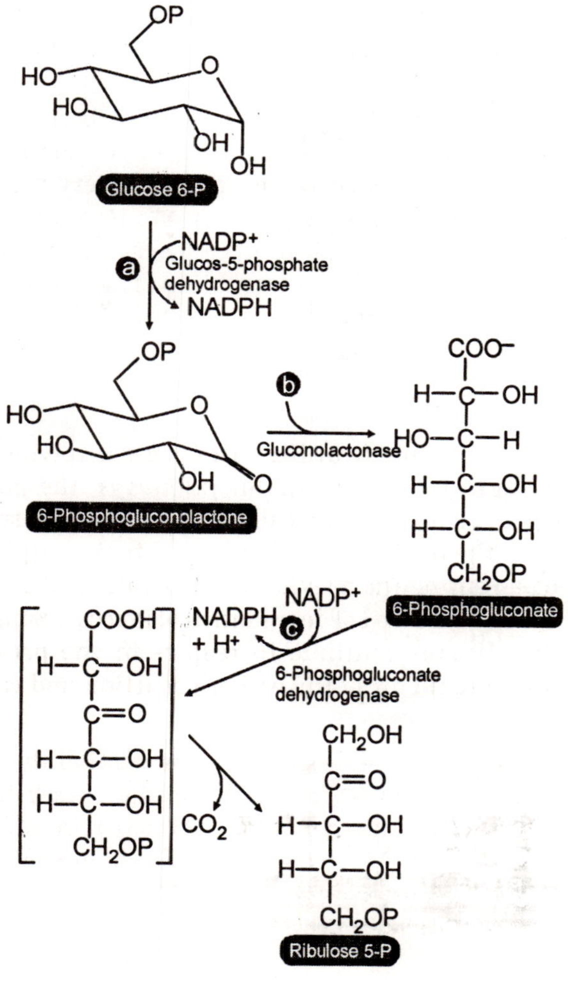

Glucose 6-P

(a) $NADP^+$ → NADPH; Glucos-5-phosphate dehydrogenase

6-Phosphogluconolactone

(b) Gluconolactonase

6-Phosphogluconate

(c) $NADP^+$ → NADPH + H^+; 6-Phosphogluconate dehydrogenase

CO_2

Ribulose 5-P

...(7.12)

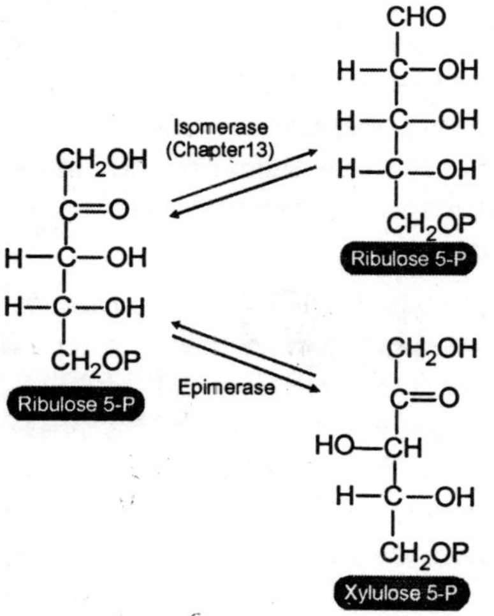

$$
\begin{array}{c}
\text{Ribulose 5-P} \xrightleftharpoons[\text{(Chapter13)}]{\text{Isomerase}} \text{Ribose 5-P (CHO, H-C-OH, H-C-OH, H-C-OH, } CH_2OP\text{)}\\
\text{Ribulose 5-P } (CH_2OH, C{=}O, H{-}C{-}OH, H{-}C{-}OH, CH_2OP) \xrightleftharpoons{\text{Epimerase}} \text{Xylulose 5-P } (CH_2OH, C{=}O, HO{-}CH, H{-}C{-}OH, CH_2OP)
\end{array}
\quad ...(7.13)
$$

Ribulose 5-P

Ribulose 5-P

Xylulose 5-P

...(7.13)

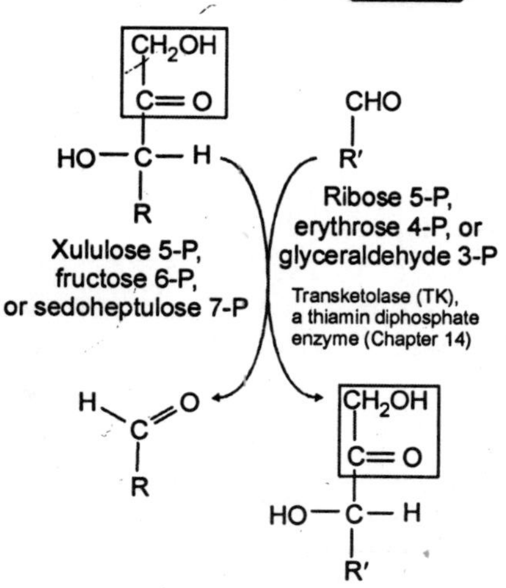

CH_2OH–C=O–HO–C–H–R (Xululose 5-P, fructose 6-P, or sedoheptulose 7-P) + CHO–R′ (Ribose 5-P, erythrose 4-P, or glyceraldehyde 3-P)

Transketolase (TK), a thiamin diphosphate enzyme (Chapter 14)

→ H–C(=O)–R + CH_2OH–C=O–HO–C–H–R′

...(7.14)

CH_2OH–C=O–HO–C–H–H–C–OH–R″ + CHO–R′

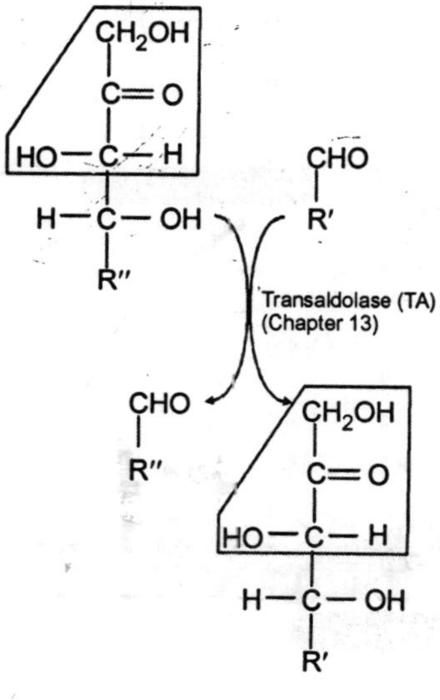

Transaldolase (TA) (Chapter 13)

→ CHO–R″ + CH_2OH–C=O–HO–C–H–H–C–OH–R′

...(7.15)

The ingenious sugar rearrangement system uses two enzymes, *transketolase* and *transaldolase*. Both catalyze chain cleavage and transfer reactions (Eqs. 7.14 and 7.15) that involve the same group of substrates. These enzymes use the two basic types of C-C bond cleavage; adjacent to a carbonyl group (α) and one carbon removed from a carbonyl group (β). Both types are needed in the pentose phosphate pathways just as they are in the citric acid cycle. The enzymes of the pentose phosphate pathway are found in the cytoplasm of both animal and plant cells. Mammalian cells appear to have an additional set that is active in the endoplasmic reticulum and plants have another set in the chloroplasts.

An Oxidative Pentose Phosphate Cycle : Putting the three enzyme systems together, we can form a cycle that oxidizes hexose phosphates. Three carbon atoms are chopped off one at a time (Fig. 7.8A) leaving a three-carbon triose phosphate as the product. Since the dehydrogenation system works only on glucose 6-P, a part of the sugar rearrangement system must be utilized between each of the three oxidation steps. Notice that a C_5 unit (ribose 5-P) is used in the first reaction with transketolase but is regenerated at the end of the sequence. This C_5 unit is the regenerating substrate for the cycle.

As indicated by the dashed arrows, it is formed readily in any quantity needed by oxidation of glucose 6-P. Before the C_5 unit that is formed in each oxidation step can be processed by the sugar rearrangement reactions, it must be isomerized from ribulose 5-P to xylulose 5-P; before the C_5 unit, produced at the end of the sequence in Fig. 7.8, can be reutilized as a regenerating substrate, it must be isomerized to ribose 5-P. Thus, the cycle is quite complex. The same C_5 substrates appear at several points in Fig. 7.8A and substrates from different parts of the cycle become scrambled and the pathway does not degrade all the hexose molecules in a uniform manner. For this reason, Zubay described the pentose phosphate pathways as a "swamp." The oxidative pentose phosphate cycle is often presented as a means for complete oxidation of hexoses to CO_2.

For this to happen the C_3 unit indicated as the product in Fig. 7.8A must be converted (through the action of aldolase, a phosphatase, and hexose phosphate isomerase) back to one-half of a molecule of glucose-6-P which can enter the cycle at the beginning. On the other hand, alternative ways of degrading the C_3 product glyceraldehyde-P are available. For example, using glycolytic enzymes, it can be oxidized to pyruvate and to CO_2 via the citric acid cycle. As a general rule, NAD^+ is associated with catabolic reactions and it is somewhat unusual to find $NADP^+$ acting as an oxidant.

However, in mammals the enzymes of the pentose phosphate pathway are specific for $NADP^+$. The reason is thought to lie in the need of NADPH for biosynthesis (Section I. On this basis, the occurrence of the pentose phosphate pathway in tissues having an unusually active biosynthetic function (liver and mammary gland) is understandable. In these tissues the cycle may operate as indicated in Fig. 7.8A with the C_3 product also being used in biosynthesis. Furthermore, any of the products from C_4 to C_7 may be withdrawn in any desired amounts without disrupting the smooth operation of the cycle. For example, the C_4 intermediate *erythrose 4-P* is required in synthesis of aromatic amino acids by bacteria and plants (but not in animals). *Ribose 5-P* is needed for formation of several amino acids and of nucleic acids by all organisms.

In some circumstances the formation of ribose 5-P may be the only essential function for the pentose phosphate pathway. Several studies of the metabolism of isotopically labeled glucose have been in accord with operation of the pentose phosphate pathway as is depicted in Fig. 7.8. However, Williams and associates proposed a modification in the sugar rearrangement sequence in liver to include the formation of arabinose 5-P (from ribose 5-P), an octulose bisphosphate, and an octulose 8-monophosphate. Many investigators argue that these additional reactions are of minor significance.

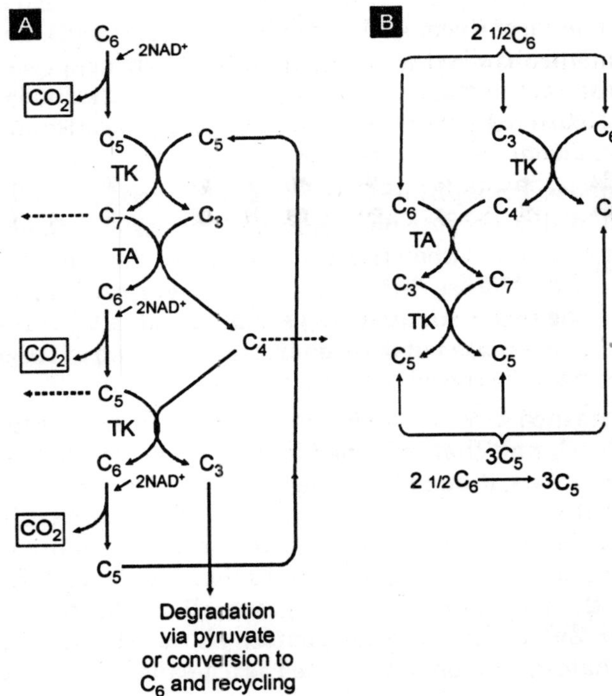

Fig. 7.8. The pentose phosphate pathways. (A) Oxidation of a hexose (C_6) to three molecules of CO_2 and a three-carbon fragment with the option of removing C_3, C_4, C_5, and C_7 units for biosynthesis (dashed arrows). (B) Non-oxidative pentose pathways: 2 1/2 $C_6 \rightarrow 3\ C_5$ or 2 $C_6 \rightarrow 3\ C_4$ or 31/2$C_6 \rightarrow 3C_7$.

The measured concentrations of pentose phosphate pathway intermediates in rat livers are close to those predicated for a near-equilibrium state from equilibrium constants measured for the individual steps of Fig. 7.8.

Most of the concentrations are in the 4- to 10-μM range but the level of erythrose 4-*P*, which is predicted to be ~ 0.2 μM, is too low to measure. In contrast to animals, the resurrection plant *Craterostigma plantaginenm* accumulates large amounts of a 2-oxo-octulose. This plant is one of a small group of angiosperms that can withstand severe dehydration and are able to rehydrate and resume normal metabolism within a few hours. During desiccation much of the octulose is converted into sucrose. The plant has extra transketolase genes which may be essential for this rapid interconversion of sugars.

Nonoxidative Pentose Phosphate Pathways

The sugar rearrangement system together with the glycolytic enzymes that convert glucose 6-*P* to glyceraldehyde 3-*P* can function to transform hexose phosphates into pentose phosphates (Fig. 7.8B; Eq. 7.16) which may be utilized for nucleic acid synthesis in erythrocytes and other cells.

$$2^1/_2\ C_6 \rightarrow 3\ C_5 \qquad ...(7.16)$$

The reader can easily show that the same enzymes will catalyze the net conversion of hexose phosphate to erythrose 4-phosphate or to sedoheptulose 7-phosphate (Eq. 7.17):

$$2\ C_6 \rightarrow 3C_4;\ 31/2\ C_6 \rightarrow 3C_7 \qquad \text{...(7.17)}$$

An investigation of metabolism of the red lipid-forming yeast *Rhodotorula gracilis* (which lacks phosphofructokinase and is thus unable to break down sugars through the glycolytic pathway) indicated that 20% of the glucose is *oxidized* through the pentose phosphate pathways while 80% is *altered* by the nonoxidative pentose phosphate pathway. However, it is not clear how the C_3 unit used in the nonoxidative pathway (Fig. 7.8B) is formed if glycolysis is blocked. A number of fermentations are also based on the pentose phosphate pathways (Section F,5).

The Entner-Doudoroff Pathway

An additional way of cleaving a six-carbon sugar chain provides the basis for the *Entner-Doudoroff pathway* which is used by *Zymomonas lindneri* and many other species of bacteria. Glucose 6-P is oxidized first to 6-phosphogluconate, which is converted by dehydration to a 2-oxo-3-deoxy derivative (Eq. 7.18 step *a*). The resulting 2-oxo-3-deoxy sugar is cleaved by an aldolase (Eq. 7.18, step *b*) to pyruvate and glyceraldehyde 3-*P*, which are then metabolized in standard ways.

6-Phosphogluconate —(a, $-H_2O$)→ COO^-–C=O–CH_2–HCOH–HCOH–CH_2O-P

2-Oxo-3-depxu-6-phosphogluconate

(b) An aldolase ↓

COO^-–C=O–CH_2 (Pyruvate) + HC=O–HCOH–CH_2O-P (Glyceraldehyde 3-P) ...(7.18)

FERMENTATION: "LIFE WITHOUT OXYGEN"

Pasteur recognized in 1860 that fermentation was not a spontaneous process but a result of life in the near absence of air. He realized that yeasts decompose much more sugar under anaerobic conditions than they do aerobically, and that the anaerobic fermentation was essential to the life of these organisms. In addition to the alcoholic fermentation of yeast, there are many other fermentations which have been attractive subjects for biochemical study. If life evolved at a time when no oxygen was available, the most primitive organisms must have used fermentations. They may be the oldest as well as the simplest ways in which cells obtain energy.

The enzymes of the glycolysis pathway are found in the small genomes of *Mycoplasma, Haemophila,* and *Methococcus*. Fermentation is also a vital process in the human body. Our muscles usually receive enough oxygen to oxidize pyruvate and to obtain ATP through aerobic metabolism, but there are circumstances in which the oxygen supply is inadequate.

During extreme exertion, after most oxygen is consumed, muscle cells produce lactate by fermentation. White muscle of fish and fowl has little aerobic metabolism and normally yields L-lactic acid as a principal end product. Likewise, a variety of tissues within the human body, including the transparent lens and cornea of our eyes, are poorly supplied with blood and depend upon fermentation of glucose to lactic acid. Red blood cells and skin and sometimes adipose tissue are also major producers of lactic acid. Of the ~ 115 g of lactic acid present in a 70-kg human body, about 29% comes from erythrocytes, 29% from skin, 17% from the brain, and 16% from skeletal muscle. Because lactic acid lowers the pH of cells it must be removed efficiently. Some of the lactic acid formed in muscle and most of the lactate formed in less aerobic tissues (*e.g.*, adipose tissue) enters the bloodstream, which normally contains 1-2 mM lactate, and is carried to the liver where it is reoxidized to pyruvate.

Part of the pyruvate is then oxidized via the citric acid cycle while a larger part is reconverted to glucose (Section J,5). This glucose may be released into the bloodstream and returned to the muscles. The overall process is known as the *Cori cycle*. Lactic acid accumulates in muscle after vigorous exercise. It is exported to the liver slowly, but if mild exercise continues the lactate may be largely oxidized within muscle via the tricarboxylic acid cycle.

Recent NMR studies have shown that lactic acid is formed rapidly dxiring muscular contraction, even when exercise is mild. During the initial 15 ms of contraction the ATP utilized is regenerated from creatine phosphate. During the remainder of the contraction (up to ~100 ms) glycogen is converted to lactic acid to provide ATP and to replenish the creatine phosphate. In the resting period following contraction most of the lactate is either dehydrogenated to pyruvate and oxidized in mitochondria or exported to other tissues. The glycogen stores in muscle are renewed by synthesis from blood glucose. Lactic acid is a convenient energy carrier and a precursor for gluconeogenesis which can be transferred between tissues easily.

Cancer cells often take advantage of this opportunity to grow rapidly using fermentation of glucose to lactic acid as a source of energy. Alcoholic fermentation allows roots of some plants to survive short periods of flooding. Ethanol does not acidify the tissues as does lactic acid, avoiding possible damage from low pH. Goldfish can also use the ethanolic fermentation for short times, excreting the ethanol.

Fermentations Based on the Embden-Meyerhof Pathway

Homolactic and Alcoholic Fermentations : The reactions by which glucose can be converted to lactate and, by yeast cells, to ethanol and CO_2 (Fig. 7.7) illustrate several features common to all fermentations. The NADH produced in the oxidation step is reoxidized in a reaction by which substrate is reduced to the final end product. The NAD alternates between oxidized and reduced forms. This coupling of oxidation steps with reduction steps in exact equivalence is characteristic of all true (anaerobic) fermentations.

The formation of ATP from ADP and P_i by substrate-level phosphorylation is also common to all fermentations. The stoichiometry is often nearly exact and simple. For example, according to the reaction of Eq. 7.19, which is outlined step-by-step in Fig. 7.7, a net total of two moles of ATP is formed per mole of glucose fermented.

Energy Relationships : If we disregard the synthesis of ATP, the equations for the lactic acid and ethanol fermentations are given by Eqs. 7.19 and 7.20.

$$\text{Glucose} \rightarrow 2\ \text{lactate}^- + 2\ \text{H}^+$$
$$\Delta G'\ (\text{pH } 7) = -196\ \text{kJ mol}^{-1}\ (-46.8\ \text{kcal mol}^{-1}) \tag{7.19}$$
$$\text{Glucose} \rightarrow 2\ \text{CO}_2 + 2\ \text{ethanol}$$
$$\Delta G^\circ = -235\ \text{kJ mol}^{-1} \tag{7.20}$$

The Gibbs energy changes are negative and of sufficient magnitude that the reactions will unquestionably go to completion. However, the synthesis of two molecules of ATP from inorganic phosphate and ADP, a reaction [Eq. 7.21) for which $\Delta G'$ is substantially positive, is coupled to the fermentation.

$$\text{ADP}^{3-} + \text{HPO}_4^{\ 2-} + \text{H}^+ \rightarrow \text{ATP}^{4-} + \text{H}_2\text{O}$$
$$\Delta G'\ (\text{pH } 7) = +\ 34.5\ \text{kJ mol}^{-1} \tag{7.21}$$

To obtain the net Gibbs energy change for the complete reaction we must add 2 × 34.5 = + 69.0 kJ to the values of $\Delta G'$ for Eqs. 7.19 and 7.20. When this is done we see that the net Gibbs energy changes are still highly negative, that the reactions will proceed to completion, and that these fermentations can serve as an usable energy source for organisms. Biochemists sometimes divide AG for the ATP synthesis in a coupled reaction sequence (in this case +69 kJ) by the overall Gibbs energy decrease for the coupled process (196 or 235 kJ mol^{-1}) to obtain an "efficiency." In the present case the efficiency would be 35% and 29% for coupling of Eq. 7.21 (for 2 mol of ATP) to Eqs. 7.19 and 7.20, respectively.

According to this calculation, nature is approximately one-third efficient in the utilization of available metabolic Gibbs energy for ATP synthesis. However, it is important to realize that this calculation of efficiency has no exact thermodynamic meaning. Furthermore, the utilization of ATP formed by a cell for various purposes is far from 100% efficient. Why are the Gibbs energy decreases for Eqs. 7.19 and 7.20 so large? No overall oxidation takes place; there is only a rearrangement of the existing bonds between atoms of the substrate. Why does this rearrangement of bonds lead to a substantial negative ΔG? An answer is suggested by an examination of the numbers of each type of bond in the substrate and in the products.

During the conversion from glucose to two molecules of lactate one C-C bond, one C-O bond, and one O-H bond are lost and one C-H bond and one C = O are gained. If we add up the bond energies for these bonds (Table 7.6) we find that the difference (ΔH) between substrate and products amounts to only about 20 kJ/mol. However, lactic acid contains a carboxyl group, and carboxyl groups have a special stability as a result of resonance. The extra resonance energy of a carboxyl group (Table 7.6) is ~ 117 kJ (28 kcal) per mole or 234 kJ/mol for two carboxyl groups. This is approximately the same as the Gibbs energy change (Eq. 7.19) for fermentation of glucose to lactate. Thus, the energy available results largely from the rearrangement of bonds by which the carboxyl groups of lactate are formed.

Likewise, the resonance stabilization of CO_2 is given by Pauling as 151 kJ/mol, again of just the right magnitude to explain ΔG in alcoholic fermentations (Eq. 7.20). On this basis we can state as a general rule that fermentations can occur when substrates consisting of largely singly bonded atoms and groups, such as the carbonyl groups that are not highly stabilized by resonance, are converted to products containing carboxyl groups or to CO_2. If we assume an efficiency of ~ 30%, the energy available will be about sufficient for synthesis of one ATP molecule for each carboxyl group or CO_2 created.

Bear in mind that generation of ATP also depends upon availability of a mechanism. It is of interest that most synthesis of ATP is linked directly to the chemical processes by which carboxyl groups or CO_2 molecules are created in a fermentation process. The most important single reaction is the oxidation of the aldehyde group of glyceraldehyde 3-P to the carboxyl group of 3-phosphoglycerate (steps *6a- 6c* and 7 in Fig. 7.7) Compare the fermentation of glucose with the complete oxidation of the sugar to carbon dioxide and water (Eq. 7.22), a process which yeast cells (as well as our own cells) carry out in the presence of air.

The overall Gibbs energy change is over 10 times greater than that for fermentation, a fact that permits the cell to form an enormously greater quantity of ATP. The

$$\text{Glucose} + 6\ O_2 \rightarrow 6\ CO_2 + 6\ H_2O$$

$$\Delta G' = -2872\ \text{kJ}\ (-686.5\ \text{kcal})\ \text{mol}^{-1} \qquad (7.22)$$

net gain in ATP synthesis, accompanying Eq. 7.22, is usually about 38 mol of ATP—19 times more than is available from fermentation of glucose. Thus, the explanation of Pasteur's observation that yeast decomposes much less sugar in the presence of air than in its absence is clear. Also, we can understand why a cell, living anaerobically, must metabolize a very large amount of substrate to grow.

Variations of the Alcoholic and Homolactic Fermentations: The course of a fermentation is often affected drastically by changes in conditions. Many variations can be visualized by reference to Fig. 7.9, which shows a number of available metabolic sequences. We have already discussed the conversion of glucose to triose phosphate and via reaction pathway *a* to pyruvate, via reaction c to lactate, and via reaction *d* to ethanol. If bisulfite is added to a fermenting culture of yeast, the acetaldehyde formed through reaction *d* is trapped as the bisulfite adduct blocking the reduction of acetaldehyde to ethanol, an essential part of the fermentation.

Yeast cells accommodate this change by using the accumulating NADH to reduce half of the triose-*P* to glycerol through pathway *b*. Two enzymes are needed, a dehydrogenase and a phosphatase, to hydrolytically cleave off the phosphate. The balanced reaction is given by Eq. 7.23:

$$\text{Glucose} \rightarrow \text{glycerol} + \text{acetaldehyde (trapped)} + CO_2\ \Delta G'\ (\text{pH } 7) = -105\ \text{kJ mol}^{-1} \qquad (7.23)$$

In this reaction only one molecule of CO_2 is produced but the overall Gibbs energy change is still adequate to make the reaction highly spontaneous. However (referring to Fig. 7.9), we see that the net synthesis of ATP is now apparently zero.

The fermentation apparently does not permit cell growth. Nevertheless, it has been used industrially for production of glycerol. Reduction of dihydroxyacetone phosphate to glycerol phosphate also occurs in insect flight muscle and apparently operates as an alternative to lactic acid formation in that tissue. There is no net gain of ATP in the conversion of free glucose to glycerol phosphate and pyruvate, but using stored glycogen in muscle as the starting material, the dismutation of triose-*P* to glycerol-*P* and pyruvate provides one ATP per glucose unit rapidly during the vigorous contraction of the powerful insect flight muscle.

During the slower recovery phase, glycerol-*P* is thought to be reoxidized after entering the mitochondria of these highly aerobic cells. Thus, the transport of glycerol-*P* into mitochondria serves as a means for transporting reducing equivalents derived from reoxidation of NADH into the mitochondria. Indeed the significance of glycerol-*P* to muscle metabolism may be more related to this function than to the rapid formation of ATP. Why does the glycolysis sequence begin with phosphorylation of glucose by ATP? The phospho groups probably provide convenient handles and doubtless assist in substrate recognition. There may be a kinetic advantage but also a danger. Unless there is adequate regulation the "turbo design," in which ATP is used at the outset to drive glycolysis, may lead to accumulation of phosphorylated intermediates and to inadequate concentrations of ATP and inorganic phosphate.

Yeast cells guard against this problem by synthesizing trehalose 6-phosphate, which acts as a feedback inhibitor of hexokinase. Trypanosomes utilize a different type of control. The enzymes that convert glucose into 3-phosphoglycerate are present in membrane-bounded organelles called *glycosomes*. Phosphoglycerate is exported from them into the cytosol where glycolysis is completed. Since inorganic phosphate is essential for ATP formation, if the P_i concentration falls too low the rate of fermentation by yeast juice is greatly decreased, an observation made by Harden and Young in 1906.

The Mixed Acid Fermentation

Enterobacteria, including *E. coli,* convert glucose to ethanol and acetic acid and either formic acid or CO_2 and H_2 derived from it. The stoichiometry is variable but the fermentation can be described in an idealized form as follows:

$$\text{Glucose} + \text{H2O} \rightarrow \text{ethanol} + \text{acetate}^- + H^+ + 2\ H_2 + 2\ CO_2$$

$$\Delta G' \text{ (pH 7)} + - 225 \text{ kJ mol}^{-1} \qquad ...(7.24)$$

The details of the process and the oxidation-reduction balance can be pictured as in Eq. 7.25. Pyruvate is cleaved by the pyruvate formatelyase reaction to acetyl-CoA and formic acid. Half of the acetyl-CoA is cleaved to acetate via acetyl-*P* with generation of ATP, while the other half is reduced in two steps to ethanol using the two molecules of NADH produced in the initial oxidation of triose phosphate (Eq. 7.25). The overall energy yield is three molecules of ATP per glucose.

The "efficiency" is thus (3 × 34.5) ÷ 225 = 46%. Some of the glucose is converted to D-lactic and to succinic acids (pathway *f*, Fig. 7.9); hence the name *mixed acid fermentation*. Table 7.1 gives typical yields of the mixed acid fermentation of *E. coli*. Among the four major products are acetate, ethanol, H_2, and CO_2, as shown in Eq. 7.25. However, at high pH formate accumulated instead of CO_2.

Glucose
2 NADH
2 ATP
2 Pyruvate
Pyruvate formate-lyase
2 HCOOH
$2CO_2$
Formic hydrogen-lyase
$2\ H_2$
2 Acetyl-CoA
ATP
NADH
NADH
Acetate
Acetaldehyde → Ethanol

...(7.25)

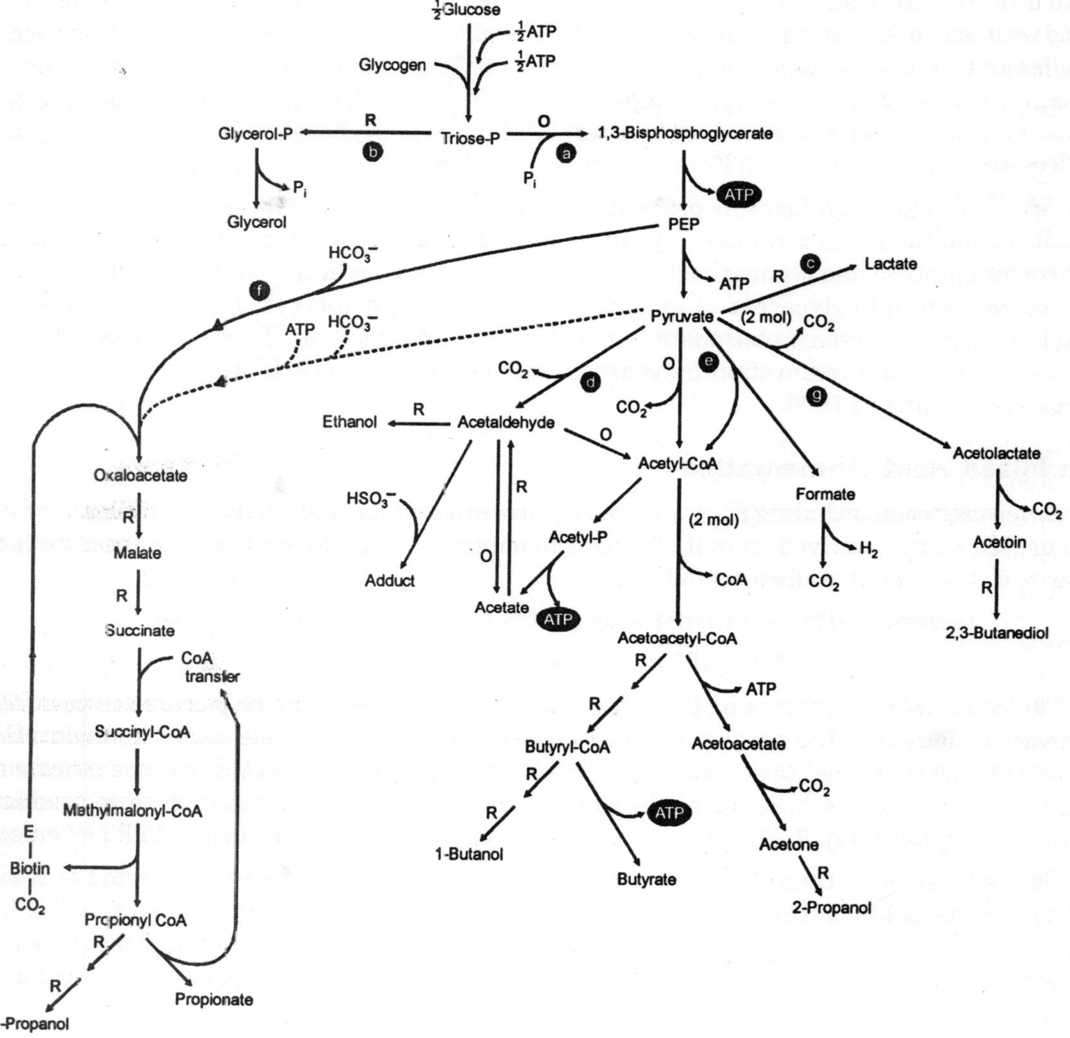

Fig. 7.9. Reaction sequences in fermentation based on the Embden-Meyerhof-Parnas pathway. Oxidation steps (producing NADH + H⁺) are marked "O"; reduction steps (using NADH + H⁺) are marked "R."

Over one-third of the glucose was fermented to lactate in both cases.

In some mixed acid fermentations (*e.g.,* that of *Shigella*) formic acid accumulates, but in other cases (e.g., with E. *coli* at pH 6) it is converted to CO_2 and H_2 (Eq. 7.25). The equilibration of formic acid with CO_2 and hydrogen is catalyzed by the *formic hydrogen-lyase* system which consists of two iron-sulfur enzymes. The selenium-containing *formate dehydrogenase* catalyzes oxidation of formate to CO_2 by NAD^+, while a membrane-bound *hydrogenase* equilibrates NADH + H^+ with NAD^+ + H_2.

Hydrogenase also serves to release H_2 from excess NADH. Krebs pointed out that an excess of NADH may arise because growth of cells requires biosynthesis of many components such as amino acids. When glucose is the sole source of carbon, biosynthetic reactions involve an excess of oxidation steps over reduction steps. The excess of reducing equivalents may be released as H_2 or may be used to form highly reduced products such as succinate.

Table 7.1. Products of the Mixed Acid Fermentation by *E. coli* at Low and High Values of pH

Product (Millimole formed from 100 mmol of glucose)	*pH 6.2*	*pH 7.8*
Acetate	36	39
Ethanol	50	51
H2	70	0.3
CO_2	88	1.7
Formate	2.4	86
Lactate	79	70
Succinate	11	15
Glycerol	1.4	0.3
Acetion	0.1	0.2
Butanediol	0.3	0.2

Among such genera as *Aerobacter* and *Serratia* part *of the* pyruvate formed is condensed with decarboxylation to form *S acetolactate,* which is decarboxylated to acetoin; pathway *g* of Fig. 7.9). The acetoin is reduced with NADH to 2,3-*butanediol,* while a third molecule of pyruvate is converted to ethanol, hydrogen, and CO_2 (Eq. 7.26). The reaction provides the basis for industrial production of butanediol, which can be dehydrated nonenzymatically to butadiene.

...(7.26)

Mixed acid fermentations are not limited to bacteria. For example, trichomonads, parasitic flagellated protozoa, have no mitochondria. They export pyruvate into the bloodstreams of their hosts and also contain particles called *hydrogenosomes* which can convert pyruvate to acetate, succinate, CO_2, and H_2. Hydrogenosomes are bounded by double membranes and have a common evolutionary relationship with both mitochondria and bacteria.

The enzyme that catalyzes pyruvate cleavage in hydrogenosomes apparently does not contain lipoate and may be related to the pyruvate-ferredoxin oxidoreductase of clostridia. The hydrogenosomes also contain an active hydrogenase. Many invertebrate animals are true facultative anaerobes, able to survive for long periods, sometimes indefinitely, without oxygen. Among these are *Ascaris*, oysters, and other molluscs.

Succinate and alanine are among the main end products of anaerobic metabolism. The former may arise by a mixed acid fermentation that also produces pyruvate. The pyruvate is converted to acetate to balance the fermentation in *Ascaris lumbricoides,* which is in effect an obligate anaerobe. However, in molluscs the pyruvate may undergo transamination with glutamate to form alanine and 2-oxoglutarate; the oxoglutarate may be oxidatively decarboxylated to succinate. The reactions depend upon the availability of a store of glutamate or of other amino acids, such as arginine, that can give rise to glutamate.

The Propionic Acid Fermentation

Propionic (propanoic) acid-producing bacteria are numerous in the digestive tract of ruminants. Within the rumen some bacteria digest cellulose to form glucose, which is then converted to lactate and other products. The propionic acid bacteria can convert either glucose or lactate into propionic and acetic acids which are absorbed into the bloodstream of the host. Usually some succinic acid is also formed. The basis of the propionic acid fermentation is conversion of pyruvate to oxaloacetate by carboxyla-tion and the further conversion through succinate and succinyl-CoA to methylmalonyl-CoA and propionyl-CoA, reactions which are almost the exact reverse of those for the oxidation of propionate in the animal body (Fig. 7.3, pathway *d*).

However, whereas the carboxylation of pyruvate to oxaloacetate in the animal body requires ATP, the propionic acid bacteria save one equivalent of ATP by using a carboxyltransferase. This enzyme donates a carboxyl group from a preformed carboxybiotin compound generated in the decarboxylation of methylmalonyl-CoA in the next to final step of the reaction sequence (Fig. 7.10). A second molecule of ATP is saved by linking directly the conversion of succinate to succinyl-CoA to the cleavage of propionyl-CoA to propionate through the use of a CoA transferase. To provide for oxidation-reduction balance, two-thirds of the glucose goes to propionate and one-third to acetate:

$$1^1/2 \text{ Glucose} \rightarrow 2 \text{ propionate}^- + \text{acetate}^- + 3\text{ H}^+ + CO_2 + H_2O$$

$$\Delta G' \text{ (pH 7)} = -465 \text{ kJ per 11/2 mol of glucose} \qquad ..(7.27)$$

More carboxyl groups and CO_2 molecules are formed in this fermentation ($2\ ^2/_3$ per glucose molecule) than in the regular lactic acid fermentation. The yield of ATP (also $2\ ^2/_3$ mol/mol of glucose fermented) is correspondingly greater and $\Delta G'$ is more negative. Using the same mechanism (Fig. 7.10), propionic acid bacteria are also able to ferment lactate, the product of fermentation by other bacteria, to propionate and acetate (Eq. 7.28). The net gain is one molecule of ATP. This reaction probably accounts for the niche in the ecology of the animal rumen that is occupied by propionic acid bacteria.

$$3\ \text{Lactate}^- \rightarrow 2\ \text{propionate}^- + \text{acetate}^- + H_2O + CO_2$$
$$\Delta G'\ (\text{pH } 7) = -171\ \text{kJ per 3 mol of lactate} \tag{7.28}$$

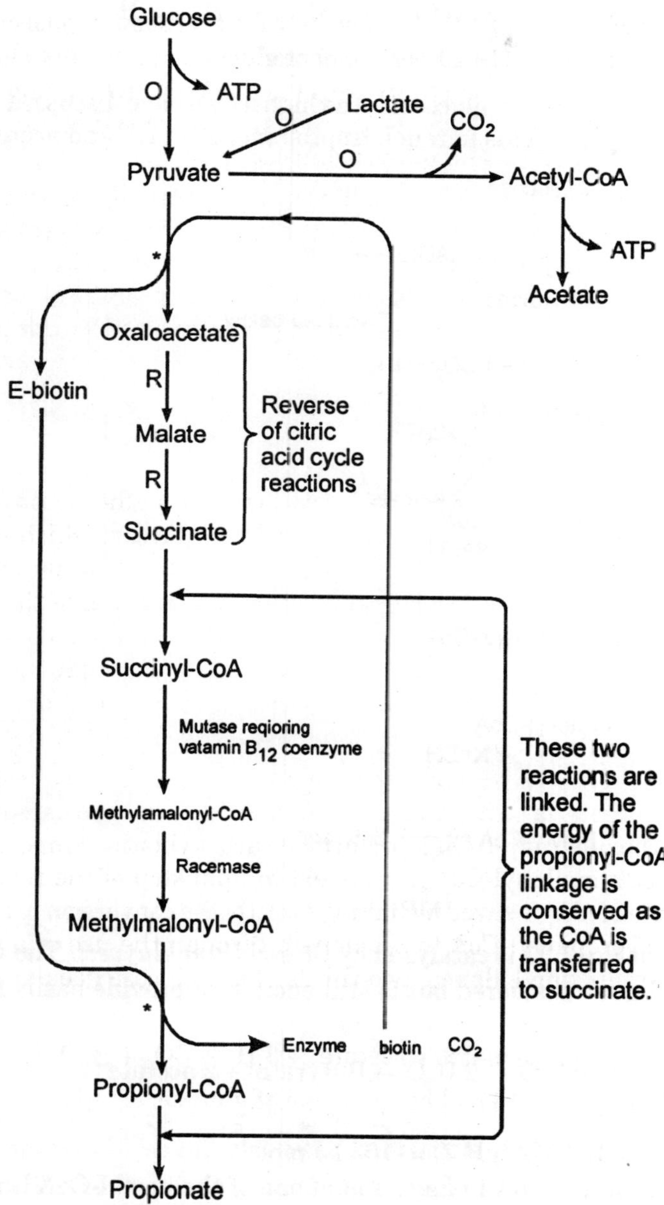

Fig. 7.10. Propionic acid fermentation of *Propionobacteria* and *Veillonella*. Oxidation steps are designated by the symbol "O" and reduction steps by "R." The two coupled reactions marked by asterisks are catalyzed by carboxyl-transferase.

Butyric Acid and Butanol-Forming Fermentations

A variety of fermentations are carried out by bacteria of the genus *Clostridium* and by the rumen organisms *Eubaderium (Butyribacterium)* and *Butyrivibrio.* For example, glucose may be converted to butyric and acetic acids together with. CO_2 and H_2 (Eqs. 7.29 and 7.30).

$$2 \text{ Glucose} + 2\ H_2O \text{ butyrate}^- + 2 \text{ acetate}^- + 4\ CO_2 + 6\ H_2 + 3\ H^+$$

$$\Delta G' \text{ (pH 7)} = -479 \text{ kJ per 2 mol of glucose} \qquad (7.29)$$

The yield of ATP (31/2 mol/mol of glucose) is the highest we have discussed giving an efficiency of 50%. Another fermentation yields butanol, isopropanol ethanol, and acetone.

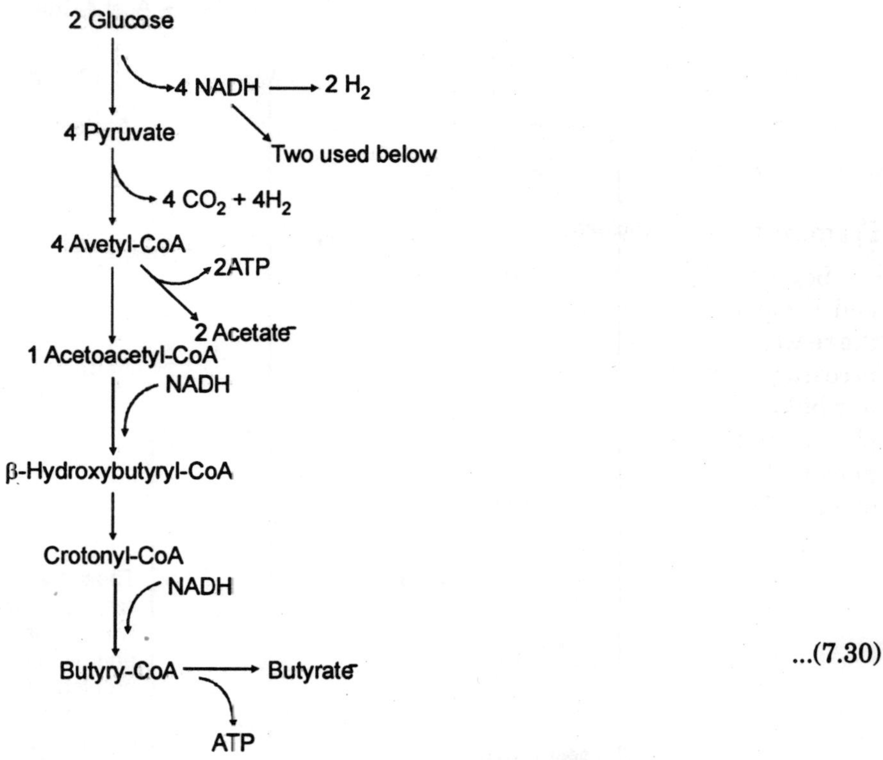

...(7.30)

The fermentation of Eq. 7.31 is catalyzed by *Clostridium kluyveri.* The value of $-\Delta G'$ is one of the lowest that we have considered but is still enough to provide easily for the synthesis of one molecule of ATP.

$$\underset{\text{Crotonate}}{2\ CH_3CH = CH\text{-}COO^-} + 2\ H_2O \rightarrow \text{Butyrate}^- + 2 \text{ acetate}^-$$

$$+ H^+ \Delta G' \text{ (pH 7)} = -105 \text{ kJ mol}^{-1} \qquad ...(7.31)$$

The energy of the butyryl-CoA linkage and of one of the acetyl-CoA linkages is conserved and utilized in the initial formation of crotonyl-CoA (Eq. 7.32). That leaves one acetyl-CoA which can be converted via acetyl-P to acetate with formation of ATP.

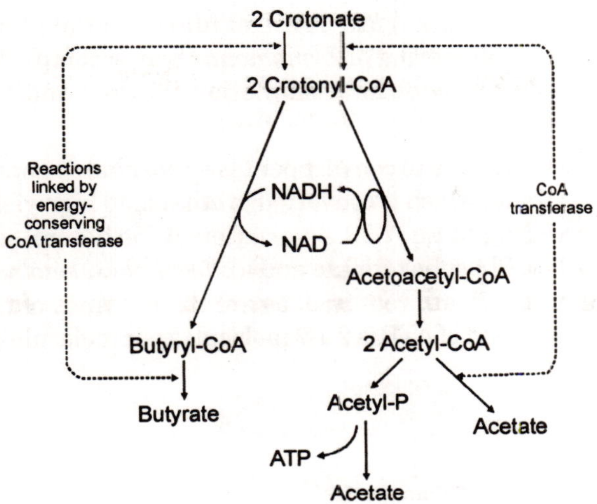

...(7.32)

Fermentations Based on the Phosphogluconate and Pentose Phosphate Pathways

Some lactic acid bacteria of the genus *Lactobacillus,* as well as *Leuconostoc niesenteroides* and *Zymomonas mobilis,* carry out the *heterolactic* fermentation (Eq. 7.33) which is based on the reactions of the pentose phosphate pathway. These organisms lack aldolase, the key enzyme necessary for cleavage of fructose 1,6-bisphosphate to the triose phosphates. Glucose is converted to ribulose 5-P using the oxidative reactions of the pentose phosphate pathway. The ribulose-phosphate is cleaved by phosphoketolase (Eq. 7.23) to acetyl-phosphate and glyceraldehyde 3-phosphate, which are converted to ethanol and lactate, respectively. The overall yield is only one ATP per glucose fermented.

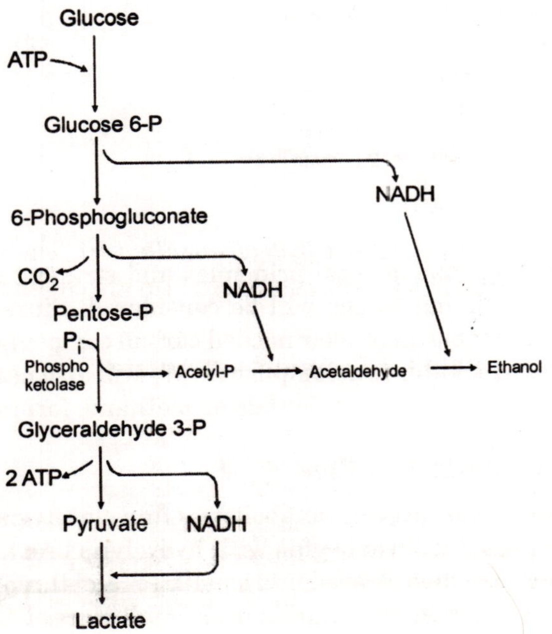

...(7.33)

This is generated in the substrate level oxidative phosphorylation catalyzed by phosphoketolase. Metabolic-engineering of *Zymomonas* was accomplished by transferring from other bacteria two operons that provide for assimilation of xylose and a complete set of enzymes for the pentose phosphate pathway.

The engineered bacteria are able to convert pentose phosphates nonoxi-datively (see Fig. 7.8) into glyceraldehyde 3-phosphate, which is converted to ethanol in high yield and with much greater synthesis of ATP than according to Eq. 7.33. A variation of the heterolactic fermentation is used by *Bifidobacterium* (Eq. 7.34). Phosphoketolase and a *phosphohexoketolase*, which cleaves fructose 6-*P* to erythrose 4-*P* and acetyl-*P*, are required, as are the enzymes of the sugar rearrangement system (Section E,3). The net yield of ATP is 2 1/2 molecules per molecule of glucose.

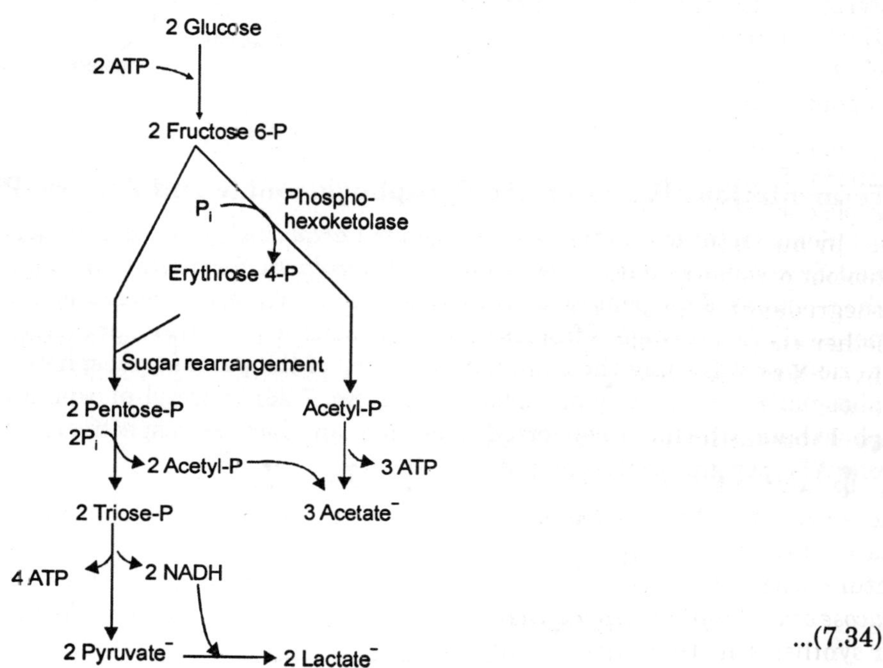

...(7.34)

Biosynthesis

In this section and sections H–K the general principles and strategy of synthesis of the many carbon compounds found in living things will be considered. Since green plants and autotrophic bacteria are able to assemble all of their needed carbon compounds from CO_2, let us first examine the mechanisms by which this is accomplished. We will also need to ask how some organisms are able to subsist on such simple compounds as methane, formate, or acetate.

Metabolic Loops and Biosynthetic Families

As was pointed out in, routes of biosynthesis (anabolism) often closely parallel pathways of biodegradation (catabolism). Thus, catabolism begins with hydrolytic breakdown of polymeric molecules; the resulting monomers are then cleaved into small two- and three-carbon fragments. Biosynthesis begins with formation of monomeric units from small pieces followed by assembly of the monomers into polymers. The mechanisms of the individual reactions of biosynthesis and biodegradation are also often closely parallel. However, in most instances, there are clear-cut differences. A first principle of biosynthesis is that *biosynthetic pathways, although related to catabolic pathways, differ from them in distinct ways and are often catalyzed by completely different sets of enzymes.* The sum of the pathways of biosynthesis and biodegradation form a continuous loop - a series of reactions that take place concurrently and often within the same part of a cell.

Metabolic loops often begin in the central pathways of carbohydrate metabolism with three- or four-carbon compounds such as phospho-glycerate, pyruvate, and oxaloacetate. After loss of some atoms as CO_2 the remainder of the compound rejoins the "mainstream" of metabolism by entering a catabolic pathway leading to acetyl-CoA and oxidation in the citric acid cycle. Not all of the loops are closed within a given species. For example, *human beings are unable to synthesize the vitamins and the "essential amino acids."* We depend upon other organisms to make these compounds, but we do degrade them.

Some metabolites, such as uric acid, are excreted by humans and are further catabolized by bacteria. From a chemical viewpoint the whole of nature can be regarded as an enormously complex set of branching and interconnecting metabolic cycles. Thus, the synthetic pathways used by autotrophs are all parts of metabolic loops terminating in oxidation back to CO_2. It is often not possible to state at what point in a metabolic loop biosynthesis has been completed and biodegradation begins. An end product X that serves one need of a cell may be a precursor to another cell component Y which is then degraded to complete the loop. The reactions that convert X to Y can be regarded as either biosynthetic (for Y) or catabolic (for X).

Key Intermediates and Biosynthetic Families

In examining routes of biosynthesis it is helpful to identify some key intermediates. One of these is *3-phosphoglycerate*. This compound is a primary product of photosynthesis and may reasonably be regarded as the starting material from which all other carbon compounds in nature are formed. Phosphoglycerate, in most organisms, is readily interconvertible with both *glucose and phosphoenolpyruvate (PEP).* Any of these three compounds can serve as the precursor for synthesis of other organic materials.

A first stage in biosynthesis consists of those reactions by which 3-phosphoglycerate or PEP arise, whether it be from CO_2, formate, acetate, lipids, or polysaccharides. The further biosynthetic pathways from 3-phosphoglycerate to the myriad amino acids, nucleotides, lipids, and miscellaneous compounds found in cells are complex and numerous. However, the basic features are relatively simple. Figure 7.11 indicates the origins of many substances including the 20 amino acids present in proteins, nucleotides, and lipids.

Among the additional key biosynthetic precursors that can be identified from this chart are *glucose 6-phosphate, pyruvate, oxaloacetate, acetyl-CoA, 2-oxoglutarate, and succinyl-CoA.* The amino acid *serine* originates almost directly from 3-phosphoglycerate. *Aspartate* arises from oxaloacetate and *glutamate* from 2-oxoglutarate. These three amino acids each are converted to "families" of other compounds. A little attention paid to establishing correct family relationships will make the study of biochemistry easier. Besides the serine, aspartate, and glutamate-oxoglutarate families, a fourth large family originates directly from pyruvate and a fifth (mostly lipids) from acetyl-CoA. The aromatic amino acids are formed from erythrose 4-*P* and PEP via the key intermediate *chorismic acid* (Fig. 7.1). Other families of compounds arise from glucose 6-*P* and from the *pentose phosphates*. These groups have been set off roughly by the boxes outlined in green in Fig. 7.11.

Harnessing the Energy of ATP for Biosynthesis

In the past it seemed reasonable to think that some biosynthetic pathways involved exact reversal of catabolic pathways. For example, it was observed that glycogen phosphorylase catalyzed elongation of glycogen branches by transfer of glycosyl groups from glucose 1-phosphate. Likewise, the enzymes needed for the β oxidation of fatty acid derivatives, when isolated from mitochondria, catalyze formation of fatty acyl-CoA derivatives from acetyl-CoA and a reducing agent such as NADH. However, reactant concentrations within cells are rarely appropriate for reversal of a catabolic sequence.

For catabolic sequences the Gibbs energy change is usually distinctly negative and reversal requires high concentrations of end products. However, the latter are often removed promptly from the cells. For example, NADH produced in degradation of fatty acids is oxidized to NAD^+ and is therefore never available in sufficient concentrations to reverse the p oxidation sequence. *Nature's answer to the problem of reversing a catabolic pathway lies in the coupling of cleavage of ATP to the biosynthetic reaction.* The concept was introduced in which one sequence for linking hydrolysis of ATP to biosynthesis was discussed. However, living cells employ several different methods of harnessing the Gibbs energy of hydrolysis of ATP to drive biosynthetic processes.

Many otherwise strange aspects of metabolism become clear if it is recognized that they provide a means for coupling ATP cleavage to biosynthesis. A few of the most important of these coupling mechanisms are summarized in this section.

Group Activation

Consider the formation of an ester (or of an amide) from a free carboxylic acid and an alcohol (or amine) by elimination of a molecule of water (Eq. 7.35). The reaction is thermodynamically unfavourable with values of ΔG′ (pH 7) ranging from ~ +10 to 30 kJ mol^{-1} depending on conditions and structures of the specific compounds. Long ago, organic chemists learned that such reactions can be made to proceed by careful removal of the water that is generated (Eq. 7.35).

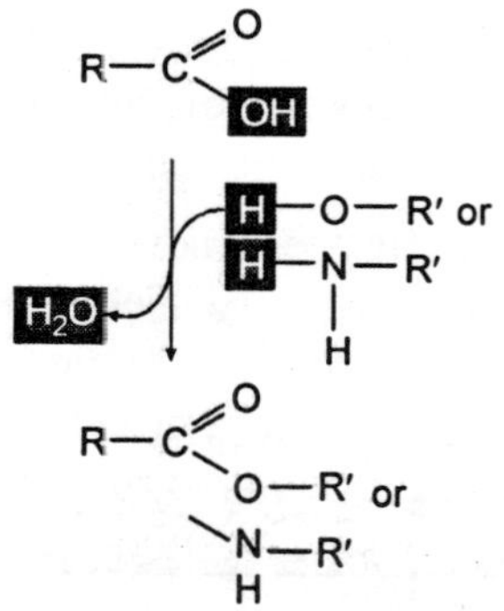

R—C(=O)—OH
H—O—R′ or
H—N(H)—R′
H_2O
R—C(=O)—O—R′ or
R—C(=O)—N(H)—R′

...(7.35)

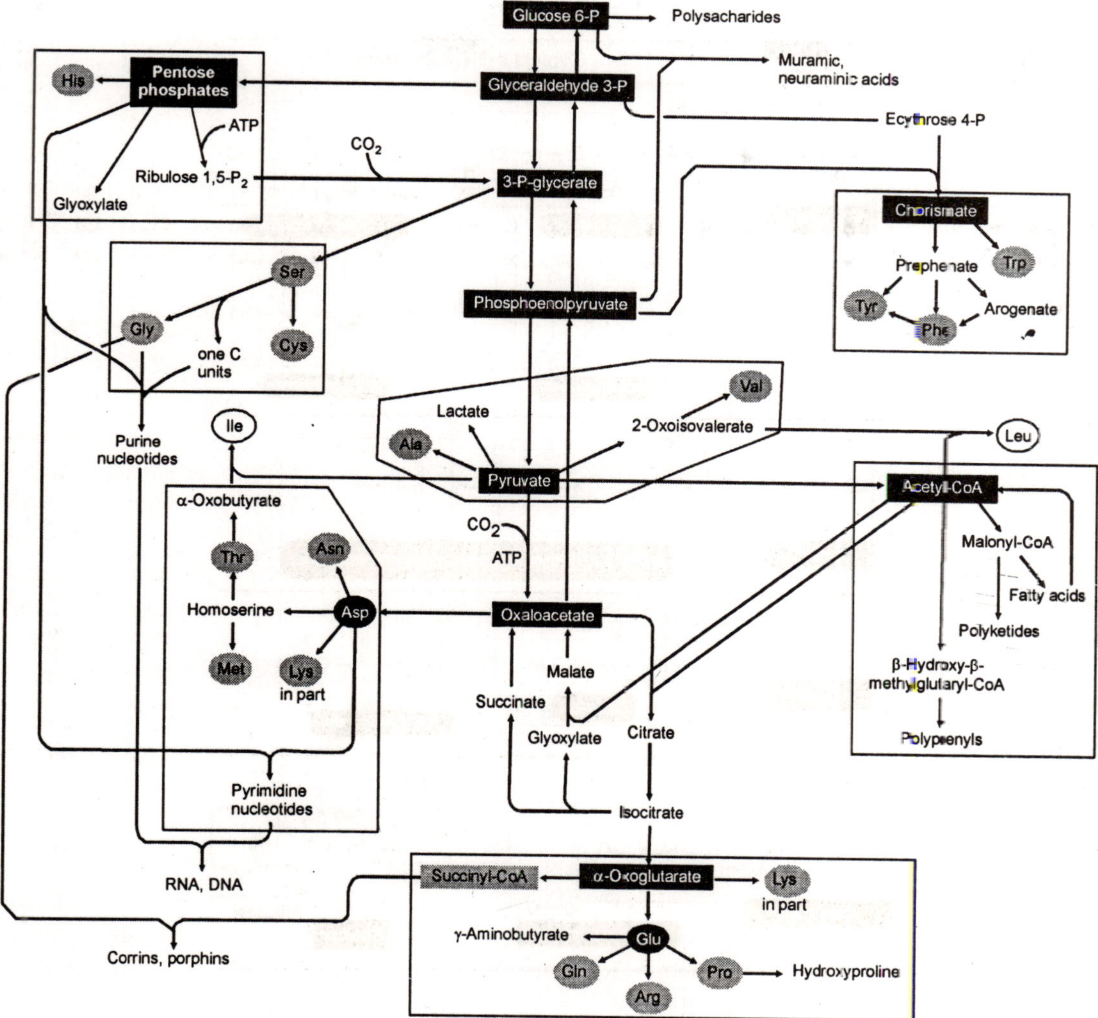

Fig. 7.11. Some major biosynthetic pathways. Some key intermediates are enclosed in boxes and the 20 common amino acids of proteins are encircled. Key intermediates for each family are in shaded boxes or elipses. Green lines trace the reactions of the glyoxylate pathway and of glucogenesis.

However, it is often better to "activate" the carboxylic acid by conversion to an acyl chloride or an anhydride:

$$R-C(=O)-Cl \qquad\qquad R-C(=O)-O-C(=O)-R$$

Acyl chloride Acid anhydride

Nucleophilic attack on the carbonyl group of such a compound results in displacement of a good leaving group, Cl^- or $R-COO^-$. Nature has followed the same approach in forming from carboxylic acids *acyl phosphates or acyl-CoA derivatives.*

Table 7.2. "Activated" Groups Used in Biosynthesis.

Group	Typical activated forms
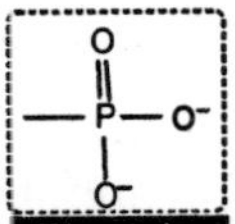 Phospho	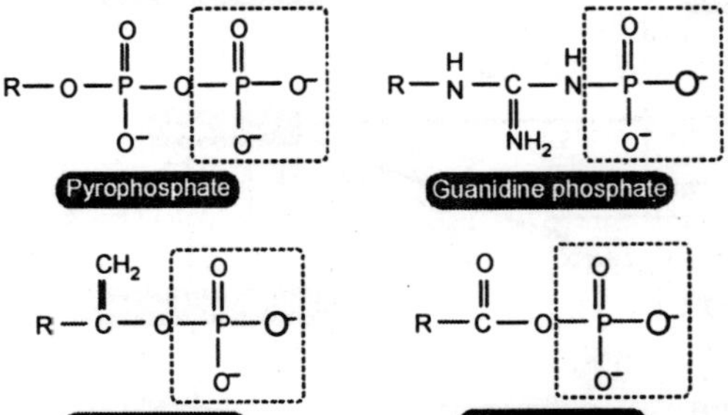 Pyrophosphate; Guanidine phosphate; Enol phosphate; Acyl phosphate
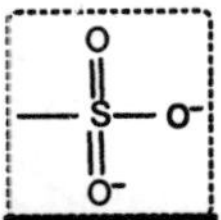 Sulfo	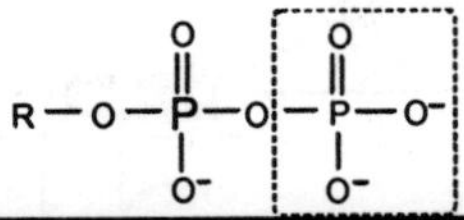 PAPS (3'-Phosphoadenosine-5'-phosphosulfate)
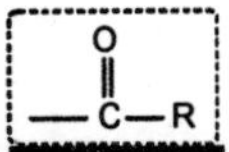 Acyl	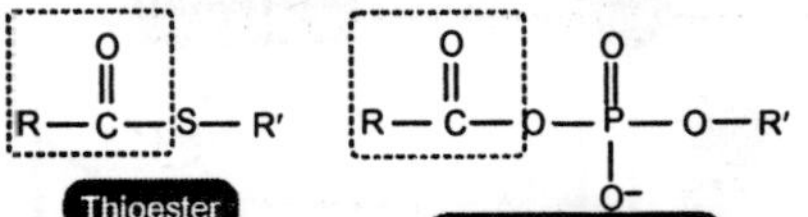 Thioester; Acyl phosphate
 Glycosyl	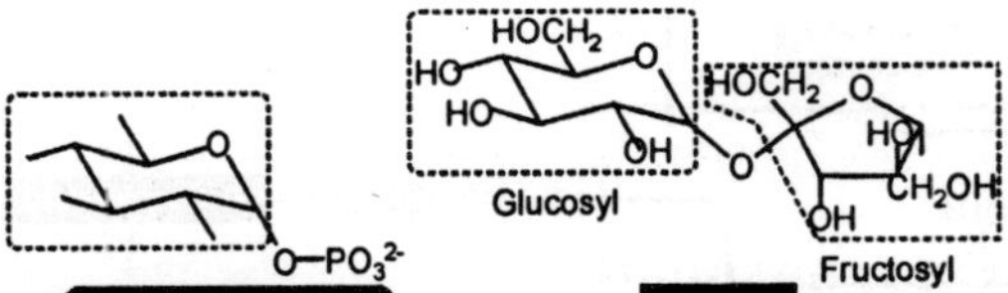 Glycosyl phosphate; Sucrose (Glucosyl, Fructosyl)
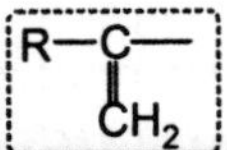 Enoyl	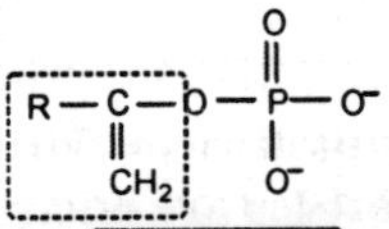 Enol phosphate
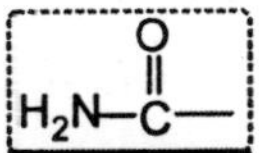 Carbamoyl	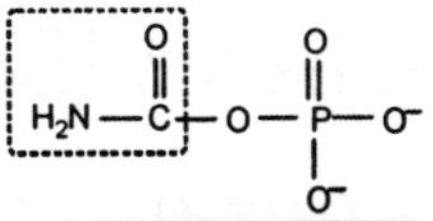 Carbamoyl phosphate
 Alkyl	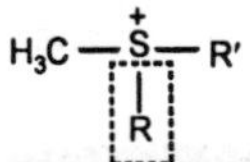 S-Adenosylmethionine

The virtue of these "activated" acyl compounds in biosynthetic reactions was considered in Table 7.1. Just as a carboxylic acid can be converted to an active acyl derivative, so other groups can be activated. ATP and other compounds with phospho groups of high group transfer potential are *active phospho* compounds. Sulfate is converted to a phospho-sulfate anhydride, an *active sulfo* derivative. Sugars are converted to compounds such as glucose 1-*P* or sucrose, which contain *active glycosyl* groups.

The group transfer potentials of the latter, though not as great as that of the phospho groups of ATP, are still high enough to make glucose 1-phosphate and sucrose effective glycosylating reagents. Table 7.2 lists several of the more important activated groups. Group activation usually takes place at the expense of ATP cleavage. As pointed out earlier, acyl phosphates play a central role in metabolism by virtue of the fact that they contain both an activated acyl group and an activated phospho group.

The high group transfer potential can be conserved in subsequent reactions *in either one group or the other* (but not in both). Thus, displacement on P by an oxygen of ADP will regenerate ATP and attack on C by an –SH will give a thioester. Several of the other compounds in Table 7.2 can also be split in two ways to yield different activated groups, *e.g.*, the phosphosulfate anhydride, enoyl phosphate, and carbamoyl phosphate.

$$\underbrace{R-\overset{O}{\overset{\|}{C}}-O}_{\text{Activated acyl group}}-\underbrace{\overset{O}{\overset{\|}{\underset{O^-}{\underset{|}{P}}}}-O^-}_{\text{Activated phospho group}} \quad \text{Acyl phosphate}$$

It is probably only through intermediates of this type that cleavage of ATP can be coupled to synthesis of activated groups. Such *common intermediates* are essential to the synthesis of ATP by substrate-level phosphorylation.

Hydrolysis of Pyrophosphate

The splitting of inorganic pyrophosphate (PP_i) into two inorganic phosphate ions is catalyzed by *pyrophosphatases* that apparently occur universally. Their function appears to be simply to remove the product PP_i from reactions that produce it, shifting the equilibrium toward formation of a desired compound. An example is the formation of *aminoacyl-tRNA* molecules needed for protein synthesis. As shown in Eq. 7.36, the process requires the use of two ATP molecules to activate one amino acid. While the "spending" of two ATPs for the addition of one monomer unit to a polymer does not appear necessary from a thermodynamic viewpoint, it is frequently observed, and there is no doubt that hydrolysis of PP_i ensures that the reaction will go virtually to completion.

Transfer RNAs tend to become saturated with amino acids according to Eq. 7.36 even if the concentration of free amino acid in the cytoplasm is low. On the other hand kinetic considerations may be involved. Perhaps the biosynthetic sequence would move too slowly if it were not for the extra boost given by the removal of PP_i. Part of the explanation for the complexity may depend on control mechanisms which are only incompletely understood. In some metabolic reactions pyrophosphate esters are formed by consecutive transfer of the terminal phospho groups of two ATP molecules onto a hydroxyl group. Such esters often react with elimination of PP_i, *e.g.*, in polymerization of prenyl units (reaction type 6B, Table 7.1; Fig. 7.1). Again, hydrolysis to P_i follows. Thus, *cleavage of pyrophosphate is a second very general method for coupling* ATP *cleavage to synthetic reactions.*

$$\text{Net: } 2\,\text{ATP} + \text{amino acid} + \text{tRNA} \longrightarrow \text{aminoacyl tRNA} + 2\,\text{ADP} + 2\text{P}_i \qquad ...(7.36)$$

Although pyrophosphatases are ubiquitous, there are organisms in which PP_i is conserved by the cell and replaces ATP in several glycolytic reactions. These include *Propionibacterium*, sulfate-reducing bacteria, the photosynthetic *Rhodospirillum,* and the parasitic *Entamoeba histolytica.* In the latter the internal concentration of PP_i is about 0.2 mM. Green plants also accumulate PP_i at concentrations of up to 0.2 mM. Apparently, pyrophosphate is not always hydrolyzed immediately.

Another mystery of metabolism is the accumulation of inorganic *polyphosphate* in chains of tens to many hundreds of phospho groups linked, as in pyrophosphate, by phosphoanhydride bonds. These polyphosphates are present in many bacteria, including *E. coli,* and also in fungi, plants, and animals. They constitute a store of energy as well as of phosphate. Various other functions have also been proposed. A polyphosphate kinase transfers a terminal phospho group from polyphosphate chains onto ADP to form ATP. This source of metabolic energy is evidently essential to the ability of *Pseudomonas aeruginosa* to form biofilms. Both endophosphatases and exophosphatases, of uncertain function, can degrade the chains hydrolytically.

An exophosphatase from *E. coii* can completely hydrolyze polyphosphate chains of 1000 units processively without release of intermediates.

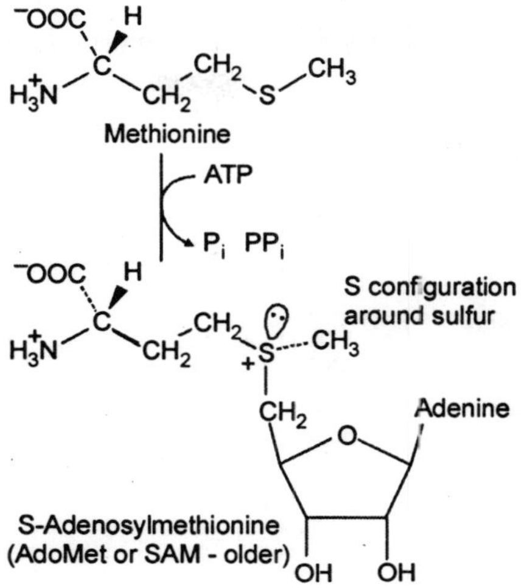

Methionine

ATP → P_i PP_i

S configuration around sulfur

Adenine

S-Adenosylmethionine (AdoMet or SAM - older)

...(7.37)

In a few instances group activation is coupled to cleavage of ATP at C-5′ presumably with formation of bound tripolyphosphate (PPP_i). The latter is hydrolyzed to P_i and PP_i and ultimately to *three* molecules of P_i. An example is the formation of S-adenosyl-methionine shown in Eq. 7.37. The reaction is a displacement on the 5′-methylene group of ATP by the sulfur atom of methionine. While the initial product may be enzyme-bound PPP_i, it is P_i and PP_i that are released from the enzyme, the P_i arising from the terminal phosphorus ($P\gamma$) of ATP. The *S*-adenosyl-methionine formed has the *S* configuration around the sulfur.

Coupling by Phosphorylation and Subsequent Cleavage by a Phosphatase

A third general method for coupling the hydrolysis of ATP to drive a synthetic sequence is to transfer the terminal phospho group from ATP to a hydroxyl group *somewhere* on a substrate. Then, after the substrate has undergone a synthetic reaction, the phosphate is removed by action of a phosphatase. For example, in the activation of sulfate (Eq. 7.38), the overall standard Gibbs energy change for steps *a* (catalyzed by *ATP sulfurylase*) and *b* is distinctly positive (+ 12 kJ mol^{-1}).

The equilibrium concentration of adenylyl sulfate formed in this group activation process is extremely low. Nature's solution to this problem is to spend another molecule of ATP to phosphorylate the 3′ –OH of adenosine phospho-sulfate. As the latter is formed, it is converted to 3′-phosphoadenosine-5′-phosphosulfate (Eq. 7.38, step *c*) by a kinase, which is often part of a bifunctional enzyme that also contains the active site of ATP sulfurylase. Since the equilibrium in this step lies far toward the right, the product accumulates in a substantial concentration (up to 1 mM in cell-free systems) and serves as the active sulfo group donor in formation of sulfate esters.

The reaction cycle is completed by two more reactions. In Eq, 7.38, step *d,* the sulfo group is transferred to an acceptor, and in step *e* the extra phosphate group is removed from adenosine 3′,5′-bisphosphate by a specific phosphatase. Since the reconversion of AMP to ADP requires expenditure of still a third high-energy linkage of ATP, the overall process makes use of three high-energy phosphate linkages for formation of one sulfate ester. An analogous use of ATP is found in photosynthetic reduction of carbon dioxide in which ATP phosphorylates ribulose 5-*P* to ribulose bisphosphate and the phosphate groups are removed later by phosphatase action on fructose bisphosphate and sedoheptulose bisphosphate. Phosphatases involved in synthetic pathways usually have a high substrate specificity and are to be distinguished from nonspecific phosphatases which are essentially digestive enzymes.

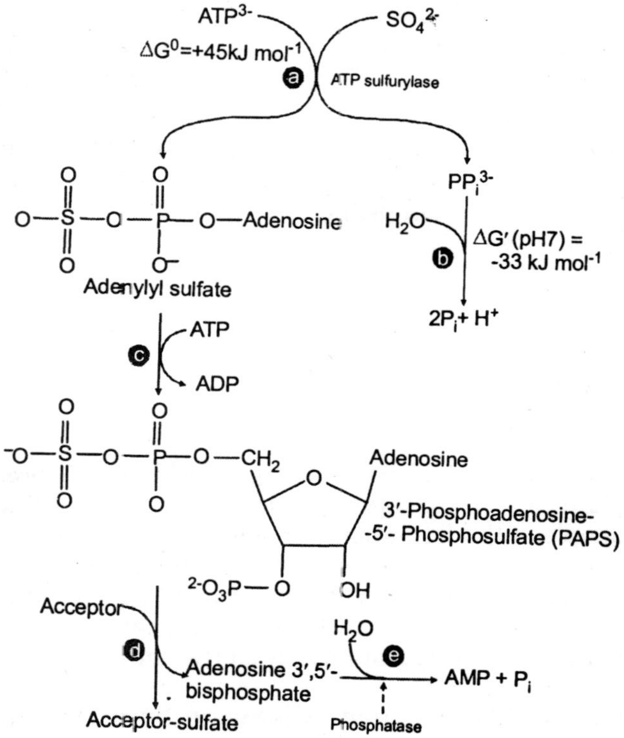

ATP^{3-} + SO_4^{2-}
ΔG^0=+45kJ mol^{-1}
(a) ATP sulfurylase

$^-O-S(=O)_2-O-P(=O)(O^-)-O-$Adenosine
Adenylyl sulfate

PP_i^{3-}
H_2O
(b) $\Delta G'$ (pH7) = -33 kJ mol^{-1}
$2P_i + H^+$

(c) ATP → ADP

$^-O-S(=O)_2-O-P(=O)(O^-)-O-CH_2$ Adenosine
$^{2-}O_3P-O$ OH
3′-Phosphoadenosine-5′- Phosphosulfate (PAPS)

Acceptor
(d) Adenosine 3′,5′-bisphosphate
Acceptor-sulfate
H_2O
(e) → AMP + P_i
Phosphatase

...(7.38)

Carboxylation and Decarboxylation: Synthesis of Fatty Acids

A fourth way in which cleavage of ATP can be coupled to biosynthesis was recognized in about 1953 when Wakil and coworkers discovered that synthesis of fatty acids in animal cytoplasm is stimulated by carbon dioxide. However, when $^{14}CO_2$ was used in the experiment no radioactivity appeared in the fatty acid formed. Rather, it was found that acetyl-CoA was carboxylated to *malonyl-CoA* in an ATP- and biotin-requiring process. The carboxyl group formed in this reaction is later converted back to CO_2 in a decarboxylation (Fig. 7.12).

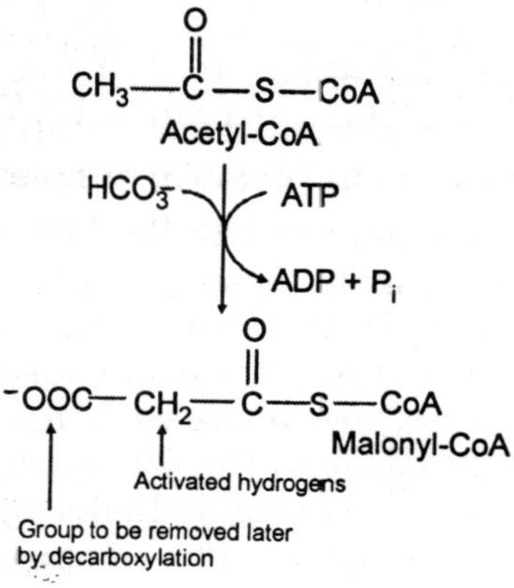

$CH_3-C(=O)-S-CoA$
Acetyl-CoA

HCO_3^- + ATP → ADP + P_i

$^-OOC-CH_2-C(=O)-S-CoA$
Malonyl-CoA

Activated hydrogens

Group to be removed later by decarboxylation

...(7.39)

We know now that in most bacteria and green plants both an acetyl group of acetyl-CoA and a malonyl group of malonyl-CoA are transferred (steps *a* and *d* of Fig. 7.12) to the sulfur atoms of the phosphopantetheine groups of a low-molecular-weight *acyl carrier protein*. The malonyl group of the malonyl-ACP is then condensed (step *f* of Fig. 7.12) with an acetyl group, which has been transferred from acetyl-ACP onto a thiol group of the enzyme (E in Eq. 7.40). The enolate anion indicated in this equation is generated by decarboxylation of the malonyl-ACP. It is this decarboxylation that drives the reaction to completion and, in effect, links C–C bond formation to the cleavage of the ATP required for the carboxylation step. A related sequence involving multifunctional proteins is used by animals and fungi.

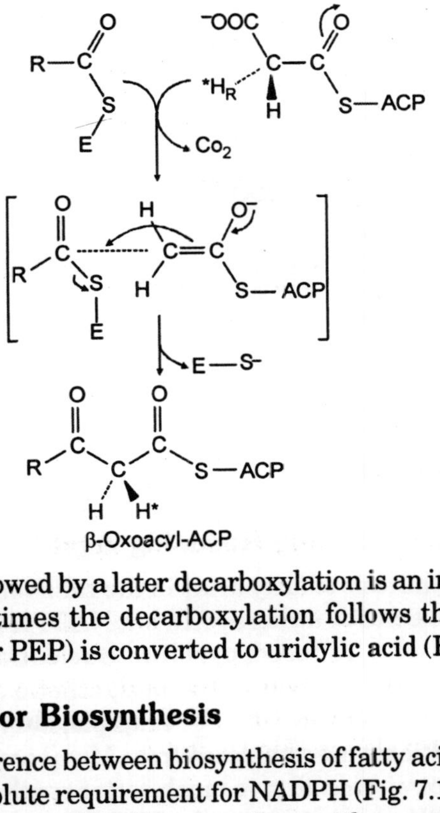

β-Oxoacyl-ACP ...(7.40)

Carboxylation followed by a later decarboxylation is an important pattern in other biosynthetic pathways, too. Sometimes the decarboxylation follows the carboxylation by many steps. For example, pyruvate (or PEP) is converted to uridylic acid (Eq. 7.41).

Reducing Agents for Biosynthesis

Still another difference between biosynthesis of fatty acids and oxidation (in mammals) is that the former has an absolute requirement for NADPH (Fig. 7.12) while the latter requires NAD^+ and flavor-proteins (Fig. 7.1). This fact, together with many other observations, has led to the generalization that *biosynthetic reduction reactions usually require NADPH rather than NADH.*

Many measurements have shown that in the cytosol of eukaryotic cells the ratio [NADPH]/[$NADP^+$] is high, whereas the ratio [NADH]/[NAD^+ is low. Thus, the NAD^+/NADH system is kept highly oxidized, in line with the role of NAD^+ as a principal biochemical oxidant, while the $NADP^+$/NADPH system is kept reduced. The use of NADPH in step *g* of Fig. 7.12 ensures that significant amounts of the β-oxoacyl-ACP derivative are reduced to the alcohol.

Another difference between β oxidation and biosynthesis is that the alcohol formed in this reduction step in the biosynthetic process has the D configuration while the corresponding alcohol in β oxidation has the L configuration.

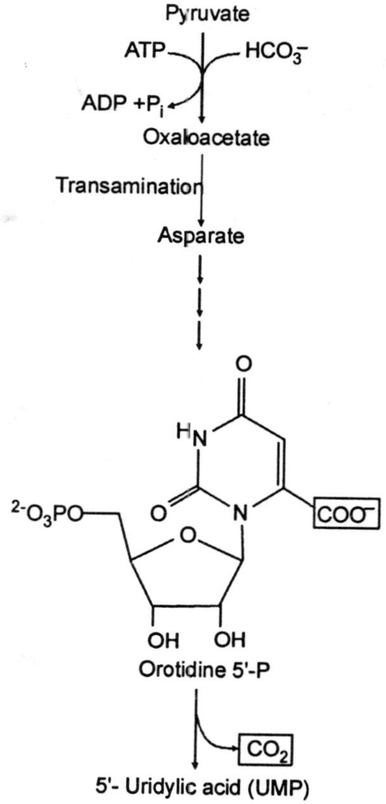

Pyruvate

ATP, HCO_3^-

ADP + P_i

Oxaloacetate

Transamination

Asparate

O

HN

$^{2-}O_3PO$ O N COO⁻

O

OH OH

Orotidine 5'-P

CO_2

5'- Uridylic acid (UMP) ...(7.41)

Reversing an Oxidative Step with a Strong Reducing Agent

The second reduction step in biosynthesis of fatty acids in the rat liver (step *i*) also required NADPH. The corresponding step in β oxidation utilizes FAD, but NADPH is a stronger reducing agent than $FADH_2$. Therefore, use of a reduced pyridine nucleotide again provides a thermodynamic advantage in pushing the reaction in the biosynthetic direction. Interesting variations have been observed among different species. For example, fatty acid synthesis in the rat requires only NADPH, but the multienzyme complexes from *Mycobacterium phlei, Euglena gradlis,* and the yeast *Sacchawmyces cerevisiae* all give much better synthesis with a mixture of NADPH and NADH than with NADPH alone. Apparently, NADPH is required in step *g* and NADH in step *i*. This seems reasonable because the equilibrium in step *i* lies far toward the product formation, and NADH at a very low concentration could carry out the reduction.

Regulation of the State of Reduction of the NAD and NADP Systems

The ratio $[NAD^+]/[NADH]$ appears to be maintained at a relatively constant value and in equilibrium with a series of different reduced and oxidized substrate pairs. Thus, it was observed that in the cytoplasm of rat liver cells, the dehydrogenations catalyzed by lactate dehydrogenase, *sn*-glycerol 3-phosphate dehydrogenase, and malate dehydrogenase are all at equilibrium with the same ratio of $[NAD^+]/[NADH]$. In one experiment rat livers were removed and frozen in less

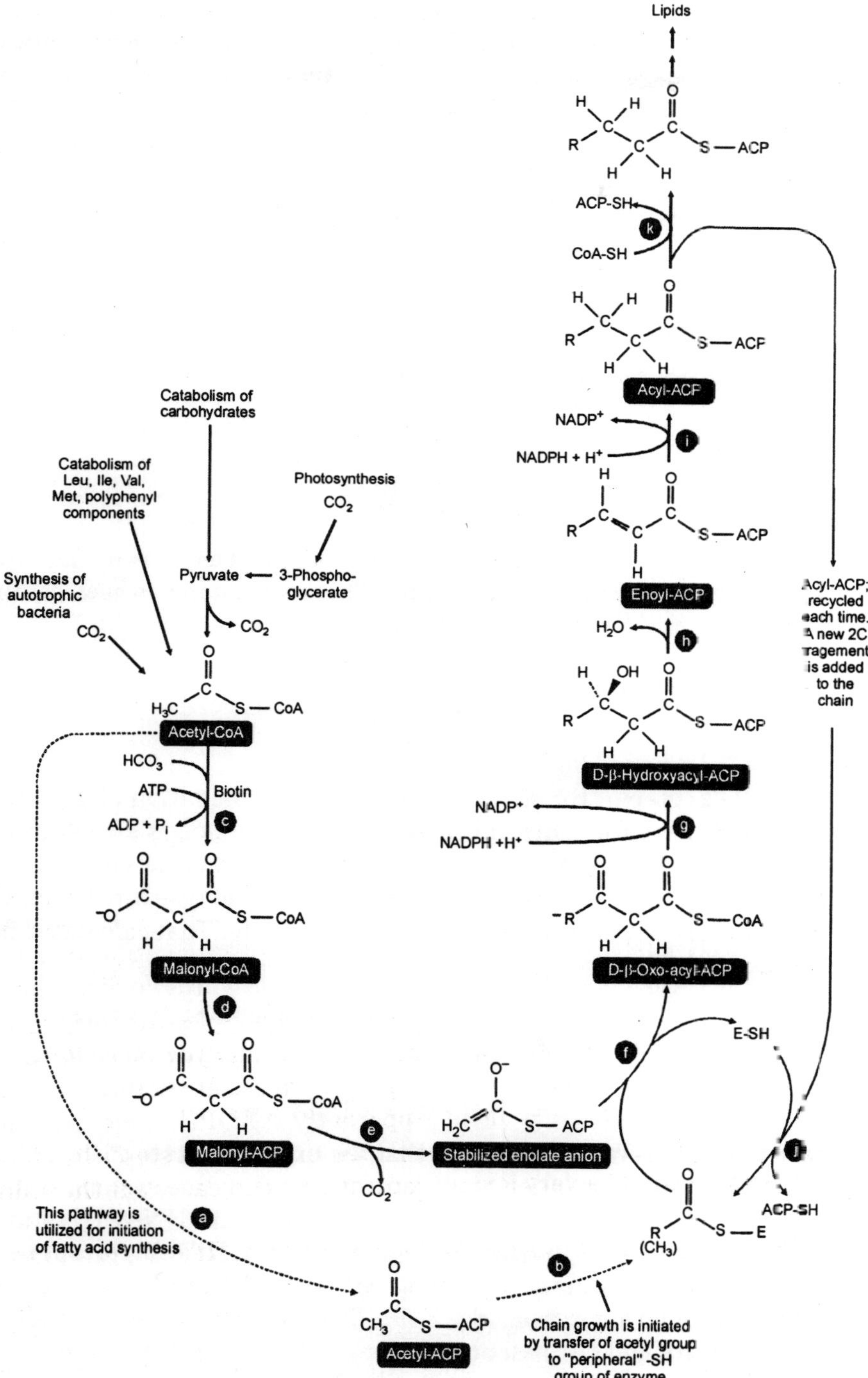

Fig. 7.12. The reactions of cytoplasmic biosynthesis of saturated fatty acids. Compare with pathway of β oxidation.

than 8 s by "freeze-clamping" (Section L,2) and the concentrations of different components of the cytoplasm determined; the ratio $[NAD^+]/[NADH]$ was found to be 634, while the ratio of [lactate]/[pyruvate] was 14.2. From these values an apparent equilibrium constant for reaction *c* of Eq. 7.42 was calculated as $K_c' = 9.0 \times 10^3$. The known equilibrium constant for the reaction (from *in*

Glyceraldehyde 3-P
Lactate
P_i
NAD^+
$NADH + H^+$
c
ADP
ATP
a
b
3-Phosphoglycerate
Slower
Pyruvate
Further metabolism

...(7.42)

vitro experiments) is 8.8×10^3 (Eq. 7.43). In a similar way it was shown that several other dehydrogenation reactions are nearly at equilibrium. This conclusion has been confirmed more recently by NMR observations.

$$K_c'(\text{pH } 7, 38°\text{C}) = \frac{[\text{lactate}]}{[\text{pyruvate}]} \times \frac{[\text{NAD}^+]}{[\text{NADH}]}$$

$$= 8.8 \times 10^3 \quad ...(7.43)$$

Now consider Eq. 7.42, step *a*, the ADP- and P_i^- requiring oxidation of glyceraldehyde 3-phosphate. Experimental measurements indicated that this reaction is also at equilibrium in the cytoplasm. In one series of experiments the measured phosphorylation state ratio [ATP]/[ADP][P_i] was 709, while the ratio [3-phosphoglycerate]/[glyceraldehyde 3-phosphate] was 55.5. The overall equilibrium constant for Eq. 7.42a is given by Eq. 7.44. That calculated from known equilibrium constants is 60.

$$K_a'(\text{ph } 7, 38°\text{C}) = \frac{[\text{ATP}]}{[\text{ADP}][\text{P}_i]} \times \frac{[\text{3-phosphoglycerate}]}{[\text{glyceraldehyde phosphate}]} \times \frac{[\text{NADH}]}{[\text{NAD}^+]}$$

$$= 709 \times 55.5 \times 1/634 = 62 \quad ...(7.44)$$

From these data Krebs and Veech concluded that the oxidation state of the NAD system is determined largely by the phosphorylation state ratio of the adenylate system. If the ATP level is high the equilibrium in Eq. 7.42 *a* will be reached at a higher $[NAD^+]/[NADH]$ ratio and lactate may be oxidized to pyruvate to adjust the [lactate]/[pyruvate] ratio. It is important not to confuse the reactions of Eq. 7.42 as they occur in an aerobic cell with the tightly coupled pair of redox reactions in the homolactate fermentation (Eq. 7.19). The reactions of steps *a* and *c* of Eq. 7.42 are essentially at equilibrium, but the reaction of step *b* may be relatively slow.

Furthermore, pyruvate is utilized in many other metabolic pathways and ATP is hydrolyzed and converted to ADP through innumerable processes taking place within the cell. Reduced NAD does not cycle between the two enzymes in a stoichiometric way and the "reducing equivalents" of NADH formed are, in large measure, transferred to the mitochondria. The proper view of the reactions of Eq. 7.42 is that the redox pairs represent a kind of *redox buffer system* that poises the NADVNADH couple at a ratio appropriate for its metabolic function. Somewhat surprisingly, within the mitochondria the ratio [NAD^+]/[NADH] is 100 times lower than in the cytoplasm.

Even though mitochondria are the site of oxidation of NADH to NAD, the intense catabolic activity occurring in the p oxidation pathway and the citric acid cycle ensure extremely rapid production of NADH. Furthermore, the reduction state of NAD is apparently buffered by the low potential of the p-hydroxybutyrate-acetoacetate couple. Mitochondrial pyridine nucleotides also appear to be at equilibrium with glutamate dehydro-genase.

How is the cytoplasmic [NADPH]/[$NADP^+$] ratio maintained at a value higher than that of [NADH]/ [NAD^+]? Part of the answer is from operation of the pentose phosphate pathway.

The reactions of Eq. 7.12, if they attained equilibrium, would give a ratio of cytosolic [NADPH]/ [$NADP^+$] > 2000 at 0.05 arm CO_2. Compare this with the ratio 1/634 for [NADH/[NAD^+] deduced from the observation on the reactions of Eq. 7.42.

Consider also the following *transhydrogenation* reaction (Eq. 7.45):

$$NADH + NADP^+ \rightarrow NAD^+ + NADPH \tag{7.45}$$

There are soluble enzymes that catalyze this reaction for which K equals ~1. Within mitochondria an energy-linked system involving the membrane shifts the equilibrium to favor NADPH. However, within the cytoplasm, the reaction of Eq. 7.45 is driven by coupling ATP cleavage to the transhydrogenation via carboxylation followed by eventual decarboxylation. One cycle that accomplishes this is given in Eq. 7.46. The first step (step *a*) is ATP-dependent carboxylation of pyruvate to oxaloacetate, a reaction that occurs only within mitochondria. Oxaloacetate can be reduced by malate dehydrogenase using NADH (Eq. 7.46, step *b*), and the resulting malate can be exported from the mitochondria.

In the cytoplasm the malate is oxidized to pyruvate, with decarboxylation, by the malic enzyme. The malic enzyme (Eq. 7.46, step *c*) is specific for $NADP^+$, is very active, and also appears to operate at or near equilibrium within the cytoplasm. On this basis, using known equilibrium constants, it is easy to show that the ratio [NADPH]/ [$NADP^+$] will be ~10^5 times higher at equilibrium than the ratio [NADH]/[NAD^+].

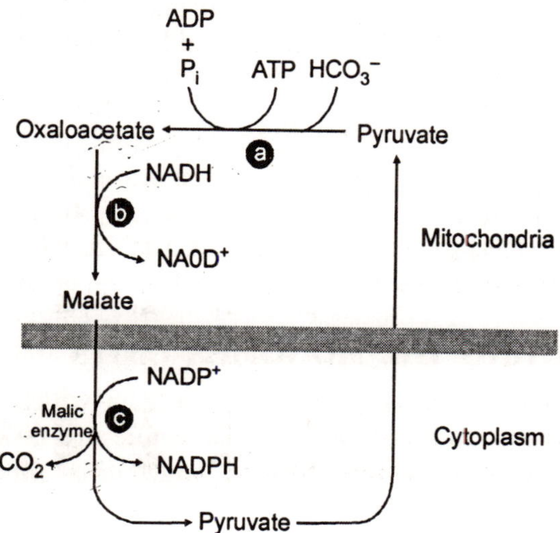

...(7.48)

Since NADPH is continuously used in biosynthetic reactions, and is thereby reconverted to $NADP^+$, the cycle of Eq. 7.46 must operate continuously. As in Eq. 7.42, a true equilibrium does not exist but steps *b* and *c* are both essentially at equilibrium. These equilibria, together with those of Eq. 7.42 for the NAD system, ensure the correct redox potential of both pyridine nucleotide coenzymes in the cytoplasm. Malate is not the only form in which C_4 compounds are exported from mitochondria. Much oxaloacetate is combined with acetyl-CoA to form citrate; the latter leaves the mitochondria and is cleaved by the ATP-dependent citrate-cleaving enzymes. This, in effect, exports both acetyl-CoA (needed for lipid synthesis) and oxaloacetate which is reduced to malate within the cytoplasm.

Alternatively, oxaloacetate may be transaminated to aspartate. The aspartate, after leaving the mitochondria, may be converted in another transamination reaction back to oxaloacetate. All of these are part of the nonequilibrium process by which C_4 compounds diffuse out of the mitochondria before completing the reaction sequence of Eq. 7.46 and entering into other metabolic processes. Note that the reaction of Eq. 7.46 leads to the *export* of reducing equivalents from mitochondria, the opposite of the process catalyzed by the malate-aspartate shuttle.

The two processes are presumably active under different conditions. While the difference in the redox potential of the two pyridine nucleotide systems is clear-cut in mammalian tissues, in *E. coli* the apparent potentials of the two systems are more nearly the same.

Reduced Ferredoxin in Reductive Biosynthesis

Both the NAD^+ and $NADP^+$ systems have standard electrode potentials $E^{\circ\prime}$ (pH 7) of –0.32 V. However, because of the differences in concentration ratios, the NAD^+ system operates at a less negative potential (–0.24 V) and the $NADP^+$ system at a more negative potential (–0.38 V) within the cytoplasm of eukaryotes. In green plants and in many bacteria a still more powerful reducing agent is available in the form of reduced ferredoxin. The value of $E^{\circ\prime}$ (pH 7) for clostridial ferredoxin is –0.41 V, corresponding to a Gibbs energy change for the two-electron reduction of a substrate ~18 kJ mol^{-1} more negative than the corresponding value of $\Delta G'$ for reduction by NADPH.

Using reduced (Fd) some photosynthetic bacteria and anaerobic bacteria are able to carry out reductions that are virtually impossible with the pyridine nucleotide system. For example, pyruvate and 2-oxoglutarate can be formed from acetyl-CoA and succinyl-*CoA,* respectively (Eq. 7.47). In our bodies the reaction of Eq. 7.47, with NAD^+ as the oxidant, goes only in the opposite direction and is essentially irreversible.

$$\text{Acetyl-CoA} \xrightarrow[\text{Fd}_{red} \;\to\; \text{Fd}_{ox}]{CO_2} \text{Pyruvate}^- + \text{CoA} \qquad \Delta G' \text{ (pH7)} = +17 \text{ kJ mol}^{-1} \quad ...(7.47)$$

CONSTRUCTING THE MONOMER UNITS

Now let us consider the synthesis of the monomeric units from which biopolymers are made. How can simple one-carbon compounds such as CO_2 and formic acid be incorporated into complex carbon compounds? How can carbon chains grow in length or be shortened? How are branched chains and rings formed?

Carbonyl Groups in Chain Formation and Cleavage

Except for some vitamin B_{12}-dependent reactions, the cleavage or formation of carbon-carbon bonds usually depends upon the participation of carbonyl groups. For this reason, carbonyl groups have a central mechanistic role in biosynthesis. The activation of hydrogen atoms β to carbonyl groups permits β condensations to occur during biosynthesis. Aldol or Claisen condensations require the participation of two carbonyl compounds. Carbonyl compounds are also essential to thiamin diphosphate-dependent condensations and the aldehyde pyridoxal phosphate is needed for most C–C bond cleavage or formation within amino acids. Because of the importance of carbonyl groups to the mechanism of condensation reactions, much of the assembly of either straight-chain or branched-carbon skeletons takes place between compounds in which the average oxidation state of the carbon atoms is similar to that in carbohydrates (or in formaldehyde, H_2CO).

The diversity of chemical reactions possible with compounds at this state of oxidation is a maximum, a fact that may explain why carbohydrates and closely related substances are major biosynthetic precursors and why the average state of oxidation of the carbon in most living things is similar to that in carbohydrates. This fact may also be related to the presumed occurrence of formaldehyde as a principal component of the earth's atmosphere in the past and to the ability of formaldehyde to condense to form carbohydrates. In Fig. 7.13 several biochemicals have been arranged according to the oxidation state of carbon.

Most of the important biosynthetic intermediates lie within ± 2 electrons per carbon atom of the oxidation state of carbohydrates. As the chain length grows, they tend to fall even closer. It is extremely difficult to move through enzymatic processes between 2C, 3C, and 4C compounds (*i.e.,* vertically in Fig. 7.13) except at the oxidation level of carbohydrates or somewhat to its right, at a slightly higher oxidation level. On the other hand, it is often possible to move horizontally with ease using oxidation-reduction reactions. Thus, fatty acids are assembled from acetate units, which lie at the same oxidation state as carbohydrates and, after assembly, are reduced. Among compounds of the same overall oxidation state, *e.g.,* acetic acid and sugars, the oxidation states of individual carbon atoms can be quite different.

Thus, in a sugar every carbon atom can be regarded as immediately derived from formaldehyde, but in acetic acid one end has been oxidized to a carboxyl group and the other has been reduced to a methyl group. Such internal oxidation-reduction reactions play an important role in the chemical manipulations necessary to assemble the carbon skeletons needed by a cell. Decarboxylation is a feature of many biosynthetic routes. Referring again to Fig. 7.13, notice that many of the biosynthetic intermediates such as pyruvate, oxoglutarate, and oxaloacetate are more oxidized than the carbohydrate level. However, their decarboxylation products, which become incorporated into the compounds being synthesized, are closer to the oxidation level of carbohydrates.

Starting with CO_2

There are three known pathways by which autotrophic organisms can use CO_2 to synthesize triose phosphates or 3-phosphoglycerate, three-carbon compounds from which all other biochemical substances can be formed. The first of these is the *reductive tricarboxylic cycle*. This is a reversal of the oxidative citric acid cycle in which reduced ferredoxin is used as a reductant in the reaction

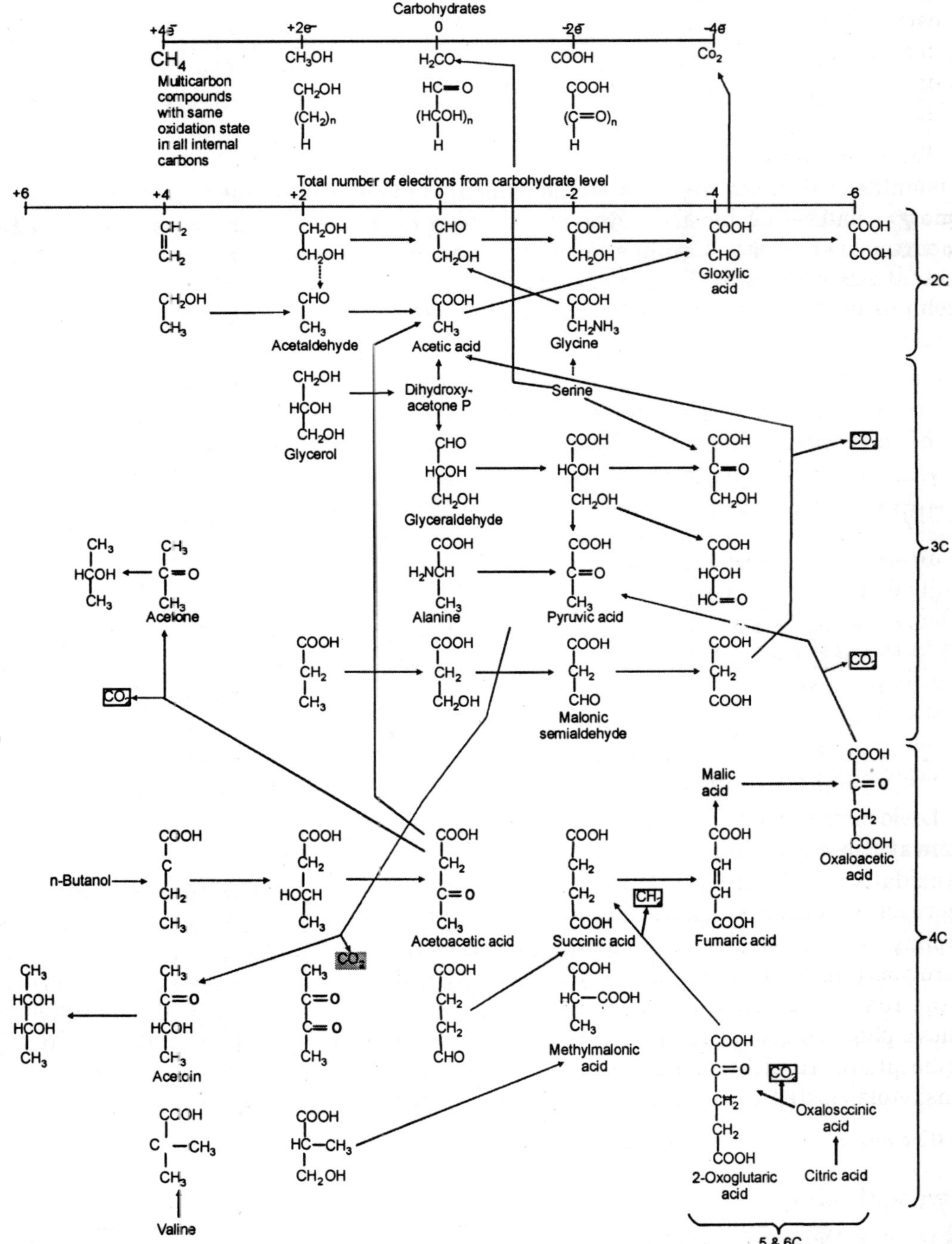

Fig. 7.13. Some biochemical compounds arranged in order of average oxidation state of the carbon atoms and by carbon-chain lengths. Black horizontal arrows mark some biological interconversions among compounds with the same chain length, while green liens show changes in chain length and are often accompanied by decarboxylation.

of Eq. 7.47 to incorporate CO_2 into pyruvate. Succinyl-CoA can react with CO_2 in the same type of reaction to form 2-oxoglutarate, accomplishing the reversal of the only irreversible step in the citric acid cycle. Using these reactions photosynthetic bacteria and some anaerobes that can generate a high ratio of reduced to oxidized ferredoxin carry out the reductive tricarboxylic acid cycle.

Together with Eq. 7.47, the cycle provides for the complete synthesis of pyruvate from CO_2. 3A quantitatively much more important pathway of CO_2 fixation is the *reductive pentose phosphate pathway* (ribulose bisphosphate cycle or *Calvin-Benson cycle;* Fig. 7.14). This sequence of reactions, which takes place in the chloroplasts of green plants and also in many chemiautotrophic bacteria, is essentially a way of reversing the oxidative pentose phosphate pathway (Fig. 7.8). The latter accomplishes the complete oxidation of glucose or of glucose 1-phosphate by $NADP^+$ (Eq. 7.48):

$$\text{Glucose 1-P}^{2-} + \text{ATP}^{4-} + 8\ \text{H}_2\text{O} + 12\ \text{NADP}^+ \rightarrow$$
$$6\ \text{CO}_2 + 12\ \text{NADPH} + \text{ADP}^{3-} + 2\text{HPO}_4^{\ 2-} + 13\ \text{H}^+$$
$$\Delta G'\ (\text{pH 7}) = -299\ \text{kJ mol}^{-1} \qquad \text{...(7.48)}$$

It would be almost impossible for a green plant to fix CO_2 using photochemically generated NADPH by an exact reversal of Eq. 7.48 because of the high positive Gibbs energy change.

To solve this thermodynamic problem the reductive pentose phosphate pathway has been modified in a way that couples ATP cleavage to the synthesis. The *reductive carboxylation* system is shown within the green shaded box of Fig. 7.14. Ribulose 5-phosphate is the starting compound and in the first step one molecule of ATP is expended to form *ribulose 1,5-bisphosphate*. The latter is carboxylated and cleaved to two molecules of 3-phosphoglycerate. This reaction was discussed. The reductive step (step *c*) of the system employs NADPH together with ATP. Except for the use of the NADP system instead of the NAD system, it is exactly the reverse of the glyceraldehyde phosphate dehydrogenase reaction of glycolysis.

Looking at the first three steps of Fig. 7.14 it is clear that in the reductive pentose phosphate pathway three molecules of ATP are utilized for each CO_2 incorporated. On the other hand, in the oxidative direction *no* ATP is generated by the operation of the pentose phosphate pathway. The reactions enclosed within the shaded box of Fig. 7.14 do not give the whole story about the coupling mechanism. A phospho group was transferred from ATP in step *a* and to complete the hydrolysis it must be removed in some future step. This is indicated in a general way in Fig. 7.14 by the reaction steps *d, e,* and *f*. Step *f* represents the action of specific phosphatases that remove phospho groups from the seven-carbon sedoheptulose bisphosphate and from fructose bisphosphate. In either case the resulting ketose monophosphate reacts with an aldose (via transketolase, step *g*) to regenerate ribulose 5-phosphate, the CO_2 acceptor.

The overall reductive pentose phosphate cycle (Fig. 7-14B) is easy to understand as a reversal of the oxida(c)tive pentose phosphate pathway in which the oxidative decarboxylation system of Eq. 7.12 is replaced by the reductive carboxylation system of Fig. 7.14A. The scheme as written in Fig. 7.14B shows the incorporation of three molecules of CO_2. The reductive carboxylation system operates three times with a net production of one molecule of triose phosphate. As with other biosynthetic cycles, any amount of any of the intermediate metabolites may be withdrawn into various biosynthetic pathways without disruption of the flow through the cycle.

The overall reaction of carbon dioxide reduction in the Calvin-Benson cycle (Fig. 7.14) becomes

$$6\ CO_2 + 12\ NADPH + 18\ ATP^{4-} + 11\ H_2O \rightarrow$$
$$\text{glucose-1-}P^{2-} + 12\ NADP^+ + 18\ ADP^{3-}\ 17HPO_4^{2-} + 6H^+ \quad (7.49)$$

The Gibbs energy change $\Delta G'$ (pH 7) is now –357 kJ mol^{-1} instead of the + 299 kJ mol^{-1} required to reverse the reaction of Eq. 7.48.

The third pathway for reduction of CO_2 to acetyl-CoA is utilized by acetogenic bacteria, by methanogens, and probably by sulfate-reducing bacteria. This *acetyl-CoA pathway* (or *Wood-Ljungdahl pathway*) involves reduction by H_2 of one of the two molecules of CO_2 to the methyl group of methyl-tetrahydromethanopterin in methanogens and of methyltetra-hydrofolate in acetogens. The pathway utilized by methanogens is illustrated. A similar process utilizing H_2 as the reductant is employed by acetogens.

In both cases a methyl corrinoid is formed and its methyl group is condensed with a molecule of carbon monoxide bound to a copper ion in a Ni-Cu cluster. The resulting acetyl group is transferred to a molecule of coenzyme A as illustrated in Eq. 7.52. The bound CO is formed by reduction of CO_2, again using H_2 as the reductant. The overall reaction for acetyl-CoA synthesis is given by Eq. 7.50. Conversion of aceryl-CoA to pyruvate via Eq. 7.47 leads into the glucogenic pathway.

$CO_2 \xrightarrow{\text{6-Electron reduction}} \rightarrow \rightarrow \rightarrow CH_3\text{–Pterin}$

↓

$CH_3\text{–CO(Corrin)}$

$CO_2 \xrightarrow{2e^-} CO$ → CO-Cu (Ni) ; Carbon monoxide dehydrogenase; + CoA-SH

↓

$$CH_3\text{–}\overset{O}{\overset{\|}{C}}\text{–S–CoA}$$

Overall:

$$2CO_2 + 4H_2 + CoASH \longrightarrow CH_3\text{–}\overset{O}{\overset{\|}{C}}\text{–S–CoA} + 3H_2O \quad ...(7.50)$$

An alternative pathway by which some acetogenic teria form acetate is via reversal of the glycine decarboxylase reaction. Methylene-THF is formed by reduction of CO_2, and together with NH_3 and CO_2 a lipoamide group of the enzyme and PLP forms glycine. The latter reacts with a second methylene-THF to form serine, which can be deaminated to pyruvate and assimilated. Methanogens may use similar pathways but ones that involve methanopterin.

Biosynthesis from Other Single-Carbon Compounds

Various bacteria and fungi are able to subsist on such one-carbon compounds as methane, methanol, methylamine, formaldehyde, and formate. Energy is obtained by oxidation to CO_2. *Methylotrophic bacteria* initiate oxidation of methane by hydro xylation and dehydrogenate the resulting methanol or exogenous methanol using the PPQ cofactor. Further dehydrogenation to formate and of formate to CO_2 via formate dehydrogenase completes the process. Some methylotrophic bacteria incorporate CO_2 for biosynthetic purposes via the ribulose bisphosphate

(Calvin-Benson) cycle but many use pathways that begin with formaldehyde (or methylene-THF). Others employ variations of the reductive pentose phosphate pathway to convert formaldehyde to triose phosphate.

In one of these, the *ribulose monophosphate cycle* or Quayle cycle, ribulose 5-*P* undergoes an aldol condensation with formaldehyde to give a 3-oxo-hexulose 6-phosphate (Eq. 7.51, step *a*). The latter is isomerized to fructose 6-*P* (Eq. 7.51, step *b*). If this equation is applied to the three C_5 sugars three molecules of fructose 6-phosphate will be formed. One of these can be phosphorylated by ATP to fructose 1,6-bisphosphate, which can be cleaved by aldolase. One of the resulting triose phosphates can then be removed for biosynthesis and the second, together with the other two molecules of fructose 6-*P*, can be recycled through the sugar rearrangement sequence of Fig. 7.8B to regenerate the three ribulose 5-*P* molecules that serve as the regenerating substrate.

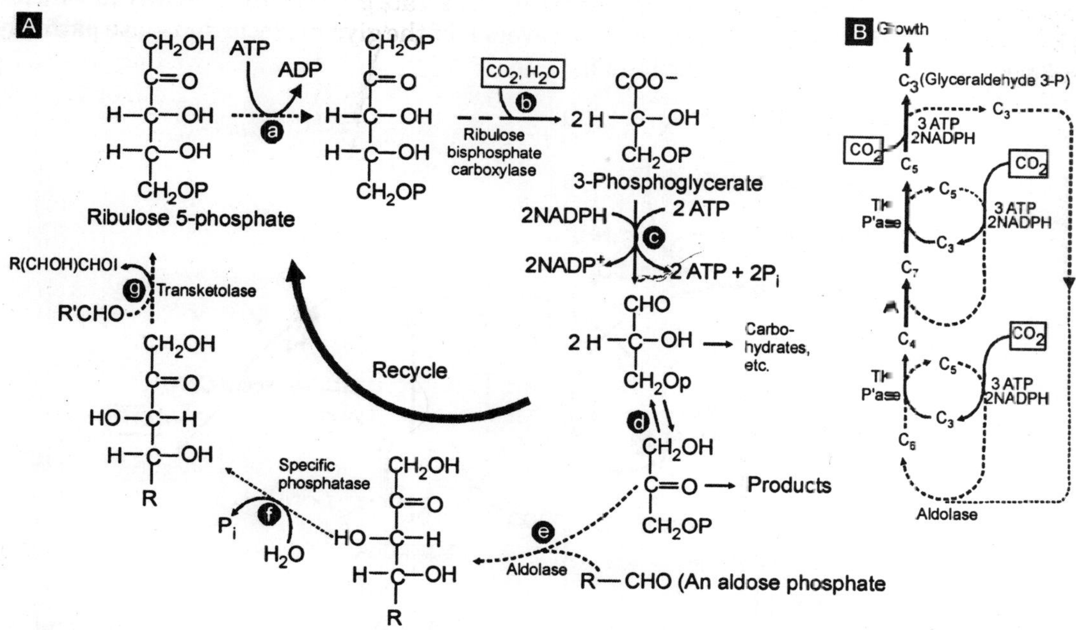

Fig. 7.14. (A) The reductive carboxylation system used in reductive pentose phosphate pathway (Calvin-Benson cycle). The essential reactions of this system are enclosed within the dashed box. Typical subsequent reactions follow. The phosphatase action completes the phosphorylation-dephosphorylation cycle. (B) The reductive pentose phosphate cycle arranged to show the combining of three CO_2 molecules to form one molecule of triose phosphate. Abbreviations are RCS, reductive carboxylation system (from above); A, aldolase, Pase, specific phosphatase; and TK, transketolase.

In bacteria, which lack formate dehydrogenase, formaldehyde can be oxidized to CO_2 to provide energy beginning with the reactions of Eq. 7.51. The resulting fructose 6-*P* is isomerized to glucose 6-*P*, which is then dehydrogenated via Eq. 7.12 to form CO_2 and the regenerating substrate ribulose 5-phosphate. A number of pseudomonads and other bacteria convert Q compounds to acetate via tetrahydrofolic acid-bound intermediates and CO_2 using the *serine pathway* shown in Fig. 7.15. This is a cyclic process for converting one molecule of formaldehyde (bound to tetrahydrofolate) plus one of CO_2 into acetate.

The regenerating substrate is *glyoxylate*. Before condensation with the "active formaldehyde" of methylene THF, the glyoxylate undergoes transamination to glycine (Fig. 7.15, step *a*). The glycine plus formaldehyde forms serine (step *b*), which is then transaminated to hydroxypyruvate, again using step *a*. Glyoxylate plus formaldehyde could have been joined in a thiamin-dependent condensation. However, as in the y-amino-butyrate shunt (Fig. 7.5), the coupled transamination step of Fig 7.15 permits use of PLP-dependent C-C bond formation. Conversion of hydroxypyruvate to PEP (Fig. 7.15) involves reduction by NADH and phosphorylation by ATP to form 3-phosphoglycerate, which is converted to PEP as in glycolysis.

The conversion of malate to acetate and glyoxylate via malyl-CoA and isocitrate lyase forms the product acetate and regenerates glyoxylate. As with other metabolic cycles, various intermediates, such as PEP, can be withdrawn for biosynthesis. However, there must be an independent route of synthesis of the regenerating substrate glyoxylate. One way in which this can be accomplished is to form glycine via the reversal of the glycine decar-boxylase pathway as is indicated by the shaded green lines in Fig. 7.15.

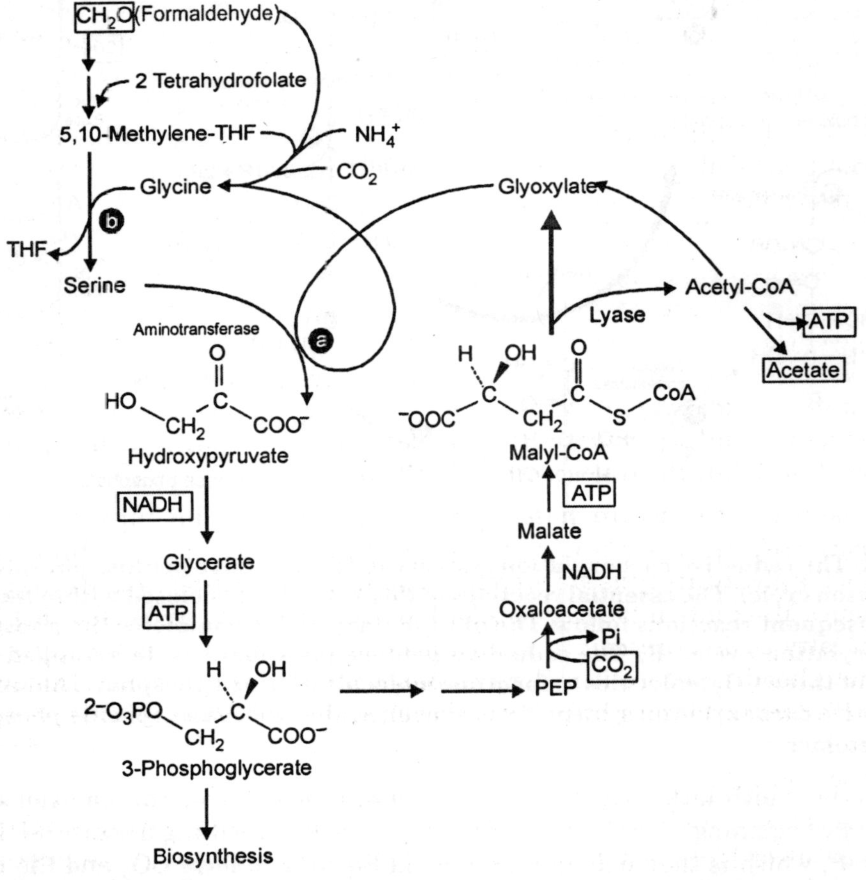

Fig. 7.15. One of the serine pathways for assimilation of one-carbon compounds.

$$\begin{array}{c} CH_2OH \\ | \\ O{=}C \\ | \\ H{-}C{-}OH \\ | \\ H{-}C{-}OH \\ | \\ CH_2{-}O{-}P \\ \text{Ribulose 5-P} \end{array} \xrightarrow[\text{(a)}]{H_2C{=}O} \begin{array}{c} CH_2OH \\ | \\ H{-}C{-}OH \\ | \\ O{=}C \\ | \\ H{-}C{-}OH \\ | \\ H{-}C{-}OH \\ | \\ CH_2{-}O{-}P \\ \text{Hexulose 6-P} \end{array} \xrightarrow[\text{Isomerase}]{\text{(b)}} \begin{array}{c} CH_2OH \\ | \\ O{=}C \\ | \\ HO{-}C{-}H \\ | \\ H{-}C{-}OH \\ | \\ H{-}C{-}OH \\ | \\ CH_2{-}O{-}P \\ \text{Fructose 6-P} \end{array} \quad ...(7.51)$$

The Glyoxylate Pathways

The reductive carboxylation of acetyl-CoA to pyruvate (Eq. 7.47) occurs only in a few types of bacteria. For most species, from microorganisms to animals, the oxidative decarboxylation of pyruvate to acetyl-CoA is irreversible. This fact has many important consequences. For example, carbohydrate is readily converted to fat; because of the irreversibility of this process, excess calories lead to the deposition of fats. However, in animals fat cannot be used to generate most of the biosynthetic intermediates needed for formation of carbohydrates and proteins because those intermediates originate largely from C_3 units. This limitation on the conversion of C_2 acetyl units to C_3 metabolites is overcome in many organisms by the *giyoxylate cycle* (Fig. 7.16), which converts *two* acetyl units into one C_4 unit.

The cycle provides a way *for organisms,* such as E.*coil*, *Saccharomyces*, *Tetmhymena,* and the nematode *Caenorhabditis* to subsist on acetate as a sole or major carbon source. It is especially prominent in plants that store large amounts of fat in their seeds (*oil seeds*). In the germinating oil seed the glyoxylate cycle allows fat to be converted rapidly to sucrose, cellulose, and other carbohydrates needed for growth. A key enzyme in the glyoxylate cycle is *isocitrate lyase*, which cleaves isocitrate to succinate and glyoxylate. The latter is condensed with a second acetyl group by the action of *malate synthase*. The L-malate formed in this reaction is dehydrogenated to the regenerating substrate oxaloacetate.

Some of the reaction steps are those of the citric acid cycle and it appears that in bacteria there is no spatial separation of the citric acid cycle and glyoxylate pathway. However, in plants the enzymes of the glyoxylate cycle are present in specialized peroxisomes known as *glyoxysomes*. The glyoxysomes also contain the enzymes for P oxidation of fatty acids, allowing for efficient conversion of fatty acids to *succinate*. This compound is exported from the glyoxysomes and enters the mitochondria where it undergoes β oxidation to oxaloacetate. The latter can be converted by PEP carboxylase or by PEP carboxykinase to PEP. An *acetyl-CoA-glyoxylate* cycle, which catalyzes oxidation of acetyl groups to glyoxylate, can also be constructed from isocitrate lyase and citric acid cycle enzymes.

Glyoxylate is taken out of the cycle as the product and succinate is recycled (Eq. 7.52). The independent pathway for synthesis of the regenerating substrate oxaloacetate is condensation of glyoxylate with acetyl-CoA (malate synthetase) to form malate and oxidation of the latter to oxaloacetate as in the main cycle of Fig. 7.16.

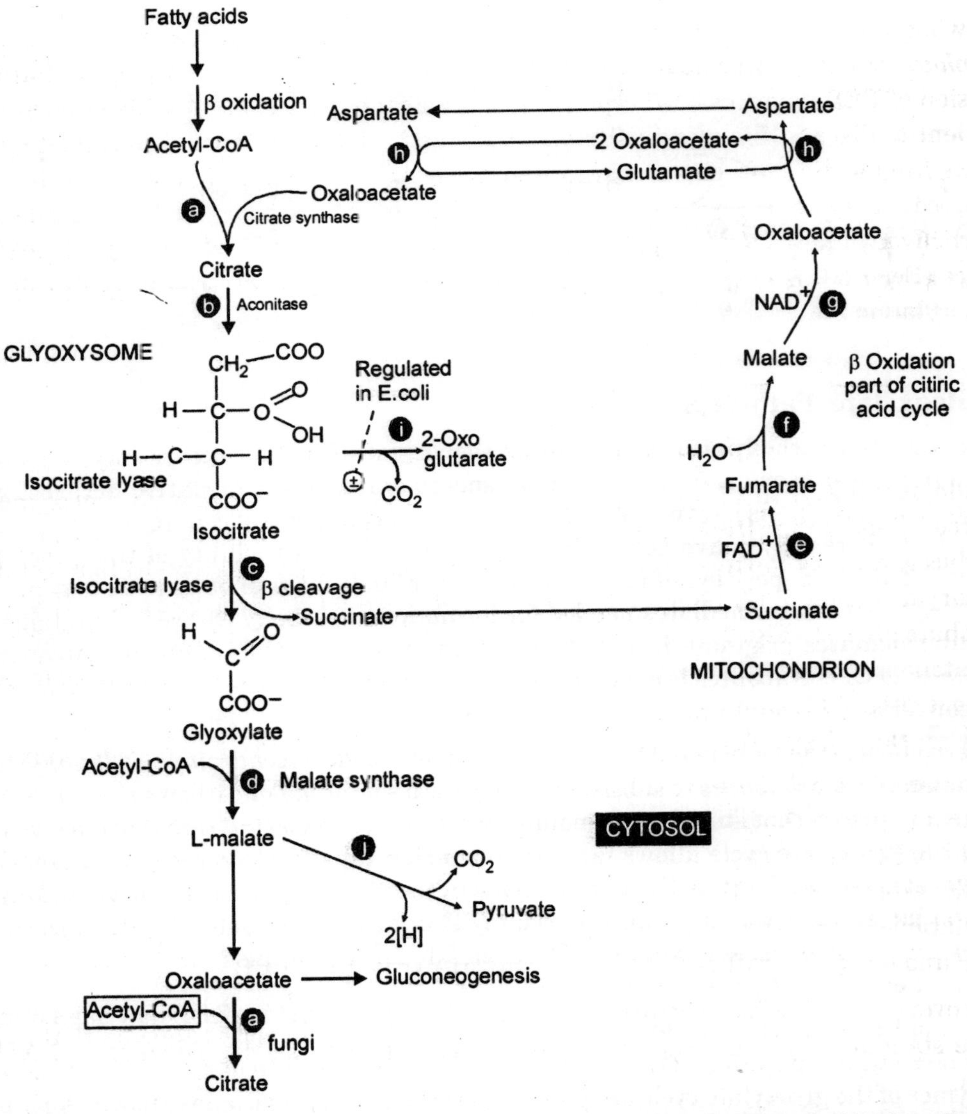

Fatty acids
β oxidation
Acetyl-CoA
Aspartate
Aspartate
h
2 Oxaloacetate
Glutamate
h
a
Oxaloacetate
Citrate synthase
Oxaloacetate
Citrate
b
Aconitase
NAD+
g
GLYOXYSOME
CH_2—COO
H—C—O
OH
H—C—C—H
Isocitrate lyase
COO^-
Isocitrate
Regulated in E.coli
i
2-Oxo glutarate
±
CO_2
Malate
β Oxidation part of citiric acid cycle
H_2O
f
Fumarate
FAD^+
e
c
Isocitrate lyase
β cleavage
Succinate
Succinate
H
C=O
COO^-
Glyoxylate
MITOCHONDRION
Acetyl-CoA
d
Malate synthase
CYTOSOL
L-malate
j
CO_2
Pyruvate
2[H]
Oxaloacetate → Gluconeogenesis
Acetyl-CoA
a
fungi
Citrate

Fig. 7.16. The glyoxylate pathway. The green line traces the pathway of labeled carbon from fatty acids or acetyl-CoA into malate and other products.

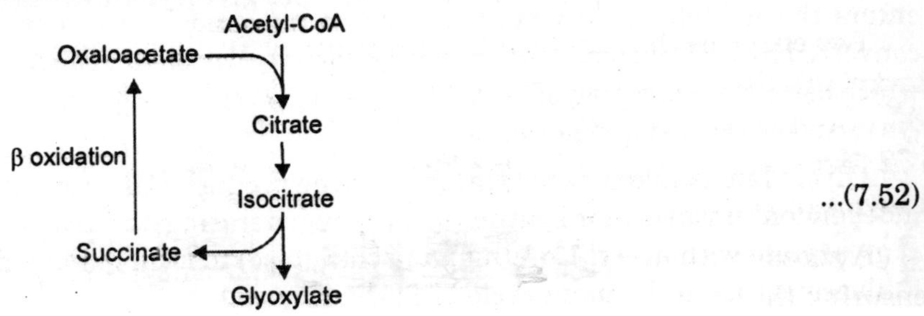

...(7.52)

Biosynthesis of Glucose from Three-Carbon Compounds

Now let us consider the further conversion of PEP and of the triose phosphates to *glucose 1-phosphate*, the key intermediate in biosynthesis of other sugars and polysaccharides. The conversion of PEP to glucose 1-*P* represents a reversal of part of the glycolysis sequence. It is convenient to discuss this along with *gluconeogenesis,* the reversal of the complete glycolysis sequence from lactic acid. This is an essential part of the Cori cycle in our own bodies, and the same process may be used to convert pyruvate derived from deamination of alanine or serine into carbohydrates. Just as with the pentose phosphate cycle, an exact reversal of the glycolysis sequence (Eq. 7.53) is precluded on thermodynamic grounds. Even at very high values of the phosphorylation state ratio $R_{p'}$ the reaction:

$$\begin{aligned}2\ \text{Lactate}^- + 3\ \text{ATP}^{4-} + 2\ H_2O &\rightarrow C_6H_{10}O_5\ (\text{glycogen})\\ &+ \text{ADP}^3 + 31 + PO_4^{2-} + H^+\\ \Delta G'\ (\text{pH 7}) &= +\ 107\ \text{kJ per glycosyl unit}\end{aligned} \quad ...(7.53)$$

would be unlikely to go to completion. The actual pathways used for gluconeogenesis (Fig. 7.17, green lines) differ from those of glycolysis (black lines) in three significant ways. First, while glycogen breakdown is initiated by the reaction with inorganic phosphate catalyzed by phosphorylase (Fig. 7.17, step *a*), the biosynthetic sequence from glucose 1-P, via uridine diphosphate glucose (Fig. 7.17, step *b;* see also Eq. 7.56), is coupled to cleavage of ATP. Second, in the catabolic process (glycolysis) fructose 6-*P* is converted to fructose 1,6-*P* through the action of a kinase (Fig. 7.17, step *c*), which is then cleaved by aldolase.

The resulting triose phosphate is degraded to PEP. In glucogenesis a phosphatase is used to form fructose *P* from fructose P_2 (Fig. 7.17, step *d*). Third, during gly-colysis PEP is converted to pyruvate by a kinase with generation of ATP (Fig. 7.17, step *e*). During glucogenesis pyruvate is converted to PEP indirectly via oxaloacetate (Fig. 7.17, steps *f* and *g*) using pyruvate carboxylase and PEP carboxykinase. This is another example of the coupling of ATP cleavage through a carboxylation-decarboxylation sequence. The net effect is to use two molecules of ATP (actually one ATP and one GTP) rather than *one* to convert pyruvate to PEP.

The overall reaction for reversal of glycolysis to form glycogen (Eq. 7.54) now has a comfortably negative standard Gibbs energy change as a result of coupling the cleavage of 7 ATP to the reaction.

$$\begin{aligned}2\ \text{Lactate}^- + 7\ \text{ATP}^{4-} + 6\ H_2O &\rightarrow \text{glycogen} + 7\ \text{ADP}^{3-} + 7\ HPO_4^{2-} + 5\ H^+\\ \Delta G'\ (\text{pH 7}) &= -31\ \text{kJ mol}^{-1}\ \text{per glycosyl unit}\end{aligned} \quad ...(7.54)$$

Two enzymes that are able to convert pyruvate directly to PEP are found in some bacteria and plants. In each case, as in the animal enzyme system discussed in the preceding paragraph, the conversion involves expenditure of two high-energy linkages of ATP. The *PEP synthase* of *E. coli* first transfers a pyrophospho group from ATP onto an imidazole group of histidine in the enzyme (Eq. 7.55).

A phospho group is hydrolyzed from this intermediate (dashed green line in Eq. 7.55, step *b*), ensuring that sufficient intermediate E-His–P is present. The latter reacts with pyruvate to form PEP. *Pyruvate-phosphate dikinase* is a similar enzyme first identified in tropical grasses and known to play an important role in the CO_2 concentrating system of the so-called "C_4 plants". The same enzyme participates in gluconeogenesis in *Acetobacter.* The reaction cycle for this enzyme is also portrayed in Eq. 7.55. In this case P_i, rather than water, is the attacking nucleophile in Eq. 7.55 and PP_i is a product. The latter is probably hydrolyzed by pyrophosphatase action, the end result being an overall reaction that is the same as with PEP synthase.

Kinetic and positional isotope exchange studies suggest that the P, must be bound to pyruvate-phosphate dikinase before the bound ATP can react with the imidazole group. Likewise, AMP doesn't dissociate until P_i has reacted to form PP_i.

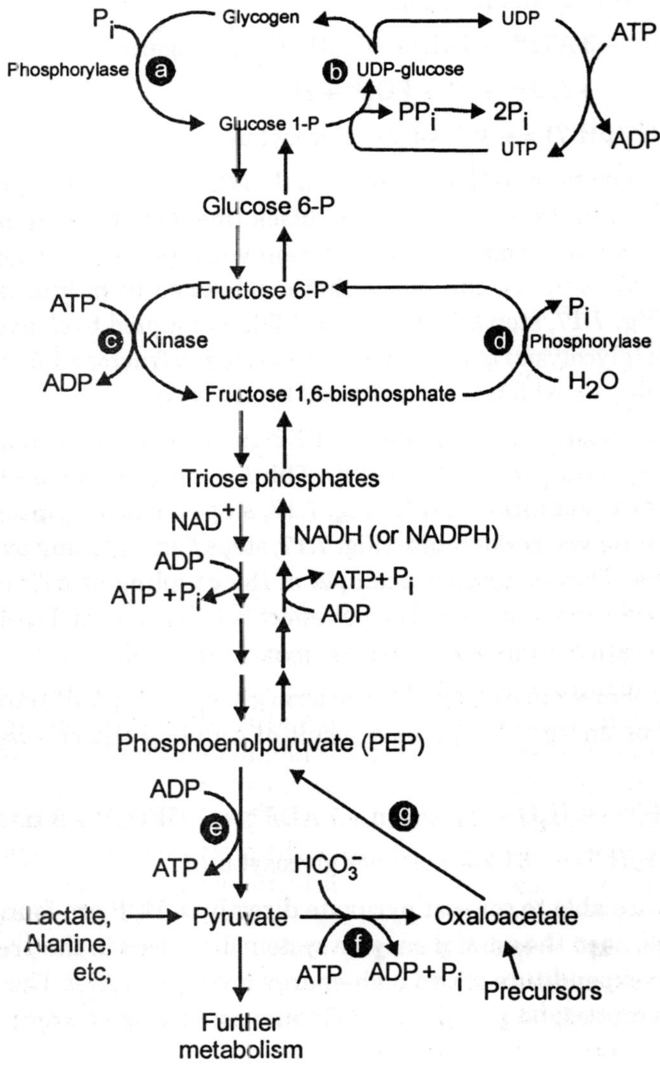

Fig. 7.17. Comparison of glycolytic pathway (left) with pathway of gluconeogenesis.

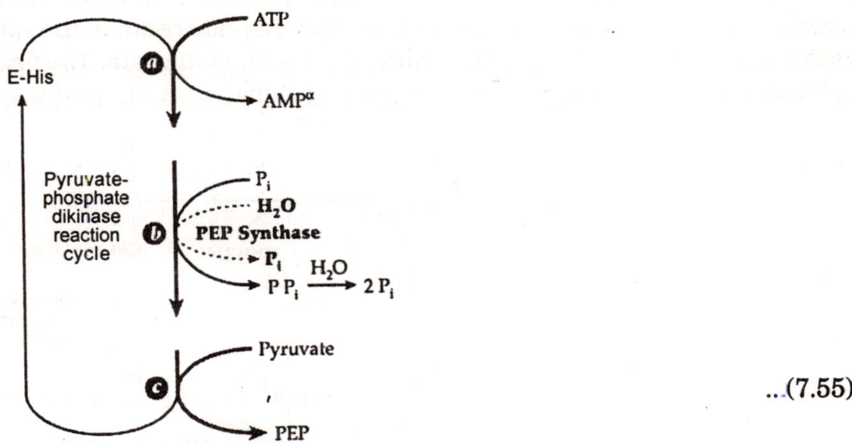

...(7.55)

Building Hydrocarbon Chains with Two-Carbon Units

Fatty acid chains are taken apart two carbon atoms at a time by β oxidation. Biosynthesis of fatty acids reverses this process by using the two-carbon acetyl unit of acetyl-CoA as a starting material. The coupling of ATP cleavage to this process by a carboxylation-decarboxylation sequence, the role of acyl carrier protein, and the use of NADPH as a reductant (Section I) have been discussed and are summarized in Fig. 7.12, which gives the complete sequence of reactions for fatty acid biosynthesis. Why does β oxidation require CoA derivatives while biosynthesis requires the more complex acyl carrier protein (ACP)? The reason may involve control.

ACP is a complex handle able to hold the growing fatty acid chain and to guide it from one enzyme to the next. In *E. coli* the various enzymes catalyzing the reactions of Fig. 7.12 are found in the cytosol and behave as independent proteins. The same is true for fatty acid syntheses of higher plants which resemble those of bacteria. It is thought that the ACP molecule lies at the center of the complex and that the growing fatty acid chain on the end of the phosphopantetheine prosthetic group moves from one subunit to the other.

The process is started by a *primer* which is usually acetyl-CoA in *E. coli.* Its acyl group is transferred first to the central molecule of ACP (step a, Fig. 7.12) and then to a "peripheral" thiol group, probably that of a cysteine side chain on a separate protein subunit (step *b*, Fig. 7.12). Next, a malonyl group is transferred (step *d*) from malonyl-CoA to the free thiol group on the ACP. The condensation (steps *e* and *f*) occurs with the freeing of the peripheral thiol group. The latter does not come into use again until the β-oxoacyl group formed has undergone the complete sequence of reduction reactions (steps *g–i).* Then the growing chain is again transferred to the peripheral -SH (step *j*) and a new malonyl unit is introduced on the central ACP. After the chain reaches a length of 12 carbon atoms, the acyl group tends to be transferred off to a CoA molecule (step *k)* rather than to pass around the cycle again. Thus, chain growth is terminated. This tendency systematically increases as the chain grows longer. In higher animals as well as in *Mycobacterium*, yeast, and *Euglena,* the *fatty acid synthase* consists of only one or two multifunctional proteins.

The synthase from animal tissues has seven catalytic activities in a single 263-kDa 2500-residue protein. The protein consists of a series of domains that contain the various catalytic activities needed for the entire synthetic sequence. One domain contains an ACP-like site with a bound 4′-phosphopantetheine as well as a cysteine side chain in the second acylation site. This synthase produces free fatty acids, principally the C_{16} palmitate. The final step is cleavage of the acyl-CoA by a thioesterase, one of the seven enzymatic activities of the synthase.

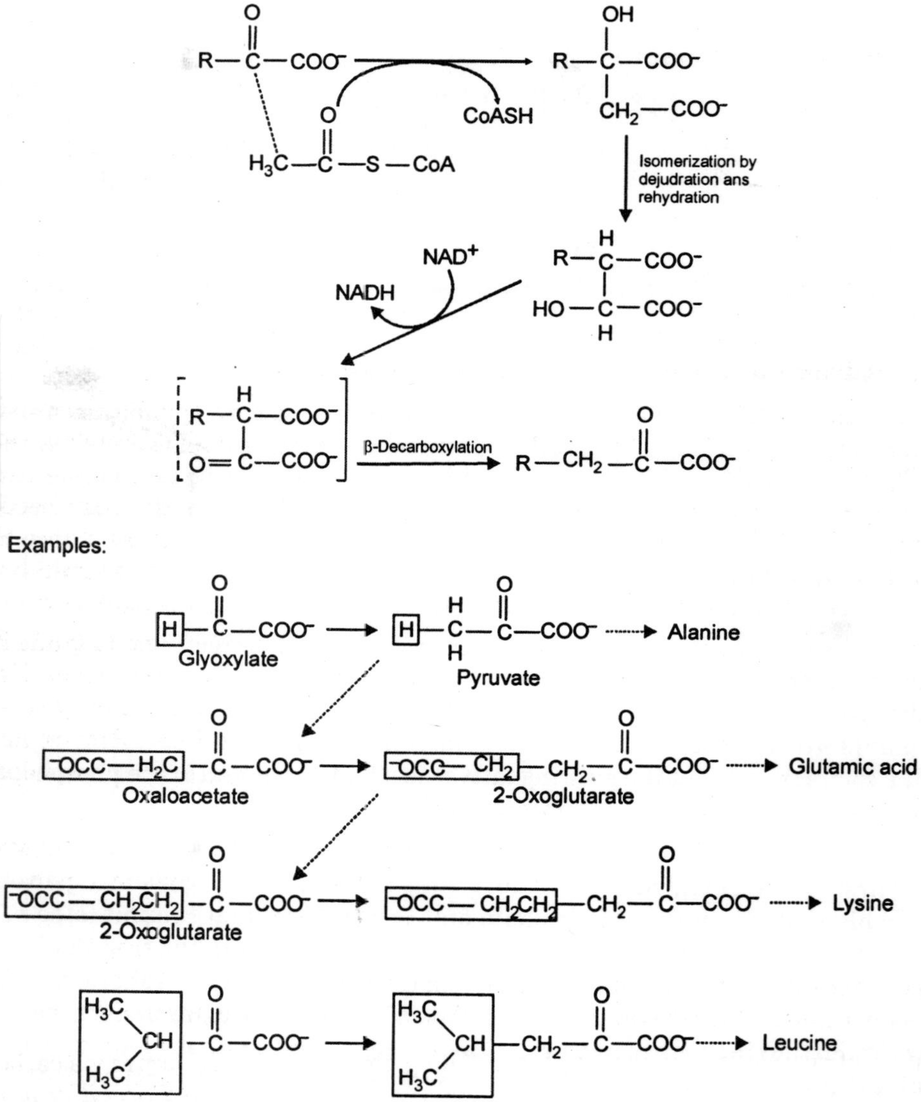

Fig. 7.18. The oxoacid chain elongation process.

The Oxoacid Chain Elongation Process

As mentioned already, glyoxylate can be converted to oxaloacetate by condensation with acetyl-CoA (Fig. 7.16) and the oxaloacetate can be decarboxylated to pyruvate. This sequence of reactions resembles that of the conversion of oxaloacetate to 2-oxoglutarate in the citric acid cycle (Fig. 7.4). Both *are examples of a frequently used general chain elongation process for α-oxo acids.* This sequence, which is illustrated in Fig. 7.18, has four steps: (1) condensation of the α-oxo acid with an acetyl group, (2) isomerization by dehydration and rehydration (catalyzed by aconitase in the case of the citric acid cycle), (3) dehydrogenation, and (4) β decarboxylation. In many cases steps 3 and 4 are combined as a single enzymatic reaction.

The isomerization of the intermediate hydroxy acid in step 2 is required because the hydroxyl group, which is attached to a tertiary carbon bearing no hydrogen, must be moved to the adjacent carbon atom before oxidation to a ketone can take place. However, in the case of glyoxylate, isomerization is not necessary-because R = H. It may be protested that the reaction of the citric acid cycle by which oxaloacetate is converted to oxo-glutarate does not follow exactly the pattern of Fig. 7.18.

The carbon dioxide removed in the decarboxylation step does not come from the part of the molecule donated by the acetyl group but from that formed from oxaloacetate. However, the end result is the same. Furthermore, there are two known citrate-forming enzymes with different stereospecificities. one of which leads to a biosynthetic pathway strictly according to the sequence of Fig. 7.18. At the bottom of Fig. 7.18 several stages of the α-oxo acid elongation process are arranged in tandem. We see that glyoxylate (a product of the acetyl-Co A-glyoxylate cycle) can be built up systematically to pyruvate, oxaloacetate, 2-oxoglutarate, and 2-oxoadipate (a precursor of lysine) using this one reaction sequence.

Methanogens elongate 2-oxoadipate by one and two carbon atoms using the same sequence to give 7- and 8-carbon dicarboxylates.

Decarboxylation as a Driving Force in Biosynthesis

Consider the relationship of the following prominent biosynthetic intermediates one to another :

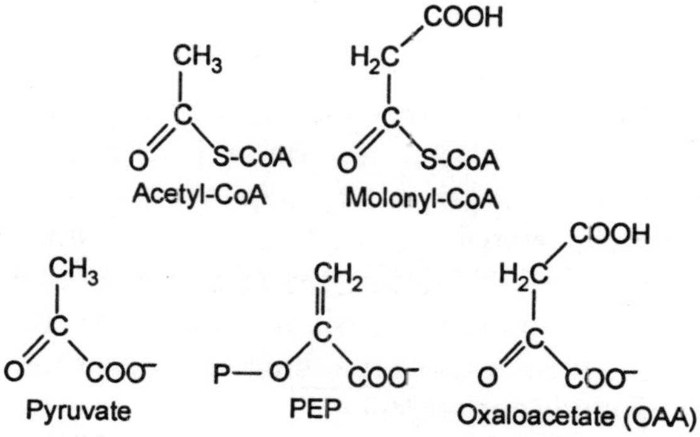

Acetyl-CoA Molonyl-CoA

Pyruvate PEP Oxaloacetate (OAA)

Utilization of acetyl-CoA for the synthesis of long-chain fatty acids occurs via carboxylation to malonyl-CoA. *We can think of the malonyl group as a β-carboxylated acetyl group.* During synthesis of a fatty acid the carboxyl group is lost, and only the acetyl group is ultimately incorporated into the fatty acid. In a similar way *pyruvate can be thought of as an a-carboxylated acetaldehyde and oxaloacetate as an α- and β-dicarboxylated acetaldehyde.*

During biosynthetic reactions these three- and four-carbon compounds also often undergo decarboxylation. Thus, they both can be regarded as "activated acetaldehyde units." Phosphoenolpyruvate is an α-carboxylated phosphoenol form of acetaldehyde and undergoes both decarboxylation and dephosphorylation before contributing a two-carbon unit to the final product. It is of interest to compare two chain elongation processes by which two-carbon units are combined. In the synthesis of fatty acids the acetyl units are condensed and then are reduced to form straight hydrocarbon chains. In the oxoacid chain elongation mechanism, the acetyl unit is introduced but is later decarboxylated. Thus, the chain is increased in length by one carbon atom at a time. These two mechanisms account for a great deal of the biosynthesis by chain extension.

However, there are other variations. For example, glycine (a carboxylated methylamine), under the influence of pyridoxal phosphate and with accompanying decarboxylation, condenses with succinyl-CoA to extend the carbon chain and at the same time to introduce an amino group. Likewise, serine (a carboxylated ethanolamine) condenses with palmitoyl-CoA in biosynthesis of sphingosine. Phosphatidylserine is decarboxylated to phosphatidylethanolamine in the final synthetic step for that phospholipid.

Stabilization and Termination of Chain Growth by Ring Formation

Biochemical substances frequently undergo cyclization to form stable five- and six-atom ring structures. The three-carbon glyceraldehyde phosphate exists in solution primarily as the free aldehyde (and its covalent hydrate) but glucose 6-phosphate exists largely as the cyclic hemiacetal. In this ring form no carbonyl group is present and further chain elongation is inhibited. When the hemiacetal of glucose 6-*P* is enzymatically isomerized to glucose 1-*P* the ring is firmly locked. Glucose 1-*P*, in turn, serves as the biosynthetic precursor of polysaccharides and related compounds, in all of which the sugar rings are stable. Ring formation can occur in lipid biosynthesis, too.

Among the *polyketides*, polyprenyl compounds, and aromatic amino acids are many substances in which ring formation has occurred by ester or aldol condensations followed by reduction and elimination processes. This is a typical sequence for biosynthesis of highly stable aromatic rings.

Branched Carbon Chains

Branched carbon skeletons are formed by standard reaction types but sometimes with addition of rearrangement steps. Compare the biosynthetic routes to three different branched five-carbon units (Fig. 7.19) The first is the use of a *propionyl group* to initiate formation of a branched-chain fatty acid. Propionyl-CoA is carboxylated to methylmalonyl-CoA, whose acyl group is transferred to the acyl carrier protein before condensation. Decarboxylation and reduction yields an acyl-CoA derivative with a methyl group in the 3-position. The second five-carbon branched unit, in which the branch is one carbon further down the chain, is an intermediate in the biosynthesis of *polyprenyl* (isoprenoid) compounds and steroids.

Three two-carbon units are used as the starting material with decarboxylation of one unit. Two acetyl units are first condensed to form acetoacetyl-CoA. Then a third acetyl unit, which has been transferred from acetyl-CoA onto an SH group of the enzyme, is combined with the acetoacetyl-CoA through an ester condensation. The thioester linkage to the enzyme is hydrolyzed to free the product *3-hydroxy-3-methylglutaryl-CoA* (HMG-CoA). This sequence is illustrated in Eq. 7.5. The thioester group of HMG-CoA is reduced to the alcohol *mevalonic acid*, a direct precursor to isopentenyl pyro-phosphate, from which the poly-prenyl compounds are formed (Fig. 7.1). The third type of carbon-branched unit is 2-oxoisovalerate, from which valine is formed by transamination.

The starting units are two molecules of pyruvate which combine in a thiamin diphosphate-dependent *α* condensation with decarboxylation. The resulting α-acetolactate contains a branched chain but is quite unsuitable for formation of an a amino acid. A rearrangement moves the methyl group to the p position and elimination of water from the diol forms the enol of the desired α-oxo acid (Fig. 7.19). The precursor of isoleucine is formed in an analogous way by condensation, with decarboxylation of one molecule of pyruvate with one of 2-oxobutyrate.

BIOSYNTHESIS AND MODIFICATION OF POLYMERS

There are three chemical problems associated with the assembly of a protein, nucleic acid, or other biopolymer. The first is to *overcome thermodynamic barriers*. The second is *to control the rate of synthesis,* and the third is to *establish the pattern or sequence in which the monomer units are linked together*. Let us look briefly at how these three problems are dealt with by living cells.

Peptides and Proteins

Activation of amino acids for incorporation into oligopeptides and proteins can occur via two routes of acyl activation. In the first of these an *acyl phosphate* (or acyl adenylate) is formed and reacts with an amino group to form a peptide linkage. The tripeptide *glutathione* is formed in two steps of this type. In the second method of activation *aminoacyl* adenylates are formed. They transfer their activated aminoacyl groups onto specific tRNA molecules during synthesis of proteins (Eq. 7.36). In other cases activated aminoacyl groups are transferred onto -SH groups to form intermediate *thioesters*. An example is the synthesis of the antibiotic *gramicidin S* formed by *Bacillus brevis.*

The antibiotic is a cyclic decapeptide with the following five-amino-acid sequence repeated twice in the ringlike molecule :

$$(\text{–D-Phe-L-Pro-L-Val-L-Orn-L-Leu-})_2$$

The soluble enzyme system responsible for its synthesis contains a large 280-kDa protein that not only activates the amino acids as aminoacyl adenylates and transfers them to thiol groups of 4′-phosphopantetheine groups covalently attached to the enzyme but also serves as a template for joining the amino acids in proper sequence. Four amino acids—proline, valine, orni thine (Orn), and leucine—are all bound. A second enzyme (of mass 100 kDa) is needed for activation of phenylalanine. It is apparently the activated phenylalanine (which at some point in the process is isomerized from L- to D-phenylalanine) that initiates polymer formation in a manner analogous to that of fatty acid elongation (Fig. 7.12).

Initiation occurs when the amino group of the activated phenylalanine (on the second enzyme) attacks the acyl group of the aminoacyl thioester by which the activated proline is held. Next, the freed imino group of proline attacks the activated valine, etc., to form the pentapeptide. Then two pentapeptides are joined and cyclized to give the antibiotic. The sequence is absolutely specific, and it is remarkable that this relatively small enzyme system is able to carry out each step in the proper sequence.

Many other peptide antibiotics, such as the bacitracins, tyrocidines, and enniatins, are synthesized in a similar way, as are depsipeptides and the immunosuppresant cyclosporin. A virtually identical pattern is observed for formation of *polyketides*, whose chemistry is considered. While peptide antibiotics are synthesized according to enzyme-controlled polymerization patterns, both proteins and nucleic acids are made by *template mechanisms*. The sequence of their monomer units is determined by genetically encoded information. A key reaction in the

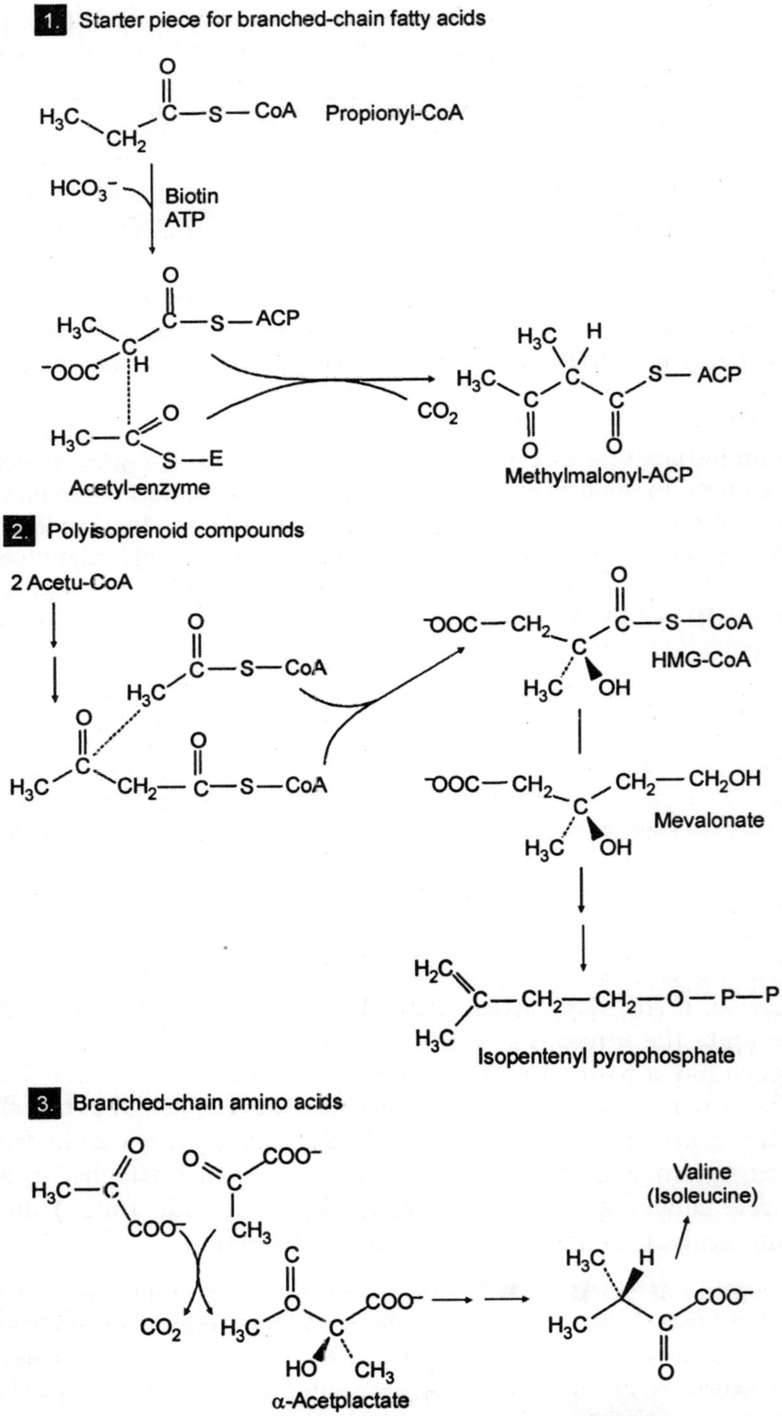

1. Starter piece for branched-chain fatty acids

$H_3C-CH_2-C(=O)-S-CoA$ Propionyl-CoA

HCO_3^- Biotin ATP

$^-OOC-CH(CH_3)-C(=O)-S-ACP$

$H_3C-C(=O)-S-E$

Acetyl-enzyme

CO_2

$H_3C-C(=O)-CH(CH_3)-C(=O)-S-ACP$

Methylmalonyl-ACP

2. Polyisoprenoid compounds

2 Acetu-CoA

$H_3C-C(=O)-S-CoA$

$H_3C-C(=O)-CH_2-C(=O)-S-CoA$

$^-OOC-CH_2-C(CH_3)(OH)-CH_2-C(=O)-S-CoA$ HMG-CoA

$^-OOC-CH_2-C(CH_3)(OH)-CH_2-CH_2OH$ Mevalonate

$H_2C=C(CH_3)-CH_2-CH_2-O-P-P$

Isopentenyl pyrophosphate

3. Branched-chain amino acids

$H_3C-C(=O)-COO^-$ $O=C(COO^-)-CH_3$

CO_2

$H_3C-C(=O)-C(OH)(CH_3)-COO^-$

α-Acetplactate

Valine (Isoleucine)

$(H_3C)_2CH-C(=O)-COO^-$

Fig. 7.19. Biosynthetic origins of three five-carbon branched structural units. Notice that decarboxylation is involved in driving each sequence.

formation of proteins is the transfer of activated aminoacyl groups to molecules of tRNA (Eq. 7.36). The tRNAs act as carriers or adapters as explained in detail in Chapter 29. Each *aminoacyl-tRNA synthetase* must recognize the correct tRNA and attach the correct amino acid to it. The tRNA then carries the activated amino acid to a ribosome, where it is placed, at the correct moment, in the active site. *Peptidyltransferase*, using a transacylation reaction, in an *insertion mechanism* transfers the C terminus of the growing peptide chain onto the amino group of the new amino acid to give a tRNA-bound peptide one unit longer than before.

Polysaccharides

Incorporation of a sugar monomer into a polysaccharide also involves cleavage of two high-energy phosphate linkages of ATP. However, the activation process has its own distinctive pattern (Eq. 7.56). Usually a sugar is first phosphorylated by a kinase or a kinase plus a phosphomutase (Eq. 7.56, step *a*). Then a nucleoside triphosphate (NuTP) reacts under the influence of a second enzyme with elimination of pyrophosphate and formation of a *glycopyranosyl ester* of the nucleoside diphosphate, more often known as a *sugar nucleotide* (Eq. 7.56, step *b*). The inorganic pyrophosphate is hydrolyzed by pyro-phosphatase while the sugar nucleotide donates the activated glycosyl group for polymerization (Eq. 7.56, step *c*). In this step the glycosyl group is transferred with displacement of the nucleoside diphosphate. Thus, the overall process involves first the cleavage of ATP to ADP and P_i, and then the cleavage of a nucleoside triphosphate to a nucleoside diphosphate plus P_i.

The nucleoside triphosphate in Eq. 7.56, step *b* is sometimes ATP, in which case the overall result is the splitting of two molecules of ATP to ADP. However, as detailed in the whole series of nucleotide "handles" serve to carry various activated glycosyl units. What determines the pattern of incorporation of sugar units into polysaccharides? Homopolysaccharides, like cellulose and the linear amylose form of starch, contain only one monosaccharide component in only one type of linkage.

A single synthetase enzyme can add unit after unit of an activated sugar (UDP glucose or other sugar nucleotide) to the growing end. However, at least two enzymes are needed to assemble a branched molecule such as that of the glycogen molecule. One is the synthetase; the second is a *branching enzyme*, a transglycosylase. After the chain ends attain a length of about ten monosaccharide units the branching enzyme attacks a glycosidic linkage somewhere in the chain. Acting much like a hydrolase, it forms a glycosyl enzyme (or a stabilized carbocation) intermediate. The enzyme does not release the severed chain fragment but transfers it to another nearby site on the branched polymer. In the synthesis of glycogen, the chain fragment is joined to a free 6-hydroxyl group of the glycogen, creating a new branch attached by an α-l,6-linkage. Other carbohydrate polymers consist of repeating oligosaccharide units. Thus, in hyaluronan units of glucuronic acid and N-acetyl-D-glucosamine alternate.

The "O antigens" of bacterial cell coats (p. 180) contain repeating subunits made up of a "block" of four or five different sugars. In these and many other cases the pattern of polymerization is established by the specificities of individual enzymes. An enzyme capable of joining an activated glucosyl unit to a growing polysaccharide will do so only if the proper structure has been built up to that point. In cases where a block of sugar units is transferred it is usually *inserted* at the nonreducing end of the polymer, which may be covalently attached to a protein. Notice that the insertion mode of chain growth exists for lipids, polysaccharides, and proteins.

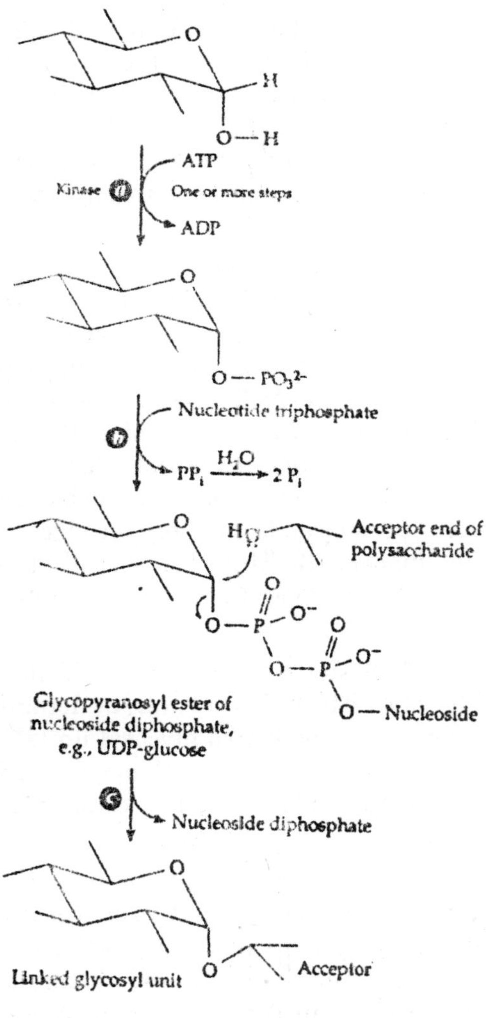

O
H
O—H
ATP
Kinase (1) One or more steps
ADP

O
$O—PO_3^{2-}$
Nucleotide triphosphate
(2)
$PP_i \xrightarrow{H_2O} 2\ P_i$

O
H O
Acceptor end of polysaccharide
O—P—O⁻ O—P—O⁻
O—Nucleoside
Glycopyranosyl ester of nucleoside diphosphate, e.g., UDP-glucose

(3) Nucleoside diphosphate

O
Linked glycosyl unit O Acceptor

...(7.56)

Nucleic Acids

The activated nucleotides are the nucleoside 5′-triphosphates. The ribonucleotides ATP, GTP, UTP, and CTP are needed for RNA synthesis and the 2′-deoxyribonucleotide triphosphates, dATP, dTTP, dGTP, and dCTP for DNA synthesis. In every case, the addition of activated monomer units to a growing polynucleotide chain is catalyzed by an enzyme that binds to the template nucleic acid. The choice of the proper nucleotide unit to place next in the growing strand is determined by the nucleotide already in place in the complementary strand, a matter that is dealt.

The chemistry is a simple displacement of pyrophosphate (Eq. 7.57). The 3′-hydroxyl of the polynucleotide attacks the phosphorus atom of the activated nucleoside triphosphate. Thus, *nucleotide chains always grow from the 5′ end, with new units being added at the 3′ end.*

$$\text{HO} \overset{3'}{—} \text{Nucleotide} \overset{5'}{—} \text{O—P(=O)(O}^-\text{)—O—P(=O)(O}^-\text{)—O—P(=O)(O}^-\text{)—OH}$$

$$\text{H—O—Polynucleotide}^{5'}$$

$$\downarrow \quad \text{PP}_i \xrightarrow{H_2O} 2\text{P}_i$$

$$\text{HO} \overset{3'}{—} \text{Nu—O—P(=O)(O}^-\text{)—O—Polynucleotide}^{5'} \qquad ...(7.57)$$

Phospholipids and Phosphate-Sugar Alcohol Polymers

Choline and ethanolamine are activated in much the same way as are sugars. For example, choline can be phosphorylated using ATP (Eq. 7.58, step *a*) and the phosphocholine formed can be further converted (Eq. 7.58, step *b*) to *cytidine diphosphate choline*. Phosphocholine is transferred from the latter onto a suitable acceptor to form the final product (Eq. 7.58, step *c*). The polymerization pattern differs from that for polysaccharide synthesis. When the sugar nucleotides react, the entire nucleoside diphosphate is eliminated (Eq. 7.56), but CDP-choline and CDP-ethanolamine react with elimination of CMP (Eq. 7.58, step *c*), leaving one phospho group in the final product. The same thing is true in the synthesis of the bacterial teichoic acids. Either CDP-glycerol or CDP-ribitol is formed first and polymerization takes place with elimination of CMP to form the alternating phosphate-sugar alcohol polymer.

Irreversible Modification and Catabolism of Polymers

While polymers are being synthesized continuously by cells, they are also being modified and torn down. Nothing within a cell is static. As discussed in everything turns over at a slower or faster rate. Hydrolases attack all of the polymers of which cells are composed, and active catabolic reactions degrade the monomers formed. Membrane surfaces are also altered, for example, by hydroxylation and glycosylation of both glycoproteins and lipid head groups. It is impossible to list all of the known modification reactions of biopolymers. They include hydrolysis, methylation, acylation, isopentenylation, phosphorylation, sulfation, and hydroxylation. Precursor molecules are cut and trimmed and often modified further to form functional proteins or nucleic acids.

Phosphotransferase reactions splice RNA transcripts to form mRNA and a host of alterations convert precursors into mature tRNA molecules . Even DNA, which remains relatively unaltered, undergoes a barrage of chemical attacks. Only because of the presence of an array of repair enzymes does our DNA remain nearly unchanged so that faithful copies can be provided to each cell in our bodies and can be passed on to new generations.

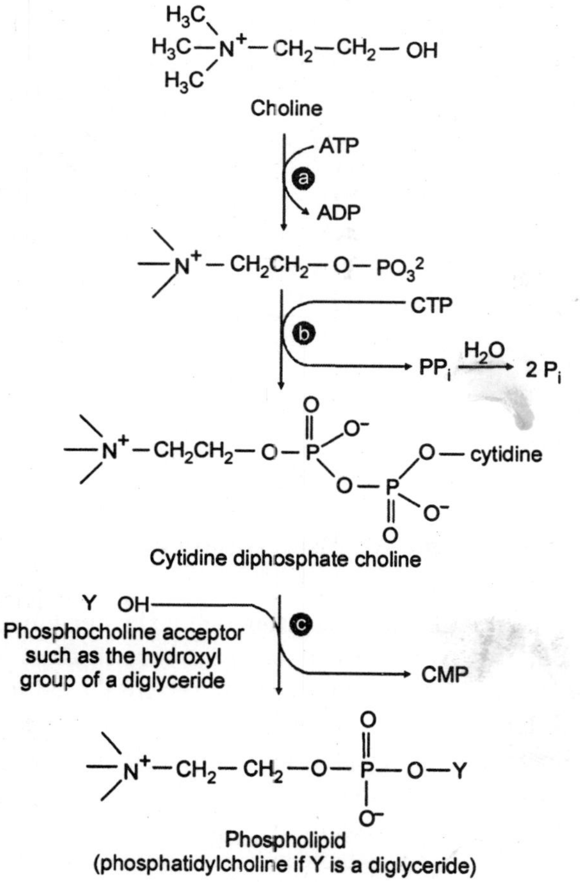

$$\begin{array}{c} H_3C\\ H_3C-N^+-CH_2-CH_2-OH\\ H_3C \end{array}$$

Choline

ATP → ⓐ → ADP

$$-N^+-CH_2CH_2-O-PO_3^{2}$$

CTP → ⓑ → $PP_i \xrightarrow{H_2O} 2\,P_i$

$$-N^+-CH_2CH_2-O-P(=O)(O^-)-O-P(=O)(O^-)-O-\text{cytidine}$$

Cytidine diphosphate choline

Y OH— Phosphocholine acceptor such as the hydroxyl group of a diglyceride → ⓒ → CMP

$$-N^+-CH_2-CH_2-O-P(=O)(O^-)-O-Y$$

Phospholipid
(phosphatidylcholine if Y is a diglyceride) ...(7.58)

REGULATION OF BIOSYNTHESIS

A simplified view of metabolism is to consider a cell as a "bag of enzymes." Indeed, much of metabolism can be explained by the action of several thousand enzymes promoting specific reactions of their substrates. These reactions are based upon the natural chemical reactivities of the substrates. However, the enzymes, through the specificity of their actions and through association with each other, channel the reactions into a selected series of metabolic pathways. The reactions are often organized as cycles which are inherently stable. We have seen that biosynthesis often involves ATP-dependent reductive reactions. *It is these reductive processes that produce the less reactive nonpolar lipid groupings and amino acid side chains so essential to the assembly of insoluble intracellular structures.*

Oligomeric proteins, membranes, microtubules, and filaments are all the natural result of aggregation caused largely by hydrophobic interactions with electrostatic forces and hydrogen bonding providing specificity. A major part of metabolism is the creation of complex molecules that aggregate spontaneously to generate structure. This structure includes the lipid-rich cytoplasmic membranes which, together with embedded carrier proteins, control the entry of substances into cells. Clearly, the cell is now much more than a bag of enzymes, containing several compartments, each of which contains its own array of enzymes and other components. Metabolite concentrations may vary greatly from one compartment to another. The reactions that modify lipids and glycoproteins provide a driving force that assists in moving membrane materials generated internally into the outer surface of cells.

Other processes, including the breakdown by lysosomal enzymes, help to recycle membrane materials. Oxidative attack on hydrophobic materials such as the sterols and the fatty acids of membrane lipids results in their conversion into more soluble substances which can be degraded and completely oxidized. The flow of matter within cells tends to occur in metabolic loops and some of these loops lead to formation of membranes and organelles and to their turnover. This flow of matter, which is responsible for growth and development of cells, is driven both by hydrolysis of ATP coupled to biosynthesis and by irreversible degradative alterations of polymers and lipid materials.

It also provides for transient formation and breakup of complexes of macromolecules, which may be very large, in response to varying metabolic needs. Anything that affects the rate of a reaction involved in either biosynthesis or degradation of any component of the cell will affect the overall picture in some way. Thus, every chemical reaction that contributes to a quantitatively significant extent to metabolism has some controlling influence. Since molecules interact with each other in so many ways, reactions of metabolic control are innumerable. Small molecules act on macro-molecules as effectors that influence conformation and reactivity. Enzymes act on each other to break covalent bonds, to oxidize, and to crosslink.

Transferases add phospho, glycosyl, methyl, and other groups to various sites. The resulting alterations often affect catalytic activities. The number of such interactions significant to metabolic control within an organism may be in the millions. Small wonder that biochemical journals are filled with a confusing number of postulated control mechanisms. Despite this complexity, some regulatory mechanisms stand out clearly. The control of enzyme synthesis through feedback repression and the rapid control of activity by feedback inhibition have been considered previously.

Under some circumstances, in which there is a constant growth rate, these controls may be sufficient to ensure the harmonious and proportional increase of all constituents of a cell. Such may be the case for bacteria during logarithmic growth a mammalian embryo growing rapidly and drawing all its nutrients from the relatively constant supply in the maternal blood. Contrast the situation in an adult. Little growth takes place, but the metabolism must vary with time and physiological state.

The body must make drastic readjustments from normal feeding to a starvation situation and from resting to heavy exercise. The metabolism needed for rapid exertion is different from that needed for sustained work. A fatty diet requires different metabolism than a high-carbohydrate diet. The necessary control mechanisms must be rapid and sensitive.

Glycogen and Blood Glucose

Two special features of glucose metabolism in animals are dominant. The first is the storage of glycogen for use in providing muscular energy rapidly. This is a relatively short-term matter but the rate of glycolysis can be intense: The entire glycogen content of muscle could be exhausted in only 20 s of anaerobic fermentation or in 3.5 min of oxidative metabolism. There must be a way to turn on glycolysis quickly and to turn it off when it is no longer needed. At the same time, it must be possible to reconvert lactate to glucose or glycogen (gluconeogenesis). The glycogen stores of the muscle must be repleted from glucose of the blood. If insufficient glucose is available from the diet or from the glycogen stores of the liver, it must be synthesized from amino acids.

Table 7.3. Some Effects of Insulin on Enzymes.

Name of Enzyme	***Type of Regulation***
A. Activity increased	
Enzymes of glycolysis	
Glucokinase	Transcription induced
Phosphofructokinase	via 2,6-fructose P_2
Pyruvate kinase	Dephosphorylation
6-Phosphofructo-2-kinase	Dephosphorylation
Enzymes of glycogen synthesis	
Glucokinase	Transcription
Glycogen synthase (muscle)	
Enzymes of lipid synthesis	
Pyruvate dehydrogenase (adipose)	Dephosphorylation (Eq. 7.9)
Acetyl-CoA carboxylase	Dephosphorylation
ATP-citrate lyase	Phosphorylation
Fatty acid synhthase	
Lipoprotein lipase	
Hydroxymethylagulataryl-CoA reductase	
B. Activity decreased	
Enzymes of gluconeogenesis	
Pyruvate carboxylase	
PEP carboxykinase	Transcription inhibited
Fructose 1,6-bisphosphate	
Glucose 6-phosphatase	
Enzymes of lipolysis	
Triglyceride lipase	
(hormone-sensitive lipase)	Dephosphorylation
Enzymes of glycogenolysis	
Enzymes of glycogenolysis	
Glycogen phosphorylase	
C. Other proteins affected by insulin	
Glucose transporter GLUT4	Redistribution
Ribosomal protein S6	Phosphorylation by $p90^{rsk}$
IGF-II receptor	Redistribution
Transferrin receptor	Redistribution
Calmodulin	Phosphorylation

The second special feature of glucose metabolism is that certain tissues, including brain, blood cells, kidney medulla, and testis, ordinarily obtain most of their energy through oxidation of glucose. For this reason, the glucose level of blood cannot be allowed to drop much below the normal 5 mM. The mechanism of regulation of the blood glucose level is complex and incompletely understood. A series of hormones are involved.

Insulin : This 51-residue cross-linked polypeptide is synthesized in the pancreatic islets of Langerhans, a tissue specialized for synthesis and secretion into the bloodstream of a series of small peptide hormones. One type of islet cells, the P cells, forms primarily insulin which is secreted in response to high (> 5mm) blood [glucose]. Insulin has a wide range of effects on metabolism. Most of these effects are thought to arise from binding to insulin receptors and are mediated by cascades. The end result is to increase or decrease activities of a large number of enzymes as is indicated in Table 7.3. Some of those are also shown in Fig. 7.20, which indicates interactions with the tricarboxylic acid cycle and lipid metabolism.

Binding of insulin to the extracellular domain of its dimeric receptor induces a conformational change that activates the intracellular tyrosine kinase domains of the two subunits. Recent studies suggest that in the activated receptor the two trans-membrane helices and the internal tyrosine kinase domains move closer together, inducing the essential autophosphosphorylation. The kinase domain of the phosphorylated receptor, in turn, phosphorylates several additional proteins, the most important of which seem to be the insulin receptor substrates IRS-1 and IRS-2. Both appear to be essential in different tissues.

Phosphorylated forms of these proteins initiate a confusing variety of signaling cascades. One of the most immediate effects of insulin is to stimulate an increased rate of uptake of glucose by muscle and adipocytes (fat cells) and other insulin-sensitive tissues. This uptake is accomplished largely by movement of the glucose transporter GLUT4 from internal "sequestered" storage vesicles located near the cell membrane into functioning positions in the membranes.

Activation of this translocation process apparently involves IRS-1 and phosphatidylinositol (PI) 3-kinase, which generates PI-3,4,5-P_3. The latter induces the translocation. However, the mechanism remains obscure. The process may also require a second signaling pathway which involves action of the insulin receptor kinase on an adapter protein known as *CAP*, a transmembrane caveolar protein *flotillin*, and a third protein *Cbl*, a known cellular protooncogene. Phosphorylated Cbl forms a complex with CAP and flotillin in a "lipid raft" which induces the exocytosis of the sequestered GLUT4 molecules. A clue to another possible unrecognized mechanism of action for insulin comes from the observation that urine of patients with non-insulin-dependent diabetes contains an unusual isomer of inositol, D-*chiro*-inositol.

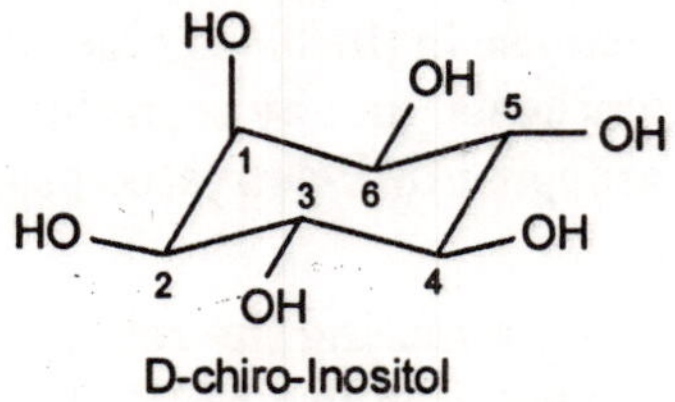

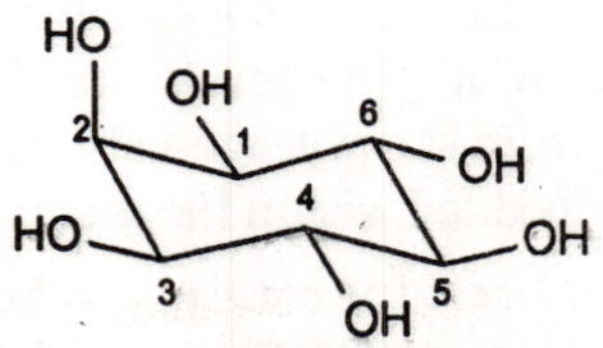

D-chiro-Inositol

myo-Inositol
(numbered as D)

Plasma of such individuals contains an antagonist of insulin action, an inositol phosphoglycan containing *myo*-inositol as a cyclic 1,2-phosphate ester and galactosamine and man-nose in a 1:1:3 ratio. This appears to be related to the glycosyl phosphatidylinositol (GPI) membrane anchors. It has been suggested that such a glycan, perhaps containing *chiro*-inositol, is released in response to insulin and serves as a second messenger for insulin. This hypothesis remains unproved. However, insulin does greatly stimulate a GPI-specific phospholipase C, at least in yeast.

Another uncertainty surrounds the possible cooperation of chromium in the action of insulin. How do the insulin-secreting pancreatic 3 cells sense a high blood glucose concentration? Two specialized proteins appear to be involved. The sugar transporter *GLUT2* allows the glucose in blood to equilibrate with the free glucose in the β cells, while *glucokinase* (hexokinase IV or D) apparently serves as the glucose sensor.

Despite the fact that glucokinase is a monomer, it displays a cooperative behavior toward glucose binding, having a low affinity at low [glucose] and a high affinity at high [glucose]. Mutant mice lacking the glucokinase gene develop early onset diabetes which is mild in heterozygotes but severe and fatal within a week of birth for homozygotes. These facts alone do not explain how the sensor works and there are doubtless other components to the signaling system. A current theory is that the increased rate of glucose catabolism in the (3 cells when blood [glucose] is high leads to a high ratio of [ATP]/[ADP] which induces closure of ATP-sensitive K^+ channels and opening of voltage-gated Ca^{2+} channels. This could explain the increase in [Ca^{2+}] within p cells which has been associated with secretion of insulin and which is thought to induce the exocytosis in insulin storage granules. The internal [Ca^{2+}] in pancreatic islet cells is observed to oscillate in a characteristic way that is synchronized with insulin secretion.

Glucagon : This 29-residue peptide is the principal hormone that counteracts the action of insulin. Glucagon acts primarily on liver cells (hepatocytes) and adipose tissue and is secreted by the a cells of the islets of Langerhans in the pancreas, the same tissue whose β cells produce insulin, if the blood glucose concentration falls much below 2 mM. Like the insulin-secreting β cells, the pancreatic ω cells contain glucokinase, which may be involved in sensing the drop in glucose concentration.

However, the carrier GLUT2 is not present and there is scant information on the sensing mechanism. Glucagon promotes an increase in the blood glucose level by stimulating breakdown of liver glycogen, by inhibiting its synthesis, and by stimulating gluconeogenesis. All of these effects are mediated by cyclic AMP through cAMP-activated protein kinase and through fructose 2,6-P_2.

Glucagon also has a strong effect in promoting the release of glucose into the bloodstream. *Adrenaline* has similar effects, again mediated by cAMP. However, glucagon affects the liver, while adrenaline affects many tissues. *Glucocorticoids* such as cortisol also promote gluconeogenesis and the accumulation of glycogen in the liver through their action on gene transcription. The release of glucose from the glycogen stores in the liver is mediated by *glucose 6-phosphatase*, which is apparently embedded within the membranes of the endoplasmic reticulum.

A labile enzyme, it consists of a 357-residue catalytic subunit, which may be associated with other subunits that participate in transport. A deficiency of this enzyme causes the very severe type la *glycogen storage disease*. Only hepatocytes have significant glucose 6-phosphatase activity.

Phosphofructo-1-Kinase in the Regulation of Glycolysis

The metabolic interconversions of glucose 1-*P*, glucose 6-*P*, and fructose 6-*P* are thought to be at or near equilibrium within most cells. However, the phosphorylation by ATP of fructose 6-*P* to fructose 1,6-P_2 catalyzed by phosphofructose-1-kinase (Fig. 7.2, step *b*; Fig. 7.17, top center) is usually far from equilibrium. This fact was established by comparing the mass action ratio [fructose 1,6-P_2] [ADP]/[fructose 6-*P*] [ATP] measured within tissues with the known equilibrium constant for the reaction. At equilibrium this mass action ratio should be equal to the equilibrium constant.

The experimental techniques for determining the four metabolite concentrations that are needed for evaluation of the mass action ratio in tissues are of interest. The tissues must be frozen very rapidly. This can be done by compressing them between large liquid nitrogen-cooled aluminum clamps. Tissues can be cooled to –80°C in less than 0.1 s in this manner. The frozen tissue is then powdered, treated with a frozen protein denaturant such as perchloric acid, and analyzed. From data obtained in this way, a mass action ratio of 0.03 was found for the phosphofructo-1-kinase reaction in heart muscle. This is much lower than the equilibrium constant of over 3000 calculated from the value of $\Delta G'$ (pH 7) = –20.1 kJ mol^{-1}. Thus, like other biochemical reactions that are nearly irreversible thermo-dynamically, this reaction is far from equilibrium in tissues. The effects of ATP, AMP, and fructose 2,6-bisphosphate on phosphofructokinase.

Fructose 2,6-P_2 is a potent allosteric activator of phosphofructokinase and a strong competitive inhibitor of fructose 1,6-bisphosphatase. It is formed from fructose 6-P and ATP by the 90-kDa bifunctional phosphofructo-2-kinase/fructose 2,6-bisphosphatase. Thus, the same protein forms and destroys this allosteric effector. Since the bifunctional enzyme is present in very small amounts, the rate of ATP destruction from the substrate cycling is small. Glucagon causes the concentration of *liver* fructose 2,6-P_2 to drop precipitously from its normal value. This, in turn, causes a rapid drop in glycolysis rate and shifts metabolism toward gluconeogenesis. At the same time, liver glycogen breakdown is inceased and glucose is released into the bloodstream more rapidly.

The effect on fructose 2,6-P_2 is mediated by a cAMP-dependent protein kinase which phosphorylates the bifunctional kinase/phosphatase in the liver. This modification greatly reduces the kinase activity and strongly activates the phosphatase, thereby destroying the fructose 2,6-P_2. The changes in activity appear to be largely a result of changes in the appropriate K_m values which are increased for fructose 6-P and decreased for fructose 2,6-P_2.

Gluconeogenesis

If a large amount of lactate enters the liver, it is oxidized to pyruvate which enters the mitochondria. There, part of it is oxidized through the tricarboxylic acid cycle. However, if [ATP] is high, pyruvate dehydrogenase is inactivated by phosphorylation (Eq. 7.9) and the amount of pyruvate converted to oxaloacetate and malate (Eq. 7.46) may increase. Malate may leave the mitochondrion to be reoxidized to oxaloacetate, which is then converted to PEP and on to glycogen (heavy green arrows in Fig. 7.20).

When [ATP] is high, phosphofructokinase is also blocked, but the fructose 1,6-bisphosphatase, which hydrolyzes one phosphate group from fructose 1,6-P_2 is active. If the glucose content of blood is low, the glucose 6-*P* in the liver is hydrolyzed and free glucose is secreted. Otherwise, most of the glucose 6-*P* is converted to glycogen. Muscle is almost devoid of glucose 6-phosphatase, the export of glucose not being a normal activity of that tissue. Gluconeogenesis in liver is strongly promoted by glucagon and adrenaline. The effects, mediated by cAMP, include stimulation of fructose 1,6-bisphosphatase and inhibition of phosphofructo-1-kinase, both caused by the drop in the level of fructose 2,6-P_2.

The conversion of pyruvate to PEP via oxaloacetate is also promoted by glucagon. This occurs primarily by stimulation of pyruvate carboxylase. However, it has been suggested that the most important mechanism by which glucagon enhances gluconeogenesis is through stimulation of mitochondrial respiration, which in turn may promote gluconeogenesis. The conversion of oxaloacetate to PEP by PEP-carboxykinase (PEPCK, Fig. 7.20) is another control point in gluconeogenesis. Insulin inhibits gluconeogenesis by decreasing transcription of the mRNA for this enzyme.

Glucagon and cAMP stimulate its transcription. The activity of PEP carboxykinase is also enhanced by Mn^{2+} and by very low concentrations of Fe^{2+}. However, the enzyme is readily inactivated by Fe^{2+} and oxygen. Any regulatory significance is uncertain. Although the regulation of gluconeogenesis in the liver may appear to be well understood, some data indicate that the process can occur efficiently in the presence of high average concentrations of fructose 2,6-P_2. A possible explanation is that liver consists of several types of cells, which may contain differing concentrations of this inhibitor of gluconeogenesis. However, mass spectroscopic studies suggest that glucose metabolism is similar throughout the liver.

Substrate Cycles

The joint actions of phosphofructokinase and fructose 1,6-bisphosphatase (see also Fig. 7.20) create a substrate cycle of the type discussed. Such cycles apparently accomplish nothing but the cleavage of ATP to ADP and P_i (ATPase activity). There are many cycles of this type in metabolism and the fact that they do not ordinarily cause a disastrously rapid loss of ATP is a consequence of the tight control of the metabolic pathways involved. In general, only one of the two enzymes is fully activated at any time. Depending upon the metabolic state of the cell, degradation may occur with little biosynthesis or biosynthesis with little degradation.

Other obvious substrate cycles involve the conversion of glucose to glucose 6-*P* and hydrolysis of the latter back to glucose (Fig. 7.20, upper left-hand corner), the synthesis and breakdown of glycogen (upper right), and the conversion of PEP to pyruvate and the reconversion of the latter to PEP via oxaloacetate and malate (partially within the mitochondria). While one might suppose that cells always keep substrate cycling to a bare minimum, experimental measurements on tissues *in vivo* have indicated surprisingly high rates for the fructose 1,6-bisphosphatase-phosphofructokinase cycle in mammalian tissues when glycolytic flux rates are low and also for the pyruvate → oxaloacetate → PEP → pyruvate cycle.

As pointed out by maintaining a low rate of substrate cycling under conditions in which the carbon flux is low (in either the glycolytic or glucogenic direction) the system is more sensitive to allosteric effectors than it would be otherwise. However, when the flux through the glycolysis pathway is high the relative amount of cycling is much less and the amount of ATP formed approaches the theoretical 2.0 per glucose. Substrate cycles generate heat, a property that is apparently put to good use by cold bumblebees whose thoracic temperature must reach at least 30°C before they can fly. The insects apparently use the fructose bisphosphatase-phosphofructokinase substrate cycle to warm their flight muscles. It probably helps to keep us warm, too.

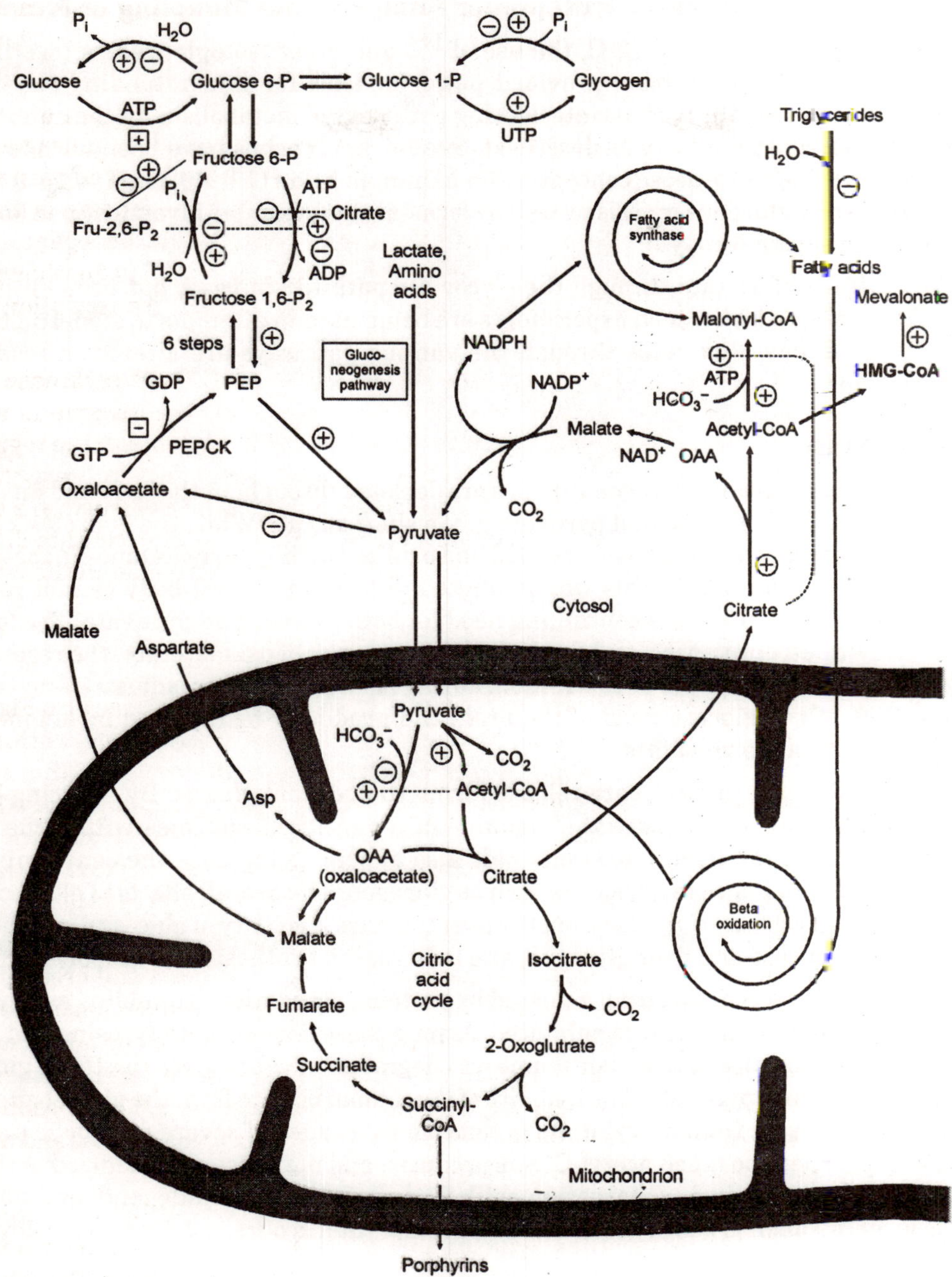

Fig. 7.20. The interlocking pathways of glycolysis, gluconeogenesis, and fatty acid oxidation and synthesis with indications of some aspects of control in hepatic tissues. (→) Reactions of glycolysis, fatty acid degradation, and oxidation by the citric acid cycle. (→) Biosynthetic pathways. Some effects of insulin via indirect action on enzymes ⊕, ⊖, or on transcription ⊞, ⊟ Effects of glucagon ⊕, ⊖.

Nuclear Magnetic Resonance, Isotopomer Analysis, and Modeling of Metabolism

As as been pointed, 3-C and 17-C, the use of ^{13}C and other isotopic tracers together with NMR and mass spectroscopy have provided powerful tools for understanding the complex interrelationships among the various interlocking pathways of metabolism. In the application of ^{13}C NMR to the citric acid cycle was described. Similar approaches have been used to provide direct measurement of the glucose concentration in human brain (1.0 ± 0.1 mM; 4.7 + 0.3 mM in plasma) and to study gluconeogenesis as well as fermentation. Similar investigations have been made using mass spectros-copy.

The metabolism of acetate through the glycoylate pathway in yeast has been observed by ^{13}C NMR. Data obtained from such experiments are being used in attempts to model metabolism and to understand how flux rates through the various pathways are altered in response to varying conditions.

The Fasting State

During prolonged fasting, glycogen supplies are depleted throughout the body and fats become the principal fuels. Both glucose and pyruvate are in short supply. While the hydrolysis of lipids provides some glycerol (which is phosphorylated and oxidized to dihydroxyacetone-*P*), the quantity of glucose precursors formed in this way is limited. Since the animal body cannot reconvert acetyl-CoA to pyruvate, there is a continuing need for both glucose and pyruvate. The former is needed for biosynthetic processes, and the latter is a precursor of oxaloacetate, the regenerating substrate of the citric acid cycle. For this reason, during fasting the body readjusts its metabolism. As much as 75% of the glucose need of the brain can gradually be replaced by ketone bodies derived from the breakdown of fats.

Glucocorticoids (*e.g.,* cortisol;) are released from the adrenal glands. By inducing enzyme synthesis, these hormones increase the amounts of a variety of enzymes within the cells of target organs such as the liver. Glucocorticoids also appear to increase the sensitivity of cell responses to cAMP and hence to hormones such as glucagon. The overall effects of glucocorticoids include an increased release of glucose from the liver (increased activity of glucose 6-phosphatase), an elevated blood glucose and liver glycogen, and a decreased synthesis of mucopolysaccharides.

The reincorporation of amino acids released by protein degradation is inhibited and synthesis of enzymes degrading amino acids is enhanced. Among these enzymes are tyrosine and alanine aminotransferases, enzymes that initiate amino acid degradation which gives rise to the glucogenic precursors fumarate and pyruvate. The inability of the animal body to form the glucose precursors pyruvate or oxaloacetate from acetyl units is sometimes a cause of severe metabolic problems. Ketosis, develops when too much acetyl-CoA is produced and not efficiently oxidized in the citric acid cycle. Ketosis occurs during starvation, with fevers, and in insulin-dependent diabetes. In cattle, whose metabolism is based much more on acetate than is ours, spontaneously developing ketosis is a frequent problem.

Lipogenesis

A high-carbohydrate meal leads to an elevated blood glucose concentration. The glycogen reserves within cells are filled. The ATP level rises, blocking the citric acid cycle, and citrate is exported from mitochondria (Fig. 7.20). Outside the mitochondria citrate is cleaved by the ATP-requiring citrate lyase to acetyl-CoA and oxaloacetate. The oxaloacetate can be reduced to malate and the latter oxidized with $NADP^+$ to pyruvate (Eq. 7.46), which can again enter the mitochondrion. In this manner acetyl groups are exported from the mitochondrion as acetyl-CoA which can be carboxylated, under the activating influence of citrate, to form malonyl-CoA, the precursor of fatty acids.

The NADPH formed from oxidation of the malate provides part of the reducing equivalents needed for fatty acid synthesis. Additional NADPH is available from the pentose phosphate pathway. Thus, excess carbohydrate is readily converted into fat by our bodies. These reactions doubtless occur to some extent in most cells, but they are quantitatively most important in the liver, in fat cells of adipose tissue, and in mammary glands. The process is also facilitated by insulin, which promotes the activation of pyruvate dehydrogenase (Eq. 7.9) and of fatty acid synthase of adipocytes. Activity of fatty acid synthase seems to be regulated by the rate of transcription of its gene, which is controlled by a transcription factor designated either as *adipocyte determination and differentiation factor-1* (*ADD-1*) or *sterol regulatory element-binding protein-1c* (*SREBP-1c*). This protein (ADD-1/SREBP-1c) may be a general mediator of insulin action.

The nuclear DNA-binding protein known as *peroxisome proliferator-activated receptor gamma* (*PPARγ*) is also involved in the control of insulin action, a conclusion based directly on discovery of mutations in persons with type II diabetes. A newly discovered hormone *resistin,* secreted by adipocytes, may also play a role. Another adipocyte hormone, *leptin*, impairs insulin action. Recent evidence suggests that both insulin and leptin may have direct effects on the brain which also influence blood glucose levels. Malonyl-CoA, which may also play a role in insulin secretion, inhibits carnitine palmitoyltransferase I (CPT I; Fig. 7.2), slowing fatty acid catabolism.